K-Theory
Proceedings of the International Colloquium, Mumbai, 2016

K-Theory
Proceedings of the International Colloquium, Mumbai, 2016

Editors

V. Srinivas,
S.K. Roushon,
Ravi A. Rao,
A.J. Parameswaran,
A. Krishna

Published for the
 **Tata Institute of Fundamental Research**
by
HINDUSTAN BOOK AGENCY

International Distribution by
American Mathematical Society, USA

HINDUSTAN BOOK AGENCY (INDIA)
P 19 Green Park Extension, New Delhi 110016

email: info@hindbook.com
http://www.hindbook.com

ISBN 978-93-86279-74-3

Cover design: Sukant Saran, TIFR

Contents

International Colloquium on K-Theory

Mumbai, 6–14 January, 2016

Preface

An international colloquium on K-Theory was held at the Tata Institute of Fundamental Research, Mumbai, from January 6, 2016 to January 14, 2016. This was the latest in the series of Colloquia held at the School of Mathematics, once every four years, since 1956. We are grateful to the IMU-CDC, the Sir Dorabji Tata Trust and Tata Institute of Fundamental Research, for financial support for hosting this event.

The organising committee of the Colloquium consisted of Professors V. Srinivas, S.K. Roushon, Ravi A. Rao, A.J. Parameswaran, A. Krishna from the School of Mathematics, TIFR.

There were thirty invited speakers and a total of 80 participants for this event. We were honoured to have a galaxy of foremost experts in the area deliver a one hour lecture each. Besides mathematicians from the School of Mathematics of the Tata Institute of Fundamental Research, several mathematicians from educational and research institutions from India and abroad attended the Colloquium.

The talks at the Colloquium covered a wide spectrum of mathematics, ranging over algebraic geometry, topology, algebraic K-theory and number theory. Algebraic theory, $\mathbb{A}^1$-homotopy theory and topological K-theory formed important sub-streams in this Colloquium.

Several branches of K-theory, like algebraic cycles, triangulated categories of motives, motivic cohomology, motivic homotopy theory, Chow groups of varieties, Euler class theory, equivariant K-theory as well as classical K-theory have developed considerably in recent years, giving rise to newer directions to the subject as well as proving results of 'classical' interest. The Colloquium brought together experts in these various branches and their talks covered this wide spectrum highlighting the interconnections giving a better perspective of the whole subject area.

The lectures were held at the Lecture Theatre AG 66 of TIFR, the same hall where Alexander Grothendieck had first announced the *Standard Conjectures on Algebraic Cycles* in the International Colloquium on Algebraic Geometry in 1968.

All lectures were videographed and the videos are available at http://conferences.math.tifr.res.in.

Our request to the speakers for contributions to the proceedings volume was warmly received and the outcome is the present memorable volume. All papers were peer reviewed by anonymous referees.

We thank the speakers for making the Colloquium a grand success. We thank all the authors for their generous contribution to this volume. We thank the anonymous referees for their splendid effort in refereeing the submitted papers.

The social programme of the Colloquium consisted of an inaugural high tea party on 6th January and a dinner at Gallops on 11th January. There was a classical Kathak dance recital by Ms. Pooja Pant and disciples on 8th January and a Hindustani vocal music concert by Pandit Arun Kashalkar on 12th January.

ix

Program of the Colloquium

Wednesday, 6 January 2016 (10:00–11:00)
Speaker   :   Spencer Bloch
Title   :   **Mixed motives and mixed Hodge structures associated to algebraic cycles**

This will be an expository talk focusing on the idea that algebraic cycles yield functions on the Tannaka group of mixed motives.

Wednesday, 6 January 2016 (11:30–12:30)
Speaker   :   Aravind Asok
Title   :   **Motivic vector bundles on projective spaces**

I will describe joint work with Mike Hopkins and Jean Fasel regarding construction of "motivic vector bundles" on P^n. By definition, a motivic vector bundle of rank r on P^n is an $\mathbb{A}^1$-homotopy class of maps from P^n to BGL_r. Such classes correspond to algebraic vector bundles on any smooth affine variety that is isomorphic in the Morel-Voevodsky $\mathbb{A}^1$-homotopy category to P^n (e.g., the variety X_n obtained as complement of the incidence hyperplane in the product of a projective space and its dual). In particular, I will explain how to construct indecomposable rank 2 bundles on X_n that "algebraize" a collection of rank 2 topological vector bundles on P^n described by Elmer Rees

Wednesday, 6 January 2016 (14:30–15:30)
Speaker   :   Arvind Nair
Title   :   **Mixed motives in A_g**

We will discuss the computation of some simple mixed motives appearing in the cohomology of the moduli space of principally polarized Abelian varieties and of its Satake compactification.

Wednesday, 6 January 2016 (16:00–17:00)
Speaker   :   Utsav Choudhury
Title   :   **Motivic Galois groups and Standard conjectures**

I will describe different constructions of motivic Galois groups due to Nori, Ayoub and Y. Andre. We will see that Nori???s motivic Galois group is canonically isomorphic to Ayoub's Galois group. Moreover, Andre's Galois group is the largest reductive quotient of the Galois group of Ayoub. In the end, we will see a reformulation of the standard conejctures using vanishing of Ayoub's Hopf d.g.a and conservative conjecture.

Thursday, 7 January 2016 (10:00–11:00)
Speaker   :   Marc Levine
Title   :   **Torsion indices of smooth projective varieties**

This is a report on a joint work with Andre Chatzistimatiou. For a smooth projective variety X of dimension d over a field k, we consider the *torsion index* $\mathrm{Tor}_k X$ of X, this being the order (possibly infinite) of the diagonal in $CH_d((X \setminus F_0) \times (X \setminus F^1))$, where F_0 is a sufficiently large dimension zero subset of X and F^1 is a sufficiently large codimension one subset of X (one takes F_0 and F^1 large enough so that the order of the diagonal remains constant under further enlargement). For X geometrically integral, we consider as well the *geometric torsion index* of X, $\mathrm{Tor}_{\bar{k}}(X_{\bar{k}})$ where $\bar{k}$ is the algebraic closure of k. The torsion index gives an upper bound for the exponent of unramified cohomology on X.

We consider the geometric torsion index of a complete intersection $X^{d_1, \ldots, d_r; n}$ of multi-degree $d_1 \geq \cdots \geq d_r \geq 2$ in $\mathbb{P}^{n+r}$ and the torsion index of a generic complete intersection in $\mathbb{P}^{n+r}$. The geometric construction of Roitman gives the upper bound

$$\mathrm{Tor}_{\bar{k}}(X_{\bar{k}}^{d_1, \ldots, d_r; n}) \leq \prod_{i=1}^{r} d_i!$$

if $\sum_i d_i \leq n + r$. Using the method of Kollár, Voisin, Colliot-Thélène-Pirutka and Totaro, we give a lower bound: Let $d_1 \geq \cdots \geq d_r \geq 2$ and $n \geq 3$ be integers such that $\sum_i d_i \leq n + r$. Let p be a prime number and suppose that

$$d_1 \geq p \cdot \left\lceil \frac{n + r + 1 - \sum_{i=2}^{r} d_i}{p + 1} \right\rceil \tag{0.1}$$

Then $p | \mathrm{Tor}(X)$ for all very general $X = X_{d_1, \ldots, d_r} \subset \mathbb{P}^{n+r}$.

Finally, we show that for the generic $X^{d_1, \ldots, d_r; n}$ in $\mathbb{P}^{n+r}$ (ie, the complete intersection defined by homogeneous equations with coefficients being independent variables $\ldots, \eta_i, \ldots$), we have the lower bound

$$\prod_{i=1}^{r} d_i!^* | \mathrm{Tor}_{k(\eta)} X^{d_1, \ldots, d_r; n}$$

Here, for a positive integer d, $d!^*$ is the least common multiple of the numbers $1, 2, \ldots, d$. For example, the generic cubic hypersurface in $\mathbb{P}^{n+1}$ with $n \geq 2$ has torsion index equal to six, while the generic quartic hypersurface in $\mathbb{P}^{n+1}$ with $n \geq 3$ has torsion index either 12 or 24.

Thursday, 7 January 2016 (11:30–12:30)
Speaker   :   **Jean Louis Colliot-Thélène**
Title   :   **Stable irrationality for some rationally connected varieties : a survey**

In 1984, the first time I came to TIFR, I described examples of stably rational varieties which are not rational. A few years later, I lectured at TIFR on developments of the Artin-Mumford method, which enables one to prove stable irrationality. A new method, which involves Chow groups, was initiated by Claire Voisin in the end of 2013. It has witnessed quick developments, with application to stable irrationality of several types of hypersurfaces of low degree. I shall survey the method and its applications.

Thursday, 7 January 2016 (14:30–15:30)
Speaker   :   **Frédéric Déglise**
Title   :   **Dimensional homotopy t-structure**

At the center of Voevodsky's theory of motivic complexes over a field is the notion of homotopy invariant sheaves with transfers. They form the heart of a canonical t-structure on motivic complexes which reflects its relation with the coniveau filtration. This t-structure is fundamental for motivic complexes, and more broadly in motivic homotopy theory. It was later extended in several direction by Fabien Morel, Déglise and lastly Joseph Ayoub in the relative case. In collaboration with Bondarko, we propose a new definition of this t-structure over an arbitrary base relying on the notion of dimension function on schemes. A new important fact is that we also construct an effective version of this t-structure, therefore staying very close to Voevodsky's original motivic complexes. During the talk, I will give the definition and state the main properties which have many common features with the perverse t-structure. Our main theorem is a characterization of the t-structure in terms of "fibers" over points of the base. Consequently, the heart of this t-structure is a convenient extension of the category of h.i. sheaves with transfers. In the end, I will give some of its main properties as well as some computations related with singularities, abelian schemes and relative curves.

Thursday, 7 January 2016 (16:00–17:00)
Speaker   :   **Ramesh Sreekantan**
Title   :   **Cycles on Abelian surfaces**

In this talk we use generalizations of classical geometric constructions of Kummer and Humbert to construct new higher Chow cycles on Abelian surfaces and K3 surfaces over p-adic local fields, generalising some work of Collino. The existence of these cycles is predicted by the poles of the local L-factor at p of the L-function of the Abelian surface. The techniques involve using some recent work of Bogomolov, Hassett and Tschinkel on the deformations of rational curves on K3 surfaces. As an application we use these cycles to prove an analogue of the Hodge-D-conjecture for Abelian surfaces

Friday, 8 January 2016 (10:00–11:00)
Speaker   :   **Shuji Saito**
Title   :   **Motives with modulus**

I will report on a joint work with Bruno Kahn and Takao Yamazki on the construction of a triangulated category of motives with modulus that extends Voevodsky's category in such a way as to encompass non-homotopy invariant phenomena

Friday, 8 January 2016 (11:30–12:30)
Speaker   :   **Marc Hoyois**
Title   :   **Cdh descent for the homotopy K-theory of tame stacks**

The cdh topology on schemes is the topology generated by Nisnevich covers and abstract blowup squares. C. Haesemeyer (in characteristic zero) and D.-C. Cisinski (in general) have shown that Weibel's homotopy invariant K-theory satisfies cdh descent. I will discuss an extension of Cisinski's result to quotient stacks of schemes by linear actions of linearly reductive algebraic groups

Friday, 8 January 2016 (14:30–15:30)
Speaker   :   **Oliver Röndigs**
Title   :   **The first motivic stable homotopy groups of spheres**

In joint work with Markus Spitzweck and Paul Arne Østvaer, we study the spectral sequence based on Voevodsky's slice filtration. This filtration on the stable homotopy category of motivic spectra over a field F measures the amount of Tate suspensions which are necessary to construct a given motivic spectrum. Work of Levine and Voevodsky shows that the slices of the motivic sphere spectrum are determined by the second page of the topological Adams-Novikov spectral sequence. We use this information to compute the first stable motivic homotopy groups of spheres over fields of characteristic zero, at least up to a completion with respect to the first algebraic Hopf map

Friday, 8 January 2016 (16:00–17:00)
Speaker   :   Wataru Kai
Title   :   A moving lemma for algebraic cycles with modulus and contravariance

The theory of algebraic cycles with modulus is an emerging branch of algebraic cycle theory. Examples are the additive higher Chow group introduced by Bloch, Esnault and Park, and the higher Chow group with modulus by Binda, Kerz and Saito. The groups are believed to correspond to relative K-groups of pairs of a smooth variety and a Cartier divisor on it. In this talk we exhibit how the contravariance of these theories can be deduced from a "moving lemma with modulus".

Saturday, 9 January 2016 (10:00–11:00)
Speaker   :   Charles Weibel
Title   :   Relative Cartier Divisors and Laurent Polynomials

This is joint work with Vivek Sadhu. If A is a subring of B (or more generally $f : X \to Y$ is affine faithful) there is a natural determinant from the relative K_0 group to the group of relative Cartier divisors, reducing to the classical case when A is normal with quotient field B. We study the behavior of relative divisors under polynomial and Laurent polynomial extensions. This is a contracted functor in Bass' sense, and the contraction is the group of global sections of the etale sheaf $f_*(Z)/Z$.

Saturday, 9 January 2016 (11:30–12:30)
Speaker   :   Bruno Kahn
Title   :   Cycles of codimension 2 and Chow-Kënneth decompositions

I will review known results and outstanding questions on algebraic cycles, concentrating on the case of codimension 2, and explain how refined Chow-Kënneth decompositions shed a light on these questions, yield a few new results and raise some new questions.

Saturday, 9 January 2016 (14:30–15:30)
Speaker   :   Kapil Paranjape
Title   :   Modular Forms and Calabi-Yau varieties

The classical result of Eichler and Shimura associates, to each eigenform of weight 2 with rational coefficients, a corresponding elliptic curve. We will discuss how this may be generalisable to other weights by using Calabi-Yau varieties of higher dimension. We will also present some examples. This talk is based on joint work with Dinakar Ramakrishnan

Saturday, 9 January 2016 (16:00–17:00)
Speaker : Masaki Hanamura
Title : Integrals of logarithmic forms on semi-algebraic sets and a generalized Cauchy formula

We consider integrals of the form

$$\int_A \frac{dz_1}{z_1} \wedge \cdots \wedge \frac{dz_n}{z_n}$$

where A is a compact semi-algebraic set of dimension n in $\mathbb{C}^n$.

(1) Under a certain condition on the dimension intersection of A with the "faces" (intersections of coordinate hyperplanes), the integral is shown to be convergent.

(2) The "Cauchy formula" for integrals as above are formulated is formulated and proven.

(3) One can apply (1) and (2) to the study of mixed Tate Hodge structures associated to mixed Tate motives.

These are meant to further part of the work of Bloch and Kriz on mixed Tate motives. This is joint work with K. Kimura and T. Terasoma.

Monday, 11 January 2016 (10:00–11:00)
Speaker : Thomas Geisser
Title : Some remarks on etale motivic cohomology

We will give some results (well-known to the experts) on motivic cohomology over algebraically closed fields, and discuss properties of duality pairings between mod m and the m-torsion of etale motivic cohomology over algebraically closed, finite and local fields.

Monday, 11 January 2016 (11:30–12:30)
Speaker : Kay Rülling
Title : Higher Chow groups with modulus and relative Milnor K-theory

For a pair (X, D) consisting of a smooth variety X over a field and an effective Catier divisor D, S. Saito and K. Kato defined in the 80's a relative version of Milnor K-theory in order to study higher dimensional geometric class field theory. Recently M. Kerz and S. Saito gave a different approach using Chow groups of zero cycles of (X, D). In this talk I will explain a link between these two approaches in the case where the support of D is a simple normal crossings divisor. Namely, in this case we construct a cycle map from the motivic complex of (X, D) in weight r, introduced by S. Saito and F. Binda, to the relative Milnor K-sheaf of (X, D) in degree r and shifted by $[-r]$. This map induces an isomorphism between the Nisnevich cohomology groups in degrees greater equal $r + \dim X$. This is joint work with S. Saito

Monday, 11 January 2016 (14:30–15:30)
Speaker : **Roy Joshua**
Title : **Equivariant Algebraic K-theory and Derived completion**

Derived completion is a technique that originated slightly over 20 years ago. However, Gunnar Carlsson may have been the first to realize the true potential of this technique, especially in the context of Algebraic K-theory. The goal of this talk is to discuss some applications of this technique, especially in the context of our joint work with Carlsson, to equivariant algebraic K-theory

Monday, 11 January 2016 (16:00–17:00)
Speaker : **Samik Basu**
Title : **Twisted Homology theories**

The study of twisted homology and cohomology generalises the definition of twisted K-theory by Karoubi and later studied by Atiyah, Segal and Freed, Hopkins and Teleman, The definition may be exteded to arbitrary commutative ring spectra. Apart from geometric considerations, the theory is amenable to detection of ring structures in module spectra and computation of invariants like topological Hochschild homology.

Tuesday, 12 January 2016 (10:00–11:00)
Speaker : **Hélène Esnault**
Title : **Chern classes of crystals**

The crystalline Chern classes of the value on the characteristic p variety of a torsion-free crystal depend only the isocrystal class, and vanish if the crystal is locally free or the isocrystal is convergent. It is conjectured that they always vanish. One deduces from it new cases of de Jong conjecture on the relation between the ètale fundamental group and the category of isocrystals. This is joint work with Atsushi Shiho, Tokyo University

Tuesday, 12 January 2016 (11:30–12:30)
Speaker : Christian Haesemeyer
Title : The K-theory of monoid algebras in mixed char-
acteristic

Given a commutative monoid M, the natural numbers act on M via dilations, that is, via power maps. Gubeladze conjectured that if M is contained
in a lattice and contains no non-trivial units, and if R is any regular commutative ring, then the dilations act nilpotently on the reduced K-theory of
the monoid algebra $R[M]$. Using trace methods, this had previously been
proved when R contains a field. In this talk on joint work with G. Cortiñas,
M. Walker and C. Weibel, we explain how to prove the conjecture in the
general case, building on our proof in positive characteristic

Tuesday, 12 January 2016 (14:30–15:30)
Speaker : Jinhyun Park
Title : Algebraic cycles and crystalline cohomology

I will talk about a description of the sheaf of big de Rham-Witt complexes
on smooth varieties in terms of certain algebraic cycle groups, called additive higher Chow groups. Its degree 0 part gives an intersection-theoretic
description of the ring structure of the ring of big Witt vectors, which is
notorious for being complicated. In addition we deduce a description of
the crystalline cohomology in terms of algebraic cycles. (joint work with
Amalendu Krishna)

Tuesday, 12 January 2016 (16:00–17:00)
Speaker : James Lewis
Title : The regulator map from Bloch's simplicial higher
Chow groups to Deligne cohomology

An explicit formula for the Bloch cycle class map from his higher Chow
groups to Deligne cohomology, was provided by Kerr/Lewis/Müller-Stach
[KLM] (Compositio Math 142) for projective algebraic manifolds, and in
the general case by Kerr-Lewis (Invent. Math. (2007)), based on a cubical
description of these groups. We provide an explicit formula for the simplicial version of these groups. This is based on joint work with Matt Kerr
and Patrick Lopatto, with an appendix by Jose Burgos Gil.

Wednesday, 13 January 2016 (10:00–11:00)
Speaker : Amnon Neeman
Title : Grothendieck duality via Hochschild homology

Hochschild cohomology was introduced in a 1945 paper by Hochschild and Grothendieck duality dates back to the early 1960s. The fact that the two have some relation with each other is very new — it came up in papers by Avramov and Iyengar [2008], Avramov, Iyengar, and Lipman [2010] and Avramov, Iyengar, Lipman and Nayak [2011]. We will review this history, and the surprising formulas that come out. We will then discuss more recent progress. The remarkable feature of all this is the role played by Hochschild homology. One example, which we will discuss in some detail, comes about as follows. The new techniques permit us to write formulas giving trace and residue maps in Grothendieck duality in terms of expressions that are very Hochschild-homological — Alonso, Jeremias and Lipman gave such a formula, but couldn't prove that it agrees with the usual formula dating back to Verdier in the 1960s. The proof that these two agree, due to Lipman and the speaker, turns out to hinge on considering the action of ordinary Hochschild homology on the various objects in the formula.

Wednesday, 13 January 2016 (11:30–12:30)
Speaker : **Anand Sawant**
Title : $\mathbb{A}^1$-**connectedness in reductive algebraic groups**

Invariants of algebraic groups such as the Whitehead group and R-equivalence have played an important role in the study of near-rationality properties of algebraic groups. Classical results about algebraic groups combined with recent works in motivic homotopy theory allow one to relate these invariants with $\mathbb{A}^1$-connectedness in certain simply connected algebraic groups. However, the picture is very different for algebraic groups that are not simply connected. We will discuss some recent results in this direction. The talk is based on joint work with Chetan Balwe.

Wednesday, 13 January 2016 (14:30–15:30)
Speaker : **Holger Reich**
Title : **Algebraic K-theory of group algebras and topological cyclic homology**

The Whiteheadgroup $Wh(G)$ and its higher analogues defined using algebraic K-theory play an important role in geometric topology. There are vanishing conjectures in the case G is torsionfree. For groups containing torsion the Farrell-Jones conjecture gives a conjectural description in terms of group homology. I will report on a new result which for example allows to detect a large direct summand inside the rationalized Whiteheadgroup of a group like Thompson's group T. The proof uses the cyclotomic trace to topological cyclic homology, Boekstedt-Hsiang-Madsen's functor C, and new general injectivity results about the assembly maps for THH and C. There are interesting relations to the Leopoldt-Schneider conjecture. The talk will report on joint work with Wolfgang Lueck (Bonn), John Rognes (Oslo) and Marco Varisco (Albany)

Wednesday, 13 January 2016 (16:00–17:00)
Speaker   :   **Andreas Rosenschon**
Title   :   **Torsion in the Lichtenbaum Chow group of arithmetic schemes**

We give an example of a smooth scheme over the spectrum of a Dedekind domain such that the torsion subgroup of a Lichtenbaum Chow group is infinite. This is joint work with V. Srinivas.

Thursday, 14 January 2016 (10:00–11:00)
Speaker   :   **Dan Edidin**
Title   :   **Strong regular embeddings of stacks and applications**

We explain how the notion of strong regular embedding can be used to compare the K-theory of a Deligne-Mumford stack to that of a regularly embedded substack. This can be applied to deduce results about K-theory of schemes in a number of interesting examples, such as the relationship between the K-theory of a hypertoric variety and its corresponding Lawrence toric variety. Although this talk is about stacks, the motivating examples come from observations about invariant rings for actions of finite groups.

Thursday, 14 January 2016 (11:30–12:30)
Speaker   :   **Amalendu Krishna**
Title   :   **Some K-cohomology of singular surfaces and applications**

We study the divisibility property of the cohomology groups of certain K-theory sheaves on singular surfaces and give some applications to algebraic cycles.

K-Theory
Copyright ©2018 Tata Institute of Fundamental Research
Publisher: Hindustan Book Agency, New Delhi, India

Relative Cartier Divisors and K-theory

Vivek Sadhu[1] and Charles Weibel

Abstract

We study the relative Picard group $\mathrm{Pic}(f)$ of a map $f : X \to S$ of schemes. If f is faithful affine, it is the relative Cartier divisor group $\mathcal{I}(f)$. The relative group $K_0(f)$ has a γ-filtration, and $\mathrm{Pic}(f)$ is the top quotient for the γ-filtration. When f is induced by a ring homomorphism $A \to B$, we show that the relative "nil" groups $N\,\mathrm{Pic}(f)$ and $NK_n(f)$ are continuous $W(A)$-modules.

Introduction

If $f : X \to S$ is a morphism of schemes, the relative Picard group $\mathrm{Pic}(f)$ was defined by Bass in [1], and fits into a natural exact sequence

$$\mathcal{O}^\times(S) \xrightarrow{f^*} \mathcal{O}^\times(X) \xrightarrow{\partial} \mathrm{Pic}(f) \longrightarrow \mathrm{Pic}(S) \xrightarrow{f^*} \mathrm{Pic}(X). \qquad (0.1)$$

The goal of this paper is to study this group as well as $N\,\mathrm{Pic}(f)$, defined to be $\mathrm{Pic}(f[t])/\mathrm{Pic}(f)$, where $f[t] : X \times \mathbb{A}^1 \to S \times \mathbb{A}^1$.

Our first observation is that when f is $\mathrm{Spec}(B) \to \mathrm{Spec}(A)$ for a commutative ring extension $A \hookrightarrow B$, $\mathrm{Pic}(f)$ is isomorphic to the relative Cartier divisor group $\mathcal{I}(f)$, defined in [13] as the group of invertible A-submodules of B under multiplication and studied in [15, 14, 16]. This definition of $\mathcal{I}(f)$ also makes sense (and we still have $\mathcal{I}(f) \cong \mathrm{Pic}(f)$) for scheme maps $f : X \to S$ for which $\mathcal{O}_S^\times \to f_*\mathcal{O}_X^\times$ is an injection of sheaves. It then follows from [16] that $\mathrm{Pic}(f)$ is a contracted functor in the sense of Bass.

We then relate $\mathrm{Pic}(f)$ to the relative group $K_0(f)$, which fits into an exact sequence

$$K_1(S) \xrightarrow{f^*} K_1(X) \xrightarrow{\partial} K_0(f) \longrightarrow K_0(S) \longrightarrow K_0(X).$$

For example, if $f : A \hookrightarrow B$ is subintegral then $K_0(f) \cong \mathrm{Pic}(f)$ (Proposition 2.5).

---

[1]Sadhu was supported by TIFR, Mumbai Postdoctoral Fellowship.

Let $\mathcal{NI}$ denote the Zariski sheaf associated to the presheaf $U \mapsto N\operatorname{Pic}(U, f^{-1}U)$ on S. In Theorem 4.1 and Theorem 4.7, we prove the following:

Theorem 0.2 *Let $f : X \to S$ be a faithful affine morphism of schemes.*

(1) *The Zariski sheaf $\mathcal{NI}$ is an étale sheaf on S. Moreover,*

$$N\operatorname{Pic}(f) \cong H^0_{\mathrm{et}}(S, \mathcal{NI}) = H^0_{\mathrm{zar}}(S, \mathcal{NI}).$$

(2) *If X and S are schemes then $H^*_{\mathrm{et}}(S, \mathcal{NI}) \cong H^*_{\mathrm{zar}}(S, \mathcal{NI})$.*
(3) *If X and S are both affine schemes then $H^q_{\mathrm{et}}(S, \mathcal{NI}) = 0$ for $q \neq 0$.*

A secondary goal of this article is to study the relative K-theory groups $K_n(f)$ associated to a morphism of schemes $f : X \to S$. By definition, $K_n(f) = \pi_n K(f)$, where $K(f)$ is the homotopy fiber of $K(S) \to K(X)$. Comparing $X \to S$ to $X[t] \to S[t]$ yields groups $NK_*(f)$.

Theorem 0.3 *For each homomorphism $f : A \to B$:*

(1) *$NK_n(f)$ is a continuous $W(A)$-module, for all n.*
(2) *$N\operatorname{Pic}(f)$ is a continuous $W(A)$-module.*
(3) *$\det : NK_0(f) \to N\operatorname{Pic}(f)$ is a $W(A)$-module homomorphism.*

(See Theorems 3.3 and 5.6, and Proposition 3.2). This implies that if $\operatorname{char}(A) = p > 0$ then both $NK_n(f)$ and $N\operatorname{Pic}(f)$ are p-groups, while if $\operatorname{char}(A) = 0$ the groups have the structure of A-modules.

We conclude with some remarks about $K_n(f)$ when n is negative. If X and S have dimension at most d, then $K_n(S) = K_n(X) = 0$ for $n < -d$ in many cases. In such cases, it follows that $K_n(f) = 0$ for $n < -d - 1$. The cohomological interpretation of the negative K-theory of a scheme in terms of the cdh-cohomology of the constant sheaf $\mathbb{Z}$ is given in [4]. In the relative situation, we prove the following (Theorem 6.2 and Theorem 6.3):

Theorem 0.4 *Let $f : X \to S$ be a finite morphism of d-dimensional noetherian schemes.*

(1) *If X and S are essentially of finite type over a field k of characteristic 0, $K_{-d-1}(f) \cong H^d_{\mathrm{cdh}}(S, f_*\mathbb{Z}/\mathbb{Z})$.*
(2) *If $\dim S = 1$, then $K_{-2}(f) \cong H^1_{\mathrm{nis}}(S, f_*\mathbb{Z}/\mathbb{Z})$ and there is an extension*

$$0 \longrightarrow H^1_{\mathrm{nis}}(S, f_*\mathcal{O}^\times_X/\mathcal{O}^\times_S) \longrightarrow K_{-1}(f) \longrightarrow H^0_{\mathrm{nis}}(S, f_*\mathbb{Z}/\mathbb{Z}) \longrightarrow 0.$$

1 Relative Pic and Invertible Submodules

In [1], Bass defined $\operatorname{Pic}(f)$ to be the abelian group generated by $[L_1, \alpha, L_2]$, where the L_i are line bundles on S and $\alpha : f^*L_1 \to f^*L_2$ is an isomorphism. The relations are:

1. $[L_1, \alpha, L_2] + [L_1', \alpha', L_2'] = [L_1 \otimes L_1', \alpha \otimes \alpha', L_2 \otimes L_2']$;
2. $[L_1, \alpha, L_2] + [L_2, \beta, L_3] = [L_1, \beta\alpha, L_3]$;
3. $[L_1, \alpha, L_2] = 0$ if $\alpha = f^*(\alpha_0)$ for some $\alpha_0 : L_1 \cong L_2$.

Remark 1.0.1 By (1), every element of $\operatorname{Pic}(f)$ has the form $[L, \alpha, \mathcal{O}_S]$. Writing $[L, \alpha]$ for $[L, \alpha, \mathcal{O}_S]$, an alternative presentation for $\operatorname{Pic}(f)$ is that it is generated by elements $[L, \alpha]$ satisfying: $[L, \alpha] + [L', \alpha'] = [L \otimes L', \alpha \otimes \alpha']$; $[L, \alpha] = 0$ if (and only if) there is an isomorphism $\alpha_0 : L \cong \mathcal{O}_S$ so that $\alpha = f^*(\alpha_0)$. It is easy to see, and observed by Bass, that the map $\operatorname{Pic}(f) \to \operatorname{Pic}(S)$ sending $[L, \alpha]$ to $[L]$ fits into an exact sequence (0.1), where $\partial(b) = [\mathcal{O}_S, b]$.

Proposition 1.1 *Bass' $\operatorname{Pic}(f)$ is the hypercohomology group $H^0(S, \mathcal{O}_S^\times \to f_*\mathcal{O}_X^\times)$.*

Proof Let C^* denote the mapping cone of $\mathcal{O}_S^\times \to f_*\mathcal{O}_X^\times$. A 0-cocyle of C^* is given by a cover $\{U_i\}$ of S, a unit b_i of $f^{-1}(U_i)$ for each i, and units a_{ij} of $U_i \cap U_j$ for each i, j satisfying the cocyle condition (so that the $\{a_{ij}\}$ define a line bundle L on S) and such that $b_i/b_j = f^\#(a_{ij})$ on each $f^{-1}(U_i \cap U_j)$. Since the $\{b_i\}$ define an isomorphism $f^*L \cong \mathcal{O}_X$, each 0-cocyle defines an element $\lambda = [L, \beta, \mathcal{O}_S]$ of $\operatorname{Pic}(f)$. A 0-coboundary is given by $a_{ij} = a_i/a_j$ and $b_i = f^\#(a_i)$ for units a_i of U_i; adding it to a cocyle does not change λ. Refining the cover does not change λ either. The result follows from the 5-lemma applied to the following diagram with exact rows (which is easily checked to be commutative):

$$H^0(S, \mathcal{O}^\times) \to H^0(X, \mathcal{O}^\times) \to H^0(S, C^*) \to H^1(S, \mathcal{O}^\times) \to H^1(X, \mathcal{O}^\times)$$

$$\cong\downarrow \qquad \cong\downarrow \qquad \downarrow \qquad \cong\downarrow \qquad \cong\downarrow$$

$$\mathcal{O}^\times(S) \longrightarrow \mathcal{O}^\times(X) \longrightarrow \operatorname{Pic}(f) \longrightarrow \operatorname{Pic}(S) \longrightarrow \operatorname{Pic}(X).$$

$\square$

Now suppose that f is faithful and affine. As observed in [16], $\mathcal{I}(f)$ is isomorphic to $H^0(S, f_*\mathcal{O}_X^\times/\mathcal{O}_S^\times)$. Thus Proposition 1.1 implies that $\mathcal{I}(f) \cong \operatorname{Pic}(f)$. Here is a more elementary proof.

Lemma 1.2 *If $f : X \to S$ is a faithful affine map, there is an isomorphism $\rho : \mathcal{I}(f) \overset{\cong}{\longrightarrow} \mathrm{Pic}(f)$, sending L to $[L, i, \mathcal{O}_S]$, where $i : f^*L \cong \mathcal{O}_X$.*

The isomorphism $f^*L \cong \mathcal{O}_X$ is well defined, because in any affine open $U = \mathrm{Spec}(A)$ of S we have $f^{-1}U = \mathrm{Spec}(B)$ with $A \subset B$; it was proven by Roberts and Singh [13] that $L \subset B$ induces $L \otimes_A B \cong B$.

Proof Since $\rho(LL') = [L \otimes L', i \otimes i', \mathcal{O}_S] = [L, i, \mathcal{O}_S] + [L', i', \mathcal{O}_S]$, ρ is a homomorphism. To define the inverse map, we use the presentation of $\mathrm{Pic}(f)$ and the observation that because $\mathcal{O}_S \to f_*\mathcal{O}_X$ is an injection, so is $L \to L \otimes f_*\mathcal{O}_X$ for every line bundle L. Given a triple $[L_1, \alpha, L_2]$, we set $L = L_2^{-1} \otimes L_1$, so that α induces an isomorphism $f^*L \cong f^*(L_2)^{-1} \otimes f^*(L_1) \cong \mathcal{O}_X$, and define $\psi([L_1, \alpha, L_2])$ to be the submodule L of $L \otimes f_*\mathcal{O}_X \cong f_*\mathcal{O}_X$. Since ψ is compatible with the relations of $\mathrm{Pic}(f)$, it descends to a homomorphism $\psi : \mathrm{Pic}(f) \to \mathcal{I}(f)$. Since $[L_1, \alpha, L_2] = [L_2^{-1} \otimes L_1, \alpha, \mathcal{O}_S]$ in $\mathrm{Pic}(f)$ and $f^*(L) = \mathcal{O}_X$ for all $L \in \mathcal{I}(f)$, ψ is an inverse to ρ.

$\square$

2   Relative K_0 and Pic

Bass gave a presentation of a relative group $K_0(f)$ associated to $f : A \to B$ in [1] and [2, VII.5]; see [29, II.2.10]. It is generated by triples $[P_1, \alpha, P_2]$, where the P_i are finitely generated projective A-modules (or vector bundles on S) and α is an isomorphism $f^*(P_1) \overset{\cong}{\longrightarrow} f^*(P_2)$, and agrees with the group $\pi_0 K(f)$ of [29, IV.1.11]. The relations are:

(1) $[P_1, \alpha, P_2] + [P_1', \alpha', P_2'] = [P_1 \oplus P_1', \alpha \oplus \alpha', P_2 \oplus P_2']$,

(2) $[P_1, \alpha, P_2] + [P_2, \beta, P_3] = [P_1, \beta\alpha, P_3]$,

(3) $[P_1, \alpha, P_2] = 0$ if $\alpha = f^*(\alpha_0)$ for some $\alpha_0 : P_1 \cong P_2$.

By (1), every element of $K_0(f)$ has the form $[P, \alpha, A^n]$.

Bass showed [2, VII.5.3] that there is an exact sequence for each $f : A \to B$:

$$K_1(A) \overset{f^*}{\longrightarrow} K_1(B) \overset{\partial}{\longrightarrow} K_0(f) \longrightarrow K_0(A) \longrightarrow K_0(B), \qquad (2.1)$$

where for $g \in GL_n(B)$ we have $\partial([g]) = [A^n, g, A^n]$. Since we do not know if the corresponding sequence is exact for a quasi-projective map $f : X \to S$, we will restrict to the affine case in this section and the next.

Lemma 2.2 (Excision) *Let $f : A \to B$ be a ring homomorphism, and let I be an ideal of A mapping isomorphically onto an ideal of B; write $\bar{f} : A/I \subset B/I$ for the induced map. Then excision holds for K_n for all $n \leq 0$: $K_n(f) \cong K_n(\bar{f})$.*

Proof It suffices to consider the case $n = 0$. Because $K_0(A, I) \cong K_0(B, I)$ [29, Ex. II.2.3] and $K_1(A, I) \to K_1(B, I)$ is onto [29, III.2.2.1], the double-relative group vanishes: $K_0(A, B, I) = 0$. Applying contraction, we also have $K_{-1}(A, B, I) = 0$. The result now follows from the exact sequence

$$K_0(A, B, I) \longrightarrow K_0(f) \longrightarrow K_0(\bar{f}) \longrightarrow K_{-1}(A, B, I).$$

$\square$

Remark The failure of Lemma 2.2 in the non-affine setting was investigated in [12, A.5–6]. For example, if X is the normalization of S and the support Y of the conductor $\mathfrak{c}$ is 1-dimensional, the obstruction is $K_0(S, X, Y) \cong H^1(Y, \mathfrak{c}/\mathfrak{c}^2 \otimes \Omega_{X/S})$.

As observed by Bass and Murthy long ago [3], the determinant $K_0(S) \to \mathrm{Pic}(S)$ induces a surjective homomorphism

$$\det : K_0(f) \to \mathrm{Pic}(f), \quad \det[P_1, \alpha, P_2] = [\det(P_1), \det(\alpha), \det(P_2)]. \quad (2.3)$$

Since $SK_0(S)$ is the kernel of $\det : K_0(S) \to \mathrm{Pic}(S)$, we write $SK_0(f)$ for the kernel of $\det : K_0(f) \to \mathrm{Pic}(f)$.

Recall [29, II.4.2] that a λ-ring $K = \mathbb{Z} \oplus \widetilde{K}$ has a *positive structure* if it contains a λ-semiring P (positive elements) including $\mathbb{N}$, such that every element of $\widetilde{K}$ can be written as a difference of positive elements, the augmentation $\epsilon : K \to \mathbb{Z}$ sends P to $\mathbb{N}$ and, if $p \in P$ has $\epsilon(p) = n$, then $\lambda^i p = 0$ for $i > n$ and $\lambda^n p$ is a unit. The *line elements* are $\{p \in P : \epsilon(p) = 1\}$; they form a subgroup of the units of K.

Proposition 2.4 *Let $f : A \to B$ be a homomorphism of commutative rings. The operations $\lambda^i[P_1, \alpha, P_2] = [\Lambda^i P_1, \Lambda^i \alpha, \Lambda^i P_2]$ give $\mathbb{Z} \oplus K_0(f)$ the structure of a λ-ring with a positive structure. The top two ideals in the γ-filtration are $F_\gamma^1 = \widetilde{K}_0$ and $F_\gamma^2 = SK_0(f)$, and the group of its line elements is $\mathrm{Pic}(f) \cong F_\gamma^1/F_\gamma^2$.*

Proof Given $f : A \to B$, choose a surjection $\pi : \mathbb{Z}[X] \to B$ from a polynomial ring $\mathbb{Z}[X]$ in many variables to B; let R be the pullback ring $R = \{(p, a) \in \mathbb{Z}[X] \times A : \pi(p) = f(a)\}$, with $\tilde{f} : R \to \mathbb{Z}[X]$ the projection.

Since $K_1(\mathbb{Z}[X]) = \pm 1$ and $K_0(\mathbb{Z}[X]) = \mathbb{Z}$, we have $K_0(\tilde{f}) \xrightarrow{\cong} \tilde{K}_0(R)$, and this map is compatible with the operations λ^i. Similarly, we have $\mathrm{Pic}(\tilde{f}) \cong \mathrm{Pic}(R)$. By Excision 2.2 for K_0 and Pic, $K_0(\tilde{f}) \cong K_0(f)$ and $\mathrm{Pic}(\tilde{f}) \cong \mathrm{Pic}(f)$. Hence $\mathbb{Z} \oplus K_0(f) \cong \mathbb{Z} \oplus \tilde{K}_0(R)$ is a λ-ring. Thus the result follows from the fact that the operations λ^i make $K_0(R)$ into a λ-ring, with $F_\gamma^2 = SK_0(R)$, and $\tilde{K}_0(R)/SK_0(R) \cong \mathrm{Pic}(R)$. $\qquad\square$

Recall (Swan [17]) that an extension $A \subset B$ is said to be *subintegral* if B is integral over A, and $\mathrm{Spec}(B) \to \mathrm{Spec}(A)$ is a bijection inducing isomorphisms on all residue fields.

Proposition 2.5 (Ischebeck) *If $f : A \hookrightarrow B$ is subintegral then $K_0(f) \cong \mathrm{Pic}(f)$, $K_n(f) = 0$ for all $n < 0$, and there is an exact sequence*

$$1 \longrightarrow B^\times/A^\times \longrightarrow K_0(f) \longrightarrow K_0(A) \longrightarrow K_0(B) \longrightarrow 0.$$

Proof When $A \subset B$ is subintegral, Ischebeck proved in [9, Prop. 7] that the natural map $K_0(A) \to K_0(B)$ is surjective and $SK_1(A) \to SK_1(B)$ is onto, so the cokernel of $K_1(A) \to K_1(B)$ is $B^\times/A^\times$. The exact sequence follows from (2.1). Finally, Ischebeck proved in [9, p. 331] that the determinant (2.3) induces an isomorphism from the kernel of $K_0(A) \to K_0(B)$ onto the kernel of $\mathrm{Pic}(A) \to \mathrm{Pic}(B)$. The result now follows from (2.1).

Replacing A and B by Laurent polynomial extensions, the Fundamental Theorem of K-theory [29, III.4.1] implies that $LK_n(f) \cong K_{n-1}(f)$ and $K_{-1}(f) \cong L\,\mathrm{Pic}(f)$. Since $A[t, 1/t] \subset B[t, 1/t]$ is subintegral, we have $L\,\mathrm{Pic}(f) = 0$ by Proposition 5.6 of [16]. This shows that that $K_n(f) = 0$ for all $n < 0$. $\qquad\square$

Given an extension $f : A \hookrightarrow B$, let $i : A \hookrightarrow {}^+A$ be the seminormalization of A in B and ${}^+f : {}^+A \hookrightarrow B$ the induced map. There is an exact sequence

$$\cdots \longrightarrow K_n(i) \longrightarrow K_n(f) \longrightarrow K_n({}^+f) \longrightarrow K_{n-1}(i) \longrightarrow \cdots.$$

Corollary 2.6 $K_n(f) \xrightarrow{\cong} K_n({}^+f)$ *for $n < 0$, and the following sequence is exact.*
$$0 \longrightarrow K_0(i) \longrightarrow K_0(f) \longrightarrow K_0({}^+f) \longrightarrow 0.$$

Proof By Proposition 2.5 and [16, Lemma 3.3], the map $K_0(i) \cong \mathrm{Pic}(i) \to \mathrm{Pic}(f)$ is an injection. Since it factors through $K_0(i) \to K_0(f)$, the latter map is an injection. Since $K_n(i) = 0$ for $n < 0$, again by Proposition 2.5, we are done.

$\square$

3 The $W(A)$-module Structure on $NK_0(f)$ and $N\operatorname{Pic}(f)$

In this section, we fix a ring homomorphism $f : A \to B$ and show that $NK_0(f)$ and $N\operatorname{Pic}(f)$ are continuous modules over the ring $W(A)$ of big Witt vectors, so that

$$NK_1(A) \longrightarrow NK_1(B) \xrightarrow{\partial} NK_0(f) \longrightarrow NK_0(A) \longrightarrow NK_0(B) \quad (3.1)$$

is a sequence of $W(A)$-modules. Recall that $(1 + tA[[t]])^\times$ is the underlying abelian group of the ring $W(A)$; a $W(A)$-module is continuous if every element is killed by one of these ideals $(1 + t^n A[[t]])^\times$.

We first recall the continuous $W(R)$-module structure on $NK_*(A)$ when R is commutative and A is an R-algebra, due to Stienstra [18]. As $NK_*(A)$ is a continuous module, it suffices to describe multiplication by $(1 - rt^m)$, $r \in R$. Setting $S = R[s]/(s^m - r)$, the inclusion $i : R \subset S$ induces a base change functor $i^* : \mathbf{P}(A[t]) \to \mathbf{P}(A \otimes_R S[t])$ and a transfer map $i_* : \mathbf{P}(A \otimes_R S[t]) \to \mathbf{P}(A[t])$. If σ denotes the S-algebra map $S[t] \to S[t]$, $\sigma(t) = st$, then the composition $F = i_* \sigma^* i^*$ is an additive self-functor of $\mathbf{P}(A[t])$. As noted in [26, 1.5], the composition $\mathbf{P}(A) \to \mathbf{P}(A[t]) \xrightarrow{F} \mathbf{P}(A[t]) \to \mathbf{P}(A)$ is $\otimes_R S$, so F induces multiplication by m on the summand $K_*(A)$ of $K_*(A[t])$; the restriction of F to $NK_*(A)$ is multiplication by $(1 - rt^m)*$. If $A \to B$ is an R-algebra map, $NK_*(A) \to NK_*(B)$ is a homomorphism of continuous $W(R)$-modules.

We can adapt these formulas to define a multiplication by $(1 - at^m)*$ on $K_0(f)$ and $NK_0(f)$ when $a \in A$: send $[P_1, \alpha, P_2]$ to $[F(P_1), F(\alpha), F(P_2)]$. It is clear from (2.1) that $(1 - at^m)*$ is compatible with the exact sequence (3.1). A priori, though, the maps $(1 - at^m)*$ do not fit together to make $NK_0(f)$ into a $W(A)$-module.

Proposition 3.2 *For any homomorphism* $f : A \to B$, $NK_0(f)$ *is a continuous* $W(A)$-*module, and* (3.1) *is an exact sequence of continuous* $W(A)$-*modules.*

Proof As in the proof of Proposition 2.4, write $B = \mathbb{Z}[X]/I$, where $\mathbb{Z}[X]$ is a polynomial ring. Let R denote the pullback ring $A \times_B \mathbb{Z}[X]$, and write $\tilde{f} : R \to \mathbb{Z}[X]$ for the quotient map. Since $NK_*(\mathbb{Z}[X]) = 0$, we have

$NK_n(\tilde{f}) \cong NK_n(R)$ for all n. Since $A = R/I$, Lemma 2.2 and [25] imply that the groups $NK_0(f) \cong NK_0(\tilde{f}) \cong NK_0(R)$ are continuous $W(R)$-modules.

Since $W(A) = W(R)/W(I)$, where $W(I) = 1 + tI[[t]]$, we are reduced to showing that $(1 - rt^m)$ acts as zero on $K_0(f)$ whenever $r \in I$. When r is in the kernel I of $R \to A$, the ring $A \otimes_R S$ is just $A[s]/(s^m)$, so $(1 - rt^m)$ and $(1 - 0t^m)$ act identically on $K_0(f[t])$. This shows that $(1 - rt^m)$ acts as zero on $K_0(f)$ and proves that the action of $W(A)$ on $K_0(f)$ is well defined and continuous. $\square$

Applying N to the determinant described in (2.3), we get an exact sequence

$$0 \longrightarrow NSK_0(f) \longrightarrow NK_0(f) \xrightarrow{\det} N\operatorname{Pic}(f) \longrightarrow 0.$$

If $[P, \alpha, A[t]^n]$ is in $NK_0(f)$ then $\det[P, \alpha, A[t]^n] = [\det(P), \det(\alpha), A[t]]$.

Theorem 3.3 *For any homomorphism* $f : A \to B$, $N\operatorname{Pic}(f)$ *is a continuous* $W(A)$-*module, and* $\det : NK_0(f) \to N\operatorname{Pic}(f)$ *is a* $W(A)$-*module homomorphism.*

Proof Since the group $NK_0(f)$ is a continuous $W(A)$-module by Proposition 3.2, it is enough to show that $NSK_0(f)$ is closed under multiplication by $W(A)$. Since every element of $W(A)$ can be written as $\prod_{m>0}(1 - a_m t^m)$, with $a_m \in A$, and for any element u of $NK_0(f)$ there is an n so that $\prod_{m \geq n}(1 - a_m t^m) * u = 0$, it is enough to show that $NSK_0(f)$ is closed under multiplication by $(1 - at^m)$ for any $a \in A$ and $m \geq 1$.

It is enough to show that $F = i_* \sigma^* i^*$ sends $SK_0(f[t])$ to itself. We now modify the argument of [5, 4.1]. Fix $u = [P, \alpha, A[t]^n]$ in $SK_0(f[t])$; By Remark 1.0.1, $\det(u) = 0$ implies that $\det(P) = A[t]$ and $\det(\alpha) \in A$. By naturality of det, $\sigma^* i^*(u) = [P \otimes S, \alpha \otimes S, S[t]^n]$, $\det(P \otimes S) = S[t]$, $\det(\alpha \otimes S) \in S$ and $F(u) = [i_*(P \otimes S), i_*(\alpha \otimes S), A[t]^n]$. By [5, Corollary 3.2] applied to $A[t] \subset S[t]$, $\det(i_*(P \otimes S)) = A[t]$ and $\det(\alpha \otimes S) = \det(\alpha)^m \in A$, so $\det(F(u)) = 0$. $\square$

Corollary 3.4 *If* $\operatorname{char}(A) = p$ *then* $N\operatorname{Pic}(f)$ *is a* p-*group. If* $\mathbb{Q} \subseteq A$ *then* $N\operatorname{Pic}(f)$ *is an* A-*module.*

Proof Any continuous $W(A)$-module has these properties; see [25, 3.3]. $\square$

4   Sheaf Properties of $N \operatorname{Pic}(f)$

When $f : X \to S$ is a faithful affine morphism of schemes, let $\mathcal{I}(f)_{\mathrm{zar}}$ denote the Zariski sheaf $f_* \mathcal{O}_X^\times / \mathcal{O}_S^\times$ on the category Sm/S of smooth schemes over S; by [16, 4.4], $\mathcal{I}(f)_{\mathrm{zar}}$ is also an étale sheaf, and $H_{\mathrm{et}}^0(S, \mathcal{I}(f)_{\mathrm{zar}}) = H_{\mathrm{nis}}^0(S, \mathcal{I}(f)_{\mathrm{zar}}) = \operatorname{Pic}(f)$. Our choice of Sm/S is dictated by the need to not only include étale extensions but be closed under product with $\mathbb{A}_S^1 \xrightarrow{\pi} S$.

Let $\pi^* \mathcal{I}(f)$ denote the restriction of $\mathcal{I}(f)_{\mathrm{zar}}$ to $Sm/\mathbb{A}_S^1$ along π. Its direct image $\pi_*(\pi^* \mathcal{I}(f))$ is the Zariski sheaf $\mathcal{I}(f)_{\mathrm{zar}} \oplus \mathcal{NI}(f)$ on Sm/S, where $\mathcal{NI}(f)$ denotes the Zariski sheaf on Sm/S associated to the presheaf $U \mapsto N \operatorname{Pic}(f \times_S U)$.

Theorem 4.1 *Let $f : X \to S$ be a faithful affine morphism of schemes. Then $\mathcal{NI}(f)$ is an étale sheaf on S. Moreover,*

$$H_{\mathrm{et}}^0(S, \mathcal{NI}(f)) = H_{\mathrm{zar}}^0(S, \mathcal{NI}(f)) = N \operatorname{Pic}(f).$$

Proof   Since $\pi^* \mathcal{I}(f)$ is an étale sheaf on $\mathbb{A}_S^1$, its direct image $\pi_* \pi^* \mathcal{I}(f)$ is an étale sheaf on S; since $\pi_* \pi^* \mathcal{I}(f) \cong \mathcal{I}(f)_{\mathrm{zar}} \oplus \mathcal{NI}(f)$, $\mathcal{NI}(f)$ is also an étale sheaf. Since

$$H_{\mathrm{et}}^0(S, \pi_* \pi^* \mathcal{I}(f)) = H_{\mathrm{et}}^0(\mathbb{A}_S^1, \pi^* \mathcal{I}(f)) = \operatorname{Pic}(f[t]) = \operatorname{Pic}(f) \oplus N \operatorname{Pic}(f),$$

we see that $H_{\mathrm{et}}^0(S, \mathcal{NI}(f)) = N \operatorname{Pic}(f)$. If S_s is a Zariski local scheme of S, this shows that the stalk $\mathcal{NI}(f)_s = H_{\mathrm{zar}}^0(S_s, \mathcal{NI}(f))$ equals $H_{\mathrm{et}}^0(S_s, \mathcal{NI}(f))$.

$\square$

Example 4.2 If f is seminormal, the sheaf $\mathcal{NI}(f)$ vanishes and $N \operatorname{Pic}(f) = 0$. This follows from Theorem 4.1 and [15, 1.5], which states that $N \operatorname{Pic}(A, B) = 0$ when A is seminormal in B.

We now modify an argument of Vorst [22] and van der Kallen [21]. Suppose that $\operatorname{Spec}(A) = \bigcup_{i=0}^r U_i$, where $U_i = \operatorname{Spec}(A_{s_i})$. Given a presheaf F of abelian groups on $\operatorname{Spec}(A)$, we write $C^\bullet(\{U_i\}, F)$ for the augmented Čech complex:

$$0 \to F(A) \xrightarrow{\epsilon} \prod_{i=0}^r F(A_{s_i}) \to \prod_{0 \le i < j \le r} F(A_{s_i s_j}) \to \cdots \to F(A_{s_0 s_1 \cdots s_r}) \to 0.$$

Given $s \in A$, we have an A-algebra map $\sigma : A[x] \to A[x]$ determined by $\sigma(x) = sx$. We write $N F(A)_{[s]}$ for the direct limit of $F(A[x]) \xrightarrow{\sigma} F(A[x]) \xrightarrow{\sigma} \cdots$. Suppose that for all $0 \le i_0 < i_1 < \cdots < i_p \le r$ and $j \le p$:

$$N F(A_{s_{i_0} \cdots s_{i_j} \cdots s_{i_p}}[x]) \cong N F(A_{s_{i_0} \cdots \hat{s}_{i_j} \cdots s_{i_p}}[x])_{[s_{i_j}]}. \tag{4.3}$$

In this situation, Vorst proved [22, 1.2] that the sequence $C^\bullet(\{U_i\}, NF)$ is always exact. He also proved that $F = NK_n$ satisfied (4.3), so that $C^\bullet(\{U_i\}, NK_n)$ is exact for all n. (See [22, 1.4] or [29, V.8.5]; the nonzero-divisor hypothesis is unnecessary by [20].)

Remark 4.4 It is easy to see (and follows from Vorst's result [22, 1.2]) that the functor $NU(A) = (A[t])^\times/A^\times$ satisfies (4.3). From the exact sequence of complexes

$$0 \longrightarrow C^\bullet(\{U_i\}, NU) \longrightarrow C^\bullet(\{U_i\}, NU(- \otimes_A B)) \longrightarrow$$
$$C^\bullet(\{U_i\}, NU(- \otimes_A B)/NU) \longrightarrow 0$$

we see that $C^\bullet(\{U_i\}, F)$ is also exact for the functor $F(A_s) = NU(B_s)/NU(A_s)$.

Lemma 4.5 $C^\bullet(\{U_i\}, N\operatorname{Pic})$ *is always an exact sequence.*

Proof By [27, Theorem 4.2], given $s \in A$ we have $N\operatorname{Pic}(A_s) \cong N\operatorname{Pic}(A)_{[s]}$ and hence $N\operatorname{Pic}(A_s[x]) \cong N\operatorname{Pic}(A[x])_{[s]}$. This implies that $N\operatorname{Pic}$ satisfies (4.3). Vorst's result shows that $C^\bullet(\{U_i\}, N\operatorname{Pic})$ is an exact sequence. $\square$

We apply these considerations to the presheaf $N\operatorname{Pic}(f): U \mapsto N\operatorname{Pic}(f|_U)$.

Lemma 4.6 *Suppose that* $\operatorname{Spec}(A) = \bigcup_{i=0}^n U_i$, *where* $U_i = \operatorname{Spec}(A_{s_i})$. *If* $f : A \hookrightarrow B$ *is a ring extension, the complex* $C^\bullet(\{U_i\}, N\operatorname{Pic}(f))$ *is exact.*

$$0 \to N\operatorname{Pic}(A, B) \to \prod_{i=0}^n N\operatorname{Pic}(A_{s_i}, B_{s_i}) \to \prod_{i_i < i_2} N\operatorname{Pic}(A_{s_{i_1} s_{i_2}}, B_{s_{i_1} s_{i_2}}) \to \cdots$$

Proof Let ^+A denote the subintegral closure of A in B, so ^+A is seminormal in B and we have $A \subset {}^+A \subset B$. By [14, Prop. 4.1], we have an exact sequence

$$1 \longrightarrow N\operatorname{Pic}(A, {}^+A) \longrightarrow N\operatorname{Pic}(A, B) \longrightarrow N\operatorname{Pic}({}^+A, B) \longrightarrow 1.$$

By Example 4.2, the third term vanishes and we have $N\operatorname{Pic}(A, {}^+A) \cong N\operatorname{Pic}(A, B)$. Thus we may assume that B is subintegral over A. In this case, Ischebeck proved [9, Prop. 7] that $N\operatorname{Pic}(A) \to N\operatorname{Pic}(B)$ is surjective. Now the result follows from Remark 4.4, Lemma 4.5 and the long exact cohomology sequences associated to

$$0 \longrightarrow C^\bullet(\{U_i\}, F) \longrightarrow C^\bullet(\{U_i\}, N\operatorname{Pic}(f)) \longrightarrow C^\bullet(\{U_i\}, N\operatorname{Pic}(f)/F) \longrightarrow 0,$$

$$0 \to C^\bullet(\{U_i\}, N\operatorname{Pic}(f)/F) \to C^\bullet(\{U_i\}, N\operatorname{Pic}) \to C^\bullet(\{f^{-1}(U_i)\}, N\operatorname{Pic}) \to 0.$$

□

Theorem 4.7 *Let $f : A \hookrightarrow B$ be an extension of rings. Then:*

$$H^q_{\text{et}}(\operatorname{Spec}(A), \mathcal{NI}) = \begin{cases} N\operatorname{Pic}(f) & \text{if } q = 0 \\ 0 & \text{if } q > 0. \end{cases}$$

Proof The case $q = 0$ is given by Theorem 4.1. By Lemma 4.6, the Čech cohomology groups $\check{H}^q(\operatorname{Spec}(A), \mathcal{NI})$ vanish for $q > 0$. Using the Cartan criterion [11, III.2.17], $H^q_{\text{et}}(\operatorname{Spec}(A), \mathcal{NI})$ equals $\check{H}^q(\operatorname{Spec}(A), \mathcal{NI}) = 0$ for $q > 0$.

□

Corollary 4.8 *Let $f : X \to S$ be a faithful affine morphism of schemes. Then*

$$H^*_{\text{et}}(S, \mathcal{NI}) \cong H^*_{\text{zar}}(S, \mathcal{NI}).$$

Proof Consider the site change map $\tau : S_{\text{et}} \to S_{\text{zar}}$. Then by Theorem 4.7, the higher direct image sheaves $R^q\tau_* \mathcal{NI}$ vanish for $q > 0$. Therefore the Leray spectral sequence degenerates, yielding the result.

□

Remark More generally, if $f : X \to S$ is any morphism of schemes then $\mathcal{O}_S^\times$ may not inject into $f_*\mathcal{O}_X^\times$. In this case, if we interpret $f_*\mathcal{O}_X^\times/\mathcal{O}_S^\times$ as the mapping cone of $\mathcal{O}_S^\times \to f_*\mathcal{O}_X^\times$ (a complex of Zariski sheaves) and use sheaf hypercohomology, then Theorem 4.1 remains valid. However, Theorem 4.7 may fail in this setting.

5 Module Structures on $NK_n(f)$

Given an exact functor $F : \mathcal{P} \to \mathcal{Q}$, the relative K-theory groups $K_n(F)$ fit into an exact sequence

$$\cdots \xrightarrow{F} K_{n+1}\mathcal{Q} \xrightarrow{\partial} K_n(F) \longrightarrow K_n\mathcal{P} \xrightarrow{F} K_n\mathcal{Q} \xrightarrow{\partial} \cdots$$

ending in $K_0\mathcal{Q} \xrightarrow{\partial} K_{-1}(F)$. Waldhausen showed that the $K_n(F)$ are the homotopy groups $\pi_{n+2}|wS_\bullet(S_\bullet F)|$ $(n \geq 0)$, where $S_n F$ denotes the category of pairs

$$(P_*, Q_*) = (P_1 \rightarrowtail P_2 \rightarrowtail \cdots \rightarrowtail P_n, Q_0 \rightarrowtail Q_1 \rightarrowtail \cdots \rightarrowtail Q_n)$$

($P_i \in \mathcal{P}$ and $Q_j \in \mathcal{Q}$), together with choices of Q_i/Q_j for $i > j$, such that $F(P_*)$ is $Q_1/Q_0 \rightarrowtail \cdots \rightarrowtail Q_n/Q_0$. (See [23, 1.5.4–7] or [29, IV.8.5.3].)

Example 5.1 If A is a ring, we write $\mathbf{P}(A)$ for the category of finitely generated projective A-modules. Given a ring homomorphism $f : A \to B$, we have an exact functor $\mathbf{P}(f) : \mathbf{P}(A) \to \mathbf{P}(B)$; by abuse, we write $K_*(f)$ for $K_*\mathbf{P}(f)$. Writing $f[t]$ for $A[t] \to B[t]$, we have $K_*(f[t]) = K_*(f) \oplus NK_*(f)$. The Fundamental Theorem of K-theory easily extends to the relative setting, yielding

$$K_*(f[t, 1/t]) \cong K_*(f) \oplus NK_*(f) \oplus NK_*(f) \oplus K_{*-1}(f).$$

Let A be a commutative ring. As in [29], we write $\mathbf{End}(A)$ for the category of pairs (P, α), where P in $\mathbf{P}(A)$ and $P \overset{\alpha}{\to} P$ is an endomorphism, and write $\mathbf{Nil}(A)$ for the full subcategory of $\mathbf{End}(A)$ consisting of all (P, α) with α nilpotent. As pointed out in [29, II.7.4], $K_*\mathbf{End}(A) \cong K_*(A) \oplus \mathrm{End}_*(A)$ and $K_*\mathbf{Nil}(A) \cong K_*(A) \oplus \mathrm{Nil}_*(A)$, where $\mathrm{End}_*(A)$ is a graded-commutative ring and $\mathrm{Nil}_*(A)$ is a graded $\mathrm{End}_*(A)$-module. By naturality, the exact functors $\mathbf{Nil}(f) : \mathbf{Nil}(A) \to \mathbf{Nil}(B)$ yield relative groups $K_*\mathbf{Nil}(f) \cong K_*(f) \oplus \mathrm{Nil}_*(f)$.

The category $\mathbf{Nil}(A)$ is equivalent to the category $\mathbf{H}_{1,t}(A[t])$ of t-primary torsion $A[t]$-modules M with $pd_{A[t]}M = 1$. Specifically, if (P, ν) is in $\mathbf{Nil}(A)$, and we write P_ν for the $A[t]$-module P on which t acts as ν, then P_ν has projective dimension 1 over $A[t]$. The Fundamental Theorem ([29, V.8.2]) implies that $\mathrm{Nil}_n(A) \cong NK_{n+1}(A)$. We also have $K\mathbf{P}(A[t]) \cong K\mathbf{H}(A[t])$ (see e.g., [29, V.3.2]).

Proposition 5.2 *There is a natural isomorphism* $\mathrm{Nil}_n(f) \cong NK_{n+1}(f)$.

Proof From the diagram of exact categories

$$\mathbf{Nil}(A) \overset{\cong}{\longrightarrow} \mathbf{H}_{1,t}(A[t]) \longrightarrow \mathbf{H}(A[t]) \overset{\cong}{\longleftarrow} \mathbf{P}(A[t]) \longrightarrow \mathbf{P}(A[t, 1/t])$$
$$\downarrow \qquad\qquad\qquad\qquad\qquad\qquad\qquad \downarrow \qquad\qquad\qquad \downarrow$$
$$\mathbf{Nil}(B) \overset{\cong}{\longrightarrow} \mathbf{H}_{1,t}(B[t]) \longrightarrow \mathbf{H}(B[t]) \overset{\cong}{\longleftarrow} \mathbf{P}(B[t]) \longrightarrow \mathbf{P}(B[t, 1/t])$$

we get a fibration sequence of K-theory spectra

$$K\mathbf{Nil}(A) \longrightarrow K(A[t]) \longrightarrow K(A[t, 1/t])$$
$$\downarrow \qquad\qquad \downarrow{\scriptstyle f[t]^*} \qquad\qquad \downarrow{\scriptstyle f[t,1/t]^*}$$
$$K\mathbf{Nil}(B) \longrightarrow K(B[t]) \longrightarrow K(B[t, 1/t]).$$

Taking vertical fibers, we see that there is a long exact sequence

$$K_{n+1}(f[t]) \to K_{n+1}(f[t, 1/t]) \to K_n\mathbf{Nil}(f) \to K_n(f[t]) \to K_n(f[t, 1/t]) \to$$

and (using Example 5.1) an isomorphism $\mathrm{Nil}_n(f) \cong NK_{n+1}(f)$.

$\square$

Lemma 5.3 *For any ring homomorphism* $f : A \to B$, $\mathrm{Nil}_*(f)$ *is a graded* $\mathrm{End}_*(A)$-*module.*

Proof A typical object in the Waldhausen category $S_n\mathbf{Nil}(f)$ is a pair

$$(\mu_*, \nu_*) = ((M_1, \mu_1) \rightarrowtail \cdots (M_n, \mu_n), (N_0, \nu_0) \rightarrowtail \cdots (N_n, \nu_n)).$$

There is a pairing $\mathbf{End}(A) \times S.\mathbf{Nil}(f) \to S.\mathbf{Nil}(f)$ of simplicial Waldhausen categories, sending $(P, \alpha) \times (\mu_*, \nu_*)$ to

$$((P \otimes M_1, \alpha \otimes \mu_1) \rightarrowtail \cdots \rightarrowtail (P \otimes M_n, \alpha \otimes \mu_n), (P \otimes N_0, \alpha \otimes \nu_1) \rightarrowtail \cdots \rightarrowtail$$
$$(P \otimes N_n, \alpha \otimes \nu_n)).$$

It induces a pairing $K_*\mathbf{End}(A) \otimes K_*\mathbf{Nil}(f) \to K_*\mathbf{Nil}(f)$. Since the tensor product $(\alpha \otimes \beta) \otimes \mu \cong \alpha \otimes (\beta \otimes \mu)$ is associative up to natural isomorphism, the two pairings

$$\mathbf{End}(A) \times \mathbf{End}(A) \times S.\mathbf{Nil}(f) \longrightarrow S.\mathbf{Nil}(f)$$

agree up to natural isomorphism, making $K_*\mathbf{Nil}(f)$ a graded $K_*\mathbf{End}(A)$-module. In particular, $\mathrm{Nil}_*(f)$ is a graded module over $\mathrm{End}_*(A)$.

$\square$

Recall that the ring $W(A)$ of big Witt vectors has underlying abelian group $(1 + tA[[t]])^\times$. Almkvist's theorem [29, II.7.4.3] states that $[P, \alpha] \mapsto \det(1 - t\alpha)$ maps $\mathrm{End}_0(A)$ isomorphically onto the subring of $W(A)$ whose underlying abelian group consists of all quotients $f(t)/g(t)$ of polynomials in $1 + tA[t]$. The intersection of the ring $\mathrm{End}_0(A)$ with the ideal $(1 + t^m A[[t]])$ of $W(A)$ is the ideal $I_m = \{1 + t^m(f/g)\}$ of $\mathrm{End}_0(A)$, and $\mathrm{End}_0(A)/I_m \cong W(A)/(1 + t^m A[[t]])$. In particular, $W(A)$ is the completion of $\mathrm{End}_0(A)$ with respect to the t-adic filtration.

We say that an $\mathrm{End}_0(A)$-module M is *continuous* if for every $x \in M$ there is an m so that $I_m \cdot x = 0$. Thus every continuous $\mathrm{End}_0(A)$-module M is also continuous as a $W(A)$-module: for every $x \in M$ we have $(1 + t^m A[[t]]) \cdot x = 0$ for some m.

The exact functors $F_n, V_n : \mathbf{Nil}(A) \to \mathbf{Nil}(A)$, defined by $F_n(P, \nu) = (P, \nu^n)$ and $V_n(Q, \nu) = (Q[t]/(t^n - \nu), t)$, commute with $\mathbf{Nil}(A) \to \mathbf{Nil}(B)$. Hence they induce exact endofunctors F_n, V_n on $S.\mathbf{Nil}(f)$ by $F_n(\mu_*, \nu_*) = (F_n(\mu_*), F_n(\nu_*))$ and $V_n(\mu_*, \nu_*) = (V_n(\mu_*), V_n(\nu_*))$. For $a \in A$ and $n > 0$, and ν in $\mathrm{Nil}_*(f)$, Almkvist's theorem associates $(1 - at^n)$ to $V_n([A, a] - [A, 0])$ and yields the product formula

$$(1 - at^n) * \nu = V_n([A, a] - [A, 0]) * \nu. \tag{5.4}$$

Stienstra proved in [18, 19] that the $\mathrm{Nil}_n(A)$ are continuous $\mathrm{End}_0(A)$-modules, and hence $W(A)$-modules. The key step [18, 2.12] was showing that the projection formula holds:

$$(V_n \alpha) * \nu = V_n(\alpha * F_n(\nu)) \quad \text{for} \quad \alpha \in \mathrm{End}_0(A) \text{ and } \nu \in \mathrm{Nil}_*(A).$$

Here is the corresponding projection formula in the relative setting; we will postpone its proof in order to get to the main result.

Lemma 5.5 *For all $\alpha \in \mathrm{End}_0(A)$ and $\beta \in \mathrm{Nil}_*(f)$,*

$$(V_n \alpha) * \beta = V_n(\alpha * F_n(\beta)).$$

Theorem 5.6 *Let $f : A \to B$ be a ring map. Then the product (5.4) makes $\mathrm{Nil}_n(f) \cong NK_{n+1}(f)$ into a continuous $W(A)$-module for every integer n.*

Proof For each $m > 0$, let $\mathbf{Nil}^m(A)$ denote the exact subcategory of all (P, ν) in $\mathbf{Nil}(A)$ such that $\nu^m = 0$. Thus we have relative groups $K_* \mathbf{Nil}^m(f)$ associated to $K_* \mathbf{Nil}^m(A) \to K_* \mathbf{Nil}^m(B)$, and $K_* \mathbf{Nil}(f)$ is the direct limit of the $K_* \mathbf{Nil}^m(f)$.

Suppose that $n \geq m$. Clearly, F_n acts as zero on $\mathbf{Nil}^m(f)$. By the projection formula 5.5, $V_n(\alpha)$ acts as zero on the image $\mathrm{Nil}_*^m(f)$ of $K_* \mathbf{Nil}^m(f) \to K_* \mathbf{Nil}(f) \to \mathrm{Nil}_*(f)$. By (5.4), $(1 - at^n)$ acts as zero on $\mathrm{Nil}_*^m(f)$. Since $\mathrm{Nil}_*(f)$ is the union of the $\mathrm{Nil}_*^m(f)$, for any $\beta \in \mathrm{Nil}_*(f)$ there is an m such that $(1 - at^n) \cdot \beta = 0$ for all $n \geq m$ and $a \in A$. This shows that $\mathrm{Nil}_*(f)$ is a continuous $\mathrm{End}_0(A)$-module, and hence a continuous $W(A)$-module.
$\square$

Proof (of Lemma 5.5) Following Stienstra [18, §6], set $R = \mathbb{Z}[y_1, y_2]$, and set $\mathbf{E} = \mathbf{End}(R; S_6)$, where S_6 is the multiplicative subset of $R[x]$ generated by x and $x^n - y_1^n y_2$. As pointed out in *loc. cit.*, there is a multi-exact pairing

$$\Theta : \mathbf{E} \times \mathbf{End}(A) \times \mathbf{Nil}(B) \longrightarrow \mathbf{Nil}(B)$$

sending (E, ω), (P, α) and (N, ν) to $(E \otimes_R (P \otimes_A N), \omega \otimes 1)$, where $P \otimes_A N$ is ragarded as an R-module by letting y_1 and y_2 act as $\alpha \otimes 1$ and $1 \otimes \nu$. As this pairing is natural in B, we may replace $\mathbf{Nil}(B)$ by $S.\mathbf{Nil}(f)$. This yields (among other things) a product

$$\Theta_* : K_0\mathbf{E} \otimes \mathrm{End}_0(A) \otimes \mathrm{Nil}_*(f) \longrightarrow \mathrm{Nil}_*(f).$$

Stienstra proves in *loc. cit.* that the elements $[R^n, \omega]$ and $[R^n, \omega']$ agree in $K_0\mathbf{E}$, where

$$\omega = \begin{pmatrix} 0 & & & y_1^n y_2 \\ 1 & & & 0 \\ & \ddots & & \vdots \\ 0 & & 1 & 0 \end{pmatrix} \quad \text{and} \quad \omega' = \begin{pmatrix} 0 & & & y_1 y_2 \\ y_1 & & & 0 \\ & \ddots & & \vdots \\ 0 & & y_1 & 0 \end{pmatrix}.$$

Therefore the two maps

$$\Theta_*([R^n, \omega], -), \Theta_*([R^n, \omega'], -) : \mathrm{End}_0(A) \otimes \mathrm{Nil}_*(f) \longrightarrow \mathrm{Nil}_*(f)$$

agree. Stienstra also observes that these maps send $[P, \alpha] \otimes \beta$ to $V_n(\alpha * F_n \beta)$ and $(V_n \alpha) * \beta$, respectively; see also [19, p.14]. The projection formula follows.

$\square$

6 Negative Relative K-theory

Let $f : X \to S$ be a morphism of schemes. Then we have a long exact sequence of negative K-groups, part of which is:

$$\cdots \longrightarrow K_{-d}(f) \longrightarrow K_{-d}(S) \longrightarrow K_{-d}(X) \longrightarrow K_{-d-1}(f) \longrightarrow K_{-d-1}(S) \longrightarrow \cdots . \tag{6.1}$$

Theorem 6.2 *Let $f : X \to S$ be a morphism of d-dimensional schemes, essentially of finite type over a field k of characteristic 0. Then for all $r > 0$:*

(1) $K_n(f) = K_n(f \times \mathbb{A}^r) = 0$ *for $n \le -d - 2$.*

(2) $K_{-d-1}(f) \cong K_{-d-1}(f \times \mathbb{A}^r)$ *("f is K_{-d-1}-regular.")*

(3) *If f is a finite map then $K_{-d-1}(f) \cong H^d_{\mathrm{cdh}}(S, f_*\mathbb{Z}/\mathbb{Z})$.*

Proof By [4, Corollary 5.9, Theorem 6.2], $K_n(S) \cong K_n(S \times \mathbb{A}^r)$ for all $n \leq -d$, $K_n(S) = 0$ for $n < -d$ and $K_{-d}(S) \cong H^d_{\mathrm{cdh}}(S, \mathbb{Z})$; the analogous assertions hold for X. The exact sequence (6.1) for S and $S \times \mathbb{A}^r$ implies the first two assertions. For (3), we have a distinguished triangle cdh sheaves on S,

$$\mathbb{Z} \longrightarrow f_*\mathbb{Z} \longrightarrow f_*\mathbb{Z}/\mathbb{Z} \longrightarrow \mathbb{Z}[1].$$

Since the cdh-cohomological dimension of S is at most d, $H^{d+1}_{\mathrm{cdh}}(S, \mathbb{Z}) = 0$. Thus the long exact sequence on cdh-cohomology ends in

$$\longrightarrow H^d_{\mathrm{cdh}}(S, \mathbb{Z}) \longrightarrow H^d_{\mathrm{cdh}}(S, f_*\mathbb{Z}) \longrightarrow H^d_{\mathrm{cdh}}(S, f_*\mathbb{Z}/\mathbb{Z}) \longrightarrow 0.$$

Since f is finite, we have $H^*_{\mathrm{cdh}}(S, f_*\mathbb{Z}) \xrightarrow{\cong} H^*_{\mathrm{cdh}}(X, \mathbb{Z})$; assertion (3) follows.

$\square$

Remark 6.2.1 Let k be a perfect field of characteristic p. Kerz and Strunk have shown in [10] that $K_n(S)$ is a p-primary torsion group for $n < -d$. Then Theorem 6.2 holds for k up to p-torsion.

If in addition k is a perfect field, over which weak resolution of singularities holds, then Theorem 6.2(1,2) holds for k. This also follows from [10]; if strong resolution of singularities holds, (1) also follows from the Geisser–Hesselholt theorem in [6] that $K_n(S) = 0$ for $n < -d$.

When S is a curve, not necessarily defined over $\mathbb{Q}$, we have a similar result.

Theorem 6.3 *Let* $f : X \to S$ *be a finite map of 1-dimensional noetherian schemes. Then* $K_{-1}(f)$ *fits into an exact sequence*

$$0 \longrightarrow H^1_{\mathrm{nis}}(S, f_*\mathcal{O}_X^\times/\mathcal{O}_S^\times) \longrightarrow K_{-1}(f) \longrightarrow H^0_{\mathrm{nis}}(S, f_*\mathbb{Z}/\mathbb{Z}) \longrightarrow 0.$$

In addition, $K_{-2}(f) \cong H^1_{\mathrm{nis}}(S, f_*\mathbb{Z}/\mathbb{Z})$ *and* $K_n(f) = 0$ *for* $n < -2$.

Proof By Thomason-Trobaugh [20, 10.8], we have a spectral sequence

$$E_2^{p,q} = H^p_{\mathrm{nis}}(S, \mathcal{K}_{-q}(f)) \Longrightarrow K_{-p-q}(f),$$

where $\mathcal{K}_n(f)$ is the Nisnevich sheafification of the presheaf $U \mapsto K_n(U, f^{-1}U)$. Each stalk $\mathcal{K}_n(f)$ is $K_n(A, B)$, where A is a hensel local ring of dimension ≤ 1. By Lemma 6.4 below, we have

$$\mathcal{K}_n(f) = \begin{cases} 0 & \text{if } n \leq -2 \\ f_*\mathbb{Z}/\mathbb{Z} & \text{if } n = -1 \\ f_*\mathcal{O}_X^\times/\mathcal{O}_S^\times & \text{if } n = 0. \end{cases}$$

Since $cd_{\mathrm{nis}}(S) \leq 1$, $E_2^{p,q} \neq 0$ only for $p = 0, 1$ and $q \leq 1$. Thus the spectral sequence degenerates to yield $K_{-2}(f) \cong H^1_{\mathrm{nis}}(S, f_*\mathbb{Z}/Z)$ and $K_n(f) = 0$ for $n < -2$.

$\square$

Lemma 6.4 *Let A be a 1-dimensional hensel local ring and $f : A \hookrightarrow B$ a finite extension. If B has r components, then*

$$K_0(f) \cong B^\times/A^\times, \quad K_{-1}(f) \cong \mathbb{Z}^{r-1} \quad \text{and} \quad K_n(f) = 0 \text{ for } n < -1.$$

Proof Since B is a finite A-algebra, B is a finite product of r hensel local rings. By [24, 2.8], $K_n(A) = K_n(B) = 0$ for $n < -1$. By a result of Drinfeld [29, III.4.4.3], we have $K_{-1}(A) = K_{-1}(B) = 0$. The result now follows from (6.1).

$\square$

Remark 6.5 A necessary condition for $K_{-1}(f) = 0$ is that the ring extension $f : A \hookrightarrow B$ is *anodal*, i.e., if every $b \in B$ such that $(b^2 - b) \in A$ and $(b^3 - b^2) \in A$ belongs to A. (See [27, 3.1].) This is because (2.3) induces a surjection $L \det : K_{-1}(f) \to L\operatorname{Pic}(f)$, and we showed in [16] that $L\operatorname{Pic}(f) = 0$ implies that $A \subset B$ is anodal. The converse does not hold, even if f is a birational extension of domains, as [27, Example 3.5] shows.

Example 6.6 Here is an example to show why we assume S affine in Proposition 2.5. For each n, the scheme $S = \mathbb{P}^1_k$ has a sheaf of algebras $\mathcal{O}_B = \mathcal{O}_S \oplus \mathcal{O}(n)$ with $\mathcal{O}(n)$ a square-zero ideal; fix $n \leq -2$ and set $X = \operatorname{Spec}(\mathcal{O}_B)$. Then $H = H^1(\mathbb{P}^1, \mathcal{O}(n))$ is nonzero and $\operatorname{Pic}(X) = \operatorname{Pic}(S) \oplus H$, $K_0(X) \cong K_0(S) \oplus H$. In particular, $K_{-1}(f) = H \neq 0$.

Acknowledgements This project was initiated while the first author was visiting Rutgers University in August 2015; he would like to thank the Math Dept. of Rutgers University for the invitation and financial support. The first author is also grateful to Jan Stienstra for sending him the manuscript [18]. The second author would like to thank TIFR for providing a great environment for doing this research.

References

[1] H. Bass and A. Roy, *Lectures on topics in Algebraic K-theory*, Tata Inst. Fund. Res. Lectures on Mathematics **41**, Tata Institute of Fundamental Research, Bombay 1967.

[2] H. Bass, *Algebraic K-theory*, Benjamin, New York, 1968.

[3] H. Bass and M. P. Murthy, *Grothendieck Groups and Groups of Abelian Group Rings*, Ann. of Math. **86** (July 1967), 16–73.

[4] G. Cortiñas, C. Haesemeyer, M. Schlichting and C. Weibel, *Cyclic homology, cdh-cohomology and negative K-theory*, Ann. of Math. **167** (2008), 549–573.

[5] B. Dayton and C. Weibel, *On the naturality of* Pic, SK_0 *and* SK_1, pp. 1–28 in Algebraic K-theory: Connections with Geometry and Topology, NATO ASI Series C **279**, Kluwer Press, 1989.

[6] T. Geisser and L. Hesselholt, *On the vanishing of negative K-groups*, Math. Ann. **348** (2010), 707–736.

[7] C. Haesemeyer, *Descent properties of homotopy K-theory*, Duke Math. J. **125** (2004), 589–620.

[8] R. Hartshorne, *Algebraic geometry*, Springer-Verlag, New York, 1977.

[9] F. Ischebeck, *Subintegral ring extensions and some K-theoretical functors*, J. Algebra **121** (1989), 323–338.

[10] M. Kerz and F. Strunk, *On the vanishing of negative homotopy K-theory*, J. Pure Appl. Algebra **221** no.7 (2017), 1641–1644.

[11] J. Milne, *Étale Cohomology*, Princeton University Press, Princeton 1980.

[12] C. Pedrini and C. Weibel, *Divisibility in the Chow group of zero-cycles on a singular surface*, in K-theory (Strasbourg, 1992), Astérisque **226** (1994), 371–409.

[13] L. G. Roberts and B. Singh, *Subintegrality, invertible modules and the Picard group*, Compositio Math. **85** (1993), 249–279.

[14] V. Sadhu, *Subintegrality, Invertible Modules and Laurent polynomial Extensions*, Proc. Indian Acad. Sci. (Math. Sci.) **125** (2015), 149–160.

[15] V. Sadhu and B. Singh, *Subintegrality, invertible modules and Polynomial Extensions*, J. Algebra **393** (2013), 16–23.

[16] V. Sadhu and C. Weibel, *Relative Cartier divisors and Laurent polynomial Extensions*, Math. Zeit. **285** no. 1-2 (2017), 353–366.

[17] R. G. Swan, *On Seminormality*, J. Algebra **67** (1980), 210–229.

[18] J. Stienstra, *Operations in the higher K-theory of endomorphisms*, Current trends in algebraic topology, Part 1 (London, Ont., 1981), 59–115, CMS Conf. Proc., 2, Amer. Math. Soc., 1982.

[19] J. Stienstra, Correction to *"Cartier-Dieudonné theory for Chow groups"*, J. Reine Angew. Math. **362** (1985), 218–220.

[20] R. W. Thomason and T. Trobaugh, *Higher algebraic K-theory of schemes and of derived categories*, The Grothendieck Festschrift, Vol. III, 247–35, Progr. Math. **88**, Birkhuser Boston, 1990.

[21] W. van der Kallen, *Descent for the K-theory of polynomial rings*, Math. Z. **191** (1986), 405–415.

[22] T. Vorst, *Localization of the K-theory of polynomial extensions*, Math. Ann. **244** (1979), 33–53.

[23] F. Waldhausen, *Algebraic K-theory of spaces*, Lecture Notes in Math. **1126** (1985), Springer-Verlag.

[24] C. Weibel, *K-theory and analytic isomorphisms*, Invent. Math. **61** (1980), 177–197.

[25] C. Weibel, *Mayer-Vietoris sequences and module structures on NK**, 494–517, Lecture Notes in Math. **854**, Springer-Verlag, 1981.

[26] C. Weibel, *Module structures in the K-theory of graded rings*, J. Algebra **105** (1987), 465–483.

[27] C. Weibel, *Pic is a contracted functor*, Invent. Math. **103** (1991), 351–377.

[28] C. Weibel, *The negative K-theory of normal surfaces*, Duke Math. J. **108** (2001), 1–35.

[29] C. Weibel, *The K-book: An Introduction to Algebraic K-Theory*, Graduate Studies in Math. **145**, Amer. Math. Soc., 2013.

VIVEK SADHU, SCHOOL OF MATHEMATICS, TATA INSTITUTE OF FUNDAMENTAL RESEARCH, 1 DR. HOMI BHABA ROAD, COLABA, MUMBAI 400005, INDIA

E-mail: viveksadhu@gmail.com, sadhu@math.tifr.res.in

CHARLES WEIBEL, MATH. DEPT., RUTGERS UNIVERSITY, NEW BRUNSWICK, NJ 08901, USA

E-mail: weibel@math.rutgers.edu

K-Theory
Copyright ©2018 Tata Institute of Fundamental Research
Publisher: Hindustan Book Agency, New Delhi, India

Naive vs. genuine $\mathbb{A}^1$-connectedness

Anand Sawant

Abstract

We show that the triviality of sections of the sheaf of $\mathbb{A}^1$-chain connected components of a space over finitely generated separable field extensions of the base field is not sufficient to ensure the triviality of the sheaf of its $\mathbb{A}^1$-chain connected components, contrary to the situation with genuine $\mathbb{A}^1$-connected components. As a consequence, we show that there exists an $\mathbb{A}^1$-connected scheme for which the Morel-Voevodsky singular construction is not $\mathbb{A}^1$-local.

1 Introduction

Let k be a field and let Sm/k denote the category of smooth, finite-type schemes over k. In the 1990's, Morel and Voevodsky [14] constructed the $\mathbb{A}^1$-*homotopy category* $\mathcal{H}(k)$ by taking a suitable localization of the category of simplicial sheaves of sets on Sm/k for the Nisnevich topology. Objects of $\mathcal{H}(k)$ are often called *spaces*. Analgous to algebraic topology, one then studies the $\mathbb{A}^1$-homotopy sheaves of a (pointed) space $(\mathcal{X}, x)$ — the sheaf of $\mathbb{A}^1$-connected components $\pi_0^{\mathbb{A}^1}(\mathcal{X})$, which is a sheaf of sets and the higher homotopy sheaves $\pi_n^{\mathbb{A}^1}(\mathcal{X}, x)$, for $n \geq 1$, which are sheaves of groups. We will use the notation and terminology of [14]. Any (not necessarily smooth) scheme X over k can be viewed as an object of the $\mathbb{A}^1$-homotopy category $\mathcal{H}(k)$ (see the conventions stated at the beginning of Section 2). Recent works in $\mathbb{A}^1$-homotopy theory have indicated that the $\mathbb{A}^1$-homotopy sheaves of schemes are often related to some of their interesting classical invariants.

The simplest of objects in classical topology are the discrete topological spaces. The analogous notion in $\mathbb{A}^1$-homotopy theory is that of $\mathbb{A}^1$-*invariant* sheaves (see Section 2 for precise definitions). In topology, the set of connected components of a topological space and the homotopy groups of a (pointed) topological space are discrete as topological spaces. Analogously, one can ask if the $\mathbb{A}^1$-homotopy sheaves of a (pointed) space $\mathcal{X}$ are $\mathbb{A}^1$-invariant. It has been shown by Morel [13, Theorem 6.1, Corollary 6.2] that the higher homotopy sheaves $\pi_n^{\mathbb{A}^1}(\mathcal{X}, x)$, for $n \geq 1$, are $\mathbb{A}^1$-invariant. In fact,

Morel shows much more — these higher $\mathbb{A}^1$-homotopy sheaves are *strongly* $\mathbb{A}^1$-*invariant* in the sense of [13, Definition 1.7]. However, $\mathbb{A}^1$-invariance of the sheaf of $\mathbb{A}^1$-connected components is not yet known; this has been conjectured by Morel [13, Conjecture 1.12]. It is worthwhile to mention that $\pi_0^{\mathbb{A}^1}$ fails to be a birational invariant of smooth, proper schemes [3, Example 4.8].

There are two notions of $\mathbb{A}^1$-connectedness in unstable $\mathbb{A}^1$-homotopy theory. The *naive* notion is that of $\mathbb{A}^1$-*chain connected components* of a space (see Definition 2.2), which is obtained by taking the Nisnevich sheafi-fication of the presheaf that associates with any smooth scheme U the set of morphisms from U to the space in question modulo the equivalence relation generated by naive $\mathbb{A}^1$-homotopies. On the other hand, the *genuine* notion is that of $\mathbb{A}^1$-*connected components* (see Definition 2.3) introduced by Morel-Voevodsky. These two notions do not coincide in general, not even for smooth projective varieties over $\mathbb{C}$ (see [3, Section 4] for the first examples). Given a scheme X over k, one can infinitely iterate the construction of $\mathbb{A}^1$-chain connected components to obtain the so-called *universal $\mathbb{A}^1$-invariant quotient* of X, which is isomorphic to $\pi_0^{\mathbb{A}^1}(X)$ provided the latter sheaf is $\mathbb{A}^1$-invariant (that is, Morel's conjecture holds for X). We recall these notions and known results about them in Section 2.

A natural question is to characterize genuine $\mathbb{A}^1$-connectedness of a scheme X over k in terms of triviality of sections of $\pi_0^{\mathbb{A}^1}(X)$ over field extensions of k. A result of Morel states that $\mathbb{A}^1$-connectedness of a scheme X over an infinite field k in the genuine sense (that is, triviality of $\pi_0^{\mathbb{A}^1}(X)$ as a sheaf) is equivalent to the triviality of $\pi_0^{\mathbb{A}^1}(X)(\mathrm{Spec}\ F)$, where F runs over all finitely generated separable field extensions of k. In this short note, we examine the analogous property for the sheaf of $\mathbb{A}^1$-chain connected components in Section 3 (see Theorem 3.2). As a consequence, we obtain an example of an $\mathbb{A}^1$-connected singular proper scheme X for which the Morel-Voevodsky singular construction $\mathrm{Sing}_*^{\mathbb{A}^1} X$ is not $\mathbb{A}^1$-local (see Example 3.6).

2   Connectedness in unstable $\mathbb{A}^1$-homotopy theory

We begin this section by setting up the notation and conventions that will be used throughout the paper.

Fix a base field k. We will henceforth denote by Sm/k the big Nisnevich site of smooth, finite-type schemes over k. We begin with the category of simplicial sheaves over Sm/k. Any scheme X over k can be seen as an object of this category as follows: consider the *functor of points* h_X of X, which is the sheaf that associates with every $U \in Sm/k$ the set of morphisms of schemes over k from U to X. Any Nisnevich sheaf $\mathcal{F}$ on Sm/k can be viewed as a simplicially constant simplicial sheaf. More precisely, one considers the simplicial sheaf in which the sheaf at every level is $\mathcal{F}$ and all the face and degeneracy maps are given by the identity map. We will always denote the simplicial sheaf corresponding to h_X for a scheme X over k by the same letter X.

A morphism $\mathcal{X} \to \mathcal{Y}$ of simplicial sheaves of sets on Sm/k is a *local weak equivalence* if it induces an isomorphism on every stalk. The *Nisnevich local injective model structure* on this category is the one in which the morphism of simplicial sheaves is a cofibration (resp. a weak equivalence) if and only if it is a monomorphism (resp. a local weak equivalence). The corresponding homotopy category is called the *simplicial homotopy category* and is denoted by $\mathcal{H}_s(k)$. The left Bousfield localization of the Nisnevich local injective model structure with respect to the collection of all projection morphisms $\mathcal{X} \times \mathbb{A}^1 \to \mathcal{X}$, as $\mathcal{X}$ runs over all simplicial sheaves, is called the $\mathbb{A}^1$-*model structure*. The associated homotopy category is called the $\mathbb{A}^1$-*homotopy category* and is denoted by $\mathcal{H}(k)$. We will denote by $*$ the trivial one-point sheaf on Sm/k. We will abuse the notation and use $*$ to also denote a set with one element, whenever there is no confusion.

Definition 2.1 A space (that is, a simplicial Nisnevich sheaf of sets on Sm/k) $\mathcal{X}$ is said to be $\mathbb{A}^1$-*local* if the projection map $U \times \mathbb{A}^1 \to U$ induces a bijection
$$\mathrm{Hom}_{\mathcal{H}_s(k)}(U, \mathcal{X}) \to \mathrm{Hom}_{\mathcal{H}_s(k)}(U \times \mathbb{A}^1, \mathcal{X}).$$

for every $U \in Sm/k$. Note that a Nisnevich sheaf $\mathcal{F}$ on Sm/k is $\mathbb{A}^1$-local if and only if it is $\mathbb{A}^1$-*invariant*, that is, if the projection map $U \times \mathbb{A}^1 \to U$ induces a bijection $\mathcal{F}(U) \to \mathcal{F}(U \times \mathbb{A}^1)$, for every $U \in Sm/k$. Following standard convention, we say that a scheme is $\mathbb{A}^1$-*rigid* if it is $\mathbb{A}^1$-local as a space.

Let $\mathcal{X}$ be a space. We now recall the *singular construction* on $\mathcal{X}$ defined by Morel-Voevodsky [14, p. 87-88]. Define $\mathrm{Sing}_*^{\mathbb{A}^1} \mathcal{X}$ to be the simplicial sheaf given by
$$(\mathrm{Sing}_*^{\mathbb{A}^1} \mathcal{X})_n = \underline{\mathrm{Hom}}(\Delta_n, \mathcal{X}_n),$$

where $\Delta_\bullet$ denotes the cosimplicial scheme
$$\Delta_n = \mathrm{Spec}\left(\frac{k[x_0, ..., x_n]}{(\sum_i x_i - 1)} \right)$$

with natural face and degeneracy maps analogous to the ones on topological simplices. There exists a natural transformation $Id \to \mathrm{Sing}_*^{\mathbb{A}^1}$ such that for any simplicial sheaf $\mathcal{X}$, the morphism $\mathcal{X} \to \mathrm{Sing}_*^{\mathbb{A}^1}(\mathcal{X})$ is an $\mathbb{A}^1$-weak equivalence. Observe that the singular construction $\mathrm{Sing}_*^{\mathbb{A}^1}$ takes naive $\mathbb{A}^1$-homotopies to simplicial homotopies.

Given a simplicial sheaf of sets $\mathcal{X}$ on Sm/k, we will denote by $\pi_0(\mathcal{X})$ the presheaf on Sm/k that associates with $U \in Sm/k$ the coequalizer of the diagram $\mathcal{X}_1(U) \rightrightarrows \mathcal{X}_0(U)$, where the maps are the face maps coming from the simplicial data of $\mathcal{X}$. We will denote by $\pi_0^s(\mathcal{X})$ the Nisnevich sheafification of the presheaf $\pi_0(\mathcal{X})$.

Definition 2.2 The sheaf of $\mathbb{A}^1$-*chain connected components* of a space $\mathcal{X}$ is defined to be

$$\mathcal{S}(\mathcal{X}) := \pi_0^s(\mathrm{Sing}_*^{\mathbb{A}^1}\mathcal{X}).$$

Thus, $\mathcal{S}(\mathcal{X})$ is the Nisnevich sheafification of the presheaf that associates with any smooth scheme U the set $\mathcal{X}(U)/\sim$, where $\sim$ is the equivalence relation generated by the image of $\mathcal{X}_1 \to \mathcal{X}_0 \times \mathcal{X}_0$, where the maps are the face maps coming from the simplicial data of $\mathcal{X}$ (in other words, $\sim$ is the equivalence relation generated by naive $\mathbb{A}^1$-homotopies).

Definition 2.3 The sheaf of $\mathbb{A}^1$-*connected components* of a space $\mathcal{X}$ is defined to be

$$\pi_0^{\mathbb{A}^1}(\mathcal{X}) := \pi_0^s(L_{\mathbb{A}^1}\mathcal{X}),$$

where $L_{\mathbb{A}^1}$ denotes an $\mathbb{A}^1$-fibrant replacement functor. A space $\mathcal{X}$ is said to be $\mathbb{A}^1$-*connected* if $\pi_0^{\mathbb{A}^1}(\mathcal{X}) \simeq *$.

Morel-Voevodsky explicitly describe an $\mathbb{A}^1$-fibrant replacement functor as follows:

$$L_{\mathbb{A}^1} = Ex \circ (Ex \circ \mathrm{Sing}_*^{\mathbb{A}^1})^{\mathbb{N}} \circ Ex,$$

where Ex denotes a simplicial fibrant replacement functor on the model category of simplicial Nisnevich sheaves of sets over Sm/k [14, §2, Lemma 2.6, p. 107]. There exists a natural transformation $Id \to L_{\mathbb{A}^1}$ which factors through the natural transformation $Id \to \mathrm{Sing}_*^{\mathbb{A}^1}$ mentioned above. For any object $\mathcal{X}$, the morphism $\mathcal{X} \to L_{\mathbb{A}^1}(\mathcal{X})$ is an $\mathbb{A}^1$-weak equivalence. A result of Morel-Voevodsky [14, §2, Corollary 3.22] describes what happens to the natural map $\mathrm{Sing}_*^{\mathbb{A}^1}\mathcal{X} \to L_{\mathbb{A}^1}(\mathcal{X})$ after applying π_0^s; we record it below for the sake of convenience.

Lemma 2.4 *The canonical map* $\mathcal{S}(\mathcal{X}) \to \pi_0^{\mathbb{A}^1}(\mathcal{X})$ *is an epimorphism, for every space* $\mathcal{X}$. *If* $\mathrm{Sing}_*^{\mathbb{A}^1}\mathcal{X}$ *is* $\mathbb{A}^1$-*local, then the map* $\mathcal{S}(\mathcal{X}) \to \pi_0^{\mathbb{A}^1}(\mathcal{X})$ *is an isomorphism.*

We will henceforth focus on a specific class of spaces, namely, sheaves of sets on the big Nisnevich site on Sm/k. We will eventually specialize to the case of schemes. Let $\mathcal{F}$ be a Nisnevich sheaf on Sm/k. By Lemma 2.4, we have a sequence of epimorphisms

$$\mathcal{F} \longrightarrow \mathcal{S}(\mathcal{F}) \longrightarrow \mathcal{S}^2(\mathcal{F}) \longrightarrow \cdots \longrightarrow \mathcal{S}^n(\mathcal{F}) \longrightarrow \cdots ,$$

where $\mathcal{S}^{n+1}(\mathcal{F})$ is defined inductively to be $\mathcal{S}(\mathcal{S}^n(\mathcal{F}))$, for every $n \in \mathbb{N}$. We define

$$L(\mathcal{F}) := \varinjlim_{n} \mathcal{S}^n(\mathcal{F}). \tag{2.1}$$

The following result was proved in [3] (see [3, Theorem 2.13, Remark 2.15, Corollary 2.18]), which shows that $L(\mathcal{F})$ is the *universal* $\mathbb{A}^1$*-invariant quotient* of $\mathcal{F}$.

Theorem 2.5 *Let* $\mathcal{F}$ *be a sheaf of sets on* Sm/k. *Then the sheaf* $L(\mathcal{F})$ *is* $\mathbb{A}^1$*-invariant. Moreover, if* $\mathcal{G}$ *is an* $\mathbb{A}^1$*-invariant sheaf, then any map* $\mathcal{F} \to \mathcal{G}$ *factors uniquely through the epimorphism* $\mathcal{F} \to L(\mathcal{F})$. *Moreover, if* $\pi_0^{\mathbb{A}^1}(\mathcal{F})$ *is* $\mathbb{A}^1$*-invariant, then the canonical map* $L(\mathcal{F}) \to \pi_0^{\mathbb{A}^1}(\mathcal{F})$ *is an isomorphism.*

We will henceforth focus on schemes over a field. In view of Theorem 2.5, it is clear that a good understanding of $L(X)$ is tantamount to understanding $\pi_0^{\mathbb{A}^1}(X)$. It is natural to ask the following question.

Question 2.6 *Let* X *be a smooth scheme over* k. *Does there exist* $n \in \mathbb{N}$ *such that* $L(X) \simeq \mathcal{S}^n(X)$?

For every scheme X over a field k, we have the following commutative diagram in which every morphism is an epimorphism

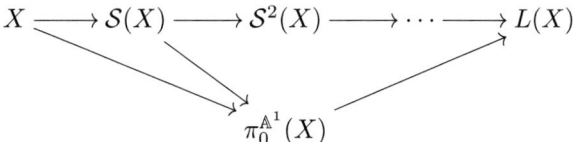

where the existence of the map $\pi_0^{\mathbb{A}^1}(X) \to L(X)$ making the diagram commute is a consequence of the $\mathbb{A}^1$-invariance of $L(X)$ (see [3, Lemma 2.8]). The morphism $\pi_0^{\mathbb{A}^1}(X) \to L(X)$ is an isomorphism if $\pi_0^{\mathbb{A}^1}(X)$ is $\mathbb{A}^1$-invariant. An affirmative answer to Question 2.6 will give a conjectural but explicit geometric description of $\pi_0^{\mathbb{A}^1}(X)$. We end this section by enlisting the known examples in which such an explicit description is available.

Examples 2.7 ($\mathbb{A}^1$-rigid varieties) For an $\mathbb{A}^1$-rigid variety X, one has isomorphisms of Nisnevich sheaves

$$X \simeq \mathcal{S}(X) \simeq \pi_0^{\mathbb{A}^1}(X).$$

Examples of $\mathbb{A}^1$-rigid varieties include $\mathbb{G}_m$, algebraic tori, abelian varieties, curves of genus ≥ 1 etc.

Examples 2.8 (Reductive algebraic groups) For any sheaf of groups G, it is known that $\pi_0^{\mathbb{A}^1}(G)$ is $\mathbb{A}^1$-invariant [7]. We therefore have $\pi_0^{\mathbb{A}^1}(G) \simeq L(G) = \varinjlim_n \mathcal{S}^n(G)$. We will now focus on the case where G is a reductive algebraic group over a field.

(a) **Isotropic groups.** Suppose that G satisfies the following *isotropy condition*: every almost k-simple component of the derived group G_{der} of G contains a k-subgroup scheme isomorphic to $\mathbb{G}_m$. Under this hypothesis, Asok, Hoyois and Wendt have shown that $\mathrm{Sing}_*^{\mathbb{A}^1} G$ is $\mathbb{A}^1$-local [1, Theorem 2.3.2]. Therefore, for G satisfying the above isotropy condition, one has

$$\mathcal{S}(G) \simeq \pi_0^{\mathbb{A}^1}(G) \simeq L(G).$$

The sections of this sheaf over fields can often be described explicitly. If G is a semisimple, simply connected group over finite field k satisfying the above isotropy hypothesis, then one has isomorphisms $W(k, G) \simeq \mathcal{S}(G)(k) \simeq G(k)/R$. This is a consequence of [1, Theorem 2.3.2] and a classical result [9, Théorème 7.2]. Here $W(k, G)$ denotes the *Whitehead group* of G and $G(k)/R$ denotes the group of *R-equivalence classes* (see [9], for example, for precise definitions).

(b) **Anisotropic groups.** Let us assume that the base field k is infinite and perfect. Suppose now that G does not satisfy the above isotropy hypothesis; that means, the derived group G_{der} of G has at least one almost k-simple factor which is anisotropic. In this case, it is known that $\mathrm{Sing}_*^{\mathbb{A}^1} G$ fails to be $\mathbb{A}^1$-local [5, Theorem 4.7]. We now assume that G is a semisimple anisotropic group. Note that one has $W(k, G) = G(k)$ in this case. A result of Borel-Tits implies in this case that $\mathcal{S}(G)(k) \simeq G(k)$ (see [4, Lemma 3.7] for details). However, one has the following result (see [4, Theorem 4.2], [5, Theorem 3.6]): if G is a semisimple, simply connected group over an infinite perfect field k which does not satisfy the above isotropy hypothesis, then one has canonical isomorphisms

$$\pi_0^{\mathbb{A}^1}(G)(k) \simeq \mathcal{S}^2(G)(k) \simeq G(k)/R.$$

It is worthwhile to mention here that we do not yet know whether $\pi_0^{\mathbb{A}^1}(G)$ agrees with $\mathcal{S}^2(G)$ as a sheaf.

(c) $\mathbb{A}^1$**-connected reductive algebraic groups.** Recall that a space $\mathcal{X}$ is said to be $\mathbb{A}^1$-connected if $\pi_0^{\mathbb{A}^1}(\mathcal{X}) \simeq *$. In [5, Theorem 5.2], $\mathbb{A}^1$-connected reductive algebraic groups have been characterized: a reductive algebraic group G over a field k of characteristic 0 is $\mathbb{A}^1$-connected if and only if G is semisimple, simply connected and R-trivial (that is $G(F)/R$ is trivial for every finitely generated separable field extension F of k).

Examples 2.9 (Proper varieties) (a) If X is a proper variety over k and if F is a finitely generated field extension of k, then one has $\mathcal{S}(X)(F) \simeq \pi_0^{\mathbb{A}^1}(X)(F)$ (see [2, Theorem 2.4.3]). One also has $\mathcal{S}(X)(F) \simeq \mathcal{S}^n(X)(F)$, for every n (see [3, Theorem 3.9, Corollary 3.10]).

(b) The case of proper schemes of dimension ≤ 1 is very easy. For reduced, proper (possibly singular) schemes X over k of dimension ≤ 1, one always has $\mathcal{S}(X) \simeq \mathcal{S}^2(X)$ ([3, Proposition 3.13]). Consequently, $\mathcal{S}(X) \simeq \pi_0^{\mathbb{A}^1}(X)$.

(c) If X is a proper, non-uniruled surface over k, then one has $\mathcal{S}(X) \simeq \pi_0^{\mathbb{A}^1}(X) \simeq \mathcal{S}^2(X)$ ([3, Theorem 3.14]).

(d) The case of smooth projective ruled surfaces is surprisingly very complicated. If X is a smooth proper rational surface, then one has $\mathcal{S}^2(X) = *$ (see Corollary 3.3). However, one has $\pi_0^{\mathbb{A}^1}(X) = *$ as well. We do not yet know if the sheaf $\mathcal{S}(X)$ for a rational surface X is trivial.

Let us now assume that the characteristic of k is 0. The case of ruled surfaces whose minimal model is of the form $\mathbb{P}^1 \times C$, where C is a smooth projective curve of genus ≥ 1 (note that such a C is $\mathbb{A}^1$-rigid) is the most complicated one. If E is a $\mathbb{P}^1$-bundle over C, then one has $\mathcal{S}(E) \simeq \pi_0^{\mathbb{A}^1}(E) \simeq C$. If X is the surface obtained by blowing up one closed point on E and when k is assumed to be algebraically closed, one has $\mathcal{S}(X) \neq \mathcal{S}^2(X)$. However, in this case one has $\mathcal{S}^2(X) \simeq \mathcal{S}^3(X)$. The details will appear in a forthcoming paper [6].

3 Naive $\mathbb{A}^1$-connectedness on field-valued points

Let X be a scheme over a field k. It is often much simpler to determine sections of the sheaf $\mathcal{S}(X)$ on smooth schemes which are the spectrum of a finitely generated separable field extension of the base field k. A result of Morel (see [12, Lemma 6.1.3]) states that a space $\mathcal{X}$ over an infinite field k is $\mathbb{A}^1$-connected (that is, $\pi_0^{\mathbb{A}^1}(\mathcal{X})$ is trivial) if and only if $\pi_0^{\mathbb{A}^1}(\mathcal{X})(\operatorname{Spec} F) = *$, for every finitely generated separable field extension F of k. The argument given by Morel also works when the base field is finite, thanks to Gabber's presentation lemma over finite fields proved in [10]. We wish to study the analogue of this result in the context of the sheaf of $\mathbb{A}^1$-chain connected components. The method used here closely follows the one employed by Morel in [12, Section 6.1] and in [11, Section 3.3].

Lemma 3.1 *Let V be an irreducible smooth scheme over k and let $W \hookrightarrow V$ be the inclusion of a dense open subscheme. Then $\mathcal{S}(V/W) \simeq *$.*

Proof Since we have epimorphisms $V \to V/W \to \mathcal{S}(V/W)$, triviality of $\mathcal{S}(V/W)$ follows from the following statement: any point $x \in V$ has an open neighbourhood U such that $\mathcal{S}(U/(W \cap U))$ is trivial.

Let $Z \hookrightarrow V$ be the closed immersion of the complement of W, with the reduced induced subscheme structure. By Gabber's presentation lemma (see [8, Theorem 3.1.1] for the case where k is infinite and [10, Theorem 1.1] for the case where k is finite), x admits an open neighbourhood U and an étale morphism $\pi : U \to \mathbb{A}^1_{V'}$, for some open subscheme V' of $\mathbb{A}^{d-1}$, where d is the dimension of V at x, such that π induces a closed immersion $Z \cap U \hookrightarrow \mathbb{A}^1_{V'}$ satisfying $Z \cap U = \pi^{-1}(\pi(Z \cap U))$ and such that $Z \cap U \to V'$ is a finite morphism. Therefore, we have an isomorphism of Nisnevich sheaves

$$U/(U \setminus Z \cap U) \xrightarrow{\sim} \mathbb{A}^1_{V'}/(\mathbb{A}^1_{V'} \setminus \pi(Z \cap U)).$$

Hence, it suffices to check that $\mathcal{S}(\mathbb{A}^1_{V'}/(\mathbb{A}^1_{V'} \setminus \pi(Z \cap U)))$ is trivial. Now, since $Z \cap U \to V'$ is a finite morphism, $Z \cap U \to \mathbb{P}^1_{V'}$ is proper. This closed immersion does not intersect the section at infinity $s_\infty : V' \to \mathbb{P}^1_{V'}$. By Mayer-Vietoris excision (see [14, §3, Lemma 1.6]), we have an isomorphism of Nisnevich sheaves

$$\mathbb{A}^1_{V'}/(\mathbb{A}^1_{V'} \setminus \pi(Z \cap U)) \xrightarrow{\sim} \mathbb{P}^1_{V'}/(\mathbb{P}^1_{V'} \setminus \pi(Z \cap U)).$$

Also observe that $\mathbb{A}^1_{V'} \to \mathbb{P}^1_{V'}/(\mathbb{P}^1_{V'} \setminus \pi(Z \cap U))$ is onto and that $\operatorname{Sing}_*^{\mathbb{A}^1}(\mathbb{A}^1_{V'}) \simeq$

$\mathrm{Sing}_*^{\mathbb{A}^1}(V')$ (since $\mathrm{Sing}_*^{\mathbb{A}^1}$ preserves $\mathbb{A}^1$-weak equivalences). Thus, the composition

$$V' \longrightarrow \mathbb{A}^1_{V'} \longrightarrow \pi_0^s(\mathrm{Sing}_*^{\mathbb{A}^1}(\mathbb{A}^1_{V'}/(\mathbb{A}^1_{V'} \setminus \pi(Z \cap U)))) \longrightarrow$$
$$\pi_0^s(\mathrm{Sing}_*^{\mathbb{A}^1}(\mathbb{P}^1_{V'}/(\mathbb{P}^1_{V'} \setminus \pi(Z \cap U))))$$

is surjective for any section $V' \to \mathbb{A}^1_{V'}$; in particular, for the zero section. But, in $\mathbb{P}^1_{V'}$, the zero section is $\mathbb{A}^1$-homotopic to the section at infinity s_∞. Since $s_\infty(V') \subseteq \mathbb{P}^1_{V'} \setminus \pi(Z \cap U)$, it follows that

$$V \to \pi_0^s(\mathrm{Sing}_*^{\mathbb{A}^1}(\mathbb{P}^1_{V'}/(\mathbb{P}^1_{V'} \setminus \pi(Z \cap U)))) = \mathcal{S}(\mathbb{P}^1_{V'}/(\mathbb{P}^1_{V'} \setminus \pi(Z \cap U)))$$

is the trivial morphism, as desired.

$\square$

Theorem 3.2 *Let k be a field and let $\mathcal{X}$ be a simplicial sheaf of sets on Sm/k. Suppose that $\mathcal{S}(\mathcal{X})(\mathrm{Spec}\ F) = *$, for every finitely generated separable field extension F of k. Then $\mathcal{S}(\mathrm{ExSing}_*^{\mathbb{A}^1}\mathcal{X}) \simeq *$. Consequently, $\mathcal{S}^2(\mathcal{X}) \simeq *$.*

Proof We need to show that for every $U \in Sm/k$, the pointed set $\mathcal{S}(\mathrm{Sing}_*^{\mathbb{A}^1}\mathcal{X})(U)$ is trivial. It suffices to show that for every morphism $U \to \mathcal{S}(\mathrm{Sing}_*^{\mathbb{A}^1}\mathcal{X})$, there is a Nisnevich cover $\mathcal{V} = \coprod V_i \to U$ such that the composite $\mathcal{V} \to U \to \mathcal{S}(\mathrm{Sing}_*^{\mathbb{A}^1}\mathcal{X})$ is trivial.

Claim: For any irreducible, smooth k-scheme V and a morphism $\phi : V \to \mathcal{X}$, the composition $V \overset{\phi}{\to} \mathcal{X} \to \mathrm{ExSing}_*^{\mathbb{A}^1}\mathcal{X} \to \mathcal{S}(\mathrm{ExSing}_*^{\mathbb{A}^1}\mathcal{X})$ is trivial.

Proof of the claim: Let $k(V)$ denote the function field of V. Since

$$\mathcal{S}(\mathcal{X})(\mathrm{Spec}\ k(V)) = \varinjlim_{W \hookrightarrow V \text{nonempty open}} \mathcal{S}(\mathcal{X})(W)$$

is trivial by hypothesis, there exists a dense open subset $W \hookrightarrow V$ such that the composite $W \to V \overset{\phi}{\to} \mathcal{X}$ is $\mathbb{A}^1$-chain homotopic to the trivial morphism. Therefore the composite of this morphism with the morphism $\mathcal{X} \to \mathrm{Sing}_*^{\mathbb{A}^1}\mathcal{X}$ is simplicially homotopic to the trivial morphism. Choose a simplicial fibrant replacement $\mathrm{Sing}_*^{\mathbb{A}^1}\mathcal{X} \to \mathrm{ExSing}_*^{\mathbb{A}^1}\mathcal{X}$. The composite $W \hookrightarrow V \overset{\phi}{\to} \mathcal{X} \to \mathrm{Sing}_*^{\mathbb{A}^1}\mathcal{X} \to \mathrm{ExSing}_*^{\mathbb{A}^1}\mathcal{X}$ continues to be simplicially homotopic to the trivial map. We denote this simplicial homotopy by $H : W \times \Delta^1 \to \mathrm{ExSing}_*^{\mathbb{A}^1}\mathcal{X}$, where $H|_{W \times \{0\}}$ is the trivial map and $H|_{W \times \{1\}}$ is induced by $\phi|_W$ (here Δ^1 denotes the simplicial 1-simplex). Consider the acyclic cofibration $V \times \{1\} \cup W \times \Delta^1 \to V \times \Delta^1$. The maps $V \times \{1\} \xrightarrow{\sim}$ $V \xrightarrow{\phi} \mathcal{X} \to Ex\mathrm{Sing}_*^{\mathbb{A}^1} \mathcal{X}$ and $H : W \times \Delta^1 \to Ex\mathrm{Sing}_*^{\mathbb{A}^1} \mathcal{X}$ clearly glue to give a map $\Phi : V \times \{1\} \cup W \times \Delta^1 \to Ex\mathrm{Sing}_*^{\mathbb{A}^1} \mathcal{X}$, which fits in the following commutative diagram.

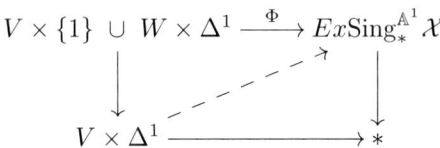

We now use the right lifting property of the projection map $Ex\mathrm{Sing}_*^{\mathbb{A}^1} \mathcal{X} \to *$ with respect to acyclic cofibrations to see that dotted arrow in the above diagram exists. It follows that $\phi : V \to Ex\mathrm{Sing}_*^{\mathbb{A}^1} \mathcal{X}$ is simplicially homotopic to a morphism $\phi' : V \to Ex\mathrm{Sing}_*^{\mathbb{A}^1} \mathcal{X}$ whose restriction to W is trivial. Thus, we get an induced morphism (of spaces) $\bar{\phi}' : V/W \to Ex\mathrm{Sing}_*^{\mathbb{A}^1} \mathcal{X}$. Now, applying Lemma 3.1 and commutativity of the diagram

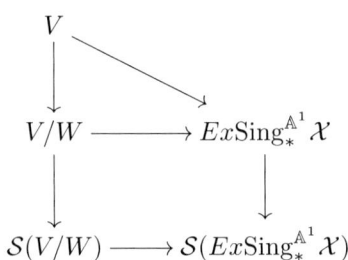

proves the claim.

We now complete the proof of the theorem using the claim. Since the natural map $Ex\mathrm{Sing}_*^{\mathbb{A}^1} \mathcal{X} \to \mathcal{S}(Ex\mathrm{Sing}_*^{\mathbb{A}^1} \mathcal{X})$ is an epimorphism, there is a Nisnevich covering $\coprod V_i \to U$, where V_i are irreducible smooth k-schemes such that every composite $V_i \to U \to \mathcal{S}(Ex\mathrm{Sing}_*^{\mathbb{A}^1} \mathcal{X})$ lifts to a morphism $V_i \to Ex\mathrm{Sing}_*^{\mathbb{A}^1} \mathcal{X}$. Since $Ex\mathrm{Sing}_*^{\mathbb{A}^1} \mathcal{X}$ is a simplicial fibrant replacement of $\mathcal{X}$, each map $V_i \to Ex\mathrm{Sing}_*^{\mathbb{A}^1} \mathcal{X}$ is represented by a map $V_i \to \mathrm{Sing}_*^{\mathbb{A}^1} \mathcal{X}$ in the simplical homotopy category $\mathcal{H}_s(k)$.

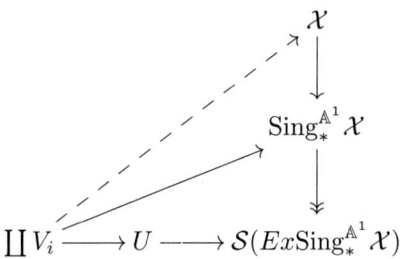

Since the sheaf at simplicial level 0 of $\mathrm{Sing}^{\mathbb{A}^1}_* \mathcal{X}$ is $\mathcal{X}_0$, and since any map from a space of simplicial dimension 0 to another space is determined by a map at the 0th simplicial level, this map factors through the monomorphism $\mathcal{X} \to \mathrm{Sing}^{\mathbb{A}^1}_* \mathcal{X}$. Thus, the theorem now follows from the claim, applied to each of the maps $V_i \to \mathcal{X}$.

□

Corollary 3.3 *Let X be a scheme over a field k such that $\mathcal{S}(X)(\mathrm{Spec}\ F) = *$, for every finitely generated separable field extension F of k. Then $\mathcal{S}^2(X) \simeq *$.*

The condition in Corollary 3.3 can be seen to hold when X is a smooth proper rational variety over a field of characteristic 0. In general, it holds when X is a smooth proper R-trivial variety over a field of characteristic 0 (see [2, Theorem 2.4.3, Corollary 2.4.9]).

Corollary 3.4 *Let G be an $\mathbb{A}^1$-connected reductive algebraic group over an infinite perfect field k. Then $\mathcal{S}^3(G) \simeq *$.*

Proof This is a straightforward consequence of Theorem 3.2 and Examples 2.8 (a), (b).

□

Remark 3.5 Note that if G in Corollary 3.4 is such that every almost k-simple factor of G contains a copy of $\mathbb{G}_m$, then one has $\mathcal{S}(G) = *$ by [1, Theorem 2.3.2].

We end this note with an example of a singular, projective scheme X for which $\mathcal{S}(X) \neq *$, but $\mathcal{S}(X)(\mathrm{Spec}\ F) = *$, for every finitely generated field extension F of k. This example was pointed out to the author by Chetan Balwe.

Example 3.6 Let k be a field and let $C \hookrightarrow \mathbb{P}^2$ be an elliptic curve over k. Let X_1 denote the blow up of $\mathbb{P}^1 \times C \hookrightarrow \mathbb{P}^1 \times \mathbb{P}^2$ at the closed point $P = ((1 : 0), Q)$, where Q is a closed point on C. Let E denote the exceptional divisor. We have an obvious morphism $\pi : X_1 \to C$, which is the composition of the blow-up morphism $\phi : X_1 \to \mathbb{P}^1 \times C$ with the projection map $\mathbb{P}^1 \times C \to C$. Let X_2 denote the plane $\{(0 : 1)\} \times \mathbb{P}^2$; so X_1 and X_2 intersect in $\{(0 : 1)\} \times C$. Let $X := X_1 \cup X_2$. Since $\mathcal{S}(X_2) = *$, it is easy to see that $\mathcal{S}(X)(\operatorname{Spec} F) = *$, for every field extension F of k. By [2, Theorem 2.4.3], we have $\pi_0^{\mathbb{A}^1}(X)(\operatorname{Spec} F) = *$, for every finitely generated field extension F of k. By [12, Lemma 6.1.3], we have $\pi_0^{\mathbb{A}^1}(X) \simeq *$.

Let U be a smooth henselian local scheme of dimension 1 over k with closed point u. Let $\gamma : U \to X_1$ be a morphism that maps u on E and the generic point of U outside E. We begin by considering naive $\mathbb{A}^1$-homotopies of U inside X_1, starting at γ. Let h be a morphism $\mathbb{A}^1 \times U \to X_1$ such that $h|_{\{0\} \times \mathbb{A}^1} = \gamma$. Since C is $\mathbb{A}^1$-rigid, the composition $\pi \circ h : \mathbb{A}^1 \times U \to C$ factors through the projection $\mathbb{A}^1 \times U \to U$. Thus, $h : \mathbb{A}^1 \times U \to X_1$ factors through $X_1 \times_C U$. By assumption, h is such that the point Q on C is in the image of $\pi \circ h$, that is, the image of the closed fiber $\mathbb{A}^1 \times \{u\}$ intersects the exceptional divisor E. Hence, $h^{-1}(P)$ is a closed subscheme of $\mathbb{A}^1 \times U$ with support contained in $\mathbb{A}^1 \times \{u\}$. Since $\phi \circ h$ can be lifted to X_1, we see that $h^{-1}(P)$ is a closed subscheme of $\mathbb{A}^1 \times U$ of codimension 1. Therefore, the support of $h^{-1}(P)$ must be exactly $\mathbb{A}^1 \times \{u\}$. Thus, h maps $\mathbb{A}^1 \times \{u\}$ into the exceptional divisor E.

Now, let $h : \mathbb{A}^1 \times U \to X$ be a morphism such that $h|_{\{0\} \times U} = \gamma$. Since $\mathbb{A}^1 \times U$ is irreducible, this implies that h factors through the inclusion of X_1 into X. The discussion in the above paragraph shows that h maps the whole closed fiber of $\mathbb{A}^1 \times U$ on E. Hence, we cannot have $\mathcal{S}(X)(U) = *$ and consequently, $\mathcal{S}(X) \neq *$. However, Theorem 3.2 implies that $\mathcal{S}^2(X) = *$. Since $\pi_0^{\mathbb{A}^1}(X) \simeq *$ as observed above, $\operatorname{Sing}_*^{\mathbb{A}^1} X$ cannot be $\mathbb{A}^1$-local in view of Lemma 2.4.

Acknowledgements   The author thanks Chetan Balwe for suggesting Example 3.6 as well as for his comments on this article and Marc Hoyois for a helpful discussion during the International Colloquium on K-theory at TIFR. The author also thanks the referee for a careful reading of the note and for comments that helped him improve the exposition.

References

[1] A. Asok, M. Hoyois and M. Wendt, *Affine representability results in $\mathbb{A}^1$-homotopy theory II: principal bundles and homoge-*

neous spaces, Preprint, arXiv: 1507:08020v3 [math.AG] (2015).

[2] A. Asok and F. Morel, *Smooth varieties up to $\mathbb{A}^1$-homotopy and algebraic h-cobordisms*, Adv. Math. **227** (2011) no. 5, 1990–2058.

[3] C. Balwe, A. Hogadi and A. Sawant, *$\mathbb{A}^1$-connected components of schemes*, Adv. Math. **282** (2015), 335–361.

[4] C. Balwe and A. Sawant, *R-equivalence and $\mathbb{A}^1$-connectedness in anisotropic groups*, Int. Math. Res. Not. **22** (2015), 11816–11827.

[5] C. Balwe and A. Sawant, *$\mathbb{A}^1$-connectedness in reductive algebraic groups*, Trans. Amer. Math. Soc., **369** (2017), no. 8, 5999-6015.

[6] C. Balwe and A. Sawant, *$\mathbb{A}^1$-connected components of ruled surfaces*, in preparation.

[7] U. Choudhury, *Connectivity of motivic H-spaces*, Alg. Geom. Topol. **14** (2014) no. 1, 37–55.

[8] J.-L. Colliot-Thélène, R. Hoobler, B. Kahn, *The Bloch-Ogus-Gabber theorem*, Algebraic K-theory (Toronto, ON, 1996), 31-94, Fields Inst. Commun. 16, Amer. Math. Soc., 1997.

[9] P. Gille, *Le problème de Kneser-Tits*, Séminaire Bourbaki. Vol. 2007/2008. Astérisque No. **326** (2009), Exp. No. 983, vii, 39–81 (2010).

[10] A. Hogadi and G. Kulkarni, *Gabber's presentation lemma for finite fields*, Preprint, arXiv: 1612:09393v2 [math.AG] (2016).

[11] F. Morel, *An introduction to $\mathbb{A}^1$-homotopy theory*, Contemporary Developments in Algebraic K-Theory, ICTP Lect. Notes, vol. XV, Abdus Salam Int. Cent. Theoret. Phys., Trieste (2004), 357 – 441 (electronic).

[12] F. Morel, *The stable $\mathbb{A}^1$-connectivity theorems*, K-Theory **35** (2005), 1–68.

[13] F. Morel, *$\mathbb{A}^1$-algebraic topology over a field*, Lecture Notes in Mathematics, **2052** (2012), Springer, Heidelberg.

[14] F. Morel and V. Voevodsky, *$\mathbb{A}^1$-homotopy theory of schemes* Inst. Hautes Études Sci. Publ. Math. **90** (1999), 45–143.

ANAND SAWANT, MATHEMATISCHES INSTITUT, LUDWIG-MAXIMILIANS UNIVERSITÄT, THERESIENSTR. 39, D-80333 MÜNCHEN, GERMANY.
E-mail: sawant@math.lmu.de

K-Theory
Copyright ©2018 Tata Institute of Fundamental Research
Publisher: Hindustan Book Agency, New Delhi, India

Cellularity of hermitian K-theory and Witt-theory

Oliver Röndigs, Markus Spitzweck and Paul Arne Østvær

Abstract

Hermitian K-theory and Witt-theory are cellular in the sense of stable motivic homotopy theory over any base scheme without points of characteristic two.

1 Introduction

The notion of a cellular object in motivic homotopy theory is intrinsically linked to the geometry of motivic spheres $S^{p,q}$ [4]. Suppose the smooth scheme X admits a filtration by closed subschemes

$$\emptyset \subset X_0 \subset \cdots \subset X_{n-1} \subset X_n = X,$$

where $X_i \smallsetminus X_{i-1}$ is a disjoint union of affine spaces $\mathbf{A}^{n_{ij}}$. Examples of such filtrations arise in the context of Białynicki-Birula decompositions for $\mathbf{G}_m$-action on smooth projective varieties [2], cf. [3] for a more recent implementation. By homotopy purity [7, Theorem 3.2.23] for Thom spaces of normal bundles of closed embeddings, there is a homotopy cofiber sequence

$$X \smallsetminus X_i \longrightarrow X \smallsetminus X_{i-1} \longrightarrow \mathbf{Th}(\mathcal{N}_i).$$

By assumption the normal bundle $\mathcal{N}_i$ is trivial. Thus the splitting $\mathbf{Th}(\mathcal{N}_i) \cong \bigvee_j S^{2n_{ij}, n_{ij}}$ and the two-out-of-three property for stably cellular objects [4, Lemma 2.5] imply inductively that X is stably cellular in the sense of [4, Definition 2.10].

In this paper we employ a similar strategy to prove cellularity for Thom spaces of direct sums of tautological sympletic bundles over quaternionic Grassmannians. This allows us to show cellularity of the motivic spectra representing hermitian K-theory and Witt-theory [5]. By a base scheme we mean any regular noetherian separated scheme of finite Krull dimension.

Theorem 1.1 *Suppose all points on the base scheme have residue characteristic unequal to two. Then hermitian K-theory* **KQ** *and Witt-theory* **KW** *are cellular motivic spectra.*

For a related antecedent result showing cellularity of algebraic K-theory, see [4, Theorem 6.2]. The proof of Theorem 1.1 exploits the geometry of quaternionic Grassmannians and the explicit model for hermitian K-theory from [9].

Recent applications of **KQ** and **KW** concern computations of stable homotopy groups of motivic spheres [6], [8], [12], and a proof of the Milnor conjecture on quadratic forms [11]. For cellular motivic spectra one has the powerful fact that stable motivic weak equivalences are detected by $\pi_{*,*}$-isomorphisms [4, Corollary 7.2]. Our main motivation for proving Theorem 1.1 is that it is being used in the computation of the slices of **KQ** in [12, Theorem 2.14]. In terms of motivic cohomology with integral and mod-2 coefficients, the result is

$$\mathsf{s}_q(\mathbf{KQ}) \cong \begin{cases} \Sigma^{2q,q}\mathbf{MZ} \vee \bigvee_{i<\frac{q}{2}} \Sigma^{2i+q,q}\mathbf{MZ}/2 & q \text{ even} \\ \bigvee_{i<\frac{q+1}{2}} \Sigma^{2i+q,q}\mathbf{MZ}/2 & q \text{ odd.} \end{cases}$$

In turn, this is an essential ingredient in our proof of Morel's π_1-conjecture in [12]. It is an interesting problem to make sense of Theorem 1.1 without any assumptions on the points of the base scheme.

This short paper is organized into Section 2 on basic properties of motivic cellular spectra, Section 3 on the geometry of quaternionic Grassmannians, and Section 4 on hermitian K-theory and Witt-theory.

2   Cellular Objects

The subcategory of cellular spectra in the motivic stable homotopy category is the smallest full localizing subcategory that contains all suspensions of the sphere spectrum, cf. [4, §2.8]. For our purposes it suffices to know four basic facts about cellular motivic spectra. First we recall part (3) of Definition 2.1 in [4].

Lemma 2.1 *The homotopy colimit of a diagram of cellular motivic spectra is cellular.*

The second fact is a specialization of [4, Lemma 2.4].

Lemma 2.2 *Let E be a motivic spectrum and let p, q be integers. Then E is cellular if and only if its (p,q)-suspension $\Sigma^{p,q}E$ is cellular.*

The third fact is a specialization of [4, Lemma 2.5].

Lemma 2.3 *If $E \to F \to G$ is a homotopy cofiber sequence of motivic spectra such that any two of E, F, and G are cellular, then so is the third.*

Finally, we recall Lemma 3.2 in [4].

Lemma 2.4 *If E_i is a cellular motivic spectrum for all $i \in I$, then $\bigvee_{i \in I} E_i$ is cellular.*

3   Quaternionic Grassmannians

The quaternionic Grassmannian $\mathbf{HGr}(r, n)$ is the open subscheme of the ordinary Grassmannian $\mathbf{Gr}(2r, 2n)$ parametrizing $2r$-dimensional subspaces of the trivial vector bundle $\mathcal{O}^{\oplus 2n}$ on which the standard symplectic form is nondegenerate. It is smooth affine of dimension $4r(n - r)$ over the base scheme. Let $\mathcal{U}_{r,n}$ be short for the tautological symplectic subbundle of rank $2r$ on $\mathbf{HGr}(r, n)$. It is the restriction to $\mathbf{HGr}(r, n)$ of the tautological subbundle of $\mathbf{Gr}(2r, 2n)$ together with the restriction to $\mathcal{U}_{r,n}$ of the standard symplectic form on $\mathcal{O}^{\oplus 2n}$.

More generally, to every symplectic bundle $(\mathcal{E}, \phi)$ one associates the quaternionic Grassmannian $\mathbf{HGr}(r, \mathcal{E}, \phi)$; it is the open subscheme of the Grassmannian $\mathbf{Gr}(2r, \mathcal{E})$ parametrizing $2r$-dimensional subspaces of the fibers of $\mathcal{E}$ on which ϕ is nondegenerate. Associated to the trivial rank $2n - 2$ symplectic bundle $(\mathcal{E}, \psi)$ is the bundle $\mathcal{F} = \mathcal{O} \oplus \mathcal{E} \oplus \mathcal{O}$ equipped with the direct sum of ψ and the hyperbolic symplectic form, i.e.,

$$\begin{bmatrix} 0 & 0 & 1 \\ 0 & \psi & 0 \\ -1 & 0 & 0 \end{bmatrix}.$$

For simplicity we write $\mathbf{HGr}(\mathcal{E})$ for $\mathbf{HGr}(r, \mathcal{E}, \psi)$ and likewise for $\mathcal{F}$.

The normal bundle N of the embedding $\mathbf{HGr}(\mathcal{E}) \subset \mathbf{HGr}(\mathcal{F})$ is the tensor product $\mathcal{U}_{\mathcal{E}}^{\vee} \otimes \mathcal{O}^{\oplus 2}$ for the dual of the tautological symplectic subbundle of rank $2r$ on $\mathbf{HGr}(\mathcal{E})$. Theorem 4.1 in [10] shows that N is naturally isomorphic to an open subscheme of $\mathbf{Gr}(2r, \mathcal{F})$ and there is a decomposition $N = N^+ \oplus N^-$; here, $N^+ = \mathbf{HGr}(\mathcal{F}) \cap \mathbf{Gr}(2r, \mathcal{O} \oplus \mathcal{E})$ and $N^- = \mathbf{HGr}(\mathcal{F}) \cap \mathbf{Gr}(2r, \mathcal{E} \oplus \mathcal{O})$ have intersection $\mathbf{HGr}(\mathcal{E})$. Thus there are natural vector bundle isomorphisms $N^+ \cong N^- \cong \mathcal{U}_{r,n-1}$ and the normal bundle $\mathcal{N}$ of N^+ in $\mathbf{HGr}(\mathcal{F})$ is isomorphic to $\pi_+^* \mathcal{U}_{r,n-1}$ for the bundle projection $\pi_+ \colon N^+ \to \mathbf{HGr}(\mathcal{E})$. Moreover, there is a vector bundle isomorphism between the restriction $\mathcal{U}_{r,n}|N^+$ of $\mathcal{U}_{r,n}$ to N^+ and $\pi_+^* \mathcal{U}_{r,n-1}$. For $r \leq n - 1$, let Y denote the complement of N^+ in $\mathbf{HGr}(\mathcal{F})$ [10, (5.1)].

Proposition 3.1 *For $m \geq 0$ the suspension spectrum of the Thom space of the vector bundle $\mathcal{U}_{r,n}^{\oplus m}$ on $\mathbf{HGr}(r, n)$ is a finite cellular spectrum. In particular, $\Sigma^{\infty}\mathbf{HGr}(r, n)_{+}$ is a cellular spectrum.*

Proof   The proof proceeds by a double induction argument on r and $n \geq r$. The base cases $\mathbf{HGr}(0, n)$ and $\mathbf{HGr}(n, n)$ are clear, so we may assume $0 < r < n$. Define the motivic space Z by the homotopy cofiber sequence

$$\mathbf{Th}(\mathcal{U}_{r,n}^{\oplus m}|Y) \longrightarrow \mathbf{Th}(\mathcal{U}_{r,n}^{\oplus m}) \longrightarrow Z. \qquad (3.1)$$

According to [13, Lemma 3.5] there is a canonical isomorphism in the motivic homotopy category

$$Z \cong \mathbf{Th}(\mathcal{U}_{r,n}^{\oplus m}|N^{+} \oplus \mathcal{N}).$$

Using the above we note $\mathcal{U}_{r,n}^{\oplus m}|N^{+} \oplus \mathcal{N} \cong \pi_{+}^{*}\mathcal{U}_{r,n-1}^{\oplus(m+1)}$ and hence there are canonical isomorphisms

$$Z \cong \mathbf{Th}(\pi_{+}^{*}\mathcal{U}_{r,n-1}^{\oplus(m+1)}) \cong \mathbf{Th}(\mathcal{U}_{r,n-1}^{\oplus(m+1)}).$$

By induction hypothesis $\Sigma^{\infty}Z$ is a finite cellular spectrum. Thus Lemma 2.3 and (3.1) reduce the proof to showing that $\Sigma^{\infty}\mathbf{Th}(\mathcal{U}_{r,n}^{\oplus m}|Y)$ is a finite cellular spectrum. To this end we recall parts of Theorem 5.1 in [10]: There exists maps

$$Y \xleftarrow{g_1} Y_1 \xleftarrow{g_2} Y_2 \xrightarrow{q} \mathbf{HGr}(r - 1, \mathcal{E}, \psi),$$

where g_i and q are Zariski locally trivial torsors over vector bundles of rank $2r - i$ and $4n - 3$, respectively. Moreover, $g_2^{*}g_1^{*}\mathcal{U}_{r,n}$ is isomorphic to $\mathcal{O}_{Y_2}^{2} \oplus q^{*}\mathcal{U}_{r-1,n}$. Invoking [7, §3.2, Example 2.3] this implies the canonical isomorphisms

$$\Sigma^{\infty}\mathbf{Th}(\mathcal{U}_{r,n}^{\oplus m}|Y) \cong \Sigma^{\infty}\mathbf{Th}(g_2^{*}g_1^{*}\mathcal{U}_{r,n}^{\oplus m}|Y) \cong \Sigma^{\infty}\mathbf{Th}(\mathcal{O}_{Y_2}^{2m} \oplus q^{*}\mathcal{U}_{r-1,n}^{\oplus m}) \cong$$
$$\Sigma^{2m,m}\Sigma^{\infty}\mathbf{Th}(\mathcal{U}_{r-1,n}^{\oplus m}).$$

Here the suspension spectrum of $\mathbf{Th}(\mathcal{U}_{r-1,n}^{\oplus m})$ is finite cellular by the induction hypothesis. This finishes the proof using Lemma 2.2.

4   Hermitian K-theory and Witt-theory

In this section we finish the proof of Theorem 1.1 stated in the introduction.

The quaternionic plane $\mathbf{HP}^1$ is the first quaternionic Grassmannian $\mathbf{HGr}(1, 2)$. In the pointed unstable motivic homotopy category, $(\mathbf{HP}^1, x_0)$ is isomorphic to the two-fold smash product of the Tate object $T \equiv \mathbf{A}^1/\mathbf{A}^1 \smallsetminus \{0\}$. It follows that the $\mathbf{A}^1$-mapping cone $\mathbf{HP}^{1+}$ of the rational point $x_0 \colon S \to \mathbf{HP}^1$ is isomorphic to $T^{\wedge 2}$. Hence the stable homotopy category of $\mathbf{HP}^{1+}$-spectra is equivalent to the standard model for the stable motivic homotopy category [9, Theorem 12.1].

Theorem 12.3 in [9] shows there is an isomorphism between hermitian K-theory $\mathbf{KQ}$ and an $\mathbf{HP}^{1+}$-spectrum $\mathbf{BO}^{\mathrm{geom}}$. For n odd, $\mathbf{BO}^{\mathrm{geom}}_{2n} = \mathbf{Z} \times \mathbf{HGr}$ [9, (12.5)]. Here $\mathbf{HGr}$ denotes the infinite quaternionic Grassmannian, i.e., the sequential colimit

$$\operatorname*{colim}_n \mathbf{HGr}(n, 2n).$$

We note that the transition maps in the colimit are defined in [9, (8.1)]. The motivic space $\mathbf{Z} \times \mathbf{HGr}$ is pointed by $(0, \mathbf{HGr}(0, 0))$. Thus $\mathbf{KQ}$ is isomorphic to the homotopy colimit

$$\operatorname*{hocolim}_{n \text{ odd}} \Sigma^{4n, 2n} \Sigma^\infty \mathbf{Z} \times \mathbf{HGr}. \tag{4.1}$$

It remains to show cellularity of (4.1). Note that $\Sigma^\infty \mathbf{Z} \times \mathbf{HGr}$ is a homotopy colimit of cellular spectra by Lemma 2.4 and Proposition 3.1. It follows that $\Sigma^{4n, 2n} \Sigma^\infty \mathbf{Z} \times \mathbf{HGr}$ is cellular according to Lemmas 2.1 and 2.2. We conclude the proof for $\mathbf{KQ}$ by applying Lemma 2.1.

Cellularity of $\mathbf{KW}$ follows from that of $\mathbf{KQ}$ via Lemma 2.1 and the description of $\mathbf{KW}$ as the homotopy colimit of the diagram

$$\mathbf{KQ} \xrightarrow{\;\eta\;} \Sigma^{-1, -1}\mathbf{KQ} \xrightarrow{\;\Sigma^{-1, -1}\eta\;} \Sigma^{-2, -2}\mathbf{KQ} \xrightarrow{\;\Sigma^{-2, -2}\eta\;} \cdots$$

given in [1, Theorem 6.5]. Here η is the first stable Hopf map induced by the canonical map $\mathbf{A}^2 \smallsetminus \{0\} \to \mathbf{P}^1$.

References

[1] A. Ananyevskiy, *On the relation of special linear algebraic cobordism to Witt groups*, Homology Homotopy Appl. **18** no.1 (2016), 204–230.

[2] A. Białynicki-Birula, *Some theorems on actions of algebraic groups*, Ann. of Math. (2), **98** (1973), 480–497.

[3] P. Brosnan, *On motivic decompositions arising from the method of Białynicki-Birula*, Invent. Math. **161(1)** (2005), 91–111.

[4] D. Dugger and D. C. Isaksen, *Motivic cell structures*, Algebr. Geom. Topol. **5** (2005), 615–652.

[5] J. Hornbostel, $\mathbf{A}^1$-*representability of Hermitian K-theory and Witt groups*, Topology, **44(3)** (2005), 661–687.

[6] F. Morel, *On the motivic π_0 of the sphere spectrum*, In Axiomatic, enriched and motivic homotopy theory, of NATO Sci. Ser. II Math. Phys. Chem. **131** (2004), 219–260. Kluwer Acad. Publ.

[7] F. Morel and V. Voevodsky, $\mathbf{A}^1$-*homotopy theory of schemes*, Inst. Hautes Études Sci. Publ. Math. **90** (2001), 1999, 45–143.

[8] K. M. Ormsby and P. A. Østvær, *Stable motivic π_1 of low-dimensional fields*, Adv. Math. **265** (2014), 97–131.

[9] I. Panin and C. Walter, *On the motivic commutative ring spectrum* BO, arXiv:1011.0650.

[10] I. Panin and C. Walter, *Quaternionic Grassmannians and Pontryagin classes in algebraic geometry*, arXiv:1011.0649.

[11] O. Röndigs and P. A. Østvær, *Slices of hermitian K-theory and Milnor's conjecture on quadratic forms*, Geom. Topol. **20(2)** (2016), 1157–1212.

[12] O. Röndigs, M. Spitzweck and P. A. Østvær, *The first stable homotopy groups of motivic spheres*, to appear in Ann. Math., arXiv:1604.00365.

[13] M. Spitzweck, *Relations between slices and quotients of the algebraic cobordism spectrum*, Homology, Homotopy Appl. **12(2)** (2010), 335–351.

OLIVER RÖNDIGS, INSTITUTE OF MATHEMATICS, UNIVERSITY OF OSNABRÜCK, OSNABRÜCK, GERMANY
E-mail: oliver.roendigs@uni-osnabrueck.de

MARKUS SPITZWECK, INSTITUTE OF MATHEMATICS, UNIVERSITY OF OSNABRÜCK, OSNABRÜCK, GERMANY
E-mail: markus.spitzweck@uni-osnabrueck.de

PAUL ARNE ØSTVÆR, DEPARTMENT OF MATHEMATICS, UNIVERSITY OF OSLO, NORWAY
E-mail: paularne@math.uio.no

K-Theory
Copyright ©2018 Tata Institute of Fundamental Research
Publisher: Hindustan Book Agency, New Delhi, India

On the η-inverted sphere

Oliver Röndigs

Abstract

The first and second homotopy groups of the η-inverted sphere spectrum over a field of characteristic not two are zero. A cell presentation of higher Witt theory is given as well, at least over the complex numbers.

1 Introduction

Let $\eta\colon \mathbf{A}^2 \smallsetminus \{0\} \to \mathbf{P}^1$ denote the first algebraic Hopf map given by sending a nonzero pair (x, y) to the line it generates in the plane. It induces an element of the same name $\eta \in \pi_{1,1}\mathbf{1}$ in the motivic stable homotopy groups of spheres. Hence η acts on the motivic stable homotopy groups of the sphere spectrum, in fact of any motivic spectrum. Inverting this action produces trivial groups in degrees one and two for the sphere spectrum.

Theorem 1.1 *Let F be a field of characteristic not two. Then*

$$\pi_{n+1,n}\mathbf{1}[\tfrac{1}{\eta}] = \pi_{n+2,n}\mathbf{1}[\tfrac{1}{\eta}] = 0$$

for every integer n.

This vanishing result enters the computation of the first line

$$\bigoplus_{n\in\mathbb{Z}} \pi_{n+1,n}\mathbf{1}$$

of motivic stable homotopy groups of spheres given in [31]. Fabien Morel identified the zero line of motivic stable homotopy groups of spheres, that is, the direct sum

$$\bigoplus_{n\in\mathbb{Z}} \pi_{n,n}\mathbf{1}$$

with the graded Milnor-Witt K-theory of the base field — see [24, Theorem 6.2.1], and the corresponding unstable statement [25, Theorem 4.9], documented in detail for perfect infinite fields in [26]. This identification implies that $\pi_{n,n}\mathbf{1}[\frac{1}{\eta}]$ is isomorphic to the Witt ring of the base field for every integer n. Impressive work of Guillou and Isaksen on the motivic Adams spectral sequence at the prime two computes $\pi_{n+k,n}\mathbf{1}_2^{\wedge}[\frac{1}{\eta}]$ for all integers n and k over the real and complex numbers [10], [11]. It implies that $\pi_{n+3,n}\mathbf{1}_2^{\wedge}[\frac{1}{\eta}]$ is nontrivial, at least over the real and the complex numbers. The next task is to determine $\pi_{n+3,n}\mathbf{1}[\frac{1}{\eta}]$ for an arbitrary field of characteristic not two, perhaps via the resolution of $\mathbf{1}[\frac{1}{\eta}]$ via connective Witt theory cKW employed in Section 5. Only the first two maps

$$\mathbf{1}[\tfrac{1}{\eta}] \longrightarrow \mathsf{cKW} \longrightarrow \mathsf{cKW} \wedge \mathsf{cKW} \tag{1.1}$$

of this resolution enter the proof of Theorem 1.1. The zero line computation of Morel cited above allows one to deduce the vanishing in Theorem 1.1 from the connectivity of the second map in (1.1), as Section 5 explains. This connectivity is obtained in Section 7, at least after inverting 2, by transferring results from topology on connective real K-theory to connective Witt theory. Proving that this transfer works proceeds by a real version of Joseph Ayoub's model for the Betti realization, to be explained in Section 3 and exploited in Section 4. The argument sketched so far would provide Theorem 1.1 only after inverting 2^1, which annihilates Witt theory for fields which are not formally real — see Section 6. The integral computation requires knowledge on 2-complete computations, supplied by results from [31] with the help of the convergence result [17, Theorem 1]. The passage from 2-inverted to integral results is explained in Section 8.

Theorem 1.1 implies that a cell presentation of connective Witt theory starts by attaching 4-cells via the generators

$$\Sigma^{3,0}\mathbf{1}[\tfrac{1}{\eta}] \longrightarrow \mathbf{1}[\tfrac{1}{\eta}]$$

of $\pi_{3,0}\mathbf{1}[\frac{1}{\eta}]$. For the complex numbers, where complete information is available by [10], a very small and complete cell presentation of connective Witt theory is obtained at the end of Section 4 (see also [14]). This cell presentation is not used in the proof of Theorem 1.1, but included for comparison with the impressive rational result [2]. Cell presentations and connectivity are discussed in Section 2. The following notation is used throughout.

---

[1] See [6, Proposition 36] for a different proof of this vanishing statement after inverting 2.

F, S	field, finite dimensional separated Noetherian base scheme
$\mathbf{Sm}_S$	smooth schemes of finite type over S
$\mathbf{SH}(S)$	the motivic stable homotopy category of S
E, $\mathbf{1} = S^{0,0}$	generic motivic spectrum, the motivic sphere spectrum
$S^{s,t}$, $\Sigma^{s,t}$	motivic (s,t)-sphere, (s,t)-suspension
$\pi_{s,t}\mathsf{E} = [S^{s,t}, \mathsf{E}]$	motivic stable homotopy groups of E
$\underline{\pi}_{s,t}\mathsf{E}$	sheaves of motivic stable homotopy groups of E
KQ, KW	hermitian K-theory, Witt-theory

The suspension convention is such that $\mathbf{P}^1 \simeq S^{2,1}$ and $\mathbf{A}^1 \smallsetminus \{0\} \simeq S^{1,1}$.

A previous version of this work was included in a preliminary version of [31], but separated for the sake of presentation. Theorem 1.1 enters only [31, Section 5], and its proof requires results from [31, Sections 2–4] — see Section 8 for details. I thank the RCN program *Topology in Norway*, and in particular the University of Oslo and Paul Arne Østvær, as well as the DFG priority program *Homotopy theory and algebraic geometry* for support. I thank Joseph Ayoub for his input regarding a real version of his model of complex Betti realization. Tom Bachmann, Jeremiah Heller, Glen Wilson, and an anonymous referee receive my thanks for helpful comments.

2   Cellularity and Connectivity

The following definitions regarding cells are modelled on [8] and [17]. *Attaching a cell* to a motivic spectrum E refers to the process of forming the pushout of

$$D^{s+1,t}\mathbf{1} \longleftarrow \supset \Sigma^{s,t}\mathbf{1} \xrightarrow{\ f\ } \mathsf{E} \qquad (2.1)$$

for some map f in the category of motivic spectra. The arrow pointing to the left in (2.1) denotes the canonical inclusion into $D^{s+1,t}\mathbf{1}$, the simplicial mapping cylinder of the map $\Sigma^{s,t}\mathbf{1} \to *$. The pushout D then consists of E, together with a cell of *dimension* $(s+1,t)$ and *weight* t. More generally, one may attach a collection of cells indexed by some set I by forming the pushout of a diagram

$$\bigvee_{i \in I} D^{s_i,t_i}\mathbf{1} \longleftarrow \supset \bigvee_{i \in I} \Sigma^{s_i,t_i}\mathbf{1} \xrightarrow{\ f\ } \mathsf{E}$$

in the category of motivic spectra. A cell *presentation* of a map $f \colon \mathsf{D} \to \mathsf{E}$ of motivic spectra consists of a sequence of motivic spectra

$$\mathsf{D} = \mathsf{D}_{-1} \xrightarrow{d_0} \mathsf{D}_0 \xrightarrow{\sim} \mathsf{D}_0' \xrightarrow{d_1} \mathsf{D}_1 \xrightarrow{\sim} \mathsf{D}_1' \to \cdots \xrightarrow{d_n} \mathsf{D}_n \xrightarrow{\sim} \mathsf{D}_n' \xrightarrow{d_{n+1}} \cdots \qquad (2.2)$$

with canonical map to the colimit $c\colon \mathsf{D} \to \mathsf{D}_\infty$, attaching maps

$$\bigvee_{i \in I_n} \Sigma^{s_i, t_i} \mathbf{1} \xrightarrow{\alpha_n} \mathsf{D}'_{n-1}$$

for every natural number n, such that D_n is obtained by attaching cells to D'_{n-1} along α_n, and a weak equivalence $w\colon \mathsf{D}_\infty \xrightarrow{\sim} \mathsf{E}$ with $w \circ c = f$. Furthermore, all arrows in diagram (2.2) labelled with $\sim$ denote acyclic cofibrations.

Remark 2.1 The most important instance is the absolute case, that is, a cell presentation of a map $* \to \mathsf{E}$. Then one speaks of a cell presentation of E. The distinction between a motivic spectrum E and a cell presentation D_∞ of it might be neglected if no confusion can arise.

Example The suspension spectrum $\Sigma^\infty \mathbf{P}^\infty$ admits a cell presentation, having precisely one cell of dimension $(2n, n)$ for every natural number n.

Let $\mathbf{SH}(S)^{\mathrm{cell}}$ denote the full subcategory of $\mathbf{SH}(S)$ of motivic spectra admitting a cell presentation. It can be identified with the homotopy category of *cellular* motivic spectra, that is, the smallest full localizing subcategory of $\mathbf{SH}(S)$ containing the spheres $S^{s,t}$, as [8, Remark 7.4] and the following statement show.

Lemma 2.2 *Let E be a motivic spectrum. If it admits a cell presentation, it is cellular. Conversely, if it is cellular, it admits a cell presentation.*

Proof The first statement follows from the definitions. For the second statement, let E be a cellular motivic spectrum which is fibrant. One constructs inductively a suitable sequence of motivic spectra. Start with

$$\mathsf{D}_0 := \bigvee_{\alpha \in \pi_{s,t}\mathsf{E}, s, t} \Sigma^{s,t} \mathbf{1} \xrightarrow{\vee \alpha} \mathsf{E}$$

and factor the canonical map $\mathsf{D}_0 \to \mathsf{E}$ as an acyclic cofibration $\mathsf{D}_0 \xrightarrow{\sim} \mathsf{D}'_0$ followed by a fibration $\mathsf{D}'_0 \to \mathsf{E}$. The latter induces a surjection on $\pi_{s,t}$ for all s, t by construction. For every s, t, choose lifts of generators of the kernel of $\pi_{s,t}(\mathsf{D}'_0 \to \mathsf{E})$ and use these to attach cells to D'_0, leading to a map $\mathsf{D}_1 \to \mathsf{E}$. As before, factor it as an acyclic cofibration, followed by a fibration $\mathsf{D}'_1 \to \mathsf{E}$. This fibration still induces a surjection on $\pi_{s,t}$ for all s, t. Iterating this procedure leads to a map

$$\mathsf{D}_\infty = \mathrm{colim}_n \mathsf{D}_n \longrightarrow \mathsf{E}$$

which induces a surjective and injective map on $\pi_{s,t}$ for all s, t. The statement on injectivity requires that $\Sigma^{s,t}\mathbf{1}$ is compact, whence any element in the kernel lifts to a finite stage and hence is killed in the next stage. Since both D_∞ and E are cellular, the map $\mathsf{D}_\infty \to \mathsf{E}$ is even a weak equivalence [8, Cor. 7.2]. $\qquad\square$

The motivic stable homotopy category $\mathbf{SH}(S)$ is equipped with the *homotopy t-structure*

$$\mathbf{SH}(S)_{\geq n} \lhook\joinrel\longrightarrow \mathbf{SH}(S) \longleftarrow\!\!\!\!\!\supset \mathbf{SH}(S)_{\leq n}$$

where $\mathbf{SH}(S)_{\geq n} = \langle \Sigma^{s,t}\Sigma^\infty X_+ | X \in \mathbf{Sm}_S, s-t \geq n \rangle$ is the full subcategory generated under homotopy colimits and extensions by the shifted motivic suspension spectra of smooth S-schemes of connectivity at least n. See [16, Section 2.1] for details, and in particular [16, Theorem 2.3] for the identification with Morel's original definition via Nisnevich sheaves of motivic stable homotopy groups in case S is the spectrum of a field.

3   Real realization

Real Betti realization will be employed in order to give a topological interpretation of connective Witt theory over the real numbers. It is defined as the homotopy-colimit preserving functor $\mathbf{SH}(\mathbb{R}) \to \mathbf{SH}$ from the motivic stable homotopy category of the real numbers to the classical motivic stable homotopy category which is determined by sending the suspension spectrum of a smooth variety X over $\mathbb{R}$ to the suspension spectrum of the topological space of real points $X(\mathbb{R})$, equipped with the real analytic topology. Another viewpoint on the real Betti realization is given by equivariant stable homotopy theory with respect to the absolute Galois group C_2 of $\mathbb{R}$: The real Betti realization is the composition of geometric fixed points and the complex Betti realization; see [13, Section 4.4] for details. Real Betti realization is given already unstably on the level of presheaves on $\mathbf{Sm}_\mathbb{R}$ with values in simplicial sets, and as such it is a simplicial functor. There is an additive variant for presheaves with values in simplicial abelian groups, or, equivalently, complexes of abelian groups.

Theorem 3.1 *Real Betti realization* $\mathbb{R}^* \colon \mathbf{SH}(\mathbb{R}) \to \mathbf{SH}$ *is strict symmetric monoidal and has a right adjoint* $\mathbb{R}_*$. *Moreover, the canonical map*

$$\mathsf{D} \wedge \mathbb{R}_*(\mathsf{E}) \longrightarrow \mathbb{R}_*\big(\mathbb{R}^*(\mathsf{D}) \wedge \mathsf{E}\big) \tag{3.1}$$

is an equivalence for all motivic spectra $\mathsf{D} \in \mathbf{SH}(\mathbb{R})$.

Proof The first statement follows from [27, Section 3.3], and also from [13, Proposition 4.8]. The second statement follows from [9, Proposition 3.2] for strongly dualizable motivic spectra. Since $\mathbf{SH}(\mathbb{R})$ is generated by strongly dualizable motivic spectra, the result follows.

□

Identifying the real Betti realization of motivic spectra which are not suspension spectra of smooth varieties is not immediate. However, Joseph Ayoub's beautiful model for the complex Betti realization given in [4, Théorème 2.67] translates to a model for the real Betti realization. In order to define this model, let $\mathcal{I}^n$ denote the pro-real analytic manifold obtained by open neighborhoods of the compact unit cube $[0,1]^n \subset \mathbb{R}^n$. Letting n vary defines, in the standard way, a cocubical pro-real analytic manifold $\mathcal{I}^\bullet$ as defined in [4, Def. A.1]. One observes that $\mathcal{I}^\bullet$ is in fact a pseudo-monoidal Σ-enriched cocubical pro-real analytic manifold [4, Def. A.29]. Let $\mathcal{R}_n$ denote the $\mathbb{R}[t_1,\ldots,t_n]$-algebra of real analytic functions defined on an open neighborhood of $[0,1]^n$.

Proposition 3.2 (Ayoub) *The* $\mathbb{R}[t_1,\ldots,t_n]$-*algebra* $\mathcal{R}_n$ *is Noetherian and regular.*

Proof The $\mathbb{C}[t_1,\ldots,t_n]$-algebra $\mathcal{R}_n \otimes_{\mathbb{R}} \mathbb{C}$ is the algebra of complex analytic functions defined on an open neighborhood of $[0,1]^n$, which is Noetherian and regular. The required statement follows by Galois descent.

□

Popescu's theorem [34, Theorem 10.1] then implies that $\mathcal{R}_n$ is a filtered colimit of smooth $\mathbb{R}[t_1,\ldots,t_n]$-algebras, and thus can be considered as an affine pro-smooth $\mathbb{R}$-variety $\mathrm{Spec}(\mathcal{R}_n)$. Hence any presheaf K on $\mathbf{Sm}_{\mathbb{R}}$ admits a value $K(\mathrm{Spec}(\mathcal{R}_n))$. Letting n vary defines a cubical object $K(\mathrm{Spec}(\mathcal{R}_\bullet))$.

Theorem 3.3 (Ayoub) *Let* $C\colon \mathbf{Sm}_{\mathbb{R}} \to \mathbf{Cx}$ *be a presheaf of complexes of abelian groups. The real Betti realization of* C *is quasi-isomorphic to the total complex* $C(\mathrm{Spec}(\mathcal{R}_\bullet))$.

In fact, as in the complex case given in [4, Théorème 2.61] it is possible to replace $\mathrm{Spec}(\mathcal{R}_n)$ in Theorem 3.3 by $\mathrm{Spec}(\mathcal{R}_n^{et})$ where $\mathcal{R}_n^{et}$ is, roughly speaking, the largest sub-algebra of $\mathcal{R}_n$ which is pro-étale over $\mathbb{R}[t_1,\ldots,t_n]$.

4 Witt theory

Let KQ denote the motivic spectrum for hermitian K-theory of quadratic forms [15]. It is defined over $\text{Spec}(\mathbb{Z}[\frac{1}{2}])$, and hence over any scheme in which 2 is invertible. Note the identification

$$\text{KQ}_{s,t} = \pi_{s,t}\text{KQ} = GW^{[-t]}_{s-2t}(F)$$

with Schlichting's higher Grothendieck-Witt groups [33, Definition 9.1]. The periodicity element is denoted $\alpha\colon S^{8,4} \to \text{KQ}$. It is denoted β in [1].

Theorem 4.1 *Suppose S is a scheme over* $\text{Spec}(\mathbb{Z}[\frac{1}{2}])$. *Then* KQ *is a cellular commutative motivic ring spectrum in* **SH**(S) *which is preserved under base change.*

Proof A ring structure is provided in [28]. Cellularity follows basically from the model given in [28]; details may be found in [30]. The statement regarding base change refers to the fact that, given a morphism $f\colon S \to S'$ of schemes over $\text{Spec}(\mathbb{Z}[\frac{1}{2}])$, there is a canonical identification $f^*(\text{KQ}) \to \text{KQ}$ of motivic spectra over S. It is induced by the corresponding canonical identification of Grassmannians which serve to model KQ by [28]. $\square$

Let KW be the motivic ring spectrum representing higher Balmer-Witt groups in the motivic stable homotopy category. It can be described as $\text{KW} = \text{KQ}[\frac{1}{\eta}]$, where $\eta\colon S^{1,1} \to S^{0,0}$ is the first algebraic Hopf map.

Corollary 4.2 *Suppose S is a scheme over* $\text{Spec}(\mathbb{Z}[\frac{1}{2}])$. *Then* KW *is a cellular motivic ring spectrum in* **SH**(S) *which is preserved under base change. It is a commutative* $1[\frac{1}{\eta}]$*-algebra.*

Over fields, one may describe the coefficients of KW as follows:

$$\text{KW}_{*,*} \cong \text{KW}_{0,0}[\eta, \eta^{-1}, \alpha, \alpha^{-1}] \cong W[\eta, \eta^{-1}, \alpha, \alpha^{-1}] \qquad (4.1)$$

Here W is the Witt ring of the base field and α is the following composition:

$$S^{8,4} \xrightarrow{\ \alpha\ } \text{KQ} \xrightarrow{\ \text{can.}\ } \text{KW}$$

The remainder of the section will be devoted to specific descriptions of the motivic spectrum KW, after suitable modifications, or over specific fields. See [32, Theorem 2.6.4] for the following statement.

Theorem 4.3 (Pfister) *Let S be the spectrum of a field. The additive group underlying the graded ring* $\mathsf{KW}_{*,*}[\frac{1}{2}]$ *is torsion-free.*

In order to state the next theorem, let KO denote the topological spectrum representing real topological K-theory. Its 0-connective cover is denoted ko. The coefficients of the topological spectrum $\mathsf{KO}[\frac{1}{2}]$ are particularly simple: A Laurent polynomial ring on a single generator α_{top} of degree 4.

Theorem 4.4 (Brumfiel) *There is an equivalence* $\mathbb{R}^*\mathsf{KW}[\frac{1}{2}] \simeq \mathsf{KO}[\frac{1}{2}]$ *of topological spectra. The unit map* $\mathsf{KW}[\frac{1}{2}] \to \mathbb{R}_*\mathbb{R}^*\mathsf{KW}[\frac{1}{2}]$ *is an equivalence in* $\mathbf{SH}(\mathbb{R})$.

Proof This follows from [19, Theorem 6.2] (but see also [2, Remark 3]) which provides a natural equivalence of classical spectra

$$\mathsf{KW}_0[\tfrac{1}{2}](X) \longrightarrow \mathsf{KO}[\tfrac{1}{2}](\mathbb{R}^*X)$$

for every finite type $\mathbb{R}$-scheme X. The Yoneda lemma then implies that the map induced by the functor $\mathbb{R}^*$ on internal simplicial sets of morphisms is an equivalence for all smooth $\mathbb{R}$-schemes, whose $\Sigma^{s,t}$-suspensions generate $\mathbf{SH}(\mathbb{R})$. Adjointness then implies that the unit map $\mathsf{KW}[\frac{1}{2}] \to \mathbb{R}_*\mathbb{R}^*\mathsf{KW}[\frac{1}{2}]$ is an equivalence in $\mathbf{SH}(\mathbb{R})$.   □

Due to the α-periodicity of its target, the unit $\mathbf{1} \to \mathsf{KW}$ is not k-connective for any $k \in \mathbb{Z}$. Let $\psi\colon \mathsf{cKW} := \mathsf{KW}_{\geq 0} \to \mathsf{KW}$ denote the 0-connective cover with respect to the homotopy t-structure. Since $\mathbf{1}$ is 0-connective, the $\mathbf{1}[\frac{1}{\eta}]$-algebra homomorphism $u\colon \mathbf{1}[\frac{1}{\eta}] \to \mathsf{KW}$ factors uniquely as a $\mathbf{1}[\frac{1}{\eta}]$-algebra homomorphism $cu\colon \mathbf{1}[\frac{1}{\eta}] \to \mathsf{cKW}$, even in a strict sense if one uses [12] whose general setup applies here. The next aim is to provide an analog of Theorem 4.4 for cKW instead of KW.

Theorem 4.5 *Suppose S is a scheme over* $\mathrm{Spec}(\mathbb{Z}[\frac{1}{2}])$. *Then* cKW *is a commutative* $\mathbf{1}[\frac{1}{\eta}]$-*algebra which is preserved under base change.*

Proof The statement regarding the multiplicative structure follows from [12]. The base change argument for connective covers is given in [16, Lemma 2.2].   □

Lemma 4.6 *The canonical map* $\mathsf{cKW} \to \mathsf{KW}$ *coincides with the canonical map* $\mathsf{cKW} \to \mathsf{cKW}[\alpha^{-1}]$ *up to canonical equivalence.*

Proof By construction, $\alpha\colon S^{8,4} \to \mathsf{KW}$ is an invertible element. Moreover, it lifts to a map $S^{8,4} \to \mathsf{cKW}$ deserving the same notation. Hence there is a canonical map $\mathsf{cKW}[\alpha^{-1}] \to \mathsf{KW}$ which induces an isomorphism on sheaves of homotopy groups, whence the statement.

$\square$

Lemma 4.6 implies that the filtration on KW given by the homotopy t-structure coincides with multiplications by powers of α on suspensions of cKW. Thus the cone of multiplication by α on cKW, henceforth denoted cKW/α, coincides with the 0-truncation of KW, or equivalently, with the motivic Eilenberg-MacLane spectrum for the sheaf of unramified Witt groups. In particular, over a field F of characteristic not two, it is a motivic spectrum whose homotopy groups are concentrated in the zero line:

$$\pi_{s,t}\mathsf{cKW}/\alpha \cong \begin{cases} W(F) & s = t \\ 0 & s \neq t \end{cases}$$

Moreover, the canonical map $\mathsf{cKW} \to \mathsf{cKW}/\alpha$ is even a map of $\mathbf{1}[\frac{1}{\eta}]$-algebras. The homotopy t-structure filtration of KW thus consists of cKW-modules, and its associated graded consists of (de)suspensions of cKW/α.

Lemma 4.7 *The canonical map $\mathbb{R}^*(\mathsf{cKW}/\alpha) \to H\mathbb{Z}[\frac{1}{2}]$ is an equivalence of topological spectra.*

Proof The algebraic Hopf map η induces the degree -2 map on the topological sphere spectrum after taking real points. Morel's Theorem, see 5.2, implies the map $\mathbf{1}[\frac{1}{\eta}] \to \mathsf{cKW}$ is 1-connective and hence induces a 1-connective map

$$\mathbb{S}[\tfrac{1}{2}] = \mathbb{R}^*(\mathbf{1}[\tfrac{1}{\eta}]) \longrightarrow \mathbb{R}^*\mathsf{cKW}$$

of topological spectra. This identifies $\pi_0\mathbb{R}^*\mathsf{cKW}$ as $\mathbb{Z}[\frac{1}{2}]$ (and $\pi_1\mathbb{R}^*\mathsf{cKW}$ as the trivial group). Since the canonical map $\mathsf{cKW} \to \mathsf{cKW}/\alpha$ is 4-connective, also $\pi_0\mathbb{R}^*\mathsf{cKW}/\alpha \cong \mathbb{Z}[\frac{1}{2}]$ (and $\pi_1\mathbb{R}^*\mathsf{cKW}/\alpha = 0$). The canonical map mentioned in the statement of the lemma is the map to the zeroth Postnikov section. It remains to show that $\pi_n\mathbb{R}^*\mathsf{cKW}/\alpha = 0$ for all $n \geq 2$. For this purpose, recall that cKW/α is the Eilenberg-MacLane spectrum associated to the sheaf of unramified Witt groups on $\mathbf{Sm}_\mathbb{R}$. Hence its real Betti realization is determined by the real Betti realization of the sheaf W of Witt groups, considered as a complex concentrated in degree zero, and the spectrum structure maps. Theorem 3.3 allows one to identify the real Betti realization of W as the complex $W(\mathrm{Spec}(\mathcal{R}_\bullet))$, which turns out to be the complex

$$\mathbb{Z} \xleftarrow{\;0\;} \mathbb{Z} \xleftarrow{\;\mathrm{id}\;} \mathbb{Z} \xleftarrow{\;0\;} \mathbb{Z} \xleftarrow{\;\mathrm{id}\;} \cdots$$

by Lemma 4.8 below. The result follows.

□

The statement of Lemma 4.7 for $\mathsf{cKW}[\frac{1}{2}]/\alpha$ is mentioned below the proof of [5, Proposition 29]. Note that the proof of Lemma 4.7 supplies an integral identification of the real points of the S^1-Eilenberg-MacLane spectrum for the sheaf of Witt groups.

Lemma 4.8 *The inclusions*

$$\mathbb{R} \lhook\joinrel\longrightarrow \mathbb{R}[t_1, \ldots, t_n] \lhook\joinrel\longrightarrow \mathcal{R}_n \lhook\joinrel\longrightarrow \mathbb{R}[\![t_1, \ldots, t_n]\!]$$

induce isomorphisms on Witt groups for every n.

Proof By construction, $\mathcal{R}_0 = \mathbb{R}$. Shrinking the cube defines a filtered system of intermediate algebras between $\mathcal{R}_n$ and the formal power series ring $\mathbb{R}[\![t_1, \ldots, t_n]\!]$ whose colimit is the ring of germs of real analytic functions at zero. The intermediate algebras are all isomorphic to $\mathcal{R}_n$. Moreover, the restriction homomorphism between any two such defines an isomorphism on Witt groups, as one concludes by a homotopy interpolating between the two cube diameters. Since the Witt group commutes with filtered colimits, the result follows.

□

Corollary 4.9 *There is an equivalence $\mathbb{R}^* \mathsf{cKW}[\frac{1}{2}] \simeq \mathsf{ko}[\frac{1}{2}]$ of topological spectra. The unit $\mathsf{cKW}[\frac{1}{2}] \to \mathbb{R}_* \mathbb{R}^* \mathsf{cKW}[\frac{1}{2}] \simeq \mathbb{R}_*(\mathsf{ko}[\frac{1}{2}])$ is an equivalence in $\mathbf{SH}(\mathbb{R})$.*

Proof This follows from Theorem 4.4, Lemma 4.7, and the compatibility of $(\mathbb{R}^*, \mathbb{R}_*)$ with the homotopy t-structure.

□

Over the complex numbers, a very explicit small cell presentation of cKW can be given, thanks to the following fantastic theorem [10], [3]. This cell presentation will not be used in the remaining sections.

Theorem 4.10 (Andrews-Miller) *Over $\mathbb{C}$, the graded ring $\pi_{*,*}\mathbf{1}[\frac{1}{\eta}]$ is isomorphic to the ring $\mathbb{F}_2[\eta, \eta^{-1}, \sigma, \mu_9]/\eta\sigma^2$ where $|\eta| = (1,1)$, $|\sigma| = (7,4)$, and $|\mu_9| = (9,5)$.*

A priori, the Andrews-Miller computation produces the homotopy groups of $1_2^{\wedge}[\frac{1}{\eta}]$ but since $\pi_{0,0}1[\frac{1}{\eta}] \cong \mathbb{Z}/2$ over $\mathbb{C}$ by Morel's Theorem 5.1, $1[\frac{1}{\eta}]$ is already 2-complete over $\mathbb{C}$. Theorem 4.10 implies the following statement on the slices of $1[\frac{1}{\eta}]$, as explained in [31, Theorem 2.34].

Theorem 4.11 *Suppose S is a scheme over* $\mathrm{Spec}(\mathbb{Z}[\frac{1}{2}])$. *There is a splitting of the zero slice of the η-inverted sphere spectrum*

$$\mathsf{s}_0(1[\tfrac{1}{\eta}]) \cong \bigvee_{1 \neq n \geq 0} \Sigma^{n,0}\mathsf{M}\mathbb{Z}/2.$$

such that the unit $1[\frac{1}{\eta}] \to \mathsf{KW}$ *induces the inclusion on every even summand.*

The remainder of this section takes place in the category of $1[\frac{1}{\eta}]$-modules over the field of complex numbers. Abbreviate $\oplus_n \pi_{n+k,n}$ as π_k.

Lemma 4.12 *The unit* $1[\frac{1}{\eta}] \to \mathsf{cKW}$ *induces an isomorphism* $\pi_{4k}1[\frac{1}{\eta}] \to \pi_{4k}\mathsf{cKW}$ *for every integer k.*

Proof The unique nontrivial element $\alpha^k \eta^{-4k} \in \pi_{4k,0}\mathsf{KW}$ comes from the unique nontrivial element in the $4k$th column of the $E^2 = E^\infty$ page of the zeroth slice spectral sequence of KW [29]. The description of the unit map $1[\frac{1}{\eta}] \to \mathsf{KW}$ on slices, Theorem 4.11, implies that this element is the image of the unique nontrivial element in the $4k$th column of the $E^2 = E^\infty$ page of the zeroth slice spectral sequence of $1[\frac{1}{\eta}]$. Another proof, which does not rely on Voevodsky's slice filtration, is given in [14, Theorem 3.2]. $\square$

Set $\mathsf{D}_1 = 1[\frac{1}{\eta}]$. Then the unit $1[\frac{1}{\eta}] = \mathsf{D}_1 \to \mathsf{cKW}$ is a 3-connective map.[2] Set D_2 to be the homotopy cofiber of

$$\sigma\eta^{-4} : \Sigma^{3,0}1[\tfrac{1}{\eta}] \longrightarrow 1[\tfrac{1}{\eta}] = \mathsf{D}_1$$

Then the canonical map $1[\frac{1}{\eta}] = \mathsf{D}_1 \to \mathsf{D}_2$ induces an isomorphism on π_m for $m \geq 4$, as a consequence of the long exact sequence of homotopy groups and Theorem 4.10. The unit $1[\frac{1}{\eta}] \to \mathsf{cKW}$ then factors over D_2 as a 7-connective map $\mathsf{D}_2 \to \mathsf{cKW}$ by Lemma 4.12.

This can be continued inductively. Suppose that there exists a sequence of cellular motivic spectra factoring the unit of cKW as

$$1[\tfrac{1}{\eta}] = \mathsf{D}_1 \longrightarrow \mathsf{D}_2 \longrightarrow \cdots \longrightarrow \mathsf{D}_n \longrightarrow \mathsf{cKW}$$

---

[2] True over every field of characteristic not two, as Theorem 8.3 shows.

where the last map is $4n - 1$-connective, and where for every $k \leq n$

$$\mathbf{1}[\tfrac{1}{\eta}] \longrightarrow \mathsf{D}_k$$

induces an isomorphism on π_m for $m \geq 4(k - 1)$. Consider the (over $\mathbb{C}$ unique nontrivial) element $\Sigma^{4n-1}\mathbf{1}[\tfrac{1}{\eta}] \to \mathsf{D}_n$ in $\pi_{4n-1}\mathsf{D}_n \cong \pi_{4n-1}\mathbf{1}[\tfrac{1}{\eta}]$ inducing an isomorphism on π_{4n-1}. Then its homotopy cofiber D_{n+1} satisfies the following two properties:

1. The map $\mathbf{1}[\tfrac{1}{\eta}] \to \mathsf{D}_{n+1}$ induces an isomorphism on π_m for $m \geq 4n$.
2. The map $\mathsf{D}_n \to \mathsf{cKW}$ factors over $\mathsf{D}_n \to \mathsf{D}_{n+1}$ as a $4n + 3$-connective map $\mathsf{D}_{n+1} \to \mathsf{cKW}$.

The first property follows from the long exact sequence of homotopy groups, the condition on D_n, and Theorem 4.10. The second property then follows from the first property, equation (4.1), the condition on D_n, and Lemma 4.12. Taking the colimit with respect to $n \to \infty$ produces the desired cell presentation of cKW by Lemma 4.12. A cell presentation for KW then follows from Lemma 4.6. Rationally this cell presentation splits by [2], even over any field of characteristic not two.

5   An Adams resolution with connective Witt theory

The section title refers to the cosimplicial diagram

$$[n] \longmapsto \mathsf{cKW}^{\wedge n+1}$$

determined by the $\mathbf{1}[\tfrac{1}{\eta}]$-algebra cKW. The starting point of this resolution of $\mathbf{1}[\tfrac{1}{\eta}]$ is the following.

Theorem 5.1 (Morel) *If F is a field, $\pi_{n,n}\mathbf{1}[\tfrac{1}{\eta}]$ is isomorphic to the Witt ring of F.*

It translates to the following statement.

Corollary 5.2 *Let F be a field of characteristic not two. The unit $cu\colon \mathbf{1}[\tfrac{1}{\eta}] \to \mathsf{cKW}$ is 1-connective.*

In other words, $\pi_{n,n}cu$ is an isomorphism and $\pi_{n+1,n}cu$ is surjective for every integer n. Let

$$\mathsf{C} \longrightarrow \mathbf{1}[\tfrac{1}{\eta}] \xrightarrow{\ cu\ } \mathsf{cKW} \qquad\qquad (5.1)$$

be the fiber of cu. Corollary 5.2 implies that it is 1-connective.

Lemma 5.3 *The canonical map*

$$\underline{\pi}_{p,q}\mathsf{C} \longrightarrow \underline{\pi}_{p,q}\mathbf{1}[\tfrac{1}{\eta}]$$

is an isomorphism if $p - q \equiv 1, 2(4)$ and surjective if $p - q \equiv 3(4)$.

Proof The cofiber sequence (5.1) induces a long exact sequence of sheaves of homotopy groups. The result then follows from vanishing $\underline{\pi}_{p,q}\mathsf{cKW} = 0$ for $p - q$ not divisible by 4, which in turn follows from the fact that higher Witt groups of a field are concentrated in degrees congruent to 0 modulo 4.

□

Smashing the cofiber sequence (5.1) with cKW produces the following cofiber sequence:

$$\mathsf{C} \wedge \mathsf{cKW} \longrightarrow \mathbf{1}[\tfrac{1}{\eta}] \wedge \mathsf{cKW} = \mathsf{cKW} \xrightarrow{\;cu \wedge \mathsf{cKW}\;} \mathsf{cKW} \wedge \mathsf{cKW} \qquad (5.2)$$

Lemma 5.4 *The connecting map*

$$\underline{\pi}_{p+1,q}\mathsf{cKW} \wedge \mathsf{cKW} \longrightarrow \underline{\pi}_{p,q}\mathsf{C} \wedge \mathsf{cKW}$$

is an isomorphism for $p - q \equiv 1, 2(4)$, and surjective for $p - q \equiv 0, 3(4)$.

Proof The cofiber sequence (5.2) induces a long exact sequence of homotopy groups. As in the proof of Lemma 5.3, the vanishing $\underline{\pi}_{p,q}\mathsf{cKW} = 0$ for $p - q$ not divisible by 4 implies the statement for $p - q \equiv 1, 2, 3(4)$. Since $cu \wedge \mathsf{cKW}$ has the multiplication as a retraction, surjectivity also holds for $p - q \equiv 0(4)$.

□

An explicit consequence of Lemma 5.4 is that $\underline{\pi}_{n+1,n}\mathsf{cKW} \wedge \mathsf{cKW} = 0$ for all integers n. In fact, C is 1-connective by Corollary 5.2, whence $\underline{\pi}_{n,n}\mathsf{C} \wedge \mathsf{cKW} = 0$.

Proposition 5.5 *For every integer n, there is an isomorphism*

$$\underline{\pi}_{n+2,n}\mathsf{cKW} \wedge \mathsf{cKW} \cong \underline{\pi}_{n+1,n}\mathbf{1}[\tfrac{1}{\eta}]$$

of sheaves of homotopy groups.

Proof Smashing the cofiber sequence (5.1) with C produces the following cofiber sequence:

$$C \wedge C \longrightarrow \mathbf{1}[\tfrac{1}{\eta}] \wedge C = C \xrightarrow{\ cu \wedge C\ } cKW \wedge C$$

Since C is 1-connective by Morel's Theorem 5.2, $C \wedge C$ is 2-connective, which implies that $\underline{\pi}_{n+1,n}C \to \underline{\pi}_{n+1,n}cKW \wedge C$ is an isomorphism. Lemma 5.4 gives that the connecting map $\underline{\pi}_{n+2,n}cKW \wedge cKW \to \underline{\pi}_{n+1,n}cKW \wedge C$ is an isomorphism. The map $\underline{\pi}_{n+1,n}C \to \underline{\pi}_{n+1,n}\mathbf{1}[\tfrac{1}{\eta}]$ is an isomorphism by Lemma 5.3. The appropriate composition provides the desired isomorphism.

$\square$

Proposition 5.6 *If* $\underline{\pi}_{n+1,n}\mathbf{1}[\tfrac{1}{\eta}] = 0$, *then there is an isomorphism*

$$\underline{\pi}_{n+3,n}cKW \wedge cKW \cong \underline{\pi}_{n+2,n}\mathbf{1}[\tfrac{1}{\eta}]$$

of sheaves of homotopy groups for every integer n.

Proof Assume that $\underline{\pi}_{n+1,n}\mathbf{1}[\tfrac{1}{\eta}] = 0$, or, equivalently, that C is 2-connective. Then $C \wedge C$ is 4-connective, which implies that $\underline{\pi}_{n+k,n}C \to \underline{\pi}_{n+k,n}cKW \wedge C$ is an isomorphism for all integers n and all integers $k \leq 3$, and in particular for $k = 2$. Lemma 5.4 implies that the connecting map $\underline{\pi}_{n+3,n}cKW \wedge cKW \to \underline{\pi}_{n+2,n}cKW \wedge C$ is an isomorphism. The canonical map $\underline{\pi}_{n+2,n}C \to \underline{\pi}_{n+2,n}\mathbf{1}[\tfrac{1}{\eta}]$ is an isomorphism by Lemma 5.3. The appropriate composition provides the desired isomorphism.

$\square$

Propositions 5.5 and 5.6 are direct manifestations of the applicability of cooperations in connective Witt theory to computations of motivic stable homotopy groups of the η-inverted sphere. The following structural result has consequences for $cKW \wedge cKW$. In order to state it, recall the real étale topology on $\mathbf{Sm}_S$ which has as stalks henselian local rings with real closed residue fields. In particular, it is finer than the Nisnevich topology. The identity functor on motivic (symmetric) spectra over S thus can be regarded as a left Quillen functor from the Nisnevich to the real étale homotopy theory, and similarly for the respective $\mathbf{A}^1$-localizations. The real étale topology is relevant because of [19, Theorem 6.6], which implies that a Nisnevich fibrant model for $cKW[\tfrac{1}{2}]$ is already real étale fibrant. Besides being crucial input for a proof of Theorem 4.4 above, this gives the base case for the following statement (which could be formulated in greater generality).

Proposition 5.7 *Let F be a field of characteristic zero. A Nisnevich fibrant* $\mathsf{cKW}[\frac{1}{2}]$-*module over F is already real étale fibrant.*

Proof The standard model structure on modules over a ring object has underlying weak equivalences and fibrations. The corresponding model structure in the case at hand is cofibrantly generated. The cofibers of these generating cofibrations are free modules $\mathsf{cKW}[\frac{1}{2}] \wedge G$, where G is a shifted motivic suspension spectrum of a smooth F-scheme. A standard argument shows that it suffices to prove that a Nisnevich fibrant replacement of $\mathsf{cKW}[\frac{1}{2}] \wedge G$ is already real étale fibrant. Since the motivic symmetric spectra G are strongly dualizable, a Nisnevich fibrant replacement is given by the internal motivic spectrum of maps from (a cofibrant model of) the dual of G to a Nisnevich fibrant model of $\mathsf{cKW}[\frac{1}{2}]$. However, since the latter is already real étale fibrant, so is the internal motivic spectrum of maps. □

Proposition 5.7 assumes that 2 is invertible and that the base field has characteristic zero. Fields of positive characteristic do not have interesting Witt theory after inverting 2, as the following short section discusses for the sake of completeness.

6   Fields of odd characteristic

Theorem 6.1 *Let F be a field of odd characteristic. Then $\mathbf{1}[\frac{1}{\eta}, \frac{1}{2}]$ is contractible.*

Proof One argument (there are simpler ones) uses the beginning of the cKW-resolution of $\mathbf{1}[\frac{1}{\eta}]$ sketched in Section 5. The motivic spectrum $\mathsf{cKW}[\frac{1}{2}]$ has trivial homotopy sheaves by [32, Theorem 2.6.4]. Hence it is contractible, and so is the motivic spectrum $C \wedge \mathsf{cKW}[\frac{1}{2}]$, where C is the homotopy fiber of $\mathbf{1}[\frac{1}{\eta}] \to \mathsf{cKW}$ described in Section 5. It follows that the canonical map $C \wedge C[\frac{1}{2}] \to C[\frac{1}{2}]$ is a weak equivalence. However, since C is 1-connective by Morel's Theorem 5.2, $C \wedge C$ is 2-connective. Hence $C[\frac{1}{2}]$ is 2-connective, so $C \wedge C[\frac{1}{2}]$ is 4-connective, and so on. Thus $C[\frac{1}{2}]$ is contractible, which implies the same for $\mathbf{1}[\frac{1}{\eta}, \frac{1}{2}]$. □

Theorem 6.1 holds for any field whose Witt group is (necessarily 2-primary) torsion. This class of fields coincides with the class of non formally real fields [32, Theorem 2.7.1].

7 Witt cooperations

All statements on Witt cooperations will be deduced from the rational case [2] and the following topological result.

Theorem 7.1 (Mahowald, Kane, Lellmann) *For every prime p, there exists a sequence $B(j)$ of connective topological spectra starting with S^0, a sequence s_j of natural numbers with $s_j \geq 4j$, and a p-local equivalence*

$$\gamma \colon \bigvee_{j \geq 0} \Sigma^{s_j} B(j) \wedge \mathsf{ko} \longrightarrow \mathsf{ko} \wedge \mathsf{ko} \tag{7.1}$$

of topological spectra.

Proof This follows from [22] for the prime 2 (which is irrelevant for the following arguments), [18], and [21] for odd primes. See also [23] and [7]. $\square$

Theorem 7.1 can be reinterpreted as follows. Fix an odd prime. Consider the motivic spectrum $\Sigma^{s_j,0} B(j) \wedge \mathsf{cKW}[\frac{1}{2}]$ over the real numbers. Here the smash product with the topological spectrum $B(j)$ can be interpreted in two equivalent ways. One way is to view the category of motivic (symmetric) spectra as enriched over the category of usual (symmetric) spectra, and to use an appropriate simplicial version of $B(j)$. Another way, to be pursued here, is to consider $B(j)$ as a constant presheaf of usual S^1-spectra, having a motivic suspension spectrum with the same notation. The real Betti realization of $\Sigma^{s_j,0} B(j) \wedge \mathsf{cKW}[\frac{1}{2}]$ is

$$\mathbb{R}^* \big(\Sigma^{s_j,0} B(j) \wedge \mathsf{cKW}[\tfrac{1}{2}] \big) \simeq \mathbb{R}^* \Sigma^{s_j,0} B(j) \wedge \mathbb{R}^* \mathsf{cKW}[\tfrac{1}{2}]) \simeq \Sigma^{s_j} B(j) \wedge \mathsf{ko}[\tfrac{1}{2}]$$

by Theorem 3.1 and Corollary 4.9. Composition with γ from (7.1) yields a map

$$\mathbb{R}^* \big(\Sigma^{s_j,0} B(j) \wedge \mathsf{cKW}[\tfrac{1}{2}] \big) \longrightarrow \mathsf{ko} \wedge \mathsf{ko}[\tfrac{1}{2}] \xrightarrow{\simeq} \mathbb{R}^* \big(\mathsf{cKW} \wedge \mathsf{cKW}[\tfrac{1}{2}] \big) \tag{7.2}$$

whose adjoint is a map $\Sigma^{s_j,0} B(j) \wedge \mathsf{cKW}[\frac{1}{2}] \to \mathbb{R}_* \mathbb{R}^* \big(\mathsf{cKW} \wedge \mathsf{cKW}[\frac{1}{2}] \big)$. Let

$$\gamma^\flat \colon \Sigma^{s_j,0} B(j) \wedge \mathsf{cKW}[\tfrac{1}{2}] \longrightarrow \mathbb{R}_* \mathbb{R}^* \big(\mathsf{cKW} \wedge \mathsf{cKW}[\tfrac{1}{2}] \big)$$

$$\simeq \mathsf{cKW} \wedge \mathbb{R}_* \mathbb{R}^* \big(\mathsf{cKW}[\tfrac{1}{2}] \big) \simeq \mathsf{cKW} \wedge \mathsf{cKW}[\tfrac{1}{2}]$$

be the composition of this adjoint map, the equivalence occurring in the projection formula (3.1) for the real Betti realization displayed in Theorem 3.1, and finally the equivalence $\mathsf{cKW}[\frac{1}{2}] \to \mathbb{R}_* \mathbb{R}^* (\mathsf{cKW}[\frac{1}{2}])$ from Corollary 4.9.

Corollary 7.2 *Let p be an odd prime. The map*

$$\gamma^{\flat}\colon \bigvee_{j\geq 0} \Sigma^{sj,0}B(j) \wedge \mathsf{cKW}[\tfrac{1}{2}] \longrightarrow \mathsf{cKW} \wedge \mathsf{cKW}[\tfrac{1}{2}]$$

induced by the p-local equivalence (7.1) is a p-local equivalence in **SH**(ℝ).

Proof The adjunction $(\mathbb{R}^{*}, \mathbb{R}_{*})$ of integral stable homotopy categories descends to the "same" adjunction of p-local stable homotopy categories. Theorem 7.1 implies that the map (7.2) is a p-local equivalence. Its adjoint $\Sigma^{sj,0}B(j) \wedge \mathsf{cKW}[\tfrac{1}{2}] \to \mathbb{R}_{*}\mathbb{R}^{*}\big(\mathsf{cKW} \wedge \mathsf{cKW}[\tfrac{1}{2}]\big)$ is computed as the image of (7.2) under $\mathbb{R}_{*}$, composed with the unit of the adjunction. Since $\mathbb{R}_{*}$ preserves p-local equivalences, it suffices to show that the unit

$$\Sigma^{sj,0}B(j) \wedge \mathsf{cKW}[\tfrac{1}{2}] \longrightarrow \mathbb{R}_{*}\mathbb{R}^{*}\Sigma^{sj,0}B(j) \wedge \mathsf{cKW}[\tfrac{1}{2}]$$

is a p-local equivalence. However, it is in fact an equivalence, by Theorem 4.9 and the projection formula from Theorem 3.1. □

Corollary 7.2 implies the same p-local equivalence over any real closed field.

Theorem 7.3 *Let F be a field of characteristic zero. Then the map*

$$cu \wedge \mathsf{cKW}[\tfrac{1}{2}]\colon \mathsf{cKW}[\tfrac{1}{2}] \longrightarrow \mathsf{cKW} \wedge \mathsf{cKW}[\tfrac{1}{2}]$$

is 3-connective. In particular, $\pi_{s,t}\mathsf{cKW}\wedge\mathsf{cKW}[\tfrac{1}{2}]$ is trivial for $1 \leq s-t \leq 3$.

Proof The source $\mathsf{cKW}[\tfrac{1}{2}]$ of the map $cu \wedge \mathsf{cKW}[\tfrac{1}{2}]$ satisfies real étale descent by [19, Theorem 6.6], and its target does so by Proposition 5.7. The statement follows by proving that the induced map

$$\pi_{s,t}\mathsf{cKW}[\tfrac{1}{2}] \longrightarrow \pi_{s,t}\mathsf{cKW} \wedge \mathsf{cKW}[\tfrac{1}{2}] \qquad (7.3)$$

on homotopy sheaves for the real étale topology is an isomorphism for all $s - t \leq 3$. Hence it suffices to prove that (7.3) is an isomorphism for all $s - t \leq 3$ after evaluating at a henselian local ring with real closed residue field. The rigidity property of Witt groups [20, Satz 3.3] provides that it suffices to evaluate (7.3) at real closed fields.

The main results of [2] imply that $cu \wedge \mathsf{cKW}_{\mathbb{Q}}\colon \mathsf{cKW}_{\mathbb{Q}} \to \mathsf{cKW} \wedge \mathsf{cKW}_{\mathbb{Q}}$ is 3-connective. More precisely, [2, Corollary 5] and [2, Theorem 5] imply that $\mathsf{cKW}_{\mathbb{Q}} \simeq \bigvee_{n\geq 0} \Sigma^{8n,4n}\mathbf{1}[\tfrac{1}{\eta}]_{\mathbb{Q}}$ whence

$$\mathsf{cKW}_{\mathbb{Q}} \wedge \mathsf{cKW}_{\mathbb{Q}} \simeq \bigvee_{n\geq 0} \Sigma^{8n,4n} \bigvee_{j=0}^{n} \mathbf{1}[\tfrac{1}{\eta}]_{\mathbb{Q}}.$$

In particular, $\pi_{s,t}\mathsf{cKW} \wedge \mathsf{cKW}_{\mathbb{Q}}$ is trivial for $1 \leq s - t \leq 3$. Hence it suffices to show that p-torsion vanishes in that range one odd prime p at a time. But this is an immediate consequence of Corollary 7.2 which holds for real closed fields.

$\square$

More precise information on Witt cooperations, at least after inverting 2, can be deduced from knowledge about the topological spectra $B(j)$. They are given as Thom spectra of maps from even parts of the May filtration on the homotopy fiber of $\Omega^2 S^3 \to S^1$ to KO; see [22, Section 2] for details.

8   The vanishing

As mentioned already in Section 1, considering connective Witt theory with 2 inverted suffices due to knowledge on $(2, \eta)$-completed groups from [31], which coincide with 2-completed groups under the following circumstances.

Theorem 8.1 (Hu-Kriz-Ormsby) *Let F be a field of finite virtual 2-cohomological dimension and of characteristic not two. Then the canonical map $\mathbf{1}_2^{\wedge} \to \mathbf{1}_{2,\eta}^{\wedge}$ is an equivalence.*

Proof The reference [17, Theorem 1] provides this statement in the case of fields of characteristic zero. Their proof goes through for fields of odd characteristic with the amendment that the motivic Eilenberg-MacLane spectrum $\mathsf{M}\mathbb{Z}/2$ admits a cell presentation of finite type also over fields of characteristic not two [31, Proposition 3.32].

$\square$

Lemma 8.2 *Let F be a field of characteristic not two. The canonical map*

$$\pi_{n+1,n}\mathbf{1} \longrightarrow \pi_{n+1,n}\mathbf{1}[\tfrac{1}{2}]$$

is injective for every natural number $n \geq 5$.

Proof Since the base field is a filtered colimit of fields of finite virtual 2-cohomological dimension, and $\pi_{n+1,n}\mathbf{1}$ commutes with filtered colimits of fields, one may assume that the base field has finite virtual 2-cohomological dimension. Consider the arithmetic square for 2 and $\mathbf{1}$. It induces a long exact sequence

$$\cdots \longrightarrow \pi_{n+2,n}\mathbf{1}_2^{\wedge}[\tfrac{1}{2}] \longrightarrow \pi_{n+1,n}\mathbf{1} \longrightarrow \pi_{n+1,n}\mathbf{1}[\tfrac{1}{2}] \oplus \pi_{n+1,n}\mathbf{1}_2^{\wedge} \longrightarrow$$
$$\pi_{n+1,n}\mathbf{1}_2^{\wedge}[\tfrac{1}{2}] \longrightarrow \cdots$$

of homotopy groups. The argument below will provide that $\pi_{n+2,n}\mathbf{1}_2^\wedge = 0$. Theorem 8.1 implies that the natural map $\pi_{s,t}\mathbf{1}_2^\wedge \to \pi_{s,t}\mathbf{1}_{2,\eta}^\wedge$ is an isomorphism for all s,t. The convergence result [31, Theorem 3.50] supplies a weak equivalence $\mathsf{sc}(\mathbf{1}) \simeq \mathbf{1}_\eta^\wedge$ between the η-completion and the completion with respect to the slice filtration of $\mathbf{1}$, hence an isomorphism $\pi_{s,t}\mathbf{1}_{2,\eta}^\wedge \cong \pi_{s,t}(\mathsf{sc}\mathbf{1})_2^\wedge$ for all s,t. Consider Milnor's derived limit exact sequence:

$$0 \longrightarrow \lim_k{}^1 \pi_{s+1,t}(\mathsf{sc}\mathbf{1})/2^k \longrightarrow \pi_{s,t}(\mathsf{sc}\mathbf{1})_2^\wedge \longrightarrow \lim_k \pi_{s,t}(\mathsf{sc}\mathbf{1})/2^k \longrightarrow 0.$$
$$(8.1)$$

The canonical map $(\mathsf{sc}\mathbf{1})/2^k \to \mathsf{sc}(\mathbf{1}/2^k)$ is an equivalence. As stated in the beginning of [31, Section 5], the computation of the first slice differential for the motivic sphere spectrum [31, Lemma 4.1] implies that $\pi_{n+1,n}\mathsf{sc}(\mathbf{1}) = 0$ for $n \geq 3$ and $\pi_{n+2,n}\mathsf{sc}(\mathbf{1}) = 0$ for $n \geq 5$. The long exact sequence of homotopy groups hence forces

$$\pi_{n+2,n}\mathsf{sc}(\mathbf{1})/2^k = 0 \quad \text{for } n \geq 5 \text{ and } k \geq 1$$

which — together with Milnor's short exact sequence (8.1) — implies the isomorphisms

$$\lim_k{}^1 \pi_{n+3,n}(\mathsf{sc}\mathbf{1})/2^k \xrightarrow{\cong} \pi_{n+2,n}(\mathsf{sc}\mathbf{1})_2^\wedge \text{ and}$$

$$\pi_{n+1,n}(\mathsf{sc}\mathbf{1})_2^\wedge \xrightarrow{\cong} \lim_k \pi_{n+1,n}(\mathsf{sc}\mathbf{1})/2^k.$$

It also implies that the canonical map

$$\pi_{n+3,n}(\mathsf{sc}\mathbf{1}) \to \pi_{n+3,n}(\mathsf{sc}\mathbf{1})/2^k$$

is surjective for all $n \geq 5$ and all $k \geq 1$. Hence the canonical map

$$\pi_{n+3,n}(\mathsf{sc}\mathbf{1})/2^{k+1} \to \pi_{n+3,n}(\mathsf{sc}\mathbf{1})/2^k$$

is surjective for all $k \geq 1$ by the commutative diagram

$$
\begin{array}{ccc}
\pi_{n+3,n}(\mathsf{sc}\mathbf{1}) & \longrightarrow & \pi_{n+3,n}(\mathsf{sc}\mathbf{1})/2^{k+1} \\
{\scriptstyle \mathrm{id}}\downarrow & & \downarrow \\
\pi_{n+3,n}(\mathsf{sc}\mathbf{1}) & \longrightarrow & \pi_{n+3,n}(\mathsf{sc}\mathbf{1})/2^k.
\end{array}
$$

In particular,

$$0 = \lim_k{}^1 \pi_{n+3,n}(\mathsf{sc}\mathbf{1})/2^k = \pi_{n+2,n}(\mathsf{sc}\mathbf{1})_2^\wedge = \pi_{n+2,n}\mathbf{1}_2^\wedge$$

for all $n \geq 5$.

Thus the map

$$\pi_{n+1,n}\mathbf{1} \longrightarrow \pi_{n+1,n}\mathbf{1}[\tfrac{1}{2}] \oplus \pi_{n+1,n}\mathbf{1}_2^{\wedge}$$

is injective for all $n \geq 5$. It remains to observe that the map $\pi_{n+1,n}\mathbf{1} \to \pi_{n+1,n}\mathbf{1}_2^{\wedge}$ is the zero map for all $n \geq 5$. In fact, the isomorphisms

$$\pi_{n+1,n}\mathbf{1}_2^{\wedge} \cong \pi_{n+1,n}\mathsf{sc}(\mathbf{1})_2^{\wedge} \cong \lim_{k} \pi_{n+1,n}(\mathsf{sc}\mathbf{1})/2^k$$

explained above show that it suffices to prove the triviality of the canonical map $\pi_{n+1,n}\mathbf{1} \to \pi_{n+1,n}\mathsf{sc}(\mathbf{1})/2^k$ for all $n \geq 5$ and all $k \geq 1$, which again follows from the vanishing of $\pi_{n+1,n}\mathsf{sc}(\mathbf{1})$ for all $n \geq 3$. $\qquad\square$

The statement in Lemma 8.2 is not optimal. The computation [31, Theorem 5.5] which is based on Theorem 8.3 below implies that $\pi_{n+1,n}\mathbf{1} = 0$ for $n \geq 3$. The map $\pi_{3,2}\mathbf{1} \to \pi_{3,2}\mathbf{1}[\tfrac{1}{2}]$ is not injective; up to isomorphism it is the projection $\mathbb{Z}/24 \to \mathbb{Z}/3$ by [31, Theorem 5.5].

Theorem 8.3 *Let F be a field of characteristic not two. Then*

$$\underline{\pi}_{n+1,n}\mathbf{1}[\tfrac{1}{\eta}] = \underline{\pi}_{n+2,n}\mathbf{1}[\tfrac{1}{\eta}] = 0$$

for every integer n.

Proof Lemma 8.2 (which holds on the level of homotopy sheaves, for example by [16, Theorem 2.7]) shows that it suffices to prove the statement for $\underline{\pi}_{n+1,n}\mathbf{1}[\tfrac{1}{\eta}, \tfrac{1}{2}]$. Thanks to Theorem 6.1, it is enough to consider fields of characteristic zero. Proposition 5.5 implies the isomorphism

$$\underline{\pi}_{n+1,n}\mathbf{1}[\tfrac{1}{\eta}, \tfrac{1}{2}] \cong \underline{\pi}_{n+2,n}\mathsf{cKW} \wedge \mathsf{cKW}[\tfrac{1}{2}]$$

for every integer n. Hence this sheaf of homotopy groups vanishes by Theorem 7.3. This in turn shows that $\mathbf{1}[\tfrac{1}{\eta}] \to \mathsf{cKW}$ is 2-connective, whence Proposition 5.6 implies an isomorphism

$$\underline{\pi}_{n+2,n}\mathbf{1}[\tfrac{1}{\eta}, \tfrac{1}{2}] \cong \underline{\pi}_{n+3,n}\mathsf{cKW} \wedge \mathsf{cKW}[\tfrac{1}{2}]$$

for every integer n. The latter sheaf again vanishes by Theorem 7.3. In order to conclude that even $\underline{\pi}_{n+2,n}\mathbf{1}[\tfrac{1}{\eta}]$ vanishes, it remains to obtain an injectivity statement similar to Lemma 8.2. Invoking again Theorem 8.1 and restriction to fields of finite virtual 2-cohomological dimension, this follows from the arithmetic square for 2 and the vanishing of

$$\pi_{n+3,n}\mathbf{1}_2^{\wedge}[\tfrac{1}{2}] \cong \pi_{n+3,n}\mathbf{1}_{2,\eta}^{\wedge}[\tfrac{1}{2}] \cong \pi_{n+3,n}\mathsf{sc}(\mathbf{1})_2^{\wedge}[\tfrac{1}{2}]$$

for suitably large n. To see this vanishing, observe that for $n \geq 8$, multiplication with η induces an isomorphism between the third, as well as the fourth, columns of the n-th and $n+1$-th slice spectral sequence for $\mathbf{1}$. On the E^2-pages, these columns consist of an infinite tower $K_*^{\mathrm{Mil}}/2$ of Milnor K-theory modulo 2, generated by suitable α_1-multiples of α_4 and α_5, respectively. Calling these generators $\widetilde{\alpha}_4$ and $\widetilde{\alpha}_5$, respectively, [11, Theorem 1.1(2)] implies that over the real numbers, the third differential maps $1 \cdot \widetilde{\alpha}_5 \in K_0^{\mathrm{Mil}}/2$ to $\rho^3 \cdot \widetilde{\alpha}_4 \in K_3^{\mathrm{Mil}}/2$. By base change, the same is true over the rational numbers, and hence over any field of characteristic zero. Using [17, Proposition 2] and the Milnor-Witt K-theory module structure within the slice spectral sequence, the E^∞-page of the n-th slice spectral sequence for $\mathbf{1}$ is finite in columns three and four over any field of finite virtual 2-cohomological dimension d. More precisely, the third column has length at most $d+3$, and the fourth column has length at most $d+1$, for $n \geq 8$. By the Milnor exact sequence, $\pi_{n+3,n}\mathsf{sc}(\mathbf{1})_2^{\wedge}[\frac{1}{2}] = 0$ for $n \geq 8$. ⊔

Theorem 8.3 shows that Theorem 7.3 already holds before inverting two. The vanishing of homotopy sheaves in Theorem 8.3 implies the vanishing of the homotopy groups

$$\pi_{n+1,n}\mathbf{1}[\tfrac{1}{\eta}] = \pi_{n+2,n}\mathbf{1}[\tfrac{1}{\eta}] = 0$$

for all integers n stated in Theorem 1.1.

References

[1] A. Ananyevskiy, *Stable operations and cooperations in derived Witt theory with rational coefficients*, Ann. K-Theory **2** (2017), no. 4, 517–560.

[2] A. Ananyevskiy, M. Levine, and I. Panin, *Witt sheaves and the η-inverted sphere spectrum*, J. Topol. **10** (2017), no. 2, 370–385.

[3] M. Andrews and H. Miller, *Inverting the Hopf map*, J. Topol., **10** (2017), no. 4, 1145–1168.

[4] J. Ayoub, *L'algèbre de Hopf et le groupe de Galois motiviques d'un corps de caractéristique nulle. I*, J. Reine Angew. Math. **693** (2014), 1–149.

[5] *On the conservativity of the functor assigning to a motivic spectrum its motive*, Duke Math. J., **167** (2018), no. 8, 1525–1571.

[6] T. Bachmann, *Motivic and real étale stable homotopy theory*, Compos. Math. **154** (2018), no. 5, 883–917.

[7] D. M. Davis, *Odd primary bo-resolutions and K-theory localization*, Illinois J. Math. **30** (1) (1986), 79–100.

[8] D. Dugger and D. C. Isaksen, *Motivic cell structures*, Algebr. Geom. Topol. **5** (2005), 615–652.

[9] H. Fausk, P. Hu, and J. P. May, *Isomorphisms between left and right adjoints*, Theory Appl. Categ. **11** (4) (2003), 107–131.

[10] B. J. Guillou and D. C. Isaksen, *The η-local motivic sphere*, J. Pure Appl. Algebra **219** (10) (2015), 4728–4756.

[11] B. J. Guillou and D. C. Isaksen, *The η-inverted ℝ-motivic sphere*, Algebr. Geom. Topol. **16** (2016), no. 5, 30053027.

[12] J. J. Gutiérrez, O. Röndigs, M. Spitzweck, and P. A. Østvær, *Motivic slices and coloured operads*, J. Topol. **5** (3) (2012), 727–755.

[13] J. Heller and K. Ormsby, *Galois equivariance and stable motivic homotopy theory*, Trans. Amer. Math. Soc. **368** (11) (2016), 8047–8077.

[14] J. Hornbostel, *Some comments on motivic nilpotence, With an appendix by Marcus Zibrowius*, Trans. Amer. Math. Soc. **370** (2018), no. 4, 30013015.

[15] J. Hornbostel, $\mathbb{A}^1$-*representability of hermitian K-theory and Witt groups*, Topology **44** (3) (2005), 661–687.

[16] M. Hoyois, *From algebraic cobordism to motivic cohomology*, J. Reine Angew. Math. **702** (2015), 173–226.

[17] P. Hu, I. Kriz, and K. Ormsby, *Convergence of the motivic Adams spectral sequence*, J. K-Theory **7** (3) (2011), 573–596.

[18] R. M. Kane, *Operations in connective K-theory*, Mem. Amer. Math. Soc. **34** (254) (1981), vi+102.

[19] M. Karoubi, M. Schlichting, and C. Weibel, *The Witt groups of real algebraic varieties*, J. Topol. **9** (2016), no. 4, 12571302.

[20] M. Knebusch, *Isometrien über semilokalen Ringen*, Math. Z. **108** (1969), 255–268.

[21] W. Lellmann, *Operations and co-operations in odd-primary connective K-theory*, J. London Math. Soc. (2) **29** (3) (1984), 562–576.

[22] M. Mahowald, *bo-resolutions*, Pacific J. Math. **92** (2) (1981), 365–383.

[23] R. J. Milgram, *The Steenrod algebra and its dual for connective K-theory*, In Conference on homotopy theory (Evanston, Ill., 1974), volume 1 of Notas Mat. Simpos. pages 127–158. Soc. Mat. Mexicana, México, (1975).

[24] Fabien Morel, *On the motivic π_0 of the sphere spectrum*, In Axiomatic, enriched and motivic homotopy theory. Proceedings of the NATO Advanced Study Institute, Cambridge, UK, September 9–20, (2002), pages 219–260. Kluwer Academic Publishers, (2004).

[25] F. Morel, $\mathbb{A}^1$-*algebraic topology*, In International Congress of Mathematicians. Vol. II, pages 1035–1059. Eur. Math. Soc., Zürich, (2006).

[26] F. Morel, $\mathbb{A}^1$-*algebraic topology over a field*, Lecture Notes in Mathematics **2052**, Springer, Heidelberg, (2012).

[27] F. Morel and V. Voevodsky, $\mathbb{A}^1$-*homotopy theory of schemes.* Inst. Hautes Étud. Sci. Publ. Math. **90** (1999), 45–143.

[28] I. Panin and C. Walter, *On the motivic commutative ring spectrum BO*, arXiv:1101.0650.

[29] O. Röndigs and P. A. Østvær, *Slices of hermitian K-theory and Milnor's conjecture on quadratic forms*, Geom. Topol. **20** (2016), 1157–1212.

[30] O. Röndigs, M. Spitzweck, and P. A. Østvær, *Cellularity of hermitian K-theory and Witt theory*, arXiv:1603.05139.

[31] O. Röndigs, M. Spitzweck, and P. A. Østvær, *The first stable homotopy groups of motivic spheres*, to appear in Ann. Math., arXiv:1604.00365.

[32] Winfried Scharlau, *Quadratic and Hermitian forms*, Grundlehren der Mathematischen Wissenschaften **270**, Springer-Verlag, Berlin, (1985).

[33] M. Schlichting, *Hermitian K-theory, derived equivalences, and Karoubi's fundamental theorem*, J. Pure Appl. Algebra **221** (2017), no. 7, 17291844.

[34] M. Spivakovsky, *A new proof of D. Popescu's theorem on smoothing of ring homomorphisms*, J. Amer. Math. Soc., **12** (2) (1999), 381–444.

INSTITUTE OF MATHEMATICS, UNIVERSITY OF OSNABRÜCK, OSNABRÜCK, GERMANY
E-mail: oliver.roendigs@uni-osnabrueck.de

K-Theory
Copyright ©2018 Tata Institute of Fundamental Research
Publisher: Hindustan Book Agency, New Delhi, India

Additive higher Chow groups and de Rham-Witt complexes

Jinhyun Park

Abstract

This article sketches some recent developments that compare the additive higher Chow groups with the big de Rham-Witt complexes. In addition, we present a clue toward a conjectural theory of crystalline cobordism via (also conjectural) descriptions of higher algebraic cobordism with modulus.

1 Introduction

This article aims to give a rapid sketch of several recent results that elucidate the relationship between the additive higher Chow groups and the big de Rham-Witt complexes, and it ends with some discussions toward a conjectural "crystalline cobordism" theory on schemes. It is based on several papers of the author, joint with Amalendu Krishna as well as the talk at the International Colloquium 2016 on K-theory at the Tata Institute of Fundamental Research.

One of the central philosophical themes in K-theory and the theory of motives is to find a suitable universal cohomology theory, the motivic cohomology theory, and to construct it via algebraic cycles. The first version of motivic cohomology with the rational coefficients was proposed by A. Beilinson [1] in terms of the eigenspace decompositions of the rational higher algebraic K-groups of D. Quillen [29] via the Adams operations. Later, S. Bloch's higher Chow groups [5] turned out to be the right motivic cohomology groups with integral coefficients on smooth varieties. Bloch proved a generalization of Grothendieck-Riemann-Roch theorem that the higher Chow groups tensored with $\mathbb{Q}$ are isomorphic to the rational motivic cohomology of Beilinson.

Beyond smooth varieties, the situation gets more complicated. For non-reduced schemes, such as $\mathrm{Spec}(k[t]/(t^2))$, where k is a field, W. van der Kallen [31] proved that there is a possible nonzero relative part, as seen in the proof of $K_2(k[t]/(t^2)) = K_2(k) \oplus \Omega^1_{k/\mathbb{Z}}$, where the absolute Kähler

1-forms are nontrivial if k has transcendental elements over $\mathbb{Z}$. But, by definition, higher Chow groups do not see nilpotent elements, so that one needs additional cycle groups to cover this relative K-group.

Additive higher Chow groups, proposed by S. Bloch and H. Esnault in [6], are possibly a way to handle this issue for schemes of the form $X \otimes_k k[t]/(t^{m+1})$. Here, codimension q cycles on $X \times \mathbb{A}^1 \times \square^{n-1}$, where $\square = \mathbb{P}^1 \setminus \{1\}$ with faces $\{0, \infty\}$, that intersect all "faces" properly and satisfy "the modulus condition" coming from the divisor $(m+1) \cdot \{t=0\}$, where $t \in \mathbb{A}^1$, are used to form a complex. Its homology is denoted by $\mathrm{TCH}^q(X, n; m)$. See §2. (N.B. This TCH should not be mistakenly seen as the topological cyclic homology.) When $m = 1$, one main theorem of *op. cit.* is that $\mathrm{TCH}^n(k, n; m) \simeq \Omega^{n-1}_{k/\mathbb{Z}}$ when $\mathrm{char}(k) \neq 2$. This result was generalized by K. Rülling that $\mathrm{TCH}^n(k, n; m) \simeq \mathbb{W}_m\Omega^{n-1}_k$, where the latter group is the big de Rham-Witt $(n-1)$-forms of k of L. Hesselholt and I. Madsen [8]. See §3.

The first part of this article, from §2 to §6, sketches how one can generalize it to isomorphisms between Zariski sheaves originating from additive higher Chow groups and Zariski sheaves of big de Rham-Witt forms for a smooth k-scheme X following mostly [18] and [19], as well as earlier results from [12], [13], [14], [16], [23], [24]. Nothing in the first part of the article is new. For simplicity, suppose the base field k is always perfect of characteristic $\neq 2$. The essential point is to sketch the proof of the following from [19]:

Theorem 1.1 *Let R be a smooth semi-local k-algebra essentially of finite type over k. Let $n, m \geq 1$ be integers. Then there are natural isomorphisms $\tau^R_{n,m} : \mathbb{W}_m\Omega^{n-1}_R \to \mathrm{TCH}^n(R, n; m)$, that form an isomorphism of restricted Witt-complexes over R.*

The definition of restricted Witt-complexes is given in §3.3. Once one gets this theorem, with some work one can show that this isomorphism extends to isomorphisms of Zariski sheaves on smooth k-varieties. However, neither proving that it is an isomorphism, nor even defining the map $\tau^R_{n,m}$ itself is trivial for such big de Rham-Witt forms.

To define $\tau^R_{n,m}$ the best way is to prove that the collection $\{\mathrm{TCH}^n(R, n; m)\}_{n,m \geq 1}$ can be equipped with restricted Witt-complex structures over R, which requires

1. a product structure $\wedge_R$,

2. a differential operator δ,

3. the Leibniz rule between δ and $\wedge_R$,

4. Frobenius and Verschiebung operators that satisfy a list of axioms,

5. a ring homomorphism from $\mathbb{W}_m(R) \to \mathrm{TCH}^1(R, 1; m)$.

Since the big de Rham-Witt forms give a universal object in the category of Witt-complexes over R, these structures automatically induce natural homomorphisms $\tau_{n,m}^R$. This is done in §4 for any smooth k-algebras essentially of finite type.

In §5, we sketch the arguments involved in the proof of Theorem 1.1.

Theorem 1.1 implies that the natural map from the Zariski sheafified additive higher Chow group on a smooth k-variety X to the Zariski sheaf of big de Rham-Witt forms on X is an isomorphism of Zariski sheaves. In other words, big de Rham-Witt complexes on a smooth variety are of algebraic cycle origin. Using the p-typicalization process inspired by the p-typical curves in [4], one can also obtain a description of the p-typical de Rham-Witt complexes of L. Illusie [10] by an object originating from algebraic cycles. It is a classical theorem of Bloch and Illusie that the crystalline cohomology of P. Berthelot [2] can be computed as Zariski hypercohomology of p-typical de Rham-Witt complexes. So, what we proved implies that the crystalline cohomology itself admits a description originating from algebraic cycles. This is discussed in §6.

In the second part of the article, §7, which is nowhere in the literature, we aim to propose a way to construct a conjectural theory of crystalline cobordism. For this, we discuss the notion of modulus for algebraic cobordism cycles over $X \times \square^n$, with $\square = \mathbb{P}^1 \setminus \{1\}$.

2   Additive higher Chow cycles

We recall the definition of additive higher Chow cycles from [16], with some possible minor changes of notations. Let k be a field. Let $\square = \mathbb{P}^1 \setminus \{1\}$, and let $B_n := \mathbb{A}^1 \times \square^{n-1}$. Let $\overline{B}_n := \mathbb{A}^1 \times (\mathbb{P}^1)^{n-1}$. We use $(t, y_1, \ldots, y_{n-1}) \in \overline{B}_n$ as the coordinates. A *face* of B_n is defined as a closed subscheme given by a set of equations of the form $y_{i_1} = \epsilon_1, \cdots, y_{i_s} = \epsilon_s$, where $\epsilon_j \in \{0, \infty\}$. We let $\iota_{n,i,\epsilon} : B_{n-1} \to B_n$ be the closed immersion $(t, y_1, \ldots, y_{n-2}) \mapsto (t, y_1, \ldots, y_{i-1}, \epsilon, y_i, \ldots, y_{n-2})$. Its image is a codimension 1 face.

Let X be an equidimensional k-scheme. For a given integral closed subscheme $Z \subset X \times B_n$, we say that Z satisfies the modulus m condition if one of the following conditions holds:

(i) $n = 1$ and $Z \cap \{t = 0\} = \emptyset$.

(ii) $n > 1$ and let $\overline{Z}$ be the Zariski closure of Z in $X \times \overline{B}_n$. Let $\nu : \overline{Z}^N \to \overline{Z} \hookrightarrow X \times \overline{B}_n$ be the normalization composed with the closed immersion. Then we have

$$(m + 1)\nu^*\{t = 0\} \leq \sum_{i=1}^{n-1} \nu^*\{y_i = 1\}.$$

One of the interesting features of this modulus condition is the following result from [13, Proposition 2.4]. A generalization of it is discussed in Lemma 7.7.

Lemma 2.1 (Containment lemma) *Let $Z_1 \subset Z_2 \subset X \times B_n$ be closed subschemes, where Z_2 satisfies the modulus m condition. Then so does Z_1.*

We define $\underline{\mathrm{Tz}}^q(X, n; m)$ to be the free abelian group on the set of *admissible* integral closed subschemes $Z \subset X \times B_n$, i.e. those which (i) intersect with all faces properly, i.e. in the right codimensions and (ii) satisfy the modulus m condition. The degenerate cycles are those obtained by pulling back via various projections $\square^{n-1} \to \square^i$ with $i < n - 1$. Modulo these, we obtain $\mathrm{Tz}^q(X, n; m)$.

The closed immersion $\iota_{n,i,\epsilon} : B_{n-1} \to B_n$ induces the face map $\partial_i^\epsilon : \mathrm{Tz}^q(X, n; m) \to \mathrm{Tz}^q(X, n - 1; m)$. This is well-defined by the containment lemma. Via the usual formalism of cubical abelian groups, one can check that $\partial := \sum_{i=1}^{n-1} (-1)^i (\partial_i^\infty - \partial_i^0)$ on $\mathrm{Tz}^q(X, n; m)$ defines a complex $(\mathrm{Tz}^q(X, \bullet; m), \partial)$ of abelian groups, called additive higher Chow complex of X with modulus m. Its homology is denoted by $\mathrm{TCH}^q(X, n; m)$ and called additive higher Chow group of X with modulus m.

3 Witt vectors, de Rham-Witt complexes and Witt-complexes

We recall the definitions of Witt vectors, big de Rham-Witt complexes and Witt-complexes from the literature. A good reference is [9].

3.1 Rings of Witt-vectors

Let R be a commutative ring with unity. Recall the definition of the rings of big Witt-vectors of R from e.g. [30, Appendix A]. A *truncation set* $S \subset \mathbb{N}$ is a nonempty subset such that if $s \in S$ and $t|s$, then $t \in S$. As a set, let $\mathbb{W}_S(R) := R^S$ and define the map $w : \mathbb{W}_S(R) \to R^S$, $a = (a_s)_{s \in S} \mapsto w(a) = (w(a)_s)_{s \in S}$, where $w(a)_s := \sum_{t|s} t a_t^{\frac{s}{t}}$. When R^S on the target of w is given the component-wise ring structure, it is known that there is a unique functorial ring structure on $\mathbb{W}_S(R)$ such that w is a ring homomorphism.

For two truncation sets $S \subset S'$, there is a restriction $\mathfrak{R} : \mathbb{W}_{S'}(R) \to \mathbb{W}_S(R)$. When $S = \{1, \ldots, m\}$, we write $\mathbb{W}_m(R) = \mathbb{W}_S(R)$. For a fixed prime number p, when $S = \{1, p, p^2, \ldots\}$ and $S_i = \{1, p, \ldots, p^{i-1}\}$, we write $W(R) = \mathbb{W}_S(R)$ and $W_i(R) = \mathbb{W}_{S_i}(R)$. They are the p-typical rings of Witt vectors.

Let $\mathbb{W}(R) := \mathbb{W}_\mathbb{N}(R)$. There is a natural bijection $\mathbb{W}(R) \simeq (1 + TR[[T]])^\times$, where T is an indeterminate and the addition of the ring $\mathbb{W}(R)$ corresponds to the multiplication of the formal power series. For a truncation set S, we have $\mathbb{W}_S(R) = (1 + TR[[T]])^\times / I_S$ for a subgroup I_S. See [30, A.7]. In case $S = \{1, \ldots, m\}$, we have an isomorphism $\gamma : \mathbb{W}_m(R) \simeq (1 + TR[[T]])^\times / (1 + T^{m+1} R[[T]])^\times$, $(a_i)_{1 \le i \le m} \mapsto \prod_{i=1}^m (1 - a_i T^i)$. The Teichmüller lift $[-]_S : R \to \mathbb{W}_S(R)$ is given by $a \mapsto 1 - aT$, which is a multiplicative map. If $S = \{1, \ldots, m\}$, we write $[-]_m$ for $[-]_S$. For each $i \ge 1$, we have the i-th Verschiebung V_i given by $V_i([a]_{\lfloor m/i \rfloor}) = (1 - aT^i)$, where for a non-negative real number x, one denotes by $\lfloor x \rfloor$ the greatest integer not bigger than x.

3.2 De Rham-Witt complexes

Let R be a $\mathbb{Z}_{(p)}$-algebra for a prime p. For each truncation set S, there is a differential graded algebra (DGA) $\mathbb{W}_S \Omega_R^\bullet$, called the big de Rham-Witt complex over R, which defines a functor on the category of truncation sets. This is an initial object in the category of V-complexes and in the category of Witt-complexes over R; a V-complex (see [30, Definition 1.1] for details) is, roughly speaking, a functor that assigns to each S a DGA, equipped with operators V_n for $n \ge 1$, called Verschiebung operators, subject to a set of relations. A Witt-complex (see [30, Definition 1.4], for instance) is a V-complex that has additional operators F_n for $n \ge 1$, called Frobenius operators, subject to further relations. For rigorous definitions, see [8], [9], or [30]. In case S is a finite truncation set, we have $\mathbb{W}_S \Omega_R^\bullet = \Omega_{\mathbb{W}_S(R)/\mathbb{Z}}^\bullet / N_S^\bullet$, where $N_S^\bullet$ is a differential graded ideal given by some generators; see [30, Proposition 1.2] for details. In case $S = \{1, 2, \ldots, m\}$, we write $\mathbb{W}_m \Omega_R^\bullet$ for this object. For a prime number p, when $S = \{1, p, p^2, \ldots\}$ and $S_i = \{1, p, \ldots, p^{i-1}\}$, we write $W\Omega_R^\bullet = \mathbb{W}_S \Omega_R^\bullet$ and $W_i \Omega_R^\bullet = \mathbb{W}_{S_i} \Omega_R^\bullet$. They are the p-typical de Rham-Witt complexes of [10].

3.3 Restricted Witt-complex over R

We recall from [8, Definition 1.1.1] that *a restricted Witt-complex over R* is a pro-system of differential graded $\mathbb{Z}$-algebras $((E_m)_{m \in \mathbb{N}}, \mathfrak{R} : E_{m+1} \to E_m)$, together with families of homomorphisms of graded rings $(F_r : E_{rm+r-1} \to E_m)_{m,r \in \mathbb{N}}$ called Frobenius maps, and homomorphisms of graded groups

$(V_r : E_m \to E_{rm+r-1})_{m,r\in\mathbb{N}}$ called Verschiebung maps, satisfying the following relations for all $n, r, s \in \mathbb{N}$:

(i) $\mathfrak{R}F_r = F_r\mathfrak{R}^r, \mathfrak{R}^rV_r = V_r\mathfrak{R}, F_1 = V_1 = \text{Id}, F_rF_s = F_{rs}, V_rV_s = V_{rs}$;

(ii) $F_rV_r = r$. When $(r, s) = 1$, then $F_rV_s = V_sF_r$ on E_{rm+r-1};

(iii) $V_r(F_r(x)y) = xV_r(y)$ for all $x \in E_{rm+r-1}$ and $y \in E_m$; (projection formula)

(iv) $F_rdV_r = d$ (where d is the differential of the DGAs).

Furthermore, there is a homomorphism of pro-rings $(\lambda : \mathbb{W}_m(R) \to E_m^0)_{m\in\mathbb{N}}$ that commutes with F_r and V_r, and we have

(v) $F_rd\lambda([a]) = \lambda([a]^{r-1})d\lambda([a])$ for all $a \in R$ and $r \in \mathbb{N}$,

where $[a]$ is the Teichmüller lift in $\mathbb{W}_m(R)$ of $a \in R$.

The system $\{\mathbb{W}_m\Omega_R^\bullet\}_{m\geq 1}$ is the initial object in the category of restricted Witt-complexes over R (see [30, Proposition 1.15]).

4   Restricted Witt structure on additive higher Chow groups

In this section, we describe various structures that turn the additive higher Chow groups into a restricted Witt-complex. The main references are [16] and [18].

4.1   Algebra structure

For higher Chow groups of X, the product structure is essentially given as follows: given two cycles V_i on $X \times \square^{n_i}$, take the pull-back $\Delta_X^*(V_1 \times V_2)$ of the concatenation via the diagonal embedding $\Delta_X : X \hookrightarrow X \times X$. The existence of this pull-back Δ_X^* is a consequence of the moving lemma for higher Chow groups.

For additive higher Chow groups, the above basic idea is still necessary, but there is one extra ingredient coming from the additional $\mathbb{A}^1$-coordinate, which is the monoidal product $\mu : \mathbb{A}^1 \times \mathbb{A}^1 \to \mathbb{A}^1$. This induces $\tilde{\mu} : X \times \mathbb{A}^1 \times \square^{n_1} \times X \times \mathbb{A}^1 \times \square^{n_2} \to X \times X \times \mathbb{A}^1 \times \square^{n_1+n_2}$. Unfortunately this $\tilde{\mu}$ is not proper so that taking push-forwards via μ is in general not possible. However, given two admissible cycles V_i on $X \times \mathbb{A}^1 \times \square^{n_i}$, one can prove that the restricted morphism $\tilde{\mu}|_{V_1\times V_2}$ is actually finite, thus proper so that one *can* define the push-forward $V_1 \times_\mu V_2 := \tilde{\mu}_*(V_1 \times V_2)$, and it is again an admissible cycle. See [18, Proposition 5.7, Lemmas 5.8, 5.10].

On the other hand, either when X is smooth projective or smooth affine essentially of finite type over k, the moving lemmas [13, Theorem 4.1] or [11] (or [18, Theorem 4.1]) show that the pull-back operator Δ_X^* exists on additive higher Chow cycle classes. Hence, one can define $V_1 \wedge_X V_2 := \Delta_X^*(\mu_*(V_1 \times V_2))$. We call it the *Pontryagin*-intersection product in [18] as an analogy of the classical Pontryagin product on homology groups of Lie groups.

One can check that it is associative, it is functorial for pull-backs, and it satisfies the projection formula. See [16, Theorem 3.19] and [18, Theorem 5.4]. As a result we have

Theorem 4.1 *Let X be a smooth affine k-algebra essentially of finite type, or smooth projective. Then $\{\mathrm{TCH}^q(X, n; m)\}_{q,n \geq 1}$ form an associative graded commutative algebra.*

4.2 Differential graded structure

The additive higher Chow groups also have differential operators among them, analogous to the differential operators on Kähler forms. These operators are also induced by correspondences. We discuss them here.

Let X be a smooth quasi-projective variety. Consider the rational map from [16, §4.1]

$$\delta : X \times \mathbb{A}^1 \times \square^{n-1} \dashrightarrow X \times \mathbb{A}^1 \times \square^n, \quad (x, t, y_1, \ldots, y_{n-1})$$

$$\longmapsto (x, t, t^{-1}, y_1, \ldots, y_{n-1})$$

This is a morphism on the open subset $X \times (\mathbb{G}_m \setminus \{1\}) \times \square^{n-1}$. For a given closed subvariety $V \subset X \times \mathbb{A}^1 \times \square^{n-1}$, consider its restriction $V^\times$ on $X \times (\mathbb{G}_m \setminus \{1\}) \times \square^{n-1}$. In case V is an admissible cycle, then one can show that $\delta(V^\times)$ is actually a closed subscheme of $X \times \mathbb{A}^1 \times \square^n$, which is admissible as an additive higher Chow cycle. Furthermore, δ is compatible with the boundary operators in the sense that $\delta\partial + \partial\delta = 0$. See [16, Proposition 4.4]. Hence we obtain an induced homomorphism $\delta : \mathrm{TCH}^q(X, n; m) \to \mathrm{TCH}^{q+1}(X, n+1; m)$.

To see that δ is a differential, i.e. $\delta^2 = 0$, we need two results. The first is called the normalization theorem ([16, Theorem A.2]):

Lemma 4.2 *Let X be a smooth quasi-projective variety. Then every cycle class in $\mathrm{TCH}^q(X, n; m)$ has a representative α such that all of its codimension 1 faces $\partial_i^\epsilon(\alpha)$ are zero for $1 \leq i \leq n-1$ and $\epsilon = 0, \epsilon$.*

The other we need is the following on the symmetric group action (see [18, Lemma 5.3]):

Lemma 4.3 *Let X be a smooth quasi-projective variety. Let $\sigma \in \mathfrak{S}_{n-1}$ be a permutation on $n-1$ letters acting naturally on $\{y_1, \ldots, y_{n-1}\}$. Let $\alpha \in \mathrm{Tz}^q(X, n; m)$ be an admissible cycle such that $\partial_i^\epsilon(\alpha) = 0$ for all $1 \leq i \leq n-1$ and $\epsilon = 0, \infty$, where the notation Tz is from §2. Then for some $\gamma \in \mathrm{Tz}^q(X, n+1; m)$, we have $\sigma \cdot \alpha - (\mathrm{sgn}(\sigma)) \cdot \alpha = \partial \gamma$.*

For a cycle representative $\alpha \in \mathrm{TCH}^q(X, n; m)$ with all face $\partial_i^\epsilon(\alpha) = 0$, one can check that $\delta(\alpha)$ also has trivial codimension 1 faces. On the other hand, by the definition of the map δ, for the permutation $\tau = (1, 2)$, we have $\tau \cdot \delta^2(\alpha) = \delta^2(\alpha)$. However, by Lemma 4.3, $\tau \cdot \delta^2(\alpha)$ is equivalent to $-\delta^2(\alpha)$. Hence, $2\delta^2(\alpha) = 0$ in $\mathrm{TCH}^{q+2}(X, n+2; m)$. Since $\mathrm{char}(k) \neq 2$, using that $\mathrm{TCH}^{q+2}(X, n+2; m)$ is $\mathbb{W}_m(k)$-module, thus 2 is divisible, we deduce $\delta^2(\alpha) = 0$. This shows δ is a differential. This is done in [18, Corollary 6.6].

To prove the graded Leibniz rule is somewhat technical, but here is a sketch from [18, §6.2]. For two given cycles $\xi \in \mathrm{Tz}^{q_1}(X, n_1; m)$ and $\eta \in \mathrm{Tz}^{q_2}(X, n_2; m)$, all of whose codimension 1 faces are trivial, we need to show that $\delta(\xi \times_\mu \eta) - \delta\xi \times_\mu \eta - (-1)^{n_1-1}\xi \times_\mu \delta\eta$ is the boundary of an admissible cycle. This can be done rather directly via a construction of certain admissible cycles denoted by $\xi \times_{\mu'} \eta$ in $\mathrm{Tz}^{q-1}(X, n_1 + n_2; m)$ from [18, Definition 6.10]. Using this and Lemma 4.3, the assertion follows. Of course, the construction of $\xi \times_{\mu'} \eta$ and checking that they satisfy the admissibility require some works. See [18, §6.2] for details. The desired Leibniz rule between $\wedge_X$ and δ comes by applying the pull-back Δ_X^*. This discussion thus summarizes into:

Theorem 4.4 *Let k be a perfect field of characteristic $\neq 2$. Let X be a smooth affine k-algebra essentially of finite type, or smooth projective. Then $\{\mathrm{TCH}^q(X, n; m)\}_{q, n \geq 1}$ form an graded commutative differential graded algebra.*

4.3   Restricted Witt structure

The last sets of operations necessary on additive higher Chow groups are those related to the restricted Witt structures. Here, the affine case is more interesting, so we suppose now that $X = \mathrm{Spec}(R)$ is a smooth affine k-variety essentially of finite type. There are three necessary operations:

1. the ring homomorphisms $\tau_R : \mathbb{W}_m(R) \to \mathrm{TCH}^1(R, n; m)$,

2. the Frobenius maps $F_r : \mathrm{TCH}^q(R, n; rm + r - 1) \to \mathrm{TCH}^q(R, n; m)$, $r \geq 1$ and

3. the Verschiebung maps $V_r : \mathrm{TCH}^q(R, n; m) \to \mathrm{TCH}^q(R, n; rm+r-1)$, $r \geq 1$.

We need to check that they satisfy the list of axioms given in §3.3.

The Frobenius F_r and the Verschiebung V_r operators for $r \geq 1$ are defined in [16, §5.1, 5.2] as follows. Consider the morphism $\phi_r : X \times \mathbb{A}^1 \times \square^{n-1} \to X \times \mathbb{A}^1 \times \square^{n-1}$ given by $(x, t, y_1, \ldots, y_{n-1}) \mapsto (x, t^r, y_1, \ldots, y_{n-1})$. This is both finite and flat. Define $F_r := (\phi_r)_*$ and $V_r := (\phi_r)^*$. For these operators, the property (i) of §3.3 is trivial. The property (ii) and (iii) follow from basic properties of push-forward and pull-back of cycles as seen in [7, Proposition 1.7], for instance. The properties (iv) and (v) require some work, where (v) also requires a ring homomorphism $\mathbb{W}_m(R) \to \mathrm{TCH}^1(R, n; m)$.

To define the map $\tau_R : \mathbb{W}_m(R) \to \mathrm{TCH}^1(R, n; m)$, one uses the description $\mathbb{W}_m(R) = (1+tR[[t]])^\times / (1+t^{m+1}R[[t]])^\times$ as a group. It is a well-known fact that every member of $(1+tR[[t]])^\times$ factors uniquely into the product of the form $\prod_{n \geq 1}(1 - a_n t^n)$, with $a_n \in R$ (see [4, §1]). Thus, each element in $(1+tR[[t]])^\times / (1+t^{m+1}R[[t]])^\times$ actually has a polynomial representative f, and for this we take the associated divisor $\mathrm{div}(f)$ on $X \times \mathbb{A}^1$. One can also show that it is independent of the choice of the polynomial representative and defines a group homomorphism $\tau_R : \mathbb{W}_m(R) \to \mathrm{TCH}^1(R, 1; m)$. See [18, Proposition 7.6]. This group homomorphism commutes with both F_r and V_r by [18, Lemma 7.7].

To show that τ_R is a ring homomorphism, [18, §7] uses again the fact that every element of $(1 + tR[[t]])^\times$ has a unique expression as the product $\prod_{n \geq 1}(1 - a_n t^n)$. By [4, Proposition (1.1)], one reduces to show that $\Gamma_{a,u} \wedge_R \Gamma_{b,v} \equiv w \Gamma_{\frac{v}{a^w b^w}, \frac{u}{w}}$ in $\mathrm{TCH}^1(R, 1; m)$, where $\Gamma_{a,n} = \mathrm{div}(1 - at^n)$ and $w = \gcd(u, v)$.

This is done in [18, Proposition 7.9] in three steps. When $u = v = 1$, one can check it directly. When $v = 1$ and $u \geq 1$ is any, we can use the projection formula (the property (iii) of §3.3) between F_r and V_r to reduce it to the first case. When $u, v \geq 1$ are any, then we use again the projection formula to reduce it to the second case. Thus τ_R is a ring homomorphism.

The property (iv) of §3.3, i.e. $F_r \delta V_r \equiv \delta$, is not immediate. We sketch the arguments from [16, §5.5]. One first proves that $r F_r \delta = \delta F_r$ as an intermediate step toward (iv). In case $\mathrm{char}(k) = 0$ or $\mathrm{char}(k) = p > 0$ with $(r, p) = 1$, we have $r F_r \delta V_r = \delta F_r V_r \equiv r \delta$. So, $r(F_r \delta V_r - \delta) \equiv 0$. By the given assumptions on the characteristic, we have $F_r \delta V_r - \delta \equiv 0$ as desired. A complexity occurs when $\mathrm{char}(k) = p > 0$ and $r = p$, for which the scheme $\phi_p^{-1}(Z)$ of an admissible closed subscheme may or may not be reduced. Such cases require somewhat technical results such as [16, Lemmas 5.10, 5.11, 5.12]. These combinations and some technical cycle computations show $F_r \delta V_r \equiv \delta$ in this case.

In the remaining case of $\mathrm{char}(k) = p > 0$ and $r \geq 1$ is any integer divisible by p, we write $r = p^s \cdot r'$, for some $(r', p) = 1$. In this case, the property (i) shows $F_r = (F_p)^s F_{r'}$, $V_r = (V_p)^s V_{r'}$ so that $F_r \delta V_r = (F_p)^s F_{r'} \delta V_{r'} (V_p)^s = (F_p)^s \delta (V_p)^s$, where the last equality holds by the first case since $(r', p) = 1$. On the other hand, by the second case we already have $F_p \delta V_p = \delta$. So inductively we have $(F_p)^s \delta (V_p)^s = \delta$. This proves (iv).

The remaining property (v) of §3.3 is checked using some basic direct calculations as done in the proof of [18, Theorem 7.10]. This discussion summarizes into the following:

Theorem 4.5 *When $X = \mathrm{Spec}(R)$ is a smooth affine k-variety essentially of finite type, where $\mathrm{char}(k) \neq 2$, the additive higher Chow groups $\{\mathrm{TCH}^q(R, n; m)\}_{q,n,m \geq 1}$ form a restricted Witt-complex over R, and the Milnor range $\{\mathrm{TCH}^n(R, n; m)\}_{n,m \geq 1}$ form a restricted sub-Witt-complex over R.*

Since the big de Rham-Witt complex $\{\mathbb{W}_m \Omega_R^\bullet\}_{m \geq 1}$ is the universal object in the category of restricted Witt-complexes over R, we now deduce: ([18, Theorem 7.11])

Theorem 4.6 *When $X = \mathrm{Spec}(R)$ is a smooth affine k-variety essentially of finite type, where $\mathrm{char}(k) \neq 2$, there is a unique homomorphism $\tau_{n,m}^R : \mathbb{W}_m \Omega_R^{n-1} \to \mathrm{TCH}^n(R, n; m)$ that defines a morphism of restricted Witt-complexes over R such that $\tau_{1,m}^R = \tau_R$.*

5   Comparison theorem

In the previous section, we sketched how one obtains the homomorphisms $\tau_{n,m}^R$ by giving a restricted Witt-complex structure on additive higher Chow groups. In this section, we sketch from [18] and [19] how one can prove that the maps $\tau_{n,m}^R$ are isomorphisms when R is a smooth semi-local k-domains essentially of finite type, leading to Theorem 1.1.

5.1   Codimension 1 cycles and Witt vectors

In the previous section, we had the ring homomorphism $\tau_R : \mathbb{W}_m(R) \to \mathrm{TCH}^1(R, 1; m)$ when R is a smooth affine k-algebra. When is it an isomorphism? It is answered in [18, Theorem 7.12]. First of all, if R is an integral domain, then τ_R is injective. Indeed, consider $K = \mathrm{Frac}(R)$ and $\iota : R \hookrightarrow K$

is the inclusion. This gives the commutative diagram

$$
\begin{array}{ccc}
\mathbb{W}_m(R) & \xrightarrow{\ \mathbb{W}_m(\iota)\ } & \mathbb{W}_m(K) \\
{\scriptstyle\tau_R}\downarrow & & {\scriptstyle\tau_K}\downarrow{\scriptstyle\simeq} \\
\mathrm{TCH}^1(R,1;m) & \longrightarrow & \mathrm{TCH}^1(K,1;m).
\end{array}
$$

Here, the map τ_K is an isomorphism by [30], and $\mathbb{W}_m(\iota)$ is injective. So, τ_R must be injective when R is an integral domain. Secondly, when R is a UFD, then τ_R is surjective. This follows from the fact that if R is a UFD then so is $R[t]$, and that every height 1 prime ideal of a UFD is a principal ideal.

One interesting feature of this isomorphism for smooth factorial affine k-scheme $\mathrm{Spec}(R)$ is that, this ring isomorphism provides a conceptual and geometric way to describe the ring structure of $\mathbb{W}_m(R)$, which is known to be very complicated. The addition $+$ of $\mathbb{W}_m(R)$ corresponds to the multiplication of $(1+tR[[t]])^\times/(1+t^{m+1}R[[t]])^\times$, but the multiplication $\cdot$ of $\mathbb{W}_m(R)$ is quite nontrivial to express in $(1+tR[[t]])^\times/(1+t^{m+1}R[[t]])^\times$.

However, under the isomorphism τ_R, the addition $+$ of $\mathbb{W}_m(R)$ corresponds to just the sum of algebraic cycles in $\mathrm{TCH}^1(R,1;m)$, while the multiplication $\cdot$ of $\mathbb{W}_m(R)$ corresponds to the Pontryagin-intersection product of $\mathrm{TCH}^1(R,1;m)$. This provides an algebro-geometric way to understand the structure of the ring of Witt vectors.

Specializing to a smooth semi-local k-algebra R essentially of finite type, a theorem of Auslander-Buchsbaum shows that such R is always a smooth UFD, so the above identification via τ_R still holds for these R.

5.2 Injectivity

The injectivity part of the maps $\tau_{n,m}^R : \mathbb{W}_m\Omega_R^{n-1} \to \mathrm{TCH}^n(R,n;m)$ is proven in [19, Proposition 4.3.1]. Two essential points are, (i) when $K = \mathrm{Frac}(R)$, the map $\mathbb{W}_m\Omega_R^{n-1} \to \mathbb{W}_m\Omega_K^{n-1}$ is injective, and (ii) the map $\tau_{n,m}^K : \mathbb{W}_m\Omega_K^{n-1} \to \mathrm{TCH}^n(K,n;m)$ is an isomorphism by [30]. Given these two, one immediately deduces the injectivity of $\tau_{n,m}^R$. Here (i) can be reduced to prove the injectivity for the absolute Kähler forms and for p-typical de Rham-Witt forms.

5.3 Surjectivity

The surjectivity proven in [19] is lengthy and complicated. We sketch these steps.

One first reduces it to the case when R is obtained by localizing at a finite set of closed points as done in [19, Lemma 4.4.1]. This is based on the basic fact in commutative algebra that any proper ideal is contained in some maximal ideal.

The second step, which is the longest and hardest part, is to prove the following from [19, Theorem 1.0.2]:

Theorem 5.1 *For every cycle class in* $\mathrm{TCH}^n(R, n; m)$, *one can find a representative cycle, each of whose irreducible components is an "sfs-cycle".*

Here sfs stands for 'smooth, finite, surjective' and defined as follows: a closed integral subvariety $V \subset X \times \mathbb{A}^1 \times \square^{n-1}$ is called an sfs-cycle if it satisfies the following properties (see [19, Definition 5.1.2, Proposition 5.1.8]):

(a) The natural map $V \to X = \mathrm{Spec}(R)$ is finite and surjective.

(b) For each $0 \leq i \leq n-1$, let $V^{(i)}$ be the projection of V to $X \times \mathbb{A}^1 \times \square^i$, where $V^{(n-1)} = V$. Here each $V^{(i)}$ is smooth over k, and we have a finite sequence of integral extensions $R \subset R[a] \subset R[a, b_1] \cdots \subset R[a, b_1, \ldots, b_{n-1}]$, where $R[a, b_1, \ldots, b_i] = k[V^{(i)}]$ and $a = t \mod I(V^{(0)})$, $b_j = y_j \mod I(V^{(i)})$. Furthermore there are monic polynomials $P(t) \in R[t]$, $Q_i \in R[a, b_1, \ldots, b_{i-1}][y_i]$ such that

$$R[a] = R[t]/(P(t)), \quad R[a, b_1, \ldots, b_i] = R[a, b_1, \ldots, b_{i-1}][y_i]/(Q_i).$$

Although the proof is quite complicated, the essential idea is still to use some classical generic projection techniques used in the proof of Chow's moving lemma. To do so, we "spread out" the given semi-local k-algebra to an affine scheme containing the closed points, and we embed this affine scheme into a sufficiently big affine space, if necessary. Due to the additional extra requirements coming from the modulus condition, the parameter space we have to choose must be bigger than we might need to do the same for higher Chow groups, but the point is that it does work. Unfortunately, this argument is way too complicated, so the interested reader should consult [19, §5 ∼ §11] for more details.

Once we have this "sfs-moving lemma", then we can easily check that for an sfs-cycle V on $X \times \mathbb{A}^1 \times \square^{n-1}$, up to replacing R by the finite extension $S = k[V]$, the cycle $[V]$ considered as a cycle over S is contained in the image of $\tau_{n,m}^S$. Unfortunately, there is *a priori* no construction of the trace map in the literature for big de Rham-Witt forms that are compatible with the push-forward of additive higher Chow cycles, so one still has some work to do. The key point is that one can instead define the notion of "traceability" as in [19, §12.1.1] recalled below:

Let $R \subset S$ be a finite extension of smooth semi-local k-algebras essentially of finite type and let $f : \operatorname{Spec}(S) \to \operatorname{Spec}(R)$ be the corresponding morphism of schemes. Let $m, n \geq 1$. Consider the diagram

$$
\begin{array}{ccc}
\mathbb{W}_m \Omega_S^{n-1} & \xrightarrow{\ \tau_{n,m}^S\ } & \mathrm{TCH}^n(S, n; m) \\
 & & \downarrow{\scriptstyle f_*} \\
\mathbb{W}_m \Omega_R^{n-1} & \xrightarrow{\ \tau_{n,m}^R\ } & \mathrm{TCH}^n(R, n; m).
\end{array}
$$

We say that a form $\omega \in \mathbb{W}_m \Omega_S^{n-1}$ *is traceable to R* if $f_* \circ \tau_{n,m}^S(\omega) \in \mathrm{Image}(\tau_{n,m}^R)$. For instance, if we already have the trace map $\mathrm{Tr}_{S/R} : \mathbb{W}_m \Omega_S^{n-1} \to \mathbb{W}_m \Omega_R^{n-1}$, then everything in $\mathbb{W}_m \Omega_S^{n-1}$ is traceable. The problem is we do not know if this is the case. However, we do have it when $n = 1$ (on the level of rings of Witt vectors) and when $m = 1$ (on the level of absolute Kähler forms). So, one may wonder if we can inductively reduce the question to these known cases. The answer is yes if $R \subset S$ is a simple extension as done in [19, Proposition 12.1.5]:

Proposition 5.2 *Let $R \hookrightarrow S$ be a simple extension of regular semi-local k-algebras essentially of finite type, and $m, n \geq 1$. Then all elements in $\mathbb{W}_m \Omega_S^{n-1}$ are traceable to R.*

The proof of Proposition 5.2 uses a double induction argument on (m, n) exploiting the fact that the maps $\{\tau_{n,m}^R\}_{n,m \geq 1}$ and $\{\tau_{n,m}^S\}_{n,m \geq 1}$ are morphisms of restricted Witt-complexes over R and S, respectively.

Given Proposition 5.2, we can finally prove that $\tau_{n,m}^R$ is surjective as done in [19, §12.2]. By Theorem 5.1, every cycle class in $\mathrm{TCH}^n(R, n; m)$ has a representative which is sfs. Let $[V] \in \mathrm{TCH}^n(R, n; m)$ be any sfs-cycle. It has trivial faces by [19, Lemma 5.1.7] and its ideal $I(V)$ inside the ring $R[t, y_1, \ldots, y_{n-1}]$ is given by the equations of the form $P(t) = 0$, $Q_1(t, y_1) = 0, \ldots, Q_{n-1}(t, y_1, \ldots, y_{n-1}) = 0$, where for $R_0 := R$, $R_1 := R[t]/(P(t)), \ldots, R_i := R_{i-1}[y_i]/(Q_i)$, the rings R_i are all smooth semi-local k-algebras such that each extension $R_{i-1} \subset R_i$ is simple. We let $f_i : \operatorname{Spec}(R_i) \to \operatorname{Spec}(R_{i-1})$ be the induced finite surjective map. Let $f := f_1 \circ \cdots \circ f_n$. Obviously we have $[Z] = f_*([Z_n])$ for some $Z_n \in \mathrm{Tz}^n(R_n, n; m)$. (As a scheme $Z_n = Z$, seen as a scheme over R_n and $f|_{Z_n} = \mathrm{Id}$.) On the other hand, we have $[Z_n] = \tau_{n,m}^{R_n}(\eta_n)$ for some $\eta_n \in \mathbb{W}_m \Omega_{R_n}^{n-1}$. Now applying Proposition 5.2 to f_n, which is a simple extension, for some $\eta_{n-1} \in \mathbb{W}_m \Omega_{R_{n-1}}^{n-1}$ we have $f_{n*} \tau_{n,m}^{R_n}(\eta_n) = \tau_{n,m}^{R_{n-1}}(\eta_{n-1})$. Applying this argument from f_{n-1} to f_1 successively, we eventually get $[Z] = f_*[Z_n] = \tau_{n,m}^R(\eta_0)$ for some $\eta_0 \in \mathbb{W}_m \Omega_R^{n-1}$ as desired. This proves the surjectivity, finishing the sketch of the proof of Theorem 1.1.

6 Connection to crystalline cohomology

6.1 The Zariski sheaf $\mathcal{TCH}^q(-, n; m)$

On the category of smooth affine k-schemes, $\mathrm{TCH}^q(-, n; m)$ is a presheaf by the moving lemma of W. Kai [11] for additive higher Chow groups of smooth affine schemes. This can be extended to a presheaf on the category of all smooth k-schemes via the "presheafification" given in [18, §4.4] as

$$\mathcal{TCH}^q(X, n; m) := \operatorname*{colim}_{A \in (X \downarrow \mathrm{SmAff}_k)^{\mathrm{op}}} \mathrm{TCH}^q(A, n; m), \qquad (6.1)$$

where $(X \downarrow \mathrm{SmAff}_k)$ is the category whose objects are k-morphisms $X \to A$ with $A \in \mathrm{SmAff}_k$ and a morphism $g : (h_1 : X \to A) \to (h_2 : X \to B)$ is a k-morphism $g : A \to B$ such that $g \circ h_1 = h_2$. This idea was inspired while Amalendu Krishna and the author were working on [17].

One can apply the above procedure also to $\mathbb{W}_m \Omega^{n-1}_{(-)}$. The natural map $\tau^R_{n,m}$ for $R \in \mathrm{SmAff}_k$ defines a morphism of Zariski sheaves $\tau^X_{n,m} : \mathbb{W}_m \Omega^{n-1}_{X_{\mathrm{Zar}}} \to \mathcal{TCH}^n(X, n; m)_{\mathrm{Zar}}$ after sheafification for any smooth k-variety X. Theorem 1.1 says that this map is stalk-wise isomorphism, thus an isomorphism of Zariski sheaves.

6.2 The Zariski sheaf $\mathcal{TCH}_{(p)}(-; p^i)$

When $p > 0$ is a prime, S. Bloch [4, §I.3] defined a projection $\pi : \mathbb{W}(R) \to W(R)$ to the p-typical ring of Witt vectors by $\pi := \sum_{n \in I(p)} \frac{\mu(n)}{n} V_n F_n$, where $I(p)$ is the set of positive integers not divisible by p and $\mu(n)$ is the Möbius function. This projection gives the idempotent $e = \pi(1) \in \mathbb{W}(R)$, that allows one to define the p-typical version of additive higher Chow groups as in [19, §3.5]:

$$\mathrm{TCH}^n_{(p)}(R; p^\infty) := e(\mathrm{TCH}^n(R; \infty)) = e(\varprojlim_m \mathrm{TCH}^n(R; n; m)),$$

$$\mathrm{TCH}^n_{(p)}(R; p^i) := \mathrm{TCH}^n_{(p)}(R; p^\infty)/p^i.$$

Using the same idea as in (6.1) in §6.1, we can again define the presheaf $\mathcal{TCH}^n_{(p)}(-; p^i)$ on all smooth k-schemes as done in *loc. cit.*

6.3 Algebraic cycles and crystalline cohomology

The crystalline cohomology of a smooth variety X over a field k of characteristic $p > 0$ is defined by P. Berthelot in [2] via a category with a Grothendieck topology, called the crystalline topos. S. Bloch [4] and L. Illusie [10] proved that the crystalline cohomology admits a computation as the hypercohomology of a complex of Zariski sheaves. In doing so, Bloch used the p-typical curves induced from certain relative K-groups of $(X[t]/(t^{m+1}), (t))$ and Illusie used the p-typical de Rham-Witt complexes. They are actually equivalent objects as proven in [10].

Because the p-typical de Rham-Witt forms $W_i \Omega^{\bullet}_{X_{\mathrm{Zar}}}$ of level i is quasi-isomorphic to $W\Omega^{\bullet}_{X_{\mathrm{Zar}}}/p^i$ by [10, Corollaire I.3.17], which is actually $\mathcal{TCH}^{\bullet}_{(p)}(-; p^i)_{\mathrm{Zar}}[1]$ by the discussion above, the theorem of Bloch-Illusie now reads:

Theorem 6.1 *Let X be a smooth k-variety over a perfect field of characteristic $p > 2$. Then there is an isomorphism*

$$\mathrm{H}^n_{\mathrm{crys}}(X/W) \simeq \varprojlim_i \mathbb{H}^{n+1}_{\mathrm{Zar}}(X, \mathcal{TCH}^{\bullet}_{(p)}(-; p^i)_{\mathrm{Zar}}).$$

Since the group on the right originates from algebraic cycles, in a sense Theorem 6.1 gives an "algebraic-cycle description" of the crystalline cohomology.

7 A remark on crystalline cobordism

In this experimental section, we give some discussions toward a construction of a conjectural theory of crystalline cobordism, which does not exist yet. For a motivation, recall that we have at least three versions of cobordism theories that correspond to the motivic cohomology (Chow group), singular cohomology, and étale cohomology, and the maps from the cobordism theories to the corresponding ordinary cohomology theories. Namely, when X is a smooth k-variety over a field of characteristic zero, we have the algebraic cobordism $\Omega^*(X)$ as constructed by M. Levine and F. Morel [21] (see also [22]) and the maps $\Omega^*(X) \to \mathrm{CH}^*(X)$, $\Omega^*(X) \otimes_{\mathbb{L}} \mathbb{Z} \to \mathrm{CH}^*(X)$, where the second map is an isomorphism. Here $\mathbb{L}$ is the Lazard ring in [20], which is a $\mathbb{Z}$-coefficient polynomial ring in countably infinite number of variables. For smooth complex algebraic varieties, we have maps involving the complex cobordism (see [28]), $\mathrm{MU}^*(X) \to \mathrm{H}^{2*}(X, \mathbb{Z})$, $\mathrm{MU}^*(X) \otimes_{\mathbb{L}} \mathbb{Z} \to \mathrm{H}^{2*}(X, \mathbb{Z})$. When X is a smooth variety over a field k of characteristic 0 or $p > 0$ and $\ell \neq p$ is a prime, G. Quick constructed and studied the theory of étale cobordism in [26], [27], where $\mathbb{L}_\ell = \mathbb{L} \otimes_{\mathbb{Z}} \mathbb{Z}_\ell$, and maps $\mathrm{MU}^*_{\acute{e}t}(X) \to \mathrm{H}^{2*}_{\acute{e}t}(X, \mathbb{Z}_\ell)$, $\mathrm{MU}^*_{\acute{e}t} \otimes_{\mathbb{L}_\ell} \mathbb{Z}_\ell \to \mathrm{H}^{2*}_{\acute{e}t}(X, \mathbb{Z}_\ell)$.

What we seek is a conjectural cobordism theory that we denote by $\mathrm{MU}^*_{\mathrm{crys}}(X)$ for a smooth variety over a field k of characteristic $p > 0$ that has natural maps

$$\mathrm{MU}^*_{\mathrm{crys}}(X) \longrightarrow \mathrm{H}^{2*}_{\mathrm{crys}}(X/W), \mathrm{MU}^*_{\mathrm{crys}}(X) \otimes_{\mathbb{L}_W} W \longrightarrow \mathrm{H}^{2*}_{\mathrm{crys}}(X/W),$$

where $\mathbb{L}_W := \mathbb{L} \otimes_{\mathbb{Z}} W$ and $W = W(k)$ is the ring of p-typical Witt vectors of k. How do we do it? We do not know it yet, but there might be at least two possible attempts one can try as a first order approximation. One way is to imitate and extend the original construction of crystalline cohomology by P. Berthelot [2], using the crystalline topos. The second way, which we attempt a bit in this last section is to give an algebro-geometric model following philosophically along the line of the previous sections of this article.

We guess that this approach entails the following two steps: (i) to give an algebro-geometric construction of higher algebraic cobordism with modulus, extending [21] and [22] as well as [3], and (ii) to specialize to the studies of "additive higher algebraic cobordism" (namely, to the pairs $(X \times \mathbb{A}^1, (m + 1)\{t = 0\})$) and follow the outline of this article.

Unfortunately, this wasn't easy and a lot of things turned out to be missing or very hard to prove. However, we will still give some ideas, which may serve as the starting points for the future development. The reader should take this just as a provisional attempt, subject to possible improvements in the future.

7.1   Algebraic cobordism

We are not going to review the entire construction of algebraic cobordism by Levine-Morel [21] and Levine-Pandharipande [22]. The interested reader should consult these original references. If the reader seeks a concise summary of some of the discussions, read the first few sections of [15].

The crucial idea of the construction of algebraic cobordism (in the sense of [22]) is that, instead of just looking at closed subvarieties on a fixed k-scheme X to form algebraic cycles, one considers all projective morphisms $[f : Y \to X]$ from smooth varieties Y, and take the free abelian group on them to obtain "cobordism cycles". On these cobordism cycles, one imposes various relations, analogous to putting the rational equivalence on algebraic cycles. A cobordism cycle $[f : Y \to X]$ gives the algebraic cycle $f(Y) \subset X$. Conversely, given an algebraic cycle $Z \subset X$, assuming resolution of singularities one obtains a cobordism cycle $[f : Y \to X]$ (not unique) with $f(Y) = Z$.

7.2   An approach for higher algebraic cobordism

The cubical version of higher Chow groups use algebraic cycles on $X \times \square^n$, with $\square = (\mathbb{P}^1 \setminus \{1\}, \{0, \infty\})$ with proper intersection with all faces. Our guess is that, if exists, higher algebraic cobordism may use cobordism cycles over the space $X \times \square^n$.

Just like higher Chow cycles have the proper intersection condition with faces, we would like to impose a suitable admissibility condition to cobordism cycles over $X \times \square^n$. Unfortunately, yet we were not able to find a convincing definition for it, and furthermore, we are yet unsure of what the correct definition of boundary maps should be.[1]

The correct theory of higher algebraic cobordism, which we do not yet have, must satisfy a list of axioms for *bigraded* oriented cohomology theories (*cf.* [21]) and must be the universal object in the category of such theories. Or, equivalently, one must prove that the above object is isomorphic to the algebraic cobordism MGL of Voevodsky.

7.3 An approach to higher algebraic cobordism with modulus

It is somewhat outrageous to discuss even further generalizations of a non-existent higher algebraic cobordism. However, for the sake of discussion, we continue our journey and discuss something about the modulus condition. We want to discuss a bit on how to think about modulus conditions for cobordism cycles over $X \times \square^n$, associated to the pair (X, D) consists of a k-scheme and an effective Cartier divisor D on it.

7.3.1 A projective normalization and modulus

We first need to discuss what we mean by modulus conditions for this cobordism context. We use the following cancellation lemma from [16, Lemma 2.12]:

Lemma 7.1 *Let $g : Y \to X$ be a dominant morphism of normal integral k-schemes. Let D be a Cartier divisor on X such that the generic points of $\mathrm{Supp}(D)$ are contained in $g(Y)$. Suppose that $g^*(D) \geq 0$ on Y. Then $D \geq 0$ on X.*

Let $f : Y \to X \times \square^n$ be a projective morphism, where Y is irreducible. Let $D \subset X$ be an effective Cartier divisor. Since f is projective, it factors into $Y \overset{\iota}{\hookrightarrow} \mathbb{P}^m_{X \times \square^n} \overset{p}{\to} X \times \square^n$, for some $m > 0$. Here p is the natural projection. The open immersion $j : X \times \square^n \hookrightarrow X \times \overline{\square}^n$ induces the open immersion $j' : \mathbb{P}^m_{X \times \square^n} \hookrightarrow \mathbb{P}^m_{X \times \overline{\square}^n}$. Let $\overline{f} : \overline{Y} \hookrightarrow \mathbb{P}^m_{X \times \overline{\square}^n} \to X \times \overline{\square}^n$ be the inclusion of the Zariski closure of Y followed by the natural projection,

---

[1]In October 2011, Amalendu Krishna, in a private communication, informed the author of one possible construction via pseudo-cubical objects as in [25], but yet he would prefer not to make it public at this stage until one can find the correct convincing definition.

and let $\nu : \overline{Y}^N \to \overline{Y} \xrightarrow{\overline{f}} X \times \overline{\square}^n$ be the normalization composed with $\overline{f}$. Note that $\overline{Y}^N$ and $\overline{Y}$ may not be smooth so that they are not cobordism cycles in the sense of Levine-Morel [21] and Levine-Pandharipande [22]. This summarizes into the following commutative diagram

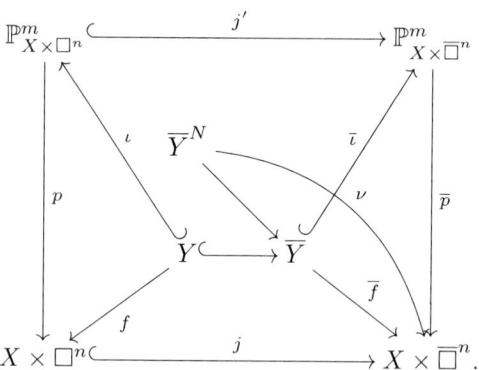

Definition 7.2 We say that $\nu : \overline{Y}^N \to X \times \overline{\square}^n$ is a *projective normalization* of $f : Y \to X \times \square^n$. This depends on the choice of the factorization $f = p \circ \iota$.

We say that the irreducible cobordism cycle $[f : Y \to X \times \square^n]$ has *modulus D with respect to the factorization $p \circ \iota$*, if we have

$$\nu^*(D \times \overline{\square}^n) \leq \nu^*(\sum_{i=1}^{n}\{y_i = 1\}).$$

Our first goal of this game is to show that, the notion of modulus condition *does not* depend on the choice of a factorization. To do so, we first prove:

Lemma 7.3 *Given two projective normalizations of $f : Y \to X \times \square^n$, there exists a common refinement projective normalization.*

More precisely, for $s = 1, 2$, let $\iota_s : Y \to \mathbb{P}^{m_s}_{X \times \square^n}$ be two closed immersions that factor f into $p_s \circ \iota_s$. Let $\overline{Y}_s$ be the Zariski closures of Y with respect to the factorization $p_s \circ \iota_s$, and $\nu_s : \overline{Y}_s^N \to X \times \overline{\square}^n$ be the projective normalizations. Then there exists another closed immersion $\iota_3 : Y \to \mathbb{P}^{m_3}_{X \times \square^n}$ that factors f, for which the associated projective normalization $\nu_3 : \overline{Y}_3^N \to X \times \overline{\square}^n$ has morphisms $q_s : \overline{Y}_3^N \to \overline{Y}_s^N$ that make the *following diagram commute:*

$$\overline{Y}_3^N \xrightarrow{q_2} \overline{Y}_2^N \qquad (7.1)$$

$$q_1 \downarrow \quad \searrow^{\nu_3} \quad \downarrow \nu_2$$

$$\overline{Y}_1^N \xrightarrow{\nu_1} X \times \overline{\square}^n.$$

Proof For simplicity, let $\star := X \times \square^n$ and $\overline{\star} := X \times \overline{\square}^n$. For the projections $p_s : \mathbb{P}_{\overline{\star}}^{m_s} \to \overline{\star}$ for $s = 1, 2$, consider the Segre embedding of the fiber product $\mathbb{P}_{\overline{\star}}^{m_1} \times_{\overline{\star}} \mathbb{P}_{\overline{\star}}^{m_2} \xrightarrow{\overline{\sigma}} \mathbb{P}_{\overline{\star}}^{m_3}$, where $m_3 := (m_1+1)(m_2+1) - 1$. On the other hand, we have the corresponding Segre embedding $\mathbb{P}_{\star}^{m_1} \times_{\star} \mathbb{P}_{\star}^{m_2} \xrightarrow{\sigma} \mathbb{P}_{\star}^{m_3}$. This gives $\iota_3 := \sigma \circ (\iota_1 \times \iota_2) : Y \hookrightarrow \mathbb{P}_{\star}^{m_3}$, that factors f into $p_3 \circ \iota_3$, where $p_3 : \mathbb{P}_{\star}^{m_3} \to \star$ is the projection. For the diagonal map $\Delta : Y \hookrightarrow \mathbb{P}_{\star}^{m_1} \times_{\star} \mathbb{P}_{\star}^{m_2}$ over $\overline{\star}$, let $\overline{Y}$ be the closure of the image of $\Delta(Y)$ in $\mathbb{P}_{\overline{\star}}^{m_1} \times_{\overline{\star}} \mathbb{P}_{\overline{\star}}^{m_2}$. But, the Zariski closure $\overline{Y}_3$ of Y for ι_3 is exactly $\overline{\sigma}(\overline{Y})$ because $\overline{\sigma}$ is a closed immersion. That means, $\overline{\sigma}| : \overline{Y} \to \overline{Y}_3$ is an isomorphism. Via the natural projections $pr_s : \mathbb{P}_{\overline{\star}}^{m_1} \times_{\overline{\star}} \mathbb{P}_{\overline{\star}}^{m_2} \to \mathbb{P}_{\overline{\star}}^{m_s}$, we have induced dominant morphisms $pr_s : \overline{Y} \to \overline{Y}_s$ (since they are projective, they must be surjective). Hence, by the universal properties of the normalizations, we have morphisms $q_s : \overline{Y}_3^N \to \overline{Y}_s^N$ that make the following diagram commutative:

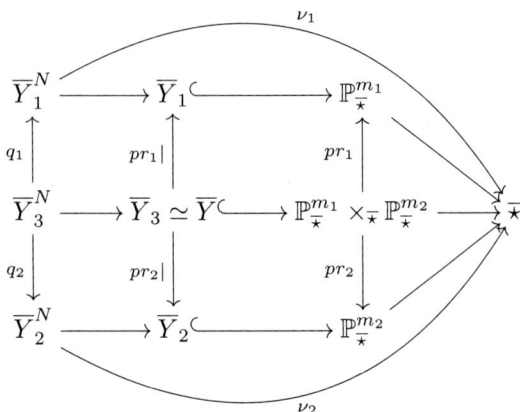

Since ν_3 is the composition of the central horizontal arrows, we obtain the diagram (7.1). This proves the lemma. $\qquad \square$

Proposition 7.4 *For $f : Y \to X \times \square^n$ as before, the modulus D condition for f as in Definition 7.2 is independent of the choice of the factorization*

$f = p \circ \iota$. *In other words, the modulus condition on Y can be stated without reference to the choice of a factorization.*

Proof For simplicity, let $E := -D \times \overline{\square}^n + \sum_{i=1}^n \{y_i = 1\}$ as a divisor on $X \times \overline{\square}^n$. For two projective normalizations $\nu_s : \overline{Y}_s^N \to X \times \overline{\square}^n$ associated to the factorizations $p_s \circ \iota_s$ for $s = 1, 2$, that Y has modulus D with respect to the factorization $p_s \circ \iota_s$ means $\nu_s^*(E) \geq 0$.

By Lemma 7.3, there exists another projective normalization $\nu_3 : \overline{Y}_3^N \to X \times \overline{\square}^n$, satisfying the commutative diagram (7.1) of the previous lemma. We claim that Y has modulus D with respect to the factorization $f = p_3 \circ \iota_3$ if and only if so does with respect to $f = p_2 \circ \iota_2$.

Suppose Y has modulus D with respect to the factorization $f = p_3 \circ \iota_3$, i.e. $\nu_3^*(E) \geq 0$. By the commutativity of the upper triangle of the diagram (7.1), we have $q_2^* \nu_2^*(E) \geq 0$. But by Lemma 7.1 this means $\nu_2^*(E) \geq 0$, so that Y has modulus D with respect to the factorization $f = p_2 \circ \iota_2$.

Conversely, if Y has modulus D with respect to ι_2, then $\nu_2^*(E) \geq 0$. This implies $q_2^* \nu_2^*(E) \geq 0$. But again by the upper triangle of the diagram (7.1), this is equivalent to that $\nu_3^*(E) \geq 0$, giving the modulus D with respect to the factorization $f = p_3 \circ \iota_3$.

But, the same argument then shows that $\nu_3^*(E) \geq 0$ if and only if $\nu_1^*(E) \geq 0$. Hence, $\nu_1^*(E) \geq 0$ if and only if $\nu_2^*(E) \geq 0$. This proves the proposition.

$\square$

For algebraic cycles, the containment lemma which first appeared in [13, Proposition 2.4] had been a useful tool in dealing with additive higher Chow cycles. By essentially the same argument, it was subsequently generalized to cycles with modulus. We want to discuss its algebraic cobordism analogue. Here, the notion of 'containment' should be generalized as follows:

Definition 7.5 Let $h_1 : W \to B$ and $h_2 : V \to B$ be morphisms of k-schemes. We say that h_1 *is contained in* h_2 if h_1 factors into $W \to V \overset{h_2}{\to} B$ for some morphism $W \to V$ of k-schemes.

Lemma 7.6 *Let $h_1 : W \to X \times \square^n$ and $h_2 : V \to X \times \square^n$ be projective morphisms with V and W smooth over k, such that h_1 is contained in h_2 in the sense of Definition 7.5. Then there exist projective normalizations $\overline{h}_1 : \overline{W} \to X \times \overline{\square}^n$ and $\overline{h}_2 : \overline{V} \to X \times \overline{\square}^n$, where $\overline{h}_1$ is contained in $\overline{h}_2$.*

Proof This is trivial: let $\star$ and $\overline{\star}$ be $X \times \square^n$ and $X \times \overline{\square}^n$ as before. Given closed immersions $\iota_1 : W \hookrightarrow \mathbb{P}_\star^{m_1}$ and $\iota_2 : V \hookrightarrow \mathbb{P}_\star^{m_2}$ that factor h_1 and h_2, we obtain $\overline{W}$ and $\overline{V}$ as the closures of W and V in $\mathbb{P}^{m_1}_{\overline{*}}$ and $\mathbb{P}^{m_2}_{\overline{*}}$. Here, the morphism $W \to V$ then obviously extends to $\overline{W} \to \overline{V}$.

$\square$

Here is a generalization of the containment lemma [13, Proposition 2.4] recalled in Lemma 2.1. The argument is essentially identical, if we replace the closed immersions by morphisms:

Lemma 7.7 (Containment lemma) *Let $h_1 : W \to X \times \square^n$ and $h_2 : V \to X \times \square^n$ be two projective morphisms, with V and W smooth over k. Suppose that h_1 is contained in h_2, and V has modulus D. Then W has modulus D, too.*

Proof Let $\overline{h}_1 : \overline{W} \to X \times \overline{\square}^n$ and $\overline{h}_2 : \overline{V} \to X \times \overline{\square}^n$ be the projective normalizations of h_1 and h_2. By Lemma 7.6, there exists $j : \overline{W} \to \overline{V}$ that factors $\overline{h}_1$ into $\overline{h}_2 \circ j$.

In case $n = 0$, the modulus condition for V means, the image of $h_1 : V \to X$ does not meet D. Since h_2 factors via h_1, the image of $h_2 : W \to X$ does not meet D either. Hence W has modulus D.

So, we now suppose $n > 0$. Consider the following commutative diagram:

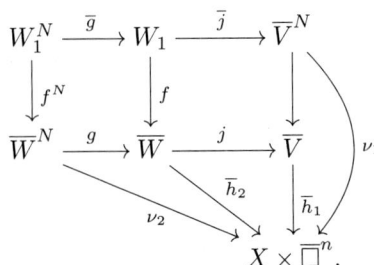

where j is the induced factorization, $W_1 = \overline{W} \times_{\overline{V}} \overline{V}^N$, the morphisms f and $\overline{j}$ are the induced ones, the morphisms g and $\overline{g}$ are the normalizations, the morphisms ν_1, ν_2 are the compositions, and f^N is induced by the universal property of normalization.

Since V satisfies the modulus condition, we have $\overline{g}^* \overline{j}^* [\nu_1^* (\sum_{i=1}^n \{y_i = 1\} - D \times \overline{\square}^n)] \geq 0$ on W_1^N. By the commutativity of the above diagram, we get $(f^N)^* [\nu_2^* (\sum_{i=1}^n \{y_i = 1\} - D \times \overline{\square}^n)] \geq 0$. But, by Lemma 7.1, this implies that $[\nu_2^* (\sum_{i=1}^n \{y_i = 1\} - D \times \overline{\square}^n)] \geq 0$ on $\overline{W}^N$, which means W has modulus D.

$\square$

7.3.2 Relation to algebraic cycles with modulus

The modulus condition we defined in this paper and the modulus condition for cycles in Binda-Saito [3] are related as follows:

Proposition 7.8 *Let $f : Y \to X \times \square^n$ be a projective morphism and Y is an irreducible k-scheme. Let $W = f(Y) \subset X \times \square^n$ be the closed subset seen as a closed subscheme with the reduced induced structure. Then f has modulus D if and only if W has modulus D in the sense of Binda-Saito.*

Proof Let $\iota : W \hookrightarrow X \times \square^n$ be the closed immersion, which is projective, with W irreducible. We have the following commutative diagram

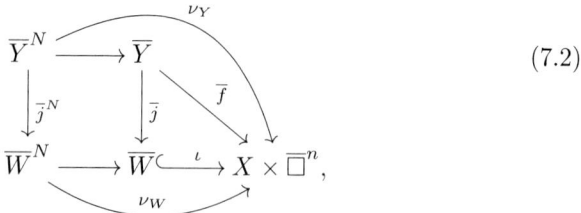

$$(7.2)$$

where $\overline{f}$ is the induced map of a projective normalization, $\overline{j}$ is the morphism given by Lemma 7.6, $\overline{Y}^N \to \overline{Y}$ and $\overline{W}^N \to \overline{W}$ are the normalizations, $\overline{j}^N$ is given by the universal property of the normalization, and ν_Y and ν_W are the compositions.

Consider the divisor $E := \sum_{i=1}^n \{y_i = 1\} - D \times \overline{\square}^n$. First suppose that W has modulus D in the sense of [3], which means $\nu_W^* E \geq 0$. This implies $\overline{j}^{N*} \nu_W^* E \geq 0$, which is equivalent to $\nu_Y^* E \geq 0$ by the commutativity of (7.2). This means Y has modulus D in the sense of this paper.

Conversely, if Y has modulus D, then we have $\nu_Y^* E \geq 0$, i.e. $\overline{j}^{N*} \nu_W^* E \geq 0$. By Lemma 7.1, this implies $\nu_W^* E \geq 0$, which means W has modulus D. $\square$

7.3.3 The external product

We do not have the notion of higher cobordism cycles with modulus yet, but at least the modulus condition we defined has the following additional basic compatibility regarding the external product:

Lemma 7.9 *Let (X_1, D_1) and (X_2, D_2) be two pairs as used previously. Let $D_1 \boxtimes D_2$ be the divisor $D_1 \times X_2 + X_1 \times D_2$ on $X_1 \times X_2$. For $s = 1, 2$, let $[f_s : Y_s \to X_s \times \square^{n_s}]$ be cobordism cycles with modulus D_s. Define $f_1 \boxtimes f_2$*

to be the cobordism cycle associated to the morphism $f_1 \times f_2 : Y_1 \times Y_2 \to X_1 \times X_2 \times \square^{n_1+n_2}$, up to the obvious exchanges of factors. Then $[f_1 \boxtimes f_2]$ has modulus $D_1 \boxtimes D_2$.

Proof To check the modulus condition, let $W \subset Y_1 \times Y_2$ be an irreducible component and let $f : W \to X_1 \times X_2 \times \square^{n_1+n_2}$ be the morphism induced by $f_1 \boxtimes f_2$. It is enough to show that W has modulus $D_1 \boxtimes D_2$.

For this, first fix projective normalizations $\overline{f}_s^N : \overline{Y}_s^N \to X_s \times \overline{\square}^{n_i}$ for $s = 1, 2$. This gives $\overline{f}_1^N \boxtimes \overline{f}_2^N : \overline{Y}_1^N \times \overline{Y}_1^N \to X_1 \times X_2 \times \overline{\square}^{n_1+n_2}$. Since k is perfect, we have $\overline{Y}_1^N \times \overline{Y}_2^N = (\overline{Y}_1 \times \overline{Y}_2)^N$ by [12, Lemma 3.1], so that there is a natural inclusion $\iota : W^N \to \overline{Y}_1^N \times \overline{Y}_2^N$.

Since f_1 has modulus D_1, we have $\overline{f}_1^{N*}(\sum_{i=1}^{n_1}\{y_i = 1\} - D_1 \times \overline{\square}^{n_1}) \geq 0$ on $\overline{Y}_1^N$. Pulling back via the projection $\overline{Y}_1^N \times \overline{Y}_2^N \to \overline{Y}_1^N$, and using commutativity, we have $(\overline{f}_1^N \boxtimes \overline{f}_2^N)^*(\sum_{i=1}^{n_1}\{y_i = 1\} - D_1 \times X_2 \times \overline{\square}^{n_1+n_2}) \geq 0$. Similarly we have $(\overline{f}_1^N \boxtimes \overline{f}_2^N)^*(\sum_{i=n_1+1}^{n_1+n_2}\{y_i = 1\} - X_1 \times D_2 \times \overline{\square}^{n_1+n_2}) \geq 0$. Taking the sum of these two inequalities and pulling back by ι, we deduce that W has modulus $D_1 \boxtimes D_2$, as desired.

$\square$

Acknowledgements The author would like to thank the organizers of the International Colloquium 2016 on K-theory at the Tata Institute of Fundamental Research for their kind invitation and hospitality. He also wishes to thank the referee for numerous suggestions that improved this paper. The author was partially supported by the National Research Foundation of Korea grant (NRF 2015R1A2A2A01004120) funded by the Korean government (MSIP).

References

[1] A. Beilinson, *Higher regulators of modular curves*, in Applications of algebraic K-theory to algebraic geometry and number theory, Part I, II (Boulder, CO, 1983), 1–34, Contemp. Math. **55** (1986), Amer. Math. Soc.

[2] P. Berthelot, *Cohomologie cristalline des schémas de caractéristique $p > 0$*, Lecture Notes in Math. **407** 604, (1974), Springer-Verlag.

[3] F. Binda and S. Saito, *Relative cycles with moduli and regulator maps*, preprint arXiv:1412.0385, (2014), to appear in J. Inst. Math. Jussieu.

[4] S. Bloch, *Algebraic K-theory and crystalline cohomology*, Publ. Math. Inst. Haut. Étud. Sci. **47** (1977), 187–268.

[5] S. Bloch, *Algebraic cycles and higher K-theory*, Adv. Math. **61** (1986), 267–304.

[6] S. Bloch and H. Esnault, *The additive dilogarithm*, Doc. Math. **Extra Vol.** Kazuya Kato's fiftieth birthday, (2003), 131–155.

[7] W. Fulton, *Intersection theory*, 2nd Edition, Ergeb. Math. Grenzgebiete, **2**, Springer, 1998.

[8] L. Hesselholt and I. Madsen, *On the K-theory of nilpotent endomorphisms*, in Homotopy methods in Algebraic Topology (Bounder, CO, 1999), edited by Greenlees, J. P. C. et al., 127–140. Contemp. Math. **271**, American Math. Soc. 2001.

[9] L. Hesselholt, *The big de Rham-Witt complex*, Acta Math. **214** (2015), 135–207.

[10] L. Illusie, *Complexe de de Rham-Witt et cohomologie cristalline*, Ann Sci. de l'Ecole Norm. Sup. 4e série, **12** (1979), 501–661.

[11] W. Kai, *A moving lemma for algebraic cycles with modulus and contravariance*, arXiv:1507.07619v1, (2015).

[12] A. Krishna and M. Levine, *Additive higher Chow groups of schemes*, J. Reine Angew. Math. **619** (2008), 75–140.

[13] A. Krishna and J. Park, *Moving lemma for additive higher Chow groups*, Algebra Number Theory **6** (2012), 293–326.

[14] A. Krishna and J. Park, *Mixed motives over $k[t]/(t^{m+1})$*, J. Inst. Math. Jussieu **11** (2012), 611–657.

[15] A. Krishna and J. Park, *Algebraic cobordism theory attached to algebraic equivalence*, J. K-theory **11** (2013), no. 1, 73–112.

[16] A. Krishna and J. Park, *DGA-structure on additive higher Chow groups*, Int. Math. Res. Not. **Vol 2015** (2015), no. 1, 1-54. doi: 10.1093/imrn/rnt193

[17] A. Krishna and J. Park, *Semitopologization in motivic homotopy theory and applications*, Alg. Geom. Topol. **15** (2015), no. 2, 823–861.

[18] A. Krishna and J. Park, *On additive higher Chow groups of affine schemes*, Doc. Math. **21** (2016), 49–89.

[19] A. Krishna and J. Park, *Algebraic cycles and crystalline cohomology*, preprint arXiv:1504.08181v4, (2015).

[20] M. Lazard, *Sur les groupes de Lie formels à un paramètre*, Bull. Soc. Math. France **83** (1955), 251–274.

[21] M. Levine and F. Morel, *Algebraic Cobordism*, Springer Monographs Math., Springer, 2007. xii+244 pp.

[22] M. Levine and R. Pandharipande, *Algebraic cobordism revisited*, Invent. Math. **176** (2009), 63–130.

[23] J. Park, *Algebraic cycles and additive dilogarithm*, Int. Math. Res. Not., **2007**, no. 18, Art. ID rnm067, 19pp. doi:10.1093/imrn/rnm067

[24] J. Park, *Regulators on additive higher Chow groups*, Amer. J. Math. **131** (2009), 257–276.

[25] I. Patchkoria, *Cubical approach to derived functors*, Homology, Homotopy, Appl. **14** (2012), no. 1, 133–158.

[26] G. Quick, *Stable étale realization and étale cobordism*, Adv. Math. **214** (2007), 730–760.

[27] G. Quick, *Torsion algebraic cycles and étale cobordism*, Adv. Math. **227** (2011), 962–985.

[28] D. Quillen, *Elementary proofs of some results of cobordism theory using Steenrod operations*, Adv. Math. **7** (1971), 29–56.

[29] D. Quillen, *Higher Algebraic K-theory. I*, in Algebraic K-theory, I: Higher K-theories (Proc. Conf. Battelle Memorial Inst., Seattle, WA, 1972), pp. 85–147. Lect. Note. Math. **341**, Springer, 1973.

[30] K. Rülling, *The generalized de Rham-Witt complex over a field is a complex of zero-cycles*, J. Algebraic Geom. **16** (2007), 109–169.

[31] W. van der Kallen, *Le K_2 des nombres duaux*, C. R. Acad. Sci. Paris Sér. A–B **273** (1971), A1204–A1207.

DEPARTMENT OF MATHEMATICAL SCIENCES, KAIST, 291 DAEHAK-RO YUSEONG-GU, DAEJEON, 34141, REPUBLIC OF KOREA (SOUTH)
E-mail: jinhyun@mathsci.kaist.ac.kr; jinhyun@kaist.edu

K-Theory
Copyright ©2018 Tata Institute of Fundamental Research
Publisher: Hindustan Book Agency, New Delhi, India

The relation between Grothendieck duality and Hochschild homology

Amnon Neeman[1]

Abstract

Grothendieck duality goes back to 1958, to the talk at the ICM in Edinburgh [13] announcing the result. Hochschild homology is even older, its roots can be traced back to the 1945 article [16]. The fact that the two might be related is relatively recent. The first hint of a relationship came in 1987 in Lipman [20], and another was found in 1997 in Van den Bergh [28]. Each of these discoveries was interesting and had an impact, Lipman's mostly by giving another approach to the computations and Van den Bergh's especially on the development of non-commutative versions of the subject. However in this survey we will almost entirely focus on a third, much more recent connection, discovered in 2008 by Avramov and Iyengar [2] and later developed and extended in several papers, see for example [3, 19].

There are two classical paths to the foundations of Grothendieck duality, one following Grothendieck and Hartshorne [15] and (much later) Conrad [10], and the other following Deligne [11], Verdier [29] and (much later) Lipman [21]. The accepted view is that each of these has its drawbacks: the first approach (of Grothendieck, Hartshorne and Conrad) is complicated and messy to set up, while the second (of Deligne, Verdier and Lipman) might be cleaner to present but leads to a theory where it's not obvious how to compute anything.

The point of this article is that the recently-discovered connection with Hochschild homology and cohomology (the one due to Avramov and Iyengar) changes this. It renders clearly superior the highbrow appoach to the subject, the one due to Deligne, Verdier and Lipman. Not only is it (relatively) easy to set up the machinery, the computations also become transparent. And in the process we learn that Grothendieck duality is not really about residues of meromorphic differential forms, it is about the local cohomology of the Hochschild homology. By a fortuitous accident, if $f : X \longrightarrow Y$ is a smooth map then the top Hochschild homology happens to be isomorphic to the

---

[1]The research was partly supported by the Australian Research Council

relative canonical bundle, and its top local cohomology is represented by meromorphic differential forms. This is the reason that, as long as we stick to smooth maps, what comes up is residues of meromorphic forms. For non-smooth, flat maps it's Hochschild homology and maps from it that we need to study.

0 Introduction

Let $f : X \longrightarrow Y$ be a morphism of noetherian schemes. At the level of derived categories there exist natural functors $\mathbf{L}f^* : \mathbf{D}_{\mathbf{qc}}(Y) \longrightarrow \mathbf{D}_{\mathbf{qc}}(X)$, its right adjoint $\mathbf{R}f_* : \mathbf{D}_{\mathbf{qc}}(X) \longrightarrow \mathbf{D}_{\mathbf{qc}}(Y)$, as well as a right adjoint for $\mathbf{R}f_*$, nowadays (following Lipman) denoted $f^\times : \mathbf{D}_{\mathbf{qc}}(Y) \longrightarrow \mathbf{D}_{\mathbf{qc}}(X)$. For general f the functor $f^\times$ can be dreadful — it can take a bounded complex of coherent sheaves, that is an object in $\mathbf{D}^b_{\mathbf{coh}}(Y) \subset \mathbf{D}_{\mathbf{qc}}(Y)$, to a truly enormous object in $\mathbf{D}_{\mathbf{qc}}(X)$. This functor $f^\times$ only behaves well under strong restrictions, the usual being that f be proper.

To remedy this one introduces a better-behaved functor $f^!$. If f is proper then $f^! = f^\times$, but for general f one traditionally does some finicky manipulations to arrive at $f^!$. And, until very recently, the recipe worked only for cohomologically bounded-below complexes. That is $f^!$ has always been viewed as a functor $f^! : \mathbf{D}^+_{\mathbf{qc}}(Y) \longrightarrow \mathbf{D}^+_{\mathbf{qc}}(X)$.

Against this background came the striking work of Avramov, Iyengar, Lipman and Nayak, see [2, 3], relating Grothendieck duality with Hochschild homology and cohomology. To give the flavor of the results let me present just one formula, and for simplicity let me give only the affine version. Suppose therefore that $X = \mathrm{Spec}(S)$, $Y = \mathrm{Spec}(R)$, assume that R and S are noetherian, and that $f : X \longrightarrow Y$ is a flat, finite-type map. In an abuse of notation we will write $f : R \longrightarrow S$ for the induced ring homomorphism, and also identify $\mathbf{D}(R) \cong \mathbf{D}_{\mathbf{qc}}(Y)$ and $\mathbf{D}(S) = \mathbf{D}_{\mathbf{qc}}(X)$. Let $S^{\mathbf{e}} = S \otimes_R S$ be the enveloping algebra. Then, for any object $N \in \mathbf{D}^+_{\mathbf{qc}}(Y) = \mathbf{D}^+(R)$, we have a canonical isomorphism

$$f^! N \cong S \otimes_{S^{\mathbf{e}}} \mathrm{Hom}_R(S, S \otimes_R N) \ .$$

In this formula the tensor products and the Hom are all derived.

The reader might find it interesting to note that, in the special case where $f : R \longrightarrow S$ is finite and étale, we recover the classical formula

$$f^! N \cong \mathrm{Hom}_R(S, N) \cong S \otimes_{S^{\mathbf{e}}} \mathrm{Hom}_R(S, S \otimes_R N) \ .$$

Of course for finite, étale maps the Homs and tensors are underived. We will revisit étale maps (not necessarily finite) in Remark 2.8.

Perhaps one needs some familiarity with the classical literature to appreciate how striking this is — assuming only that f is flat we have produced a formula for $f^!$, which took a mere paragraph to state, and is clearly free of auxiliary choices and functorial. And although the left-hand-side was defined on the assumption that N is bounded below — after all we only knew $f^!$ on the bounded-below derived category — the right-hand-side makes sense for any N. In fact the formula tells us the surprising fact that if N is an object in $\mathbf{D}^+(R)$ then $S \otimes_{S^e} \mathrm{Hom}_R(S, S \otimes_R N)$ must belong to $\mathbf{D}^+(S)$. We know, from the complicated classical construction, that $f^!$ takes $\mathbf{D}^+(R)$ to $\mathbf{D}^+(S)$, but S is not of finite Tor-dimension over S^e and we have no reason to expect an expression of the form $S \otimes_{S^e} M$ to be bounded below. The derived tensor product tends to introduce lots of negative cohomology.

In joint work with Iyengar and Lipman we revisited these results, and along the way developed a useful new natural transformation $\psi(f) : f^\times \longrightarrow f^!$, see [19]. Hints of ψ may be found in Lipman [21, Exercise 4.2.3(d)], but without the naturality properties that make it so valuable. With all these unexpected new tools it was becoming clear that the time may have come to revisit the foundations of Grothendieck duality. In this article we sketch what has come out of this.

Finally we should tell the reader the structure of this survey. The early sections, §2 and §3, survey recent results that can be found elsewhere in the literature. The results are new, meaning new in this generality — there are older avatars, what's unusual here is that the theory is developed in the unbounded derived category. The results might be innovative but we still omit the proofs. With the exception of Proposition 3.3, where the argument is included, the proofs are all to be found in recent preprints available electronically.

In §4 and §5 this changes. Special cases of the results are known, with what turn out to be artificial boundedness restrictions. We give a general treatment — both to show that the results are true more generally, and to illustrate the power of the new techniques. Because of this our treatment is complete, with proofs. The reader interested in the highlights is advised to read the statement (not proof) of Lemma 4.3, as well as Corollary 5.7 and Example 5.8.

The final sections, §6 and §7, are again "soft", with no proofs presented. They review the history and suggest open problems.

1 Conventions

In this article we consider schemes X and the corresponding derived categories $\mathbf{D}_{\mathbf{qc}}(X)$, whose objects are complexes of sheaves of $\mathcal{O}_X$–modules with quasicoherent cohomology. Since abelian categories never come up, whenever there is a possible ambiguity our functors should be assumed derived — thus we will write f^* for $\mathbf{L}f^*$, f_* for $\mathbf{R}f_*$, Hom for RHom and $\otimes$ for the derived tensor product $\otimes^{\mathbf{L}}$. For simplicity, in §2, §3, §4, §5 and §6 we will assume that our schemes are noetherian and morphisms of schemes are separated and of finite type — occasionally, but not always, we will explicitly remind the reader of these standing assumptions. Unless we specifically say otherwise all derived categories will be unbounded. For a morphism of schemes $f : X \longrightarrow Y$ we let $f^* \dashv f_* \dashv f^\times$ be the adjoint functors which, back in §0, we referred to as $\mathbf{L}f^* \dashv \mathbf{R}f_* \dashv f^\times$.

2   The formal theory

In this section and the next we sketch the current state of the formal theory, without worrying about who proved what and when.

Let $f : X \to Y$ be a morphism of schemes. The functor $f^* : \mathbf{D_{qc}}(Y) \to \mathbf{D_{qc}}(X)$ is a strict monoidal functor, meaning it respects the tensor product. Therefore for any pair of objects $E \in \mathbf{D_{qc}}(Y)$ and $F \in \mathbf{D_{qc}}(X)$ we have a natural map

$$f^*[E \otimes f_*F] \xrightarrow{\ \sim\ } f^*E \otimes f^*f_*F \xrightarrow{\ \mathrm{id}\otimes\varepsilon\ } f^*E \otimes F$$

where the first map is the natural isomorphism, and $\varepsilon : f^*f_* \longrightarrow \mathrm{id}$ is the counit of the adjunction $f^* \dashv f_*$. By adjunction we obtain a natural map $p(E,F) : E \otimes f_*F \longrightarrow f_*(f^*E \otimes F)$. The map $p(E,F)$ is known to be an isomorphism, usually called the *projection formula*. This leads us to

Definition 2.1 Let $f \cdot X \longrightarrow Y$ be a morphism of schemes and let E, F be objects in $\mathbf{D_{qc}}(Y)$. The map $\chi(f,E,F) : f^*E \otimes f^\times F \longrightarrow f^\times(E \otimes F)$ is defined by applying the adjunction $f_* \dashv f^\times$ to the composite

$$f_*(f^*E \otimes f^\times F) \xrightarrow{\ p(E,f^\times F)^{-1}\ } E \otimes f_*f^\times F \xrightarrow{\ \mathrm{id}\otimes\varepsilon'\ } E \otimes F \ ,$$

where the first map is the inverse of the isomorphism in the projection formula, while $\varepsilon' : f_*f^\times \longrightarrow \mathrm{id}$ is the counit of the adjunction $f_* \dashv f^\times$.

The first result in the theory is

Theorem 2.2 *The map $\chi(f,E,F)$ is an isomorphism whenever*

(i) *f is arbitrary, but E is a perfect complex.*

(ii) *E and F are arbitrary, but f is proper and of finite Tor-dimension.*

Next recall the base-change maps. Given a commutative square of morphisms of schemes

$$
\begin{array}{ccc}
W & \xrightarrow{\;u\;} & X \\
{\scriptstyle f}\downarrow & & \downarrow{\scriptstyle g} \\
Y & \xrightarrow{\;v\;} & Z
\end{array}
$$

there is a canonical isomorphism of functors $\alpha : f^*v^* \longrightarrow u^*g^*$. Consider the composite

$$
f^*v^*g_* \xrightarrow{\;\alpha g_*\;} u^*g^*g_* \xrightarrow{\;u^*\varepsilon\;} u^* \;,
$$

where $\varepsilon : g^*g_* \longrightarrow \mathrm{id}$ is the counit of the adjunction $g^* \dashv g_*$. Adjunction gives us a base-change map $\beta : v^*g_* \longrightarrow f_*u^*$; this map is not always an isomorphism, but there are important situations in which it is. This leads us to

Definition 2.3 Assume we are in a situation where the base-change map $\beta : v^*g_* \longrightarrow f_*u^*$ is an isomorphism; for this article the important case where this happens is when the square

$$
\begin{array}{ccc}
W & \xrightarrow{\;u\;} & X \\
{\scriptstyle f}\downarrow & & \downarrow{\scriptstyle g} \\
Y & \xrightarrow{\;v\;} & Z
\end{array}
$$

is cartesian and the map v is flat. In this scenario consider the composite

$$
f_*u^*g^\times \xrightarrow{\;\beta^{-1}g^\times\;} v^*g_*g^\times \xrightarrow{\;v^*\varepsilon'\;} v^*
$$

where the first map is the inverse of the isomorphism β while $\varepsilon' : g_*g^\times \longrightarrow \mathrm{id}$ is the counit of the adjunction $g_* \dashv g^\times$. The (second) base change map $\Phi : u^*g^\times \longrightarrow f^\times v^*$ corresponds to this composite under the adjunction $f_* \dashv f^\times$.

One can wonder when the base-change map Φ is an isomorphism. The best result to date says

Theorem 2.4 *Let the notation be as in the case of Definition 2.3 which interests us in this article — that is we assume the square cartesian and v flat. Let E be an object in $\mathbf{D}_{\mathrm{qc}}(Z)$. Then the base-change map $\Phi(E) : u^*g^\times(E) \longrightarrow f^\times v^*(E)$ is an isomorphism provided g is proper and one of the conditions below holds:*

(i) *E belongs to $\mathbf{D}_{\mathrm{qc}}^+(Z) \subset \mathbf{D}_{\mathrm{qc}}(Z)$.*

(ii) $E \in \mathbf{D_{qc}}(Z)$ is arbitrary, but the map $f : W \longrightarrow Y$ is of finite Tor-dimension.

Now one proceeds as follows: given any morphism $f : X \longrightarrow Y$ we factor it as $X \xrightarrow{u} \overline{X} \xrightarrow{p} Y$ with u an open immersion and p proper, and then define $f^! : \mathbf{D_{qc}}(Y) \longrightarrow \mathbf{D_{qc}}(X)$ by the formula $f^! = u^* p^\times$. One of the consequences of Theorem 2.4 is that $f^!$ is well-defined, meaning that it is canonically independent of the choice of factorization. And we have the following theorem.

Theorem 2.5 *The assignment, taking a morphism of schemes $f : X \longrightarrow Y$ to the functor $f^! : \mathbf{D_{qc}}(Y) \longrightarrow \mathbf{D_{qc}}(X)$, satisfies a long list of compatibility properties. We list some highlights.*

2.5.1 *Let $X \xrightarrow{f} Y \xrightarrow{g} Z$ be composable morphisms of schemes. There is a map $\rho(f,g) : (gf)^! \longrightarrow f^! g^!$, which has the property that the two ways of using ρ to go from $(hgf)^!$ to $f^! g^! h^!$ are equal.*

2.5.2 *The two functors $f^\times, f^! : \mathbf{D_{qc}}(Y) \longrightarrow \mathbf{D_{qc}}(X)$ are related by a natural transformation $\psi(f) : f^\times \longrightarrow f^!$. The ψ is compatible with composition, in the obvious sense that the square below commutes*

$$
\begin{array}{ccc}
(gf)^\times & \xrightarrow{\delta(f,g)} & f^\times g^\times \\
{\scriptstyle \psi(gf)} \downarrow & & \downarrow {\scriptstyle \psi(f)\psi(g)} \\
(gf)^! & \xrightarrow{\rho(f,g)} & f^! g^!
\end{array}
$$

where $\rho(f,g)$ is the map of 2.5.1, while $\delta(f,g) : (gf)^\times \longrightarrow f^\times g^\times$ is the canonical isomorphism.

2.5.3 *The map $\rho(f,g)$ is an isomorphism if f is of finite Tor-dimension or if either gf or g is proper. The map $\psi(f)$ is an isomorphism whenever f is proper.*

2.5.4 *Given a pair of object $E, F \in \mathbf{D_{qc}}(Y)$ then there is a way to mimick the construction in Definition 2.1 with $f^!$ in place of $f^\times$. More precisely: there is a map $\sigma(f,E,F) : f^* E \otimes f^! F \longrightarrow f^!(E \otimes F)$ so that the natural square commutes*

$$
\begin{array}{ccc}
f^* E \otimes f^\times F & \xrightarrow{\chi(f,E,F)} & f^\times(E \otimes F) \\
{\scriptstyle \mathrm{id} \otimes \psi(f)} \downarrow & & \downarrow {\scriptstyle \psi(f)} \\
f^* E \otimes f^! F & \xrightarrow{\sigma(f,E,F)} & f^!(E \otimes F)
\end{array}
$$

Furthermore we have the analog of Theorem 2.2, that is $\sigma(f, E, F)$ *is an isomorphism if one of the conditions below holds*

(i) f *is arbitrary, but* E *is a perfect complex.*

(ii) E *and* F *are arbitrary, but* f *is of finite Tor-dimension.*

2.5.5 *The base-change map* Φ *of Definition 2.3 also has an* $(-)^!$ *analog. Precisely: given a cartesian square as in Definition 2.3, there is a base-change map* $\theta : u^* g^! \longrightarrow f^! v^*$.

2.5.6 *There is an analog of Theorem 2.4 for* $(-)^!$ *in place of* $(-)^\times$. *Precisely: the map* $\theta(E) : u^* g^!(E) \longrightarrow f^! v^*(E)$ *is an isomorphism as long as one of the following holds*

(i) E *belongs to* $\mathbf{D}_{\mathbf{qc}}^+(Z) \subset \mathbf{D}_{\mathbf{qc}}(Z)$.

(ii) $E \subset \mathbf{D}_{\mathbf{qc}}(Z)$ *is arbitrary, but the map* $f : W \longrightarrow Y$ *is of finite Tor-dimension.*

The full list of compatibility properties is quite long, and in any case it is clearer and more compact to present it in a 2-category formulation. For this paper we content ourselves with what's in Theorem 2.5.

Remark 2.6 In the introduction we mentioned that people have traditionally preferred $f^!$ to $f^\times$ because it is "better behaved". Theorem 2.5 allows us to make this more precise. If we compare 2.5.4 with Theorem 2.2 we see that

(i) If f is proper then $\sigma(f, E, F)$ and $\chi(f, E, F)$ agree up to canonical isomorphism. To see this observe that, when f is proper, then the vertical maps in the commutative square of 2.5.4 are isomorphisms by 2.5.3.

(ii) The maps $\sigma(f, E, F)$ and $\chi(f, E, F)$ are defined for every triple f, E, F, but $\sigma(f, E, F)$ is an isomorphism more often. If E is perfect then both are isomorphisms. But for non-perfect E the result 2.5.4(ii) says that $\sigma(f, E, F)$ is an isomorphism whenever f is of finite Tor-dimension, whereas Theorem 2.2(ii) guarantees that $\chi(f, E, F)$ is an isomorphism only if f is proper as well as of finite Tor dimension.

The same pattern repeats itself for the base-change maps Φ and θ. They are defined for every cartesian square with flat horizontal morphisms, and coincide if the vertical maps are proper — if our Theorem 2.5 were less pared down this could be shown to follow from the general structure, the reader can see the introduction to [23] for the fullblown formalism. But if we ask ourselves when Φ and θ induce isomorphisms, the conditions on θ are less restrictive than on Φ. Precisely: when we compare Theorem 2.4 with 2.5.6 we discover

(iii) Assume E is bounded below. Then $\Phi(E) : u^* g^\times E \longrightarrow f^\times v^* E$ is an isomorphism if g is proper, while $\theta(E) : u^* g^! E \longrightarrow f^! v^* E$ is an isomorphism unconditionally.

(iv) Let E be arbitrary. Then $\Phi(E)$ is an isomorphism as long as g is proper and f is of finite Tor-dimension, while $\theta(E)$ is an isomorphism whenever f is of finite Tor-dimension (no need for any properness).

Remark 2.7 If $f : X \longrightarrow Y$ is an open immersion then the square

$$
\begin{array}{ccc}
X & \xrightarrow{\ \mathrm{id}\ } & X \\
{\scriptstyle \mathrm{id}}\downarrow & & \downarrow{\scriptstyle f} \\
X & \xrightarrow{\ f\ } & Y
\end{array}
$$

is cartesian. By 2.5.6(ii) we have that $f^! = \mathrm{id}^* f^! \xrightarrow{\ \theta\ } \mathrm{id}^! f^* = f^*$ is an isomorphism. Given any morphism $g : Y \longrightarrow Z$, the map $\rho(f,g) : (gf)^! \longrightarrow f^! g^!$ is an isomorphism by 2.5.3 — after all the open immersion f is of finite Tor-dimension. Combining these isomorphisms gives

(i) If $X \xrightarrow{\ f\ } Y \xrightarrow{\ g\ } Z$ are composable morphisms of schemes, with f an open immersion, then we have a canonical isomorphism $(gf)^! \xrightarrow{\ \rho\ } f^! g^! \xrightarrow{\ \theta\ } f^* g^!$.

If $g : Y \longrightarrow Z$ happens to be a proper morphism then $\psi(g) : g^\times \longrightarrow g^!$ is an isomorphism by 2.5.3, which we may combine with the isomorphism of (i) to deduce a canonical isomorphism $(gf)^! \cong f^* g^! \cong f^* g^\times$. The compatibilities of Theorem 2.5 force upon us the formula for $(gf)^!$. That is: any time we can factor a map $X \longrightarrow Z$ as a composite $X \xrightarrow{\ f\ } Y \xrightarrow{\ g\ } Z$, with f an open immersion and g proper, then $(gf)^!$ must be given by the formula $(gf)^! \cong f^* g^\times$.

Remark 2.8 In passing we observe that the formula of Remark 2.7(i) generalizes, we need only assume f étale. Suppose $f : X \longrightarrow Y$ is an étale morphism of noetherian schemes. Consider the following diagram

$$
\begin{array}{ccccc}
X & \xrightarrow{\ \Delta\ } & X \times_Y X & \xrightarrow{\ \pi_1\ } & X \\
& & {\scriptstyle \pi_2}\downarrow & & \downarrow{\scriptstyle f} \\
& & X & \xrightarrow{\ f\ } & Y
\end{array}
$$

Now we will apply our lemmas to the situation of Construction 3.2. We remind the reader: $f : X \longrightarrow Y$ is a finite-type, flat morphism of noetherian schemes, and we formed the diagram

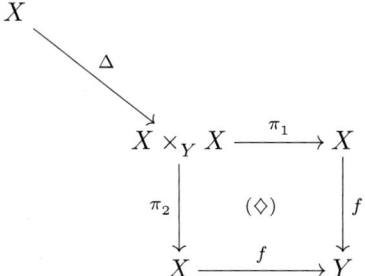

where the square is cartesian, π_1 and π_2 are the first and second projections, and $\Delta : X \longrightarrow X \times_Y X$ is the diagonal inclusion. We assert:

Proposition 5.6 *Let $A, B, C \in \mathbf{D_{qc}}(X)$ be objects and assume that A is f-perfect. Applying Construction 5.4 to the morphism $\pi_2 : X \times_Y X \longrightarrow X$ and the objects $\pi_1^* A \in \mathbf{D_{qc}}(X \times_Y X)$ and $B, C \in \mathbf{D_{qc}}(X)$, we obtain morphisms*

$$\mathcal{H}om(\pi_1^* A, \pi_2^\times B) \otimes \pi_2^* C \xrightarrow{\mathcal{H}om(\pi_1^* A, \chi) \circ \alpha} \mathcal{H}om\left[\pi_1^* A, \pi_2^\times (B \otimes C)\right]$$

$$\pi_1^* A \otimes \pi_2^* \mathcal{H}om(B, C) \xrightarrow{\beta \circ [\mathrm{id} \otimes \gamma]} \mathcal{H}om\left[\mathcal{H}om(\pi_1^* A, \pi_2^\times B), \pi_2^\times C\right].$$

We assert that, if $L \in \mathbf{D_{qc}}(X \times_Y X)$ is any object supported on the diagonal, then the functors $L \otimes (-)$ and $\mathcal{H}om(L, -)$ take both maps to isomorphisms, as do the functors Δ^* and $\Delta^\times$.

Proof Let us first observe that the statement about Δ^* and $\Delta^\times$ follows from the assertion about $L \otimes (-)$ and $\mathcal{H}om(L, -)$. Since Δ is a closed immersion the functor Δ_* is conservative, and to show that Δ^* and $\Delta^\times$ take the two maps to isomorphisms is equivalent to showing that the composites $\Delta_* \Delta^*$ and $\Delta_* \Delta^\times$ take them to isomorphisms. But it is standard that $\Delta_* \Delta^*(-) \cong \Delta_* \mathcal{O}_X \otimes (-)$ and $\Delta_* \Delta^\times(-) \cong \mathcal{H}om(\Delta_* \mathcal{O}_X, -)$. Since $\Delta_* \mathcal{O}_X$ is supported on the diagonal this reduces us to the statements about $L \otimes (-)$ and $\mathcal{H}om(L, -)$.

Now let $\mathcal{L} \subset \mathbf{D_{qc}}(X \times_Y X)$ be the full subcategory of all objects L so that $L \otimes (-)$ and $\mathcal{H}om(L, -)$ take both maps of the Proposition to isomorphisms. Clearly $\mathcal{L}$ is a localizing subcategory. We wish to show that $\mathcal{L}$ contains the category $\mathbf{D_{qc, \Delta}}(X \times_Y X)$, that is the full subcategory of

$\mathbf{D_{qc}}(X \times_Y X)$ of objects supported on the diagonal. But the subcategory $\mathbf{D_{qc,\Delta}}(X \times_Y X)$ is generated by the objects inside it which are compact in the larger $\mathbf{D_{qc}}(X \times_Y X)$; this theorem was first proved in Thomason and Trobaugh [27], and for a more general, modern proof which works for sufficienty nice algebraic stacks the reader can see Hall and Rydh [14, Theorems A, B and 4.10(2)]. This means that any localizing subcategory, containing the compact objects K supported on the diagonal, will contain all of $\mathbf{D_{qc,\Delta}}(X \times_Y X)$. It therefore suffices to show that every compact K, supported on the diagonal, belongs to $\mathcal{L}$. Hence we let K be a compact object supported on the diagonal, and wish to show that $K \otimes (-)$ and $\mathcal{H}om(K, -) \cong K^\vee \otimes (-)$ take both maps in the Proposition to isomorphisms.

Now the object $A \in \mathbf{D_{qc}}(X)$ is assumed f–perfect, and flat basechange tells us that $\pi_1^* A$ is π_2–perfect. Consider the composable morphisms $X \xrightarrow{\Delta} X \times_Y X \xrightarrow{\pi_2} X$; because the composite id $= \pi_2 \Delta$ is proper we may apply Lemma 5.2, and because it is quasi-affine Lemma 5.5 also applies. More precisely: with this pair of composable morphisms apply Lemma 5.2 to the π_2–perfect object $\pi_1^* A \in \mathbf{D_{qc}}(X \times_Y X)$ and to the perfect complexes $K, K^\vee \in \mathbf{D_{qc}}(X \times_Y X)$ supported on the image of Δ, and we learn that $\pi_{2*}(\pi_1^* A \otimes K)$ and $\pi_{2*}(\pi_1^* A \otimes K^\vee)$ are perfect in $\mathbf{D_{qc}}(X)$. But then Lemma 5.5 allows us to conclude that $K \otimes (-)$ and $\mathcal{H}om(K, -) \cong K^\vee \otimes (-)$ take both morphisms of the Proposition to isomorphisms.

$\square$

Corollary 5.7 *Let $f : X \longrightarrow Y$ be a finite-type, flat map of noetherian schemes, and let the notation be as in Construction 3.2. For objects $A, C \in \mathbf{D_{qc}}(X)$ and $B \in \mathbf{D_{qc}}(Y)$, where A is f–perfect, we have isomorphisms*

$$\mathcal{H}om(A, f^! B) \otimes C \cong \Delta^* \mathcal{H}om\left[\pi_1^* A, \pi_2^\times (f^* B \otimes C)\right]$$

$$\Delta^\times\left[\pi_1^* A \otimes \pi_2^* \mathcal{H}om(f^* B, C)\right] \cong \mathcal{H}om\left[\mathcal{H}om(A, f^! B), C\right].$$

Proof The classical isomorphism $\Delta^\times \mathcal{H}om(E, F) \cong \mathcal{H}om(\Delta^* E, \Delta^\times F)$, coupled with the fact that $\Delta^\times \psi(\pi_2) : \Delta^\times \pi_2^\times \longrightarrow \Delta^\times \pi_2^!$ is an isomorphism by Lemma 3.1, tell us that for any $E \in \mathbf{D_{qc}}(X \times_Y X)$ and any $G \in \mathbf{D_{qc}}(X)$ the map $\Delta^\times \mathcal{H}om\left[E, \psi(\pi_2)\right] : \Delta^\times \mathcal{H}om(E, \pi_2^\times G) \longrightarrow \Delta^\times \mathcal{H}om(E, \pi_2^! G)$ is an isomorphism. By [19, Proposition A.3(ii)] we deduce that $\Delta^* \mathcal{H}om\left[E, \psi(\pi_2)\right] : \Delta^* \mathcal{H}om(E, \pi_2^\times G) \longrightarrow \Delta^* \mathcal{H}om(E, \pi_2^! G)$ is also an isomorphism.

Now we turn to the proof of the Corollary. With the notation of the Corollary, as a first step we prove

- There is a natural isomorphism

$$\Delta^* \left[\mathcal{H}om(\pi_1^* A, \pi_2^\times f^* B) \right] \cong \mathcal{H}om(A, f^! B).$$

The isomorphism of • comes from the following string of isomorphisms

$$
\begin{aligned}
\Delta^* \left[\mathcal{H}om(\pi_1^* A, \pi_2^\times f^* B) \right] &\cong \Delta^* \left[\mathcal{H}om(\pi_1^* A, \pi_2^! f^* B) \right] \\
&\cong \Delta^* \mathcal{H}om(\pi_1^* A, \pi_1^* f^! B) \\
&\cong \Delta^* \pi_1^* \mathcal{H}om(A, f^! B) \\
&\cong \mathcal{H}om(A, f^! B)
\end{aligned}
$$

The first isomorphism is because the functor $\Delta^* \mathcal{H}om(\pi_1^* A, -)$ takes the map $\psi(\pi_2) : \pi_2^\times f^* B \longrightarrow \pi_2^! f^* B$ to an isomorphism. The second isomorphism is because $\theta(\Diamond) : \pi_1^* f^! \longrightarrow \pi_2^! f^*$ is an isomorphism, see 2.5.6(ii). The third isomorphism in by Lemma 4.3, and the last isomorphism is because $\Delta^* \pi_1^* \cong \mathrm{id}^*$.

With the preliminaries out of the way, apply Proposition 5.6 to the objects $A, f^* B, C \in \mathbf{D_{qc}}(X)$, where A is given to be f–perfect. The Proposition tells us that the functors Δ^* and $\Delta^\times$ take the maps below to isomorphisms

$$\mathcal{H}om(\pi_1^* A, \pi_2^\times f^* B) \otimes \pi_2^* C \xrightarrow{\quad (1) \quad} \mathcal{H}om\left[\pi_1^* A, \pi_2^\times (f^* B \otimes C) \right]$$

$$\pi_1^* A \otimes \pi_2^* \mathcal{H}om(f^* B, C) \xrightarrow{\quad (2) \quad} \mathcal{H}om\left[\mathcal{H}om(\pi_1^* A, \pi_2^\times f^* B), \pi_2^\times C \right].$$

And the first isomorphism of the Corollary is by applying Δ^* to the map (1) while the second isomorphism is by applying $\Delta^\times$ to the map (2). Let us take these one step at a time, we begin be applying Δ^* to (1). We obtain isomorphisms

$$
\begin{aligned}
\Delta^* \mathcal{H}om\left[\pi_1^* A, \pi_2^\times (f^* B \otimes C) \right] &\cong \Delta^* \left[\mathcal{H}om(\pi_1^* A, \pi_2^\times f^* B) \otimes \pi_2^* C \right] \\
&\cong \Delta^* \left[\mathcal{H}om(\pi_1^* A, \pi_2^\times f^* B) \right] \otimes [\Delta^* \pi_2^* C] \\
&\cong \mathcal{H}om(A, f^! B) \otimes C
\end{aligned}
$$

The first isomorphism is just Δ^* applied to (1). The second isomorphism is because Δ^* respects the tensor product. The third isomorphism is the tensor product of the isomorphism in • with the isomorphism $\Delta^* \pi_2^* C \cong \mathrm{id}^* C = C$.

A similar analysis works for $\Delta^\times$ applied to the map (2), which gives us the first isomorphism below

$$\Delta^\times\Big[\pi_1^*A \otimes \pi_2^*\mathcal{H}om(f^*B, C)\Big] \;\cong\; \Delta^\times\mathcal{H}om\Big[\mathcal{H}om(\pi_1^*A, \pi_2^\times f^*B)\,,\, \pi_2^\times C\Big]$$

$$\cong\; \mathcal{H}om\Big[\Delta^*\mathcal{H}om(\pi_1^*A, \pi_2^\times f^*B)\,,\, \Delta^\times\pi_2^\times C\Big]$$

$$\cong\; \mathcal{H}om\Big[\mathcal{H}om(A, f^!B)\,,\, C\Big]$$

The second isomorphism comes from the formula

$$\Delta^\times\mathcal{H}om(E, F) \cong \mathcal{H}om(\Delta^*E, \Delta^\times F).$$

The third isomorphism is the functor $\mathcal{H}om(-, -)$ applied to the isomorphism in • and the isomorphism $\Delta^\times\pi_2^\times C \cong \mathrm{id}^\times C = C$.

$\square$

Example 5.8 Let us work out what Corollary 5.7 says in the affine case: that is $f : R \longrightarrow S$ will be a finite-type, flat homomorphism of noetherian rings and, by abuse of notation, we will also write $f : \mathrm{Spec}(S) \longrightarrow \mathrm{Spec}(R)$ for the induced map of noetherian schemes. We have the usual equivalences $\mathbf{D}(R) \cong \mathbf{D_{qc}}(\mathrm{Spec}(R))$ and $\mathbf{D}(S) \cong \mathbf{D_{qc}}(\mathrm{Spec}(S))$, and $f^* : \mathbf{D}(R) \longrightarrow \mathbf{D}(S)$, $f_* : \mathbf{D}(S) \longrightarrow \mathbf{D}(R)$, $f^\times : \mathbf{D}(R) \longrightarrow \mathbf{D}(S)$ and $f^! : \mathbf{D}(R) \longrightarrow \mathbf{D}(S)$ are the affine versions of the standard functors of Grothendieck duality. Put $S^{\mathrm{e}} = S \otimes_R S$. In this affine case, to say that an object $A \in \mathbf{D}(S)$ is f–perfect means that A must have bounded cohomology which is finite as S–modules, and $f_*A \in \mathbf{D}(R)$ has finite Tor-dimension. Let $A \in \mathbf{D}(S)$ be an f–perfect complex, and let $B \in \mathbf{D}(R)$ and $C \in \mathbf{D}(S)$ be arbitrary. Then the formulas of Corollary 5.7 come down to

$$\mathrm{Hom}_S(A, f^!B) \otimes_S C \;\cong\; S \otimes_{S^{\mathrm{e}}} \mathrm{Hom}_R(A, B \otimes_R C)\,,$$
$$\mathrm{Hom}_{S^{\mathrm{e}}}\big[S, A \otimes_R \mathrm{Hom}_R(B, C)\big] \;\cong\; \mathrm{Hom}_S\big[\mathrm{Hom}_S(A, f^!B), C\big]\,.$$

where the Homs and tensors are all derived. The reader can find special cases of these formulas in Avramov, Iyengar, Lipman and Nayak [3].

The reader might note that we have already met special cases of the first of these formulas. If we put $A = C = S$ then the formula specializes to

$$f^!B \;\cong\; \mathrm{Hom}_S(S, f^!B) \otimes_S S \;\cong\; S \otimes_{S^{\mathrm{e}}} \mathrm{Hom}_R(S, S \otimes_R B)$$

of the Introduction, and if we further specialize to $B = R$ we recover the formula $f^!R \cong S \otimes_{S^{\mathrm{e}}} \mathrm{Hom}_R(S, S)$ of Remark 3.7.

6   A historical review

Grothendieck first mentioned that he knew how to prove a relative version of the Serre duality theorem in his ICM talk in Edinburgh in 1958, see [13]. The first published version was Hartshorne [15]; roughly speaking the construction of $f^!$ given in [15] is by gluing local data, not an easy thing to do in the derived category. Three and a half decades later Conrad [10] expanded and filled in details missing in [15]. The presentation of the subject given here is entirely different in spirit — it is based on early observations by Deligne [11] and Verdier [29], filled in and expanded greatly in Lipman [21]. This second construction is much more global and functorial, the usual objection to it is that it's difficult to compute anything.

Now it is time to say what's different here from the classical literature. Let us begin with the observation that, until the late 1980s, no one really understood how to handle unbounded derived categories. For the first two decades of the subject the functor f^*, which involves a derived tensor product, was treated as a functor $f^* : \mathbf{D}^-_{\mathbf{qc}}(Y) \longrightarrow \mathbf{D}^-_{\mathbf{qc}}(X)$, while the functor f_*, which involves injective resolutions, was classically viewed as a functor $f_* : \mathbf{D}^+_{\mathbf{qc}}(X) \longrightarrow \mathbf{D}^+_{\mathbf{qc}}(Y)$. A careful reader will note that, being defined on different categories, these functors are not honest adjoints — there is no counit of adjunction $f^* f_* \longrightarrow \mathrm{id}$, and a classical version of the treatment of §2 would have had to be more delicate. Luckily for us we live in modern times and can give the clean presentation of the projection formula and the base-change maps of §2.

The article that brought modernity to this discipline was Spaltenstein[26], it taught us how to take injective and flat resolutions of unbounded complexes. Spaltenstein's article made it clear how to define the adjoint functors $f^* : \mathbf{D}_{\mathbf{qc}}(Y) \longrightarrow \mathbf{D}_{\mathbf{qc}}(X)$ and $f_* : \mathbf{D}_{\mathbf{qc}}(X) \longrightarrow \mathbf{D}_{\mathbf{qc}}(Y)$. The natural question to arise was how much of Grothendieck duality could be developed in the unbounded derived category. The existence of a right adjoint $f^\times : \mathbf{D}_{\mathbf{qc}}(Y) \longrightarrow \mathbf{D}_{\mathbf{qc}}(X)$ for f_* was discovered soon after, the author even showed in [24] that it is possible to obtain this adjoint easily and very formally using Brown representability. At the time the author was promoting the point of view that the right way to approach all these classical results was to employ systematically the techniques of homotopy theory, like Brown representability — at the time this was still a novel idea. So Lipman challenged the author to try to use the techniques of homotopy theory to extend Verdier's base-change theorem [29] to the unbounded derived category. Instead of a proof the author found a counterexample, see [24,

Example 6.5]. There exists a cartesian square of noetherian schemes

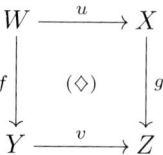

with v flat (even an open immersion) and g proper (even a closed immersion), and such that the base-change map $\Phi(\lozenge) : u^* g^\times \longrightarrow f^\times v^*$ is *not* an isomorphism. As an aside we note that the schemes in question are all affine.

This counterexample had the unfortunate effect of stifling the theory, for the next twenty years it put people off trying to develop the functor $f^!$ in the unbounded derived category. For example see Lipman's book [21] — Lipman makes a real effort to give the results in the greatest generality in which they were known at the time, and for the functor $f^!$ he works almost entirely with bounded-below complexes. Drinfeld and Gaitsgory [12] generalized a version of the theory to DG schemes, and if the structure sheaf has negative cohomology then the category $\mathbf{D}^+_{\mathbf{qc}}(X)$ does not make much sense. To finesse the issue they work in the category of Ind-coherent sheaves instead of the derived category.

In early 2013 I happened to run into Lipman at MSRI and he told me about exciting recent work, joint with Avramov, Iyengar and Nayak, which found a strange connection between Grothendieck duality and Hochschild homology and cohomology. In this survey we have already met this connection in Theorem 3.5 and Example 5.8, see also Remarks 3.6 and 3.7. Theorem 3.5 taught us about this bizarre new map from Hochschild homology to the dualizing complex $f^! \mathcal{O}_Y$, and when f is smooth and of relative dimension d this map happens to give an isomorphism of $f^! \mathcal{O}_Y$ with a shift of the relative canonical bundle. And in §5 we saw that the formulas of §3 are only the tip of the iceberg, there and many more weird and wonderful ones — we presented two of them, together with proofs, in Example 5.8. The formulas of Example 5.8 are not new, special cases may be found in [2, 3]. What was new in §5 is that we gave them as special cases of results that hold in the unbounded derived category. Back in 2013, when Lipman told me about the work, no one knew how to define $f^!$ on the unbounded derived category.

Let us observe more carefully the second formula of Example 5.8, and for simplicity let's put $B = R$. We remind the reader, the formula is

$$\mathrm{Hom}_{S^\bullet}(S, A \otimes_R C) \quad \cong \quad \mathrm{Hom}_S\big[\mathrm{Hom}_S(A, f^! R), C\big] \, .$$

If we fix A and consider the expression on the right as a functor in C then it is clearly representable — the right hand side has the form $\mathrm{Hom}_S(P, -)$, where P happens to be the expression $\mathrm{Hom}_S(A, f^! R)$. The isomorphism means that, as a functor in C, the expression $\mathrm{Hom}_{S^\bullet}(S, A \otimes_R C)$ is also representable, in particular it commutes with products — which is far from obvious. The challenge Lipman gave me was to try to use Brown representability to prove these formulas.

There is such a proof, and Iyengar and I are working on writing it up. But this survey is about another direction our research took: in trying to understand better these mysterious formulas we developed the natural transformation $\psi(f) : f^\times \longrightarrow f^!$ — early hints of it may be found in Lipman [21, Exercise 4.2.3(d)]. What was new were the naturality and functoriality properties of ψ, see [19] for some illustrations of their value. Because at the time $f^!$ was defined only on the bounded-below derived category our results imposed artificial boundedness restrictions, and it was a natural challenge to try to remove them. Working in the category of Ind-coherent sheaves, as in Drinfeld and Gaitsgory, is clearly wrong for this problem — the formulas of [3] live in the derived category. The article [23] was written to address this problem, in it Grothendieck duality is developed in the unbounded derived category, and we gave a brief summary of some of the results in §2. In §3 and §5 we gave illustrations of how one can approach the unbounded versions of the formulas of [2, 3, 19] using the techniques surveyed in this paper — we proved the formula $f^! = \Delta^* \pi_2^\times f^*$ for unbounded complexes in Proposition 3.3, while Example 5.8 showed us how to derive the reduction formulas of Avramov and Iyengar. These formulas occur in [2, 3, 19], but with unnatural boundedness hypotheses.

The reader might be puzzled. We mentioned that, twenty years ago, I produced a counterexample [24, Example 6.5] to the unbounded version of Verdier's base-change theorem. There exists a cartesian square of schemes

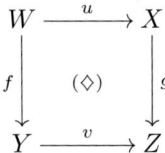

with v an open immersion and g proper, and such that the base-change map $\Phi(\Diamond) : u^* g^\times \longrightarrow f^\times v^*$ is not an isomorphism. So what has changed in two decades? What's new is Theorem 2.4(ii): it tells us that, as long as we further assume that f is *of finite Tor-dimension,* the problem goes away and $\Phi(\Diamond)$ is an isomorphism. When we compare two compactifications of

X we end up with cartesian squares of the form

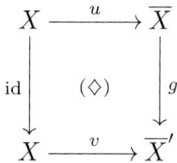

and the identity id $: X \longrightarrow X$ is of finite Tor-dimension. Thus the cartesian squares that come up in the proof that $f^! = u^* p^\times$ is independent of the factorization of $f : X \longrightarrow Y$ as $X \xrightarrow{u} \overline{X} \xrightarrow{p} Z$ all have base-change maps which are isomorphisms.

The place where the old counterexample rears its ugly head is when it comes to composition. The counterexample gave a commutative square (actually, even cartesian) and hence we have $gu = vf$. Now u and v are open immersions while f and g are proper, hence $u^! = u^*$, $v^! = v^*$, $f^! = f^\times$ and $g^! = g^\times$. On the other hand $u^! g^! = u^* g^\times$ is not isomorphic to $f^! v^! = f^\times v^*$. We have already mentioned that, in the old counterexample, the base-change map $u^* g^\times \longrightarrow f^\times v^*$ is not an isomorphism, but even more is true, the functors are not isomorphic. The 2–functor $(-)^!$ is genuinely only oplax — there are natural maps $\rho(f, v) : (vf)^! \longrightarrow f^! v^!$ and $\rho(u, g) : (gu)^! \longrightarrow u^! g^!$, but clearly they cannot both be isomorphisms. As it happens, in this particular example $\rho(u, g)$ is an isomorphism while $\rho(f, v)$ isn't.

The situation is not hopeless: 2.5.3 gives useful criteria for $\rho(f, g)$ to be an isomorphism, and in §3, §4 and §5 we illustrated how this can be applied to obtain unbounded versions of the results of [2, 3, 19]. The illustrations of §3 also showed how, with all these new methods, the abstract nonsense approach to the subject pioneered by Deligne, Verdier and Lipman can produce explicit computational formulas simply and easily. The technicalities are different: the "residual complexes" of Grothendieck are replaced by the more standard tools of Hochschild homology.

In passing let me note that Hochschild homology is a K–theoretic invariant, and its appearance raises the question whether more sophisticated K–theoretic invariants might shed even more light on the subject of Grothendieck duality. This is a volume on K–theory and its applications to algebraic geometry, and it seems the appropriate place to raise this question.

7   Generalizations

In March 2016 I received from the journal four referees' reports on my article [23]. Mostly the referees' comments were simple enough to address,

but there were two difficult issues. One referee wondered what happens if we relax the noetherian hypothesis, while another suggested that I develop the entire theory in the generality of stacks. This led me to think more carefully about these points. The noetherian hypothesis seems indispensable, at least for this approach to the theory — some of the lemmas have non-noetherian versions, but there is a point at which the argument runs into a brick wall without the noetherian assumption. But I'm happy to report that, under relatively mild hypotheses, everything generalizes to noetherian algebraic stacks.

In fact the stacky version is cleaner to state. Algebraic stacks naturally form a 2-category, as do triangulated categories. The clean way to think about the theory is to view $(-)^*$, $(-)^\times$ and $(-)^!$ as 2-functors from [suitably restricted] algebraic stacks to triangulated categories, with some relations among them. These relations can be phrased in terms of the existence of certain natural transformations relating these functors, and the assertion that certain pairs of composites of natural transformations agree. For example: it turns out that $(-)^*$ has the the structure of a monoid, meaning there is a pseudonatural transformation $(-)^* \times (-)^* \longrightarrow (-)^*$, and $(-)^\times$ and $(-)^!$ are oplax modules over it. The map ψ turns out to be an oplax natural transformation $\psi : (-)^\times \longrightarrow (-)^!$ which is a module homomorphism. Anyway: the reader can find a thorough discussion in the introduction to [23].

This led to an expository conundrum in writing the current survey — it was unclear what was the right generality for the results. Avramov, Iyengar, Lipman and Nayak work with noetherian schemes, but allow the morphisms to be essentially of finite type (rather than the more restrictive finite type), and sometimes of finite Tor-dimension (rather than flat). But the methods they use don't work for noetherian stacks — at least not yet — because one doesn't yet know that a morphism of noetherian stacks which is essentially of finite type has a Nagata compactification. Nayak [22] proved the existence of Nagata compactifications for morphisms of *noetherian schemes* essentially of finite type, but so far no one has generalized this even to algebraic spaces. In other words: the results in this paper generalize in more than one direction, and at present I do not know a common generalization that covers everything that can be proved by the methods.

The compromise I made was to present the arguments in the intersection of the known cases, that is finite-type, flat maps of noetherian schemes, and leave to the reader the various generalizations. But I did make an effort to give proofs that are easily adaptable, so that the extension to (for example) algebraic stacks is straightforward.

When I gave the talk at TIFR, which amounted to a brief summary of this survey, Geisser, Kahn, Saito and Weibel raised the question of what portion of the ideas might be transferrable to the six-functor formalism. Because shortly after giving the talk I received the referees' reports, with the questions about the non-noetherian and stacky versions of Grothendieck duality, I haven't yet had the opportunity to think about this other question. The six-functor situation is another place where one defines functors like $f^!$ using good factorizations of f, and it is eminently sensible to ask if there might be an analog of the fancy, unbounded version of the base-change theorem and of its consequences.

The question is natural enough and I would like to come back to it when I have more time. In the interim I record it for others to study.

References

[1] Leovigildo Alonso Tarrío, Ana Jeremías López, and Joseph Lipman, *Bivariance, Grothendieck duality and Hochschild homology, II: The fundamental class of a flat scheme-map*, Adv. Math. **257** (2014), 365–461.

[2] Luchezar L. Avramov and Srikanth B. Iyengar, *Gorenstein algebras and Hochschild cohomology*, Michigan Math. J. **57** (2008), 17–35, Special volume in honor of Melvin Hochster.

[3] Luchezar L. Avramov, Srikanth B. Iyengar, Joseph Lipman, and Suresh Nayak, *Reduction of derived Hochschild functors over commutative algebras and schemes*, Adv. Math. **223** (2010), no. 2, 735–772.

[4] Paul Balmer, Ivo Dell'Ambrogio, and Beren Sanders, *Grothendieck-Neeman duality and the Wirthmüller isomorphism*, Compositio Math. **152** (2016), no. 8, 1740–1776.

[5] David J. Benson, Srikanth B. Iyengar, and Henning Krause, *Local cohomology and support for triangulated categories*, Ann. Sci. École Norm. Sup. (4) **41** (2008), 573–619.

[6] ———, *Stratifying modular representations of finite groups*, Ann. of Math. (2) **174** (2011), no. 3, 1643–1684.

[7] ———, *Stratifying triangulated categories*, J. Topol. **4** (2011), no. 3, 641–666.

[8] ———, *Colocalizing subcategories and cosupport*, J. Reine Angew. Math. **673** (2012), 161–207.

[9] ———, *Module categories for group algebras over commutative rings*, J. K-Theory **11** (2013), no. 2, 297–329, With an appendix by Greg Stevenson.

[10] Brian Conrad, *Grothendieck duality and base change*, Lecture Notes in Mathematics, vol. 1750, Springer-Verlag, 2000.

[11] Pierre Deligne, *Cohomology à support propre en construction du foncteur $f^!$*, Residues and Duality, Lecture Notes in Mathematics, vol. 20, Springer–Verlag, 1966, pp. 404–421.

[12] Vladimir Drinfeld and Dennis Gaitsgory, *On some finiteness questions for algebraic stacks*, Geom. Funct. Anal. **23** (2013), no. 1, 149–294.

[13] Alexandre Grothendieck, *The cohomology theory of abstract algebraic varieties*, Proc. Internat. Congress Math. (Edinburgh, 1958), Cambridge Univ. Press, New York, 1960, pp. 103–118.

[14] Jack Hall and David Rydh, *Perfect complexes on algebraic stacks*, Compos. Math. **153** (2017), no. 11, 2318–2367.

[15] Robin Hartshorne, *Residues and duality*, Lecture notes of a seminar on the work of A. Grothendieck, given at Harvard 1963/64. With an appendix by P. Deligne. Lecture Notes in Mathematics, No. 20, Springer-Verlag, 1966.

[16] Gerhard P. Hochschild, *On the cohomology groups of an associative algebra*, Ann. of Math. (2) **46** (1945), 58–67.

[17] Reinhold Hübl and Pramathanath Sastry, *Regular differential forms and relative duality*, Amer. J. Math. **115** (1993), no. 4, 749–787.

[18] Luc Illusie, *Généralités sur les conditions de finitude dans les catégories dérivées*, Théorie des intersections et théorème de Riemann-Roch, Springer-Verlag, 1971, Séminaire de Géométrie Algébrique du Bois-Marie 1966–1967 (SGA 6, Exposé I), pp. 78–159. Lecture Notes in Mathematics, Vol. 225.

[19] Srikanth B. Iyengar, Joseph Lipman, and Amnon Neeman, *Relation between two twisted inverse image pseudofunctors in duality theory*, Compos. Math. **151** (2015), no. 4, 735–764.

[20] Joseph Lipman, *Residues and traces of differential forms via Hochschild homology*, Contemporary Mathematics, vol. 61, Ame. Math. Soc. 1987.

[21] _____ , *Notes on derived functors and Grothendieck duality*, Foundations of Grothendieck duality for diagrams of schemes, Lecture Notes in Mathematics, vol. 1960, Springer, 2009, pp. 1–259.

[22] Suresh Nayak, *Compactification for essentially finite-type maps*, Adv. Math. **222** (2009), no. 2, 527–546.

[23] Amnon Neeman, *An improvement on the base-change theorem and the functor $f^!$*, submitted.

[24] ———, *The Grothendieck duality theorem via Bousfield's techniques and Brown representability*, Jour. Amer. Math. Soc. **9** (1996), 205–236.

[25] ———, *Traces and residues*, Indiana Univ. Math. J. **64** (2015), no. 1, 217–229.

[26] Nicolas Spaltenstein, *Resolutions of unbounded complexes*, Compositio Math. **65** (1988), no. 2, 121–154.

[27] Robert W. Thomason and Thomas F. Trobaugh, *Higher algebraic K–theory of schemes and of derived categories*, The Grothendieck Festschrift (a collection of papers to honor Grothendieck's 60'th birthday), vol. 3, Birkhäuser, 1990, pp. 247–435.

[28] Michel Van den Bergh, *Existence theorems for dualizing complexes over non-commutative graded and filtered rings*, J. Algebra **195** (1997), no. 2, 662–679.

[29] Jean-Louis Verdier, *Base change for twisted inverse images of coherent sheaves*, vol. Collection: Algebraic Geometry, Tata Inst. Fund. Res., 1968, pp. 393–408.

AMNON NEEMAN, CENTRE FOR MATHEMATICS AND ITS APPLICATIONS, MATHEMATICAL SCIENCES INSTITUTE, JOHN DEDMAN BUILDING, THE AUSTRALIAN NATIONAL UNIVERSITY, CANBERRA, ACT 0200 AUSTRALIA

E-mail: `Amnon.Neeman@anu.edu.au`

K-Theory
Copyright ©2018 Tata Institute of Fundamental Research
Publisher: Hindustan Book Agency, New Delhi, India

Chow's moving lemma with modulus

Wataru Kai

Abstract

This is a report on the author's work about moving techniques of algebraic cycles with modulus. Joint work with R. Iwasa on Chern classes with modulus is also outlined. This article is aimed at a quick overview of the main ideas. Details will appear elsewhere.

1 Introduction

Bloch, Esnault [2] and Park [14] introduced additive higher Chow groups. It is expected to give a cycle-theoretic description of the K-groups of $X \otimes_k k[t]/(t^{m+1})$, up to small torsion, where X is a smooth variety over a field k. See Rülling's article [15] for an interesting computation of certain parts of the additive higher Chow group for the spectrum of a field.

By 2014, Binda and Saito [3] proceeded to introduce the higher Chow group with modulus $\mathrm{CH}^q(X|D, n)$, which may describe the relative K-groups $K_n(X, D)$ of a smooth variety X and an effective Cartier divisor D.

$$\bigoplus_q \mathrm{CH}^q(X|D, n) \longrightarrow \bigoplus_q \mathrm{CH}^q(X, n)$$

$$\left.\vphantom{\bigoplus}\right\downarrow ? \qquad\qquad \cong \Big| \text{ up to } \otimes \mathbb{Q}$$

$$\cdots \to K_{n+1}(D) \longrightarrow K_n(X, D) \longrightarrow K_n(X) \longrightarrow K_n(D) \to \cdots$$

One wishes to realize the dotted connection in the above diagram. Basically, in order to compare two things, it is often useful to establish parallel properties for both. In this article we consider the next property satisfied by the K-groups:

The groups $K_n(X, D)$ are contravariant with respect to commutative diagrams

$$
\begin{array}{ccc}
X' & \longrightarrow & X \\
\cup & & \cup \\
D' & \longrightarrow & D.
\end{array}
\tag{1.1}
$$

127

The group $\mathrm{CH}^q(X|D,n)$ is defined as the n-th homology group of the cycle complex with modulus $z^q(X|D,\bullet)$. The reader finds its definition in §2. In fact, our result does not say that the groups $\mathrm{CH}^q(X|D,n)$ are contravariant; instead, we assert that the *Nisnevich hypercohomology groups* of the complexes are contravariant. Denote by $z^q(-|D,\bullet)_{\mathrm{Nis}}$ the Nisnevich sheaf of complexes $(U \to X) \mapsto z^q(U|D\times_X U,\bullet)$. The fact that it is a sheaf follows e.g. from étale descent of closed subschemes. Let k be a fixed base field.

Theorem 1.1 ([7]) *Suppose we are given a diagram* (1.1) *in which* X, X' *are equidimensional k-schemes of finite type, D, D' are effective Cartier divisors and $X \setminus D$ is smooth. Let q be a non-negative integer and n be an integer. Then we can define a canonical pull-back map*

$$\mathbf{H}^n_{\mathrm{Nis}}(X, z^q(-|D,\bullet)_{\mathrm{Nis}}) \to \mathbf{H}^n_{\mathrm{Nis}}(X', z^q(-|D',\bullet)_{\mathrm{Nis}}).$$

The pull-back map will be defined using Moving Lemma 2.8 below.

During the writing of this article, Chern class maps $K_n(X,D) \to \mathbf{H}^{-n}_{\mathrm{Nis}}(X, z^i(-|D,\bullet)_{\mathrm{Nis}})$ have been constructed for $n \geq 0$ and $i \geq 1$ when $X \setminus D$ is smooth in joint work with R. IWASA. The contravariant functoriality gave a working platform; see §4.3.

2 Definitions and the statement of Moving Lemma

We write $\square^n := (\mathbb{P}^1 \setminus \{\infty\})^n$, and $F_n := \sum_i (\mathbb{P}^1)^{i-1} \times \{\infty\} \times (\mathbb{P}^1)^{n-i}$ (a Cartier divisor on $(\mathbb{P}^1)^n$). The letter X denotes an equidimensional k-scheme of finite type and D an effective Cartier divisor on it.

Definition 2.1 A closed subset V of $(X \setminus D) \times \square^n$ is said to satisfy the *modulus condition* if: (When $n = 0$) the closure $\overline{V}$ of V in X does not meet D (i.e. $\overline{V} = V$); (When $n > 0$) if $\overline{V}$ is the closure of V in $X \times (\mathbb{P}^1)^n$ and $\overline{V}^N$ is its normalization, the inequality of Cartier divisors $D|_{\overline{V}^N} \leq F_n|_{\overline{V}^N}$ holds, where $(-)|_{\overline{V}^N}$ denotes the pull-back of a divisor to $\overline{V}^N$.

Here, the normalization of a scheme is the disjoint union of the normalizations of its reduced irreducible components.

By definition, the *faces* of $\square^n$ are, if it has coordinates $t_1, \ldots, t_n$, the closed subschemes $\{t_i = 0\}, \{t_i = 1\}$ and their intersections. We denote by $\partial_{i,0}$ the embedding of the codimension 1 face $\{t_i = 0\}$,

$$\partial_{i,0} \colon \square^{n-1} \to \square^n, \ (t_1, \ldots, t_{n-1}) \mapsto (t_1, \ldots, \overset{i}{0}, t_i, \ldots, t_{n-1}),$$

and by $\partial_{i,1}$ the embedding of the face $\{t_i = 1\}$.

Definition 2.2 Let $\underline{z}^q(X|D, n)$ be the group of codimension q cycles on $(X \setminus D) \times \square^n$ whose irreducible components V satisfy:

- (*Face condition*) For every face F of $\square^n$, the two subsets V and $(X \setminus D) \times F$ of $(X \setminus D) \times \square^n$ intersect properly.

- V satisfies the modulus condition.

The association $n \mapsto \underline{z}^q(X|D, n)$ gives rise to a cubical abelian group. In other words, for all codimension 1 faces $\partial \colon \square^{n-1} \hookrightarrow \square^n$ and all degeneracy maps $s \colon \square^n \to \square^{n-1}$ (a map collapsing an axis), we have canonical pull-back maps $\partial^* \colon \underline{z}^q(X|D, n) \to \underline{z}^q(X|D, n-1)$ and $s^* \colon \underline{z}^q(X|D, n-1) \to \underline{z}^q(X|D, n)$ satisfying the relations one naturally expects.

The associated non-degenerate complex is denoted by $z^q(X|D, \bullet)$, so by definition the group $z^q(X|D, n)$ is the quotient of $\underline{z}^q(X|D, n)$ by the sum of subgroups $s^*\underline{z}^q(X|D, n-1)$ where s runs through the degeneracy maps $\square^n \to \square^{n-1}$. The differential map is given by $\sum_{i=1}^n (-1)^{i-1}(\partial_{i,1}^* - \partial_{i,0}^*)$.

By a general fact on cubical abelian groups, the complex $z^q(X|D, \bullet)$ is naturally a direct summand of the naively defined complex $\underline{z}^q(X|D, \bullet)$. Indeed, it is isomorphic to the subcomplex defined by $\bigcap_{i=1}^n \ker(\partial_{i,0}^*)$ in each degree.

Replacing X with étale X-schemes, we get a Nisnevich sheaf $\underline{z}^q(X|D, n)_{\mathrm{Nis}}$ on X (or also denoted by $\underline{z}^q(-|D, n)_{\mathrm{Nis}}$) and its non-degenerate companions.

Remark 2.3 When $D = \emptyset$, the definitions recover Bloch's higher Chow theory. In particular, the complex $z^q(X|\emptyset, \bullet)$ is Bloch's (cubical) cycle complex $z^q(X, \bullet)$ found e.g in [1].

Definition 2.4 Suppose we are given a finite set w of irreducible constructible subsets of $X \setminus D$ and a map of sets $e \colon w \to \mathbb{N} = \{0, 1, 2, \dots\}$. Then we denote by $\underline{z}_{w,e}^q(X|D, n)$ the subgroup of $\underline{z}^q(X|D, n)$ consisting of cycles whose components V satisfy:

- For every face F of $\square^n$ and every element $W \in w$, the following holds:

$$\mathrm{codim}_{W \times F}(V \cap (W \times F)) \geq q - e(W).$$

Remark 2.5 Every constructible subset (of a Noetherian topological space) is written as the disjoint union of finitely many locally closed subsets. So in Definition 2.4, we could have restricted ourselves to those sets w consisting of irreducible *locally closed* subsets. But allowing constructible subsets often helps keep text succinct.

One sees $\{\underline{z}^q_{w,e}(X|D,n)\}_n$ forms a cubical abelian subgroup of $\{\underline{z}^q(X|D,n)\}_n$. We denote its associated non-degenerate complex by $z^q_{w,e}(X|D,\bullet)$. We have a natural injection $z^q_{w,e}(X|D,\bullet) \hookrightarrow \underline{z}^q_{w,e}(X|D,\bullet)$. Note that for every w and functions $e \leq e'$, we have $z^q_{w,e}(X|D,n) \subset \underline{z}^q_{w,e'}(X|D,n)$. We have $\underline{z}^q(X|D,n) = \underline{z}^q_{w,q}(X|D,n)$ where q is the constant function with value q.

Example 2.6 The subcomplexes $\underline{z}^q_{w,e}(X|D,n)$ are important because they often appear as the domains of definition of partially defined operations of cycles.

Suppose we are given a commutative square (1.1) in which X, X' are equidimensional k-schemes and D, D' are effective Cartier divisors. Write $f \colon X' \to X$ for the given map. Denote by $\{\underline{z}^q_f(X|D,n)\}_n$ the cubical subgroup of $\{\underline{z}^q(X|D,n)\}_n$ consisting of cycles V (say in degree n) such that its inverse image as a closed subset $(f \times \mathrm{id}_{\square^n})^{-1}(|V|)$ has codimension $\geq q$ and it intersects all faces properly; thus if $X \setminus D$ is regular, Serre's Tor formula defines a pulled-back cycle $(f \times \mathrm{id}_{\square^n})^*V$ on $(X' \setminus D') \times \square^n$ that intersects all faces properly. It automatically satisfies the modulus condition for D', so $(f \times \mathrm{id}_{\square^n})^*V \in \underline{z}^q(X'|D',n)$.

There is a finite set w of irreducible constructible subsets of $X \setminus D$ together with a function $e\colon w \to \mathbb{N}$ such that the equality $\underline{z}^q_f(X|D,n) = \underline{z}^q_{w,e}(X|D,n)$ holds for every n. Indeed, let $Z^k \subset X \setminus D$ be the constructible subset consisting of points at which the fiber of f has dimension k. Then a cycle V in $\underline{z}^q(X|D,n)$ belongs to $\underline{z}^q_f(X|D,n)$ if and only if the inequality

$$\dim(V \cap (Z^k \times F)) + k \leq \dim(X' \times F) - q$$

holds for all faces F of $\square^n$ and all non-negative integers k. Let $Z^k = \bigcup_\lambda Z^k_\lambda$ be the irreducible decomposition. Then the inequality tells us that if we set $w = \{Z^k_\lambda\}_{k,\lambda}$ and $e(Z^k_\lambda) = q + (\dim(Z^k_\lambda) + k - \dim X')$, then we have $\underline{z}^q_f(X|D,n) = \underline{z}^q_{w,e}(X|D,n)$. With this choice, we have a map of cubical abelian groups (and in particular, of associated complexes)

$$f^* \colon \{\underline{z}^q_{w,e}(X|D,n)\}_n \to \{\underline{z}^q(X'|D',n)\}_n.$$

Example 2.7 More generally, we may take a cycle $\alpha \in \underline{z}^p(X' \setminus D', m)$ in Bloch's cubical cycle complex, and consider the (generalized) intersection product [16, V.C).7]

$$(\alpha \times \square^n) \underset{f \times \mathrm{id}_{\square^{m+n}}}{\cup} (V \times \square^m)$$

(the notation is sloppy about the isomorphism $\square^m \times \square^n \cong \square^n \times \square^m$) which gives a well-defined cycle in $\underline{z}^{p+q}(X'|D', m+n)$ as long as the closed subset

$(|\alpha| \times \square^n) \cap (f \times \mathrm{id}_{\square^{m+n}})^{-1}(|V| \times \square^m)$ of $(X' \setminus D') \times \square^{m+n}$ has codimension $\geq p + q$, intersects the faces properly, and $X \setminus D$ is regular.

Suppose $X \setminus D$ is regular and denote by $\underline{z}^q_\alpha(X|D, n)$ the subgroup of $\underline{z}^q(X|D, n)$ consisting of cycles V satisfying these conditions; we get maps $(n \geq 0)$

$$(\alpha \times \square^n) \underset{f \times \mathrm{id}_{\square^{m+n}}}{\cup} ((-) \times \square^m) \colon \underline{z}^q_\alpha(X|D, n) \to \underline{z}^{p+q}(X'|D', m+n).$$

If $\alpha \in \underline{z}^p(X' \setminus D', m)$ vanishes by the differential of $\underline{z}^p(X' \setminus D', \bullet)$, these maps give a map of complexes (in the index n). We leave it to the reader as an exercise to find a finite set w and a function $e \colon w \to \mathbb{N}$ such that $\underline{z}^q_\alpha(X|D, n) = \underline{z}^q_{w,e}(X|D, n)$. (*Hint:* Consider the dimensions of fibers of $|\alpha| \cap ((X' \setminus D') \times F') \to X$, where F' runs through the faces of $\square^m$. Answers can be found e.g. in [8, Lem. 4.7], [12, Part I Ch.II Lem.3.5.2] and [13, Lem.7.4.4].)

In view of Example 2.6, Theorem 1.1 follows from the next

Theorem 2.8 (Moving Lemma) *Let X be an equidimensional k-scheme of finite type and D be an effective Cartier divisor on it. Assume $X \setminus D$ is smooth. Then for every finite set w of irreducible constructible subsets of $X \setminus D$ and every function $e \colon w \to \mathbb{N}$, the inclusion*

$$\underline{z}^q_{w,e}(X|D, \bullet)_{\mathrm{Nis}} \hookrightarrow \underline{z}^q(X|D, \bullet)_{\mathrm{Nis}}$$

is a quasi-isomorphism of complexes of Nisnevich sheaves on X.

In particular, the inclusion induces a quasi-isomorphism on the non-degenerate complexes, cf. discussions following Definition 2.2.

The theorem roughly says that every cycle can be moved, up to quasi-isomorphism, to a cycle which has a better intersection property with prescribed subsets.

If k is a finite field, the proof can be reduced to the case where k is an infinite field by a trace (norm) argument. So we assume k is infinite and sketch a proof in the next section.

3   Sketch of the proof

Assume k is infinite. Let the notation be as in Moving Lemma 2.8 and suppose we are given a cycle $V \in \underline{z}^q(X|D, n)$. We are asked to move it to improve its intersection properties with $W \times F$, for $W \in w$ and faces $F \subset \square^n$.

3.1   Review of the non-modulus case

There is an established way of moving cycles (on affine smooth varieties) originally due to Chow [4] and applied to higher Chow groups by Bloch [1] and Levine [12, I.II.§3.4]. We review this process for Bloch's (cubical) cycle complex $\underline{z}^q(X, \bullet)$ of a smooth affine variety X. In this case we can show the quasi-isomorphism $\underline{z}_{w,e}^q(X, \bullet) \subset \underline{z}^q(X, \bullet)$ without the need of Nisnevich topology.

(1) **The case** $X = \mathbb{A}^d$. We move a given cycle $V \in \underline{z}^q(\mathbb{A}^d, n)$ using the parallel translation automorphism $\mathbb{A}^d \times \square^n \to \mathbb{A}^d \times \square^n$; $(\boldsymbol{x}, \boldsymbol{z}) \mapsto (\boldsymbol{x} + \boldsymbol{v}, \boldsymbol{z})$ with a chosen vector $\boldsymbol{v} \in \mathbb{A}^d$. Let us denote by $V + \boldsymbol{v}$ the cycle obtained after this moving. One verifies that if $\boldsymbol{v}$ is sufficiently general, then the support $|V + \boldsymbol{v}|$ satisfies all the desired intersection properties so that $V + \boldsymbol{v} \in \underline{z}_{w,e}^q(\mathbb{A}^d, n)$.

We form a homotopy cycle $\tilde{V} \in \underline{z}^q(\mathbb{A}^d, n+1)$ connecting V and $V + \boldsymbol{v}$. It is a cycle in $\mathbb{A}^d \times \square^n \times \square^1$ whose fiber over $t \in \square^1$ ("time $= t$") is $V + t\boldsymbol{v}$. One checks if V already happens to be in $\underline{z}_{w,e}^q(\mathbb{A}^d, n)$, then we have $\tilde{V} \in \underline{z}_{w,e}^q(\mathbb{A}^d, n+1)$ provided $\boldsymbol{v}$ is sufficiently generally for this V. Thus we get a homotopy inverse of the inclusion

$$\underline{z}_{w,e}^q(\mathbb{A}^d, \bullet) \hookrightarrow \underline{z}^q(\mathbb{A}^d, \bullet)$$

and prove the quasi-isomorphism.

Remark 3.1 The homotopy inverse is only partially defined because the vector $\boldsymbol{v}$ is chosen cycle-wise. It is sufficient for showing the quasi-isomorphism. One could as well set $\boldsymbol{v}$ to be the generic point of $\mathbb{A}^d$ and use a so-called specialization argument to get back to the original field k from the function field of $\mathbb{A}^d$.

(2) **Reduction to** $\mathbb{A}^d$. Now X is an arbitrary smooth affine equidimensional scheme. By (the proof of) Noether's Normalization Theorem, if we choose an arbitrary affine closed embedding $X \hookrightarrow \mathbb{A}^N$ and a sufficiently general linear projection $\mathbb{A}^N \twoheadrightarrow \mathbb{A}^d$ with $d = \dim X$, the composite $p \colon X \to \mathbb{A}^d$ is finite and surjective (in fact flat by the smoothness of X and $\mathbb{A}^d$). We have push-forward and pull-back operations of cycles, $p_* \colon \underline{z}^q(X, n) \rightleftarrows \underline{z}^q(\mathbb{A}^d, n) \colon p^*$. For a given cycle $V \in \underline{z}^q(X, n)$, we consider the tautological equality

$$p^* p_* V = (p^* p_* V - V) + V.$$

The first term on the right hand side is called the *residual cycle* (of the operation $p^* p_*$, as the author understands).

If p is chosen sufficiently generally for the given V, it can be shown that any component of V does not appear as a component of the residual cycle, and that it has an improved intersection property: if $V \in \underline{z}^q_{w,e}(X,n)$, then

$$p^* p_* V - V \in \underline{z}^q_{w,e-1}(X,n)$$

where the function $e-1 \colon w \to \mathbb{N}$ is defined by $W \mapsto \max\{e(W)-1, 0\}$. On the other hand, the cycle $p_* V$ is on $\mathbb{A}^d \times \square^n$, so by the previous step, there is a way of moving it to a good position. Pulling back this movement, we can move the cycle $p^* p_* V$ into a good position, too. Thanks to the tautology $V = p^* p_* V - (p^* p_* V - V)$, it follows that every given cycle $V \in \underline{z}^q_{w,e}(X,n)$ can be moved into $\underline{z}^q_{w,e-1}(X,n)$ up to homotopy. In this way we can prove that the inclusion $\underline{z}^q_{w,e-1}(X,\bullet) \subset \underline{z}^q_{w,e}(X,\bullet)$ is a quasi-isomorphism.

Repeated application of this argument shows that for every function e, the complex $\underline{z}^q_{w,e}(X,\bullet)$ is quasi-isomorphic to the subcomplex $\underline{z}^q_{w,0}(X,\bullet)$, where 0 is the constant function with value 0. This completes the proof.

$\square$

3.2 Our treatment (moving with modulus)

Let X be an affine equidimensional scheme and D be an effective Cartier divisor. Assume $X \setminus D$ is smooth. We want to prove that the inclusion $\underline{z}^q_{w,e}(X|D,\bullet) \subset \underline{z}^q(X|D,\bullet)$ is a quasi-isomorphism Nisnevich locally on X. By the presence of the modulus condition, we have to modify the above process in some places.

$(1)'$ **The case $X = \mathbb{A}^d$** Now we are given a divisor $D = (u)$ on $\mathbb{A}^d$. Instead of the parallel translation, we use the *translation with modulus*: we define our homotopy cycle $\tilde{V}$ to be the pull-back of the given cycle $V \in \underline{z}^q(X|D,n)$ by the morphism

$$\Phi = \Phi_{\boldsymbol{v},s} \colon \mathbb{A}^d \times \square^n \times \square^1 \to \qquad \mathbb{A}^d \times \square^n$$
$$(\boldsymbol{x}, \boldsymbol{z}, t) \mapsto (\boldsymbol{x} + tu^s \boldsymbol{v}, \boldsymbol{z}).$$

where $\boldsymbol{v}$ is a sufficiently general vector and $s \geq 1$ is a sufficiently large integer. We are moving the cycle at the speed $u^s \boldsymbol{v}$ depending on the point; the closer the point is to the divisor, the slower is the moving speed. Intersection properties are improved by this moving as in Step (1).

In order for $\tilde{V} := \Phi^*(V)$ to work as a homotopy, we have to verify the modulus condition on $\tilde{V}$. Checking it involves compactifying $\square^n \times \square^1$ into $(\mathbb{P}^1)^{n+1}$. In particular, the "time" axis $\square^1$ is compactified. It amounts to considering the behavior of the moving cycle when the time tends to the infinity $t \to \infty$. This is a new issue raised by the modulus condition, and for dealing with it I needed the translation with modulus.

The next is our key observation. It settles Step $(1)'$.

Lemma 3.2 *There is an integer $s(V) \geq 1$ such that for every vector $\boldsymbol{v} \in \mathbb{A}^d$ and every $s \geq s(V)$, the cycle $\tilde{V} := \Phi^*_{\boldsymbol{v},s}(V)$ satisfies the modulus condition.*

Remark 3.3 In fact, we will need rather the following variant. Let B be the spectrum of a discrete valuation ring and $\eta, b \in B$ be its generic and closed points. Let $\mathbb{A}^d_B$ be an affine space over B. Suppose we are given a finite set w of irreducible constructible subsets of $\mathbb{A}^d_\eta$ and a function $e \colon w \to \mathbb{N}$. Then the inclusion $\underline{z}^q_{w,e}(\mathbb{A}^d_B|\mathbb{A}^d_b, \bullet) \subset \underline{z}^q(\mathbb{A}^d_B|\mathbb{A}^d_b, \bullet)$ is a quasi-isomorphism of complexes of abelian groups (at least if the residue field at b is infinite).

$(2)'$ **Reduction to $\mathbb{A}^d$.** Now let us assume X is affine, $D = (u)$ is principal and $X \setminus D$ is smooth. We wish to employ finite surjective maps $p \colon X \to \mathbb{A}^d_B$ and operations p_*, p^* as in Step (2). In order for the operations to preserve the modulus condition, we would like to have a Cartesian diagram

$$
\begin{array}{ccc}
X & \xrightarrow{\;p\;} & \mathbb{A}^d \\
\cup \uparrow & \square & \uparrow \cup \\
D = (u) & \longrightarrow & \exists D_0
\end{array}
$$

for some divisor D_0.

We apply the next theorem, a variant of [13, Th. 10.2.2], to the morphism $u \colon X \to \mathbb{A}^1$ (to be precise, after modifying u a little in a manner depending on the cycle to move).

Theorem 3.4 (Noether's Normalization Theorem over a Dedekind base) *Let B be a one-dimensional Noetherian normal scheme and let $f \colon X \to B$ be an equidimensional morphism of relative dimension d. Let $x \in X$ be a point and $b = f(x) \in B$ its image, and assume the residue field of b is infinite.*

Then there exist Nisnevich neighborhoods $(X', x') \to (X, x)$, $(B', b') \to (B, b)$, a commutative diagram

$$
\begin{array}{ccc}
(X', x') & \longrightarrow & (X, x) \\
f' \downarrow & & \downarrow f \\
(B', b') & \longrightarrow & (B, b)
\end{array}
$$

and a finite surjective B'-morphism $X' \to \mathbb{A}^d_{B'}$.

Therefore, up to Nisnevich localization of X, we have a Nisnevich neighborhood (B, b) of $(\mathbb{A}^1, 0)$ and a finite surjective morphism $p \colon X \to \mathbb{A}^{d-1}_B$ over u, so that we have $p^*(\mathbb{A}^{d-1}_b) = D$. Then the operations p_*, p^* preserve modulus condition.

Moreover, the finite surjective map p is in fact obtained by first embedding X into some well-chosen large affine space $\mathbb{A}^N_B$ and then by taking sufficiently general linear projection $\mathbb{A}^N_B \to \mathbb{A}^{d-1}_B$. So we have an enormous choice of the finite surjective map p. By choosing a sufficiently general such p and considering the decomposition $V = p^* p_* V - (p^* p_* V - V)$ as in Step (2), we can reduce the problem to $\mathbb{A}^{d-1}_B$.

This completes the proof of Moving Lemma. $\qquad\qquad\qquad\square$

4 Application (joint work with R. Iwasa)

In addition to the computation of the additive higher Chow group of regular local k-algebras in a certain range due to Krishna-Park [10, 11], Moving Lemma 2.8 allows us to take one small step towards the construction of Chern classes with modulus.

Notation 4.1 Let X be an equi-dimensional k-scheme and D be an effective Cartier divisor on X. We assume $X \setminus D$ is smooth. Let E be a vector bundle of rank r on X, and

$$p \colon \mathbb{P}(E) \to X$$

be its associated projective bundle. Let $\xi \in \mathrm{CH}^1(\mathbb{P}(E))$ be the first Chern class of the tautological line bundle $\mathcal{O}(1)$ on $\mathbb{P}(E)$.

Let us begin with the projective bundle formula for the higher Chow group with modulus. The following projective bundle formula had been known by Krishna-Levine-Park [8, 9].

$$\mathrm{CH}^q(\mathbb{P}(E)|p^*D, n) \xleftarrow[\cong]{\xi^i \cup p^*(-)} \bigoplus_{i=0}^{r-1} \mathrm{CH}^{q-i}(X|D, n).$$

The existence of the map depends on a moving lemma for the (classical) higher Chow group. Our Moving Lemma 2.8 enhances the map to a map in the derived category (in the Nisnevich topology), cf. Example 2.7

$$\bigoplus_{i=0}^{r-1} z^{q-i}(X|D, \bullet)_{\mathrm{Nis}} \xrightarrow{\xi^i \cup p^*(-)} p_* z^q(\mathbb{P}(E)|p^*D, \bullet)_{\mathrm{Nis}}.$$

The previously known projective bundle formula implies that it is an isomorphism in the derived category. It has the flavor of a "fake" projective
bundle formula. One could ask if the "genuine" projective bundle formula
holds, namely, if the composite

$$\bigoplus_{i=0}^{r-1} z^{q-i}(X|D, \bullet)_{\mathrm{Nis}} \xrightarrow[\sim]{\xi^i \cup p^*(-)} p_* z^q(\mathbb{P}(E)|p^*D, \bullet)_{\mathrm{Nis}}$$

$$\longrightarrow \mathrm{R}p_* z^q(\mathbb{P}(E)|p^*D, \bullet)_{\mathrm{Nis}}$$

is a quasi-isomorphism. The author does not know the answer.

4.1 Chern classes of relative vector bundles

Now suppose that the vector bundle E is given a trivialization $\phi\colon E|_D \cong
\mathcal{O}_D^r$ over the divisor. We would like to call such a pair (E, ϕ) a *relative
vector bundle*. Of course, the relative vector bundles are precisely the
$\mathrm{GL}_r(I_D) := \ker[\mathrm{GL}_r(\mathcal{O}_X) \to \mathrm{GL}_r(\mathcal{O}_D))]$-torsors, where we write I_D for
the defining ideal for D. Relative vector bundles determine elements in
$K_0(X, D)$ (e.g. [17, Ch.II, §2.10] in the affine case). Jointly with Ryomei
Iwasa, the author has constructed Chern classes of relative vector bun-
dles in $\mathbf{H}^0_{\mathrm{Nis}}(X, z^i(-|D, \bullet)_{\mathrm{Nis}})$ $(1 \le i \le r)$. The "fake" projective bundle
formula plays a role.

A key step in the construction is to lift the power $\xi^r \in \mathrm{CH}^r(\mathbb{P}(E))$ to
what we call ξ^r_{rel},

$$\xi^r_{\mathrm{rel}} \in \mathbf{H}^0_{\mathrm{Nis}}(X, p_* z^r(\mathbb{P}(E)|p^*D, \bullet)_{\mathrm{Nis}}).$$

(Be aware that there is a natural map from the latter group to the for-
mer, because we know $\mathrm{CH}^r(\mathbb{P}(E)) = \mathbf{H}^0_{\mathrm{Nis}}(X, \mathrm{R}p_* z^r(\mathbb{P}(E), \bullet)_{\mathrm{Nis}})$, Bloch's
localization theorem for the higher Chow.)

Why should this lifting occur? Well, imagine that there were a motivic
cohomology theory defined for all schemes which well approximates the K-
theory, and that the (Nisnevich) higher Chow group with modulus of pairs
(X, D) were to coincide with the relative motivic cohomology of pairs. In
particular, there should be an exact sequence (where the rightmost term
does not have a concrete definition)

$$\mathbf{H}^0_{\mathrm{Nis}}(\mathbb{P}(E), z^r(-|p^*D, \bullet)) \to H^{2r}(\mathbb{P}(E), \mathbb{Z}(r)) \to \text{``}H^{2r}(\mathbb{P}(E|_D), \mathbb{Z}(r))\text{''}.$$

The vector bundle E becomes trivial over D, so the projective bundle
$\mathbb{P}(E|_D)$ is trivial. It suggests that the element ξ^r in the middle should
vanish in the right hand term. Therefore it should *lift* to the left term.

The fact that the *trivialization is given* suggests that the lift should be *canonical*.

The construction of the lift ξ_{rel}^r is sketched in the next subsection. It involves taking open coverings and defining cycles by equations, but in doing so we only need open coverings of X. So our construction gives an element in $\mathbf{H}_{\mathrm{Nis}}^0(X, p_* z^r(\mathbb{P}(E)|p^* D, \bullet)_{\mathrm{Nis}})$.

The "fake" projective bundle formula tells that this last group is isomorphic to

$$\bigoplus_{i=0}^{r-1} \mathbf{H}_{\mathrm{Nis}}^0(X, z^{r-i}(X|D, \bullet)_{\mathrm{Nis}}).$$

So we find a unique set of elements

$$c_i(E, \phi) \in \mathbf{H}_{\mathrm{Nis}}^0(X, z^i(X|D, \bullet)_{\mathrm{Nis}}) \ (1 \le i \le r)$$

satisfying

$$\xi_{\mathrm{rel}}^r + \xi^{r-1} \cup p^* c_1(E, \phi) + \cdots + p^* c_r(E, \phi) = 0$$

in the two isomorphic groups.

Definition 4.2 We call the elements thus obtained

$$c_i(E, \phi) \in \mathbf{H}_{\mathrm{Nis}}^0(X, z^i(X|D, \bullet)_{\mathrm{Nis}})$$

the *Chern classes* of the relative vector bundle (E, ϕ).

4.2 Construction of ξ_{rel}^r

Now we move on to sketch the construction of the lifting

$$\xi_{\mathrm{rel}}^r \in \mathbf{H}_{\mathrm{Nis}}^0(X, p_* z^r(\mathbb{P}(E), p^* D, \bullet)_{\mathrm{Nis}})$$

of the power ξ^r. In fact, the Zariski topology suffices for this. Whereas this construction is obsolete in logical terms now that Chern class maps have been given on the entire $K_{\ge 0}(X, D)$ (next subsection), it carries most of the main ideas and is more intuitive; that's why the author opted to present this older version which is not in the literature too.

Take an open cover $X = \bigcup_i U_i$ and trivializations $E_{U_i} \cong \mathcal{O}_{U_i}^r$ which coincide with the given ϕ over D. Thus, transition matrices on the overlaps $U_i \cap U_j$ look like $\begin{pmatrix} 1 + I_D & I_D & \\ I_D & 1 + I_D & \\ & & \ddots \end{pmatrix}$ where I_D is the defining ideal for D.

The chosen trivialization gives rise to isomorphisms $\mathbb{P}(E_{U_i}) \cong \mathbb{P}_{U_i}^{r-1}$. Let $Z_i^{(1)}, \ldots, Z_i^{(r)}$ be the standard homogeneous coordinates and $H_i^{(1)}, \ldots, H_i^{(r)}$

be the standard hyperplanes of this projective space. Also, by the given datum ϕ, we have $\mathbb{P}(E_D) \cong \mathbb{P}_D^{r-1}$; let us denote by $Z_D^{(a)}$ and $H_D^{(a)}$ for its a-th coordinate and hyperplane.

We define a map of (Zariski) presheaves $\bigoplus_i \mathbb{Z}[U_i] \to p_* z^1(-, 0)$ by assigning the hyperplane $H_i^{(1)}$ on each U_i. There arise "error terms" $H_i^{(1)} - H_j^{(1)}$ on the overlaps $U_{ij} := U_i \cap U_j$. We can cancel them by the cycles $H_{ij}^{(1)} := \mathrm{div}(1 + t\left(\frac{Z_i^{(1)}}{Z_j^{(1)}} - 1\right))$ on $\mathbb{P}(E_{U_{ij}}) \times \square^1$ where t is the coordinate of $\square^1$.

We keep cancelling higher and higer error terms by giving cycles $H_{i_0 \ldots i_n}^{(1)} \in z^1(\mathbb{P}(E_{U_{i_0 \ldots i_n}}), n)$. They form a map of complexes, representing the cycle class ξ,

$$\xi^{(1)}: \quad \begin{array}{ccccccc} \cdots \to & \bigoplus_{i,j,k} \mathbb{Z}[U_{ijk}] & \to & \bigoplus_{i,j} \mathbb{Z}[U_{ij}] & \to & \bigoplus_i \mathbb{Z}[U_i] & \to 0 \\ & \downarrow & & \downarrow & & \downarrow & \\ \cdots \to & p_* z^1(-, 2) & \to & p_* z^1(-, 1) & \to & p_* z^1(-, 0) & \to 0. \end{array}$$

For example, on U_{ijk}, it is given by the divisor of the rational function on $\mathbb{P}(E_{U_{ijk}}) \times \square^2$:

$$1 + t_1 \left(\frac{\left(1 + t_2\left(\frac{Z_j^{(1)}}{Z_k^{(1)}} - 1\right)\right) \cdot \left(1 + t_2\left(\frac{Z_i^{(1)}}{Z_j^{(1)}} - 1\right)\right)}{1 + t_2\left(\frac{Z_i^{(1)}}{Z_k^{(1)}} - 1\right)} - 1 \right)$$

where (t_1, t_2) is the coordinate of $\square^2$. Probably the reader sees the pattern.

By giving $H_i^{(a)}$ as the initial datum ($2 \le a \le r$), we can give different representatives $\xi^{(2)}, \ldots, \xi^{(r)}$ of ξ. The cup product of Čech cohomology classes gives a representative $\xi^{(1)} \cup \cdots \cup \xi^{(r)}$ for the power ξ^r, provided the intersection products involved are all well-defined. Unfortunately this is usually not the case. But it turns out that there is an open neighborhood X^* of $D \subset X$ over which the intersection products are well-defined; then if we write $U_{ijk\ldots}^* := U_{ijk\ldots} \cap X^*$, we have a well-defined map of complexes

$$\xi^{(1)} \cup \cdots \cup \xi^{(r)}: \quad \mathbb{Z}[U_\bullet^*] \longrightarrow p_* z^r(-, \bullet). \tag{4.1}$$

The next assertion is a technical key point.

Lemma 4.3 *The map (4.1) factors through the subcomplex* $p_* z^r(-|p^* D, \bullet)$ *consisting of cycles with modulus.*

Proof We give a sketch. See [6, Prop.4.6] for a complete proof of a more general result treating matrices in $\mathbf{X}_r(\mathcal{O}_X, I_D)$ (see the next subsection), not just in $\mathrm{GL}_r(I_D)$.

By Binda-Saito [3, Lem.4.1], for a normal local ring A and a principal ideal I, the divisor of a regular function $f(t_1, \ldots, t_n) = \sum_{\underline{\nu}} a_{\underline{\nu}} t^{\underline{\nu}}$ on $\mathrm{Spec}(A) \times \square^n$ with $a_{(0,\ldots,0)} \in A^*$ satisfies the modulus condition if and only if f is admissible, i.e. $a_{\underline{\nu}} \in I^{\max\{\nu_1, \ldots, \nu_n\}}$ for all multi-indices $\underline{\nu}$.

Using this criterion, one verifies that the cycle

$$H^{(a)}_{i_0 \ldots i_n} \in z^1(\mathbb{P}(E_{U_{i_0 \ldots i_n}}), n))$$

satisfies the modulus condition when restricted to the open set

$$[\mathbb{P}(E_{U_{i_0 \ldots i_n}}) \setminus (\cup_{0 \leq k \leq n} H^{(a)}_{i_k})] \times \square^n.$$

For example, the cycle $H^{(1)}_{ijk}$ restricted to the open set

$$[\mathbb{P}(E_{U_{ijk}}) \setminus \{Z^{(1)}_j Z^{(1)}_k = 0\}] \times \square^2$$

contains the divisor of the function

$$1 + t_2 \left(\frac{Z^{(1)}_i}{Z^{(1)}_k} - 1 \right) + t_1 t_2 \left(\frac{Z^{(1)}_i}{Z^{(1)}_j} + \frac{Z^{(1)}_j}{Z^{(1)}_k} - \frac{Z^{(1)}_i}{Z^{(1)}_k} - 1 \right)$$
$$+ t_1 t_2^2 \left(\frac{Z^{(1)}_i}{Z^{(1)}_j} - 1 \right) \left(\frac{Z^{(1)}_j}{Z^{(1)}_k} - 1 \right).$$

This polynomial is admissible because, by the form of the transition matrices, the functions $\frac{Z^{(1)}_i}{Z^{(1)}_j}, \frac{Z^{(1)}_j}{Z^{(1)}_k}$ and $\frac{Z^{(1)}_i}{Z^{(1)}_k}$ are congruent to 1 modulo I_D. The general case follows by easy induction.

Now the Čech data for the cup product $\xi^{(1)} \cup \cdots \cup \xi^{(r)}$ are intersections of r cycles. By the observation just made, the a-th member of each intersection satisfies the modulus condition when restricted to $\mathbb{P}(E_{U_{ijk\ldots}}) \setminus (\cup_{l \in \{i,j,k,\ldots\}} H^{(a)}_l)$. This open set contains $\mathbb{P}(E_{U_{ijk\ldots}})_D \setminus H^{(a)}_D$. It follows (by *Containment Lemma*, e.g. [9, Prop.2.4]) that the intersection product satisfies the modulus condition over an open subset containing $\cup_{a=1}^r \left[\mathbb{P}(E_{U_{ijk\ldots}})_D \setminus H^{(a)}_D \right]$ which is $\mathbb{P}(E_{U_{ijk\ldots}})_D$. So it satisfies the modulus condition over the entire $\mathbb{P}(E_{U_{ijk\ldots}})$. This last step of the argument provides another explanation as to why the cycle class ξ can satisfy the modulus condition only after being raised to the r-th power. This verifies the assertion of Lemma 4.3.

$\square$

Set $X^\circ := X \setminus D$ and $U^\circ_{ijk...} := X^\circ \cap U_{ijk...}$. The cup product may not be well-defined over this open subset because proper intersection is not assured. We note, however, that the modulus condition is vacuous over this region. Therefore we may apply the classical moving lemma of Chow to modify the cycles $\xi^{(a)}$ (say into $\xi^{(a)\circ}$) so that the cup product $\xi^{(1)\circ} \cup \cdots \cup \xi^{(r)\circ} \colon \mathbb{Z}[U^\circ_\bullet] \to p_* z^r(-, \bullet)$ is well-defined.

Combining the maps on $U^*_{ijk...}$ and $U^\circ_{ijk...}$ (and cancelling the difference on $U^*_{ijk...} \cap U^\circ_{ijk...}$), we get a map in the derived category of Zariski sheaves

$$\xi^r_{\mathrm{rel}} \colon \mathbb{Z}[X] \longrightarrow p_* z^r(-|p^*D, \bullet).$$

This concludes the desired construction of

$$\xi^r_{\mathrm{rel}} \in \mathbf{H}^0_{\mathrm{Zar}}(X, p_* z^r(-|p^*D, \bullet)_{\mathrm{Zar}}),$$

and hence of the Chern classes $c_i(E, \phi)$.

4.3   Latest development

The author thinks that the arguments so far are not enough to define maps $c_i \colon K_0(X, D) \to \mathbf{H}^0_{\mathrm{Nis}}(X, z^i(-|D, \bullet)_{\mathrm{Nis}})$; not all elements in $K_0(X, D)$ are written by relative vector bundles; nor are they powerful enough to capture higher relative K-groups $K_n(X, D)$. To put it another way, the simplicial presheaves $\mathrm{BGL}_r(I_D)$ are not powerful enough to compute relative K-groups.[1]

Satisfactory Chern class maps

$$K_n(X, D) \to \mathbf{H}^{-n}_{\mathrm{Nis}}(X, z^i(-|D, \bullet)_{\mathrm{Nis}}) \quad (n \geq 0, i \geq 1)$$

have been constructed while this manuscript was being prepared:

Theorem 4.4 (Iwasa-Kai [6]) *For pairs (X, D) of k-schemes of finite type X and Cartier divisors D with $X \setminus D$ smooth, there are functorial maps for $n \geq 0$ and $i \geq 1$,*

$$\mathsf{C}_{n,i} \colon K_n(X, D) \to \mathbf{H}^{-n}_{\mathrm{Nis}}(X, z^i(-|D, \bullet)_{\mathrm{Nis}}).$$

They coincide with Bloch's Chern classes if $D = \emptyset$. They are group homomorphisms for $n > 0$ and satisfy the Whitney sum formula for $n = 0$.

---

[1]Nonetheless, Iwasa [5] has proved that if we go *pro* over the multiples of D, then the pro relative K-groups $\{K_n(X, mD)\}_m$ ($n \geq 1$) are described in terms of $\{\mathrm{BGL}_r(I^m_D)\}_m$, with r large enough compared to n and the stability range of the non-unital ring I_D. This, combined with our present construction, gives *pro Chern classes* from the pro relative K-groups to the pro relative motivic cohomology.

Of course, the maps are consistent with the Chern classes of relative vector bundles $c_i(E, \phi)$ we have just considered.

The new ingredient is the homotopy theory of simplicial presheaves. Let $I_D \subset \mathcal{O}_X$ be the ideal defining D. Just as the usual K-groups are computed via the homotopy theory of $\mathrm{BGL}_r(\mathcal{O}_X)$, the relative K-groups $K_{\geq 0}(X, D)$ are computed via the *relative Volodin space* $\mathbf{X}_r(\mathcal{O}_X, I_D)$ which is a subsheaf of $\mathrm{BGL}_r(\mathcal{O}_X)$ consisting of triangular matrices modulo the ideal
$$\begin{pmatrix} 1 + I_D & \mathcal{O}_X & \mathcal{O}_X & \\ I_D & 1 + I_D & \mathcal{O}_X & \\ I_D & I_D & 1 + I_D & \ddots \end{pmatrix}$$
and their conjugates by permutation. The fact that $\mathbf{X}_r(\mathcal{O}_X, I_D)$ is much larger than $\mathrm{BGL}_r(I_D)$ and not representable in the category of schemes with Cartier divisors requires some extra work to be done. See *op. cit.* for details.

Acknowledgements The author is very grateful to the organizers and TIFR for inviting him and for their hospitality during the colloquium. Especially, part of the computation needed in §4 was done during his stay, on the boat. He also thanks the referee for corrections and for the encouragement to complete §4 in its current form. During the work, the author was supported by JSPS KAKENHI Grant (15J02264) and by the Program for Leading Graduate Schools, MEXT, Japan.

References

[1] S. Bloch: *Some notes on elementary properties of higher chow groups, including functoriality properties and cubical chow groups*, Notes available on his web page.

[2] S. Bloch, H. Esnault, *An additive version of higher Chow groups*, Ann. Scient. Éc. Norm. Sup. **4** (36) (2003), 463–477.

[3] F. Binda, S. Saito, *Relative cycles with modulus and regulator maps*, J. Inst. Math. Jussieu (2017), 1–61. doi:10.1017/S1474748017000391

[4] W.L. Chow, *On equivalence classes of cycles in an algebraic variety*, Ann. of Math. **64** (3) (1956), 450–479.

[5] R. Iwasa, *Homology pro stability for pro unital rings, and relative algebraic K-theory*, arXiv:1610.04998v1 [math.KT], 2016.

[6] R. Iwasa, W. Kai, *Chern classes with modulus*, arXiv:1611.07882v1 [math.KT], 2016.

[7] W. Kai, *A moving lemma for algebraic cycles with modulus and contravariance*, arXiv:1507.07619v3 [math.AG], 2015.

[8] A. Krishna, M. Levine, *Additive higher Chow groups of schemes*, J. Reine Angew. Math. **619** (2008), 75–140.

[9] A. Krishna, J. Park, *A module structure and a vanishing theorem for cycles with modulus*, Math. Res. Lett. **24** (2017), no. 4, 1147–1176.

[10] A. Krishna, J. Park, *On additive higher Chow groups of affine schemes*, Documenta Math. **21**, (2016) 49–89.

[11] A. Krishna, J. Park, *Algebraic cycles and crystalline cohomology*, arXiv:1504.08181 [math.AG], 2015.

[12] M. Levine: *Mixed Motives*, Mathematical Surveys and Monographs **57**, Amer. Math. Soc. 1998.

[13] M. Levine, *Chow's moving lemma and the homotopy coniveau tower*, K-theory **37** (1-2) (2006), 129–209.

[14] J. Park, *Regulators on additive higher Chow groups*, Amer. J. Math. **131** (1) (2009), 257–276.

[15] K. Rülling, *The generalized de Rham-Witt complex over a field is a complex of zero-cycles*, J. Algebraic Geometry **16**, (2007), 109–169; Erratum, J. Algebraic Geometry **16**, (2007), 793–795.

[16] J.P. Serre, *Algèbre Locale, Multiplicités*, Lecture Notes in Mathematics **11**, second corrected printing of the 3rd edition (1975), Springer, 1997.

[17] C.A. Weibel, *The K-book: an introduction to algebraic K-theory*, Graduate Studies in Mathematics **145**, Amer. Math. Soc. 2013.

MATHEMATICAL INSTITUTE, TOHOKU UNIVERSITY. 6-3 ARAMAKI AZA-AOBA, AOBA-KU, 980-8578 SENDAI, JAPAN.
E-mail: kaiw@m.tohoku.ac.jp

K-Theory
Copyright ©2018 Tata Institute of Fundamental Research
Publisher: Hindustan Book Agency, New Delhi, India

Multiplicative properties of the multiplicative group

Bruno Kahn

Abstract

We give a few properties equivalent to the Bloch-Kato conjecture
(now the norm residue isomorphism theorem).

Introduction

The Bloch-Kato conjecture, now called the norm residue isomorphism theorem, was finally proven by Voevodsky in 2011 [19], using key inputs from Rost. The proof has many ramifications and involves a combination of sophisticated motivic techniques, including motivic Steenrod operations, and results of a more combinatorial kind like the existence of norm varieties.

This state of the art gives some interest to the issue of finding a more elementary proof. In this direction, one can consider the early work of Thomason on inverting the Bott element in algebraic K-theory [15] as a "stable" version of the conjecture; Levine later gave a motivic version of Thomason's theorem in [9]. I wondered how close to the norm residue isomorphism theorem the latter work takes us; the result is the following theorem, which was obtained in 2009.

Theorem 1 *Let k be an infinite perfect field and let l be a prime number invertible in k. If $l = 2$, assume that k is non-exceptional in the sense of Harris-Segal: the Galois group of the extension $k(\mu_{2\infty})/k$ is torsion-free. Then the following statements are equivalent:*

(i) *The Beilinson-Lichtenbaum conjecture holds modulo l over k.*

(ii) *For all $n \geq 1$ and all $i > 0$, $H_i(\mathbb{G}_m \otimes K_n^M/l) = 0$. Here the tensor product is taken in $\mathbf{DM}^{\mathrm{eff}}$.*

(iii) *For any $n \geq 2$, any function field K/k, any semi-local K-algebra A and any ideals $I, J \subset A$ with $I \cap J = 0$, the map*

$$K_n^M(A)/l \to K_n^M(A/I)/l \oplus K_n^M(A/J)/l$$

is injective.

(iv) *Same as (iii), for A the coordinate ring of $\hat{\Delta}^q_{K,S}$ for all $q \geq 2$ and all $\emptyset \neq S \subseteq [0,q]$ and I, J defined by sets of vertices.*

Here are some explanations on the notation. We assume the reader familar with Voevodsky's category $\mathbf{DM}^{\mathrm{eff}}$ of effective motivic complexes [17, 11, 1]; in (ii) and later, H_i is relative to its homotopy t-structure. The Beilinson-Lichtenbaum conjecture is recalled at the end of §3: it is equivalent to the Bloch-Kato conjecture by [2, 14]. If A is a commutative semi-local ring, we write $K^M_*(A)$ for the Milnor ring of A in the naïve sense, i.e. the quotient of the tensor algebra $T(A^*)$ by the two-sided ideal generated by elements $a \otimes (1-a)$ with $a, 1-a \in A^*$. We shall write K^M_n for the associated Nisnevich sheaf on the category $\mathbf{Sm}$ of smooth separated k-schemes of finite type.

In (iv), we write $\hat{\Delta}^*_K$ for the cosimplicial K-scheme whose q-th term $\hat{\Delta}^q_K$ is the semi-localisation of $\Delta^q_K = \operatorname{Spec} K[t_0, \ldots, t_q]/(\sum t_i = 1)$ at its vertices. If $i \in [0,q]$ (resp. $S \subseteq [0,q]$), we write $\hat{\Delta}^q_{K,i}$ for the i-th face of $\hat{\Delta}^q_K$ and $\hat{\Delta}^q_{K,S} = \bigcup_{i \in S} \hat{\Delta}^q_{K,i}$. We shall also write ∂_i for the inclusion $\hat{\Delta}^q_{K,i} \hookrightarrow \hat{\Delta}^q_K$ (i-th face map), and $\hat{\Delta}^q_{K,[0,q]} =: \partial\hat{\Delta}^q_K$.

Of course, all statements in Theorem 1 are true since the first one is. The game we shall play here, however, is to forget about this fact and prove the equivalences without using it. Statement (ii) explains the title of this note. It is possible that such vanishing holds in more generality, which would be one possible direction of attack for a more elementary proof of [19]. The scant evidence in this direction is a remarkable theorem of Sugiyama [13, Prop. A.1] that the tensor product of Nisnevich sheaves of $\mathbf{Q}$-vector spaces with transfers is exact. The most appealing leads are of course (iii) and (iv), because of their seemingly elementary nature. When I came up with Theorem 1, I tried to prove either of these statements by using the techniques of Guin and Nesterenko-Suslin in [3, 12], but was not successful.

(Added in November 2017.) When I sent this paper to Voevodsky in June 2017, he answered:

> I cannot say that I knew this particular result, but I have encountered some facts of a similar nature and even tried to prove some of them. Without any success... It is strange that the existing proof is the only one known.

> I am, BTW, partially in connection with my current interests, very interested in the elimination of the non-constructive elements from the proof of the BK or, at least, from the proof of the Merkurjev-Suslin theorem about K_2/l.

> The main such element is the use of the axiom of choice or

rather of the existence of well-ordering on any set quite early in the proof.

I am very interested in finding a proof that avoids this part of the argument.

(....)

I am sure that I can formalize constructively the statement of the BK. I can also formalize constructively most of my mathematics such as the motivic Steenrod operations.

This was a few months before his death on September 30th, 2017. It will take time for many of us to recover from it.

1   Proof of (i) $\Rightarrow$ (iii)

Recall that the Bloch-Kato conjecture is a special case of the Beilinson-Lichtenbaum conjecture; the statement thus follows from:

Proposition 1.1 *We assume the Bloch-Kato conjecture holds modulo l. Let A be a semi-local k-algebra. Let I, J be two ideals of A such that $I \cap J = 0$. If $n \geq 2$, the homomorphism*

$$K_n^M(A)/l \to K_n^M(A/I)/l \oplus K_n^M(A/J)/l$$

is injective.

Proof   By Kerz [8, Th. 1.2], the norm residue homomorphism

$$K_n^M(A)/l \to H_{\text{ét}}^n(A, \mu_l^{\otimes n})$$

is bijective for $n \geq 1$. By the usual transfer argument [8, Def. 5.5], we may assume that $\mu_l \subset k$. Recall that étale cohomology with finite coefficients verifies closed Mayer-Vietoris, as a consequence of proper base change (for closed immersions!). Consider the diagram

$$
\begin{array}{ccc}
K_{n-1}^M(A/I) \otimes \mu_l \oplus K_{n-1}^M(A/J) \otimes \mu_l & \longrightarrow & H_{\text{ét}}^{n-1}(A/I, \mu_l^{\otimes n}) \oplus H_{\text{ét}}^{n-1}(A/J, \mu_l^{\otimes n}) \\
\downarrow{\scriptstyle \alpha} & & \downarrow \\
K_{n-1}^M(A/I+J) \otimes \mu_l & \longrightarrow & H_{\text{ét}}^{n-1}(A/I+J, \mu_l^{\otimes n}) \\
& & \downarrow{\scriptstyle \partial} \\
K_n^M(A)/l & \longrightarrow & H_{\text{ét}}^n(A, \mu_l^{\otimes n}) \\
\downarrow{\scriptstyle b} & & \downarrow \\
K_n^M(A/I)/l \oplus K_n^M(A/J)/l & \longrightarrow & H_{\text{ét}}^n(A/I, \mu_l^{\otimes n}) \oplus H_{\text{ét}}^n(A/J, \mu_l^{\otimes n})
\end{array}
$$

where the horizontal maps are norm residue isomorphisms and ∂ is the boundary map for the long exact sequence corresponding to the closed covering $\operatorname{Spec} A = \operatorname{Spec}(A/I) \cup \operatorname{Spec}(A/J)$. The two squares obviously commute, and all horizontal maps are isomorphisms since $n \geq 2$. But a is surjective, hence $\partial = 0$, hence b is injective.

$\square$

Remark 1.2 This proof does not work for $n = 1$. In fact the conclusion is false: the short exact sequence

$$0 \to A^* \to (A/I)^* \oplus (A/J)^* \to (A/I + J)^* \to 0$$

yields a long exact sequence

$$0 \to {}_lA^* \to {}_l(A/I)^* \oplus {}_l(A/J)^* \xrightarrow{\rho} {}_l(A/I + J)^*$$
$$\to A^*/l \to (A/I)^*/l \oplus (A/J)^*/l \to (A/I + J)^*/l \to 0$$

so $\operatorname{Coker} \rho$ is finite but may be nontrivial if $A/I + J$ is too disconnected.

2   Motivic cohomology and Milnor K-theory

For $n \geq 0$, the n-th motivic complex of Suslin and Voevodsky may be defined as

$$\mathbf{Z}(n) = C_*(\mathbb{G}_m^{\wedge n})[-n]$$

where $\mathbb{G}_m^{\wedge n}$ denotes the direct summand of $L((\mathbf{A}^1 - 0)^n)$ given by sections trivial at $(\mathbf{A}^1 - 0)^i \times \{1\} \times (\mathbf{A}^1 - 0)^{n-i-1}$ $(0 \leq i < n)$ and C_* is the Suslin complex [11, Th. 15.2]. We have the following basic results:

Theorem 2.1 *[14], [8, Th. 1.1] We have* $\mathbf{Z}(0) = \mathbf{Z}$, $\mathbf{Z}(1) \simeq \mathbb{G}_m[-1]$, $H^i(\mathbf{Z}(n)) = 0$ *for* $i > n$ *and* $H^n(\mathbf{Z}(n)) = K_n^M$.

3   Inverting the motivic Bott element, after Thomason and Levine

Assume that k contains a primitive l-th root of unity: the Nisnevich sheaf μ_l is then constant, cyclic of order l. From the exact triangle

$$\mu_l[0] \to \mathbf{Z}/l(1) \to \mathbb{G}_m/l[-1] \xrightarrow{+1} \qquad (3.1)$$

and the isomorphism $\mathbf{Z}/l(n) \otimes \mathbf{Z}/l(1) \xrightarrow{\sim} \mathbf{Z}/l(n+1)$, we get a map in $\mathbf{DM}^{\mathrm{eff}}$:

$$\mathbf{Z}/l(n) \otimes \mu_l \to \mathbf{Z}/l(n + 1) \qquad (3.2)$$

hence another map

$$\mathbf{Z}/l(n) \to \mathbf{Z}/l(n+1) \otimes \mu_l^{-1}$$

which becomes an isomorphism after sheafifying for the étale topology. Let $i \le n$: iterating, we get a commutative diagram in **HI**, the heart of the homotopy t-structure of $\mathbf{DM}^{\mathrm{eff}}$:

$$
\begin{array}{ccccccc}
H^i(\mathbf{Z}/l(n)) & \to & H^i(\mathbf{Z}/l(n+1)\otimes\mu_l^{-1}) & \to & H^i(\mathbf{Z}/l(n+2)\otimes\mu_l^{-2}) & \to \cdots \\
\downarrow \psi_n^i & & \downarrow \psi_{n+1}^i & & \downarrow \psi_{n+2}^i & \\
H^i(R\alpha_*\alpha^*\mathbf{Z}/l(n)) \xrightarrow{\sim} & H^i(R\alpha_*\alpha^*(\mathbf{Z}/l(n+1)\otimes\mu_l^{-1})) \xrightarrow{\sim} & H^i(R\alpha_*\alpha^*(\mathbf{Z}/l(n+2)\otimes\mu_l^{-2})) \xrightarrow{\sim} \cdots
\end{array}
$$

where α is the projection $\mathbf{Sm}_{\text{ét}} \to \mathbf{Sm}_{\text{Nis}}$. We have:

Theorem 3.1 *[9, Th. 1.1] Assume that k is non exceptional if $l = 2$. Then the direct limit of the above diagram is a (vertical) isomorphism.*

(For $l = 2$, Levine assumes either char $k > 0$ or that k contains a square root of -1, but the hypothesis he actually uses is that k is not exceptional.)

The Beilinson-Lichtenbaum conjecture is the statement that ψ_n^i is an isomorphism for all (i, n) such that $i \le n$. Hence Theorem 3.1 implies:

Proposition 3.2 *Under the assumption of Theorem 3.1, the Beilinson-Lichtenbaum conjecture holds modulo l if and only if the map*

$$H^i(\mathbf{Z}/l(n)) \otimes \mu_l \to H^i(\mathbf{Z}/l(n+1))$$

is an isomorphism for any (i, n) such that $i \le n$.

$\square$

4 Reformulation of Proposition 3.2

Proposition 4.1 a) *For all $n \ge 0$, the objects $\mathbb{G}_m \otimes K_n^M/l$ and $\mathbf{Z}(n)[n] \otimes \mathbb{G}_m/l$ of $\mathbf{DM}^{\mathrm{eff}}$ are concentrated in cohomological degrees ≤ 0 (for the homotopy t-structure), and we have isomorphisms*

$$H^0(\mathbb{G}_m \otimes K_n^M/l) \simeq H^0(\mathbf{Z}(n) \otimes \mathbb{G}_m/l[n]) \simeq K_{n+1}^M/l.$$

b) *Assume that k is non exceptional if $l = 2$. Then the following statements are equivalent:*

(i) *The Beilinson-Lichtenbaum conjecture holds modulo l.*

(ii) *For all $n \geq 1$, $\mathbf{Z}(n) \otimes \mathbb{G}_m/l \xrightarrow{\sim} K^M_{n+1}/l[-n]$ in $\mathbf{DM}^{\mathrm{eff}}$.*

(iii) *For all $n \geq 1$, $\mathbb{G}_m \otimes K^M_n/l \xrightarrow{\sim} K^M_{n+1}/l[0]$ in $\mathbf{DM}^{\mathrm{eff}}$.*

(iv) *For all $n \geq 2$, the image of $K^M_n/l[0]$ under the localisation functor $\nu_{\leq 0} : \mathbf{DM}^{\mathrm{eff}} \to \mathbf{DM}^\circ$ of [6, (4.5)] is 0, where $\mathbf{DM}^\circ$ is the category of birational motivic sheaves of [6].*

(v) *For any function field K/k, any $n \geq 2$ and any $q \geq 0$, we have*
$$H_q(K^M_n/l(\hat{\Delta}^*_K)) = 0.$$

Proof a) follows from Theorem 2.1, the isomorphism $\mathbf{Z}(1) \simeq \mathbb{G}_m[-1]$ and the right t-exactness of $\otimes$ [6, comment after (5.2)]. b) We reduce to $\mu_l \subset k$. Let C_n be the cone of (3.2), so that $C_n \simeq \mathbf{Z}(n) \otimes \mathbb{G}_m/l[-1]$. In view of a) and Proposition 3.2, (i) is equivalent to saying that C_n is concentrated in degree $n+1$ and that the map
$$K^M_{n+1}/l \simeq H^{n+1}(\mathbf{Z}/l(n+1)) \to H^{n+1}(C_n)$$
is an isomorphism. This shows that (i) $\iff$ (ii).

The identity
$$\mathbf{Z}(n) \otimes \mathbb{G}_m/l \simeq \mathbb{G}_m \otimes \mathbf{Z}(n-1) \otimes (\mathbb{G}_m/l)[-1]$$
shows that (ii) $\iff$ (iii) by induction on n (note that (ii) and (iii) are identical for $n = 1$).

By [6, Prop. 4.2.5], the statement in (iv) is equivalent to K^M_n/l being divisible by $\mathbf{Z}(1)$ in $\mathbf{DM}^{\mathrm{eff}}$, which is implied by (iii). Conversely, if $K^M_n/l \simeq C(1)$ for some $C \in \mathbf{DM}^{\mathrm{eff}}$, Voevodsky's cancellation theorem [18] shows that $C \simeq \underline{\mathrm{Hom}}(\mathbf{Z}(1), K^M_n/l) = \underline{\mathrm{Hom}}(\mathbb{G}_m, K^M_n/l)[1] = (K^M_n/l)_{-1}[1] = K^M_{n-1}/l[1]$ (compare [7, Prop. 4.3 and Rk. 4.4]).

For (iv) $\iff$ (v), we use [6, Rk. 4.6.3] (see also [5, Rk. 2.2.6]): let $i^\circ : \mathbf{DM}^\circ \to \mathbf{DM}^{\mathrm{eff}}$ be the inclusion. For any $C \in C(\mathbf{PST})$ which is $\mathbf{A}^1$-invariant and satisfies Nisnevich excision, and for any connected $Y \in \mathbf{Sm}(k)$ with function field K, one has a quasi-isomorphism
$$(i^\circ \nu_{\leq 0} C_{\mathrm{Nis}})(Y) \simeq R\Gamma(\hat{\Delta}^*_K, C). \tag{4.1}$$

For any $\mathcal{F} \in \mathbf{HI}$, one has $H^q_{\mathrm{Nis}}(X, \mathcal{F}) = 0$ for $q \neq 0$ for any smooth semi-local k-scheme X as a consequence of [16, Th. 4.37]. Therefore, the right hand side of (4.1) for $C = \mathcal{F}[0]$ is quasi-isomorphic to the complex associated to the simplicial abelian group
$$\mathcal{F}(\hat{\Delta}^*_K),$$
which shows the equivalence of (iv) and (v) by taking $\mathcal{F} = K^M_n/l$. This concludes the proof.

$\square$

Remark 4.2 In Proposition 4.1 b), (ii) is also (trivially) true for $n = 0$, but not (iii) (see (3.1)).

5   Elementary lemmas on Milnor K-groups

Let A be a commutative semi-local ring, and let I be an ideal of A. We write $(1 + I)^* = (1 + I) \cap A^* = \mathrm{Ker}(A^* \to (A/I)^*)$.

Lemma 5.1 *Assume that $|A/\mathfrak{m}| > 2$ for all maximal ideals $\mathfrak{m}$ of A. Then, with the above notation:*

  (i) $A^* \to (A/I)^*$ *is surjective.*

  (ii) *Let $\bar{a} \in A/I$ be such that $\bar{a}, 1 - \bar{a} \in (A/I)^*$. Then there exists $a \in A$ such that $a \mapsto \bar{a}$ and $a, 1 - a \in A^*$.*

  (iii) *Let J be another ideal of A, with image $\bar{J} \subset A/I$. Then $(1 + J)^* \to (1 + \bar{J})^*$ is surjective.*

Proof Let R be the Jacobson radical of A, so that $1 + R \subset A^*$. Assume first $R = 0$: then A is a finite product of fields and the three statements are obvious (the cardinality hypothesis is used in (ii)). The general case follows from chasing in the commutative square

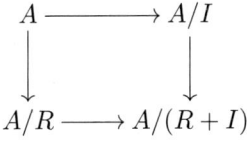

$\square$

Lemma 5.2 *Keep the assumption of Lemma 5.1. With the above notation, $K_*^M(A) \to K_*^M(A/I)$ is surjective with kernel the ideal generated by $(1+I)^*$.*

Proof The first assertion follows from Lemma 5.1 (i). To prove the second one, let us construct a surjective section to the surjection

$$\frac{K_*^M(A)}{\{(1 + I)^*\} K_*^M(A)} \twoheadrightarrow K_*^M(A/I).$$

It suffices to show that the surjective ring homomorphism $T((A/I)^*) \twoheadrightarrow K_*^M(A)/\{(1 + I)^*\} K_*^M(A)$ extending the identity map in degree 1 kills the Steinberg relations: this follows from Lemma 5.1 (ii).

$\square$

Proposition 5.3 *Keep the assumption of Lemma 5.1, and let I, J be two ideals of A. Then the sequence*

$$K_*^M(A) \to K_*^M(A/I) \oplus K_*^M(A/J) \to K_*^M(A/I + J) \to 0$$

is exact.

Proof Let $\bar{I}$ be the image of I in A/J. Consider the commutative diagram

$$\begin{array}{ccccccc}
\{(1+I)^*\}K_*^M(A) & \longrightarrow & K_*^M(A) & \longrightarrow & K_*^M(A/I) & \longrightarrow & 0 \\
\downarrow & & \downarrow & & \downarrow & & \\
\{(1+\bar{I})^*\}K_*^M(A/J) & \longrightarrow & K_*^M(A/J) & \longrightarrow & K_*^M(A/I+J) & \longrightarrow & 0.
\end{array}$$

By Lemma 5.2, the rows are exact and the middle and right vertical maps are surjective; by Lemma 5.1 (iii), the left vertical map is also surjective. The claim now follows from a diagram chase.

$\square$

6   End of proof of Theorem 1

Lemma 6.1 *Let* **sLoc** *be the category of semi-local K-schemes. Let F be a contravariant functor from* **sLoc** *to abelian groups. Suppose that, for any $X \in$ **sLoc** *and any closed cover $X = Z_1 \cup Z_2$, the sequence*

$$0 \to F(X) \to F(Z_1) \oplus F(Z_2) \to F(Z_1 \cap Z_2)$$

is exact. (Here, $Z_1 \cap Z_2 := Z_1 \times_X Z_2$ is the scheme-theoretic intersection.) Then, for any closed cover $X = Z_1 \cup \cdots \cup Z_r$, the sequence

$$0 \to F(X) \to \bigoplus_{j=1}^r F(Z_j) \to \bigoplus_{j<k} F(Z_j \cap Z_k)$$

is exact.

Proof Of course this lemma is much more general and the point is to spell out its proof. Let $Y = Z_1 \cup \ldots Z_{r-1}$. By hypothesis, the sequence

$$0 \to F(X) \to F(Y) \oplus F(Z_r) \to F(Y \cap Z_r)$$

is exact and, by induction on r, the map

$$F(Y \cap Z_r) \to \bigoplus_{j<r} F(Z_j \cap Z_r)$$

is injective. The conclusion follows by chasing in the diagram

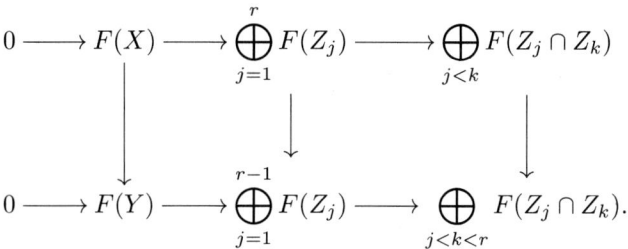

Lemma 6.2 *(See also [10, Lemma 2.4]) Let* $A_* = (A_n)_{n \geq 0}$ *be a simplicial abelian group. Let* (A_n^0) *be the normalised complex of* A: $A_r^0 = \bigcap_{i>0} \mathrm{Ker}(\partial_i : A_r \to A_{r-1})$. *For* $q > 0$, *consider the commutative diagram of complexes*

$$
\begin{array}{ccccc}
A_{q+1}^0 & \xrightarrow{\partial_0} & A_q^0 & \xrightarrow{\partial_0} & A_{q-1}^0 \\
\downarrow{\scriptstyle\alpha} & & \downarrow{\scriptstyle\beta} & & \downarrow{\scriptstyle\gamma} \\
A_{q+1} & \longrightarrow & \bigoplus_{i=0}^{q} A_q & \longrightarrow & \bigoplus_{0 \leq j < k \leq q} A_{q-1}
\end{array}
\tag{6.1}
$$

where α is inclusion, $\beta(x) = (x, 0, \ldots, 0)$, $\gamma(x)_{j,k} = \begin{cases} x & si\ (j,k) = (0,1) \\ 0 & else, \end{cases}$

$a = (\partial_i)_{0 \leq i \leq q}$ *and the (i, j, k)-component of b is*

$$
\begin{cases}
0 & if\ i \neq j, k \\
\partial_{k-1} & if\ i = j \\
-\partial_j & if\ i = k.
\end{cases}
$$

Then this diagram induces an injection on homology.

Proof Obvious.

    □

Proof (End of proof of Theorem 1) We saw in §1 that (i) $\Rightarrow$ (iii); we have (i) $\Longleftrightarrow$ (ii) by the equivalence (i) $\Longleftrightarrow$ (iii) in Proposition 4.1 b). Obviously, (iii) $\Rightarrow$ (iv). It remains to show that (iv) $\Rightarrow$ (i).

Suppose that (iv) holds in Theorem 1. In view of Proposition 5.3 and (the proof of) Lemma 6.1, we get for all $q > 0$ an exact sequence

$$0 \to K_n^M(\partial\hat{\Delta}_K^q)/l \to \bigoplus_{j=0}^{q} K_n^M(\hat{\Delta}_{K,j}^q)/l \to \bigoplus_{j<k} K_n^M(\hat{\Delta}_{K,j,k}^q)/l.$$

But $K_n^M(\hat{\Delta}_K^q) \to K_n^M(\partial\hat{\Delta}_K^q)$ is surjective, hence the bottom row of (6.1) is exact for $A_* = K_*^M(\hat{\Delta}_K^*)/l$. By Lemma 6.2, Condition (v) of Proposition 4.1 b) holds, and therefore so does its Condition (i). This concludes the proof.

$\square$

Remark 6.3 For $A_* = K_*^M(\hat{\Delta}_K^*)/l$, the homology group of the bottom row of (6.1) may be reinterpreted in a more suggestive way: it is

$$\text{Coker}\left(H_{\text{Zar}}^0(\hat{\Delta}_K^{i+1}, \mathcal{F}) \to H_{\text{oc}}^0(\partial\hat{\Delta}_K^{i+1}, \mathcal{F})\right)$$

where oc denotes the open-closed topology introduced in [4].

References

[1] A. Beilinson and V. Vologodsky, *A DG guide to Voevodsky's motives*, Geom. Funct. Anal. **17** (2008), 1709–1787.

[2] T. Geisser and M. Levine, *The Bloch-Kato conjecture and a theorem of Suslin-Voevodsky*, J. Reine Angew. Math. **530** (2001), 55–103.

[3] D. Guin, *Homologie du groupe linéaire et K-théorie de Milnor des anneaux*, J. Alg. **123** (1989), 27–59.

[4] B. Kahn, *The Geisser-Levine method revisited and algebraic cycles over a finite field*, Math. Ann. **324** (2002), 581–617.

[5] B. Kahn and M. Levine, *Motives of Azumaya algebras*, J. Inst. Math. Jussieu **9** (2010), 481–599.

[6] B. Kahn and R. Sujatha, *Birational motives, II: triangulated birational motives*, Int. Math. Res. Not. IMRN 2017, no. 22, 6778–6831.

[7] B. Kahn and T. Yamazaki, *Voevodsky's motives and Weil reciprocity*, Duke Math. J. **162** (2013), 2751–2796.

[8] M. Kerz, *The Gersten conjecture for Milnor K-theory*, Invent. Math. **175** (2009), 1–33.

[9] M. Levine, *Inverting the motivic Bott element*, K-Theory **19** (2000), 1–28.

[10] M. Levine, *Techniques of localization in the theory of algebraic cycles*, J. Alg. Geom. **10** (2001), 299–363.

[11] C. Mazza and V. Voevodsky, C. Weibel *Lecture notes on motivic cohomology*, Clay Math. Monographs **2**, AMS, Clay Math. Inst., 2006.

[12] Yu. Nesterenko and A. Suslin, *Homology of the general linear group over a local ring, and Milnor's K-theory* (Russian), Izv. Akad. Nauk SSSR Ser. Mat. **53** (1989), 121–146; translation in Math. USSR-Izv. **34** (1990), 121–145.

[13] R Sugiyama, *Motivic homology of a semiabelian variety over a perfect field*, Doc. Math. **19** (2014), 1061–1084.

[14] A. Suslin and V. Voevodsky, *Bloch-Kato conjecture and motivic cohomology with finite coefficients*, in The arithmetic and geometry of algebraic cycles (Banff, AB, 1998), 117–189, NATO Sci. Ser. C Math. Phys. Sci., **548**, Kluwer, 2000.

[15] R. Thomason, *Algebraic K-theory and étale cohomology*, Ann. Sci. Éc. Norm. Sup. **18** (1985), 437–552.

[16] V. Voevodsky, *Cohomological theory of presheaves with transfers*, *in* Cycles, transfers, and motivic homology theories, Ann. of Math. Stud. **143**, Princeton Univ. Press, 2000, 87–137.

[17] V. Voevodsky, *Triangulated categories of motives over a field*, *in* Cycles, transfers, and motivic homology theories, Ann. of Math. Stud. **143**, Princeton Univ. Press, 2000, 188–238.

[18] V. Voevodsky, *Cancellation theorem*, Doc. Math. 2010, Extra volume: Andrei A. Suslin sixtieth birthday, 671–685.

[19] V. Voevodsky, *On motivic cohomology with $\mathbf{Z}/l$-coefficients*, Annals of Math. **174** (2011), 401–438.

BRUNO KAHN, IMJ-PRG, CASE 247, 4 PLACE JUSSIEU, 75252 PARIS CEDEX 05, FRANCE

E-mail: `bruno.kahn@imj-prg.fr`

K-Theory
Copyright ©2018 Tata Institute of Fundamental Research
Publisher: Hindustan Book Agency, New Delhi, India

Explicit Chow-Lefschetz decompositions for Kummer manifolds

Reza Akhtar and Roy Joshua[1]

Abstract

Let X be a pseudo-smooth projective variety over a field k (that is, a quotient of a smooth projective variety over k by the action of a finite group) whose singularities are isolated k-rational points. Let $f : Y \to X$ be the morphism obtained by blowing up these points on X. Assume further that Y is pseudo-smooth, and that the components of the exceptional divisor are projective spaces. Then, without invoking the theory of finite-dimensional motives or assuming any of the standard conjectures, we show that a Chow-Künneth decomposition on either X or Y gives rise, by means of an explicit construction, to a Chow-Künneth decomposition on the other. We use these constructions to show that various properties (among them Murre's Conjectures and being of Lefschetz type) hold for X if and only if they hold for Y. The main examples of interest to us are *Kummer manifolds*: these are obtained by taking the quotient of an abelian variety by the involution $a \mapsto -a$, and then blowing up the singular locus. We give several further applications of our construction to this particular class of examples.

1 Introduction and summary of results

Let $X \mapsto H^*(X)$ be a Weil cohomology theory on varieties over some algebraically closed field. According to the standard conjectures of Grothendieck formulated in [GR], one expects — among other things — that if X has dimension d, then the Künneth components of the diagonal class $[\Delta_X] \in H^d(X \times X)$ should lie in the subgroup $A^d(X \times X)$ of $H^{\dim X}(X \times X)$ generated by algebraic cycles. Moreover, for any smooth hyperplane section $W \subseteq X$, the so-called Hard Lefschetz Theorem should hold: specifically, if $L : H^i(X) \to H^{i+2}(X)$ is the Lefschetz operator, then composing with the iterated operator L^{d-i} should define an isomorphism $A^i(X) \to A^{d-i}(X)$

---

[1]The second author was supported by the NSF

for all i. A detailed exposition of the standard conjectures may be found in the comprehensive survey article of Kleiman [Kl].

A related but stronger set of conjectures was formulated by Jacob Murre in [Mu]. Murre conjectured that the Künneth components of the diagonal class of each such X should actually be defined in the category of (rational) Chow motives, and that these projectors should act on the (rational) Chow groups of X in a prescribed manner. The first of Murre's conjectures — the existence of a Chow-Künneth decomposition — has been verified for certain classes of varieties (curves, surfaces, abelian varieties, and various other special cases); however, it remains wide open in the general case. The existence of a Chow-Künneth decomposition for abelian varieties was first demonstrated by Shermenev [Sh], although it is a later construction of the same by Deninger and Murre [DM] that lends itself most readily to applications. Künnemann [Ku] used the Deninger-Murre construction to prove that the Hard Lefschetz Theorem holds for abelian varieties at the level of Chow motives. In fact, Künnemann proved much more, constructing Lefschetz, Lambda, and *-operators for abelian varieties, and showing that various identities among these, which hold in the setting of Kähler geometry, actually hold at the level of Chow groups. More significantly, he showed that if a variety is of *Lefschetz type* (see Section 2.3), then many expected properties — including the Hard Lefschetz Theorem and the existence of projectors appropriately refining the Chow-Künneth decomposition — follow immediately.

Let A denote an abelian variety over an algebraically closed field of characteristic different from 2. Its associated *Kummer variety* K_A is the quotient of A by the involution $a \mapsto -a$. If A has dimension $d > 0$, then K_A has 2^{2d} singular points, which are precisely the images of the 2-torsion points of A under the quotient map $A \to K_A$. Blowing up these points yields a smooth variety K'_A which we call the *Kummer manifold*. Even though K_A is a singular variety, it is pseudo-smooth (i.e. the quotient of a smooth variety by the action of a finite group scheme), so basic methods of intersection theory may be used to study its Chow groups with $\mathbb{Q}$-coefficients (see [F, Example 1.7.6]). In earlier work [AJ1], the authors of the present article used the Chow-Künneth decomposition for A constructed by Deninger and Murre to construct an explicit Chow-Künneth decomposition for K_A. Although the *existence* of such a decomposition follows from the theory of finite-dimensional motives (see [GP]), we gave several applications for our construction which would not have been possible from the abstract theory. This work was continued in [AJ2], in which we used Künnemann's Lefschetz algebra structure on the Chow groups of an abelian variety to establish one for Kummer varieties. Once again, the existence of such a decomposition was established in [KMP] under the assumption of parts of the standard conjectures (which are known to hold for Kummer manifolds in characteristic 0 by work of Arapura [Ar]); however, our construction has no dependence on characteristic and furthermore lends itself to several applications.

In the present article, we use our constructions for K_A to exhibit an explicit Chow-Künneth decomposition for the Kummer manifold K'_A and also an explicit Lefschetz algebra structure on its Chow groups. The following is a technical result, which, combined with our earlier results (see [AJ1] and [AJ2]), provides the Chow-Künneth decomposition for the Kummer manifolds.

Theorem *(See Theorems 3.5 and 5.1) Let X denote a pseudo-smooth variety of dimension d over a field k and Y the variety obtained by blowing up a finite number of k-rational points on X. Suppose further that Y is pseudo-smooth, and that the (respective) exceptional divisors of the blow-up at each point are isomorphic to $\mathbb{P}^{d-1}$.*
   Then:

- *If either X or Y has a Chow-Künneth decomposition, then this can be used to construct (explicitly) a Chow-Künneth decomposition on the other (cf. (3.7) and Corollary 3.13.)*
- *If the Chow-Künneth decomposition (so constructed) on either X or Y satisfies Poincaré duality (respectively, Murre's Conjecture **B**, **B'**, **C**, **D**) then the same is true for the other variety.*
- *Y is of Lefschetz type if and only if X is of Lefschetz type.*

When combined with the results of [AJ1] and [AJ2], we may then conclude:

Corollary *Let A denote an abelian variety of dimension $d > 0$ over an algebraically closed field of characteristic different from 2 and K'_A its Kummer manifold. Then K'_A has a Chow-Künneth decomposition satisfying Poincaré duality and Murre's conjecture **B'**; moreover, K'_A has Lefschetz type in the sense of Definition 2.4. If $d \leq 4$, then K'_A also satisfies Murre's conjecture **B**.*

We also give several other applications of our construction. The first concerns powers of the relation of algebraic equivalence (on algebraic cycles). For a pseudo-smooth variety V, let $LCH^p_{\mathbb{Q}}(V)$ denote the subgroup of $CH^p_{\mathbb{Q}}(V)$ consisting of cycles algebraically equivalent to zero. For $r \geq 1$, we denote by L^{*r} the rth power of (the equivalence relation) L, as defined by Hiroshi Saito [Sai].

Theorem *(See Proposition 6.3) Let A denote an abelian variety of dimension $d > 0$ over an algebraically closed field of characteristic different from 2. Let $X = K_A$ and $Y = K'_A$. Let $[\Delta_X] = \sum_{i=0}^{2d} \pi_i^X$ denote the Chow-Künneth decomposition for X constructed in [AJ1] and $[\Delta_Y] = \sum_{i=0}^{2d} \pi_i^Y$ the Chow-Künneth decomposition for Y constructed in (3.7). Define filtrations on $CH_{\mathbb{Q}}^*(X)$ and $CH_{\mathbb{Q}}^*(Y)$ by $F^r CH_{\mathbb{Q}}^*(X) = \sum_{i=0}^{2d-r} \pi_i^X \bullet CH_{\mathbb{Q}}^p(X)$ and $F^r CH_{\mathbb{Q}}^*(Y) = \sum_{i=0}^{2d-r} \pi_i^Y \bullet CH_{\mathbb{Q}}^p(Y)$. Then:*

(i) *For $r \geq 1$, $F^r CH_{\mathbb{Q}}^d(X) = L^{*r} CH_{\mathbb{Q}}^d(X)$ and $F^r CH_{\mathbb{Q}}^d(Y) = L^{*r} CH_{\mathbb{Q}}^d(Y)$.*
(ii) *For $r > d$, $L^{*r} CH_{\mathbb{Q}}^*(X) = 0$ and $L^{*r} CH_{\mathbb{Q}}^*(Y) = 0$.*

As another application, we prove a Hard Lefschetz Theorem for Chow groups of Kummer manifolds in the case that the base field is the algebraic closure of a finite field of characteristic different from 2.

Theorem *(See Corollary 6.5) Let Y denote the Kummer manifold associated to an abelian variety of dimension $d > 0$ over an algebraic closure of a finite field of characteristic different from 2, and let L_Y denote the Lefschetz operator as constructed in the present paper. Then for $2p \leq d$, the map $CH_{\mathbb{Q}}^p(Y) \to CH_{\mathbb{Q}}^{d-p}(Y)$ defined by $c \mapsto L_Y^{d-2p} \bullet c$ is an isomorphism.*

Most of our arguments rely on the following fundamental fact about the structure of the Chow groups of blow-ups of the sort we are considering. Suppose $f : Y \to X$ is the morphism describing the blow-up of a pseudo-smooth variety X at a point, such that Y is pseudo-smooth and the exceptional divisor Z is isomorphic to the projective space over the ground field. This is a strong assumption, but it guarantees that the cohomology of the exceptional divisors is generated by algebraic cycles; this may be viewed as the underlying reason why our strategy works. In this case, $CH_{\mathbb{Q}}^*(Y \times Y)$ is the internal direct sum of two subgroups, which we call A and B: A consists of cycles pulled back from $X \times X$ via $f \times f$, while B consists of cycles supported on $Z \times Y \cup Y \times Z$. With respect to the non-commutative ring structure given by the composition of correspondences on $CH_{\mathbb{Q}}^*(Y \times Y) = CH^*(Y \times Y) \otimes \mathbb{Q}$, A is a ring and B is a two-sided ideal of $CH_{\mathbb{Q}}^*(Y \times Y)$; furthermore, A and B are nearly orthogonal to each other. Using these properties —and the crucial fact that every cycle in B can be written as a sum of external products of cycles on Y — we can, starting with a Chow-Künneth decomposition on X, construct Chow-Künneth projectors on Y, and then use these to construct the appropriate operators necessary for the exhibition of a Lefschetz algebra structure on $CH_{\mathbb{Q}}^*(Y \times Y)$. We also show that if a Chow-Künneth decomposition or Lefschetz algebra structure is known for Y, then pushing forward all the relevant cycles to X will establish the analogous results there.

The paper is organized as follows. We begin in Section 2 by providing definitions and results from intersection theory. In Section 3, we establish some important structural results concerning the Chow groups of a blow-up, and then give our main construction involving Chow-Künneth decompositions. The remainder of the paper is devoted to various applications of our explicit constructions. The first application, to Murre's Conjectures, appears in Section 4. In Section 5, we study Lefschetz decompositions in the context of blow-ups. Both of these are discussed in a fairly general setting. We conclude in section 6 with various specialized applications to Kummer varieties and manifolds.

Acknowledgements   The present work was prompted by helpful exchanges with several colleagues; in particular, we thank Donu Arapura, Michel Brion, Igor Dolgachev, and Charles Vial for helpful discussions. We thank the referee for a careful reading, and for several suggestions which helped improve this article.

2   Preliminaries

2.1   Correspondences and Murre's Conjectures

Let k denote a field. For convenience, we refer to the quotient of a smooth variety by the action of a finite group (scheme) as a *pseudo-smooth variety*. It is well-known that the basic machinery of intersection theory and the usual formalism for correspondences extends naturally from smooth varieties to pseudo-smooth varieties, provided one uses rational coefficients. We may thus define the category $\mathcal{M}_k(\mathbb{Q})$ of (rational) Chow motives of pseudo-smooth projective varieties in the same way as for smooth projective varieties (see for example, [Sch]). Throughout this article, we use the notation $CH^i(X)$ for the Chow groups of (an algebraic scheme) X and write $CH^i_{\mathbb{Q}}(X) = CH^i(X) \otimes \mathbb{Q}$. It is worth noting that if a finite group G acts on a smooth variety X, then the machinery of equivariant intersection theory allows us to identify the equivariant Chow groups $CH^*_G(X)_{\mathbb{Q}}$ with $CH^*_{\mathbb{Q}}(X/G)$. Thus, the extension of the usual formalism of correspondences to pseudo-smooth varieties (see [F, Example 1.7.6]) can also be derived from the analogous theory in the equivariant context.

Since we will make use of many projection maps in the sequel, we reserve the symbol p for these, with the superscript indicating the domain and the subscript the range. For example, if k is a field and X, Y, Z are pseudo-smooth varieties over k, $p_{13}^{XYZ} : X \times Y \times Z \to X \times Z$ is the map $(x, y, z) \mapsto (x, z)$. A subscript of $\emptyset$ indicates the structure morphism; for example, $p_{\emptyset}^{XY}$ is the structure morphism $X \times Y \to \operatorname{Spec} k$. Given cycles $\alpha \in CH^i(X)$ and $\beta \in CH^j(Y)$, we refer to their exterior product $\alpha \times \beta = p_1^{XY*}\alpha \cdot p_2^{XY*}\beta$ as a *product cycle* on $X \times Y$ of *type* (i, j); by abuse of terminology, we sometimes also refer to linear combinations of such elements as product cycles.

Now suppose X, Y, and Z are pseudo-smooth varieties over k, with $\gamma \in CH^*(X \times Y)$ and $\delta \in CH^*(Y \times Z)$. The composition $\delta \bullet \gamma \in CH^*(X \times Z)$ is defined by

$$\delta \bullet \gamma = p_{13}^{XYZ}{}_* (p_{12}^{XYZ*}\gamma \cdot p_{23}^{XYZ*}\delta).$$

Composition of correspondences is associative; we will use this fact freely without explicit mention in the sequel. If $s : X \times Y \to Y \times X$ is the exchange of factors, we define the *transpose* of $\alpha \in CH^*(X \times Y)$ by $\alpha^t := s^*(\alpha)$. We write Δ_X for the diagonal in $X \times X$ and Γ_f for the graph of a morphism f between (pseudo-smooth) varieties. Since $[\Delta_X] \bullet \gamma = \gamma = \gamma \bullet [\Delta_X]$ for $\gamma \in CH^*(X \times X)$, the operation $\bullet$ makes $CH^*(X \times X)$ into a (noncommutative) ring with unit element $[\Delta_X]$; furthermore, $CH^{\dim X}(X \times X)$ is a subring of $CH^*(X \times X)$.

We say that a variety X of dimension d has a *Chow-Künneth decomposition* (or CK-decomposition for short) if the diagonal class $[\Delta_X] \in CH_{\mathbb{Q}}^d(X \times X)$ has a decomposition into mutually orthogonal idempotents, each of which maps onto the appropriate Künneth component under the cycle map. More precisely, we require there exist $\pi_i \in CH_{\mathbb{Q}}^d(X \times X)$, $0 \le i \le 2d$, such that:

1. $[\Delta_X] = \displaystyle\sum_{i=0}^{2d} \pi_i$;

2. $\pi_i \bullet \pi_i = \pi_i$ for all i, and $\pi_i \bullet \pi_j = 0$ for $i \ne j$;

3. If H^* is a Weil cohomology theory, then for each i, the image of π_i under the cycle map $cl_X : CH_{\mathbb{Q}}^d(X \times X) \to H^{2d}(X \times X; \mathbb{Q})$ is the $(2d - i, i)$ Künneth component of the diagonal class.

We say that a CK-decomposition as above satisfies *Poincaré duality* if $\pi_{2d-i} = \pi_i{}^t$ for $0 \le i \le 2d$.

Finally, we recall the conjectures of Murre, formulated in [Mu] for smooth varieties:

Murre's Conjectures

Let X denote a pseudo-smooth projective variety. Then

A. X has a CK-decomposition $[\Delta_X] = \displaystyle\sum_{i=0}^{2d} \pi_i$.

B. If $i < j$ or $i > 2j$, then π_i acts as 0 on $CH^j_{\mathbb{Q}}(X)$.

B'. If $i < j$ or $i > j + \dim X$, then π_i acts as 0 on $CH^j_{\mathbb{Q}}(X)$.

C. If we define

$$F^0 CH^j_{\mathbb{Q}}(X) = CH^j_{\mathbb{Q}}(X) \text{ and } F^k CH^j_{\mathbb{Q}}(X) = \mathrm{Ker}\ \pi_{2j+1-k\,*}\big|_{F^{k-1}CH^j_{\mathbb{Q}}(X)}$$

for $k > 0$, then the resulting filtration is independent of the particular choice of projectors π_i.

D. For any filtration as defined in **C**, $F^1 CH^j_{\mathbb{Q}}(X)$ is the subgroup of cycles in $CH^j_{\mathbb{Q}}(X)$ homologically equivalent to zero.

2.2 Intersection theory on pseudo-smooth varieties

One can define pullback maps, pushforward maps, and intersection products in the context of pseudo-smooth varieties, and many basic results (including, in particular, the projection formula) carry over from the smooth case into this setting, provided one uses rational coefficients; see [dBN] for details. For this reason, we use rational coefficients throughout this section, even though many of the results (appropriately rephrased) hold with integral coefficients in the smooth case. In the interest of making our proofs more concise, we will work with correspondences as much as possible; however, it will occasionally serve intuition better to argue directly using pullback and pushforward maps. To this end, we record the following "dictionary" (cf. [F, Proposition 16.1.1] and [F, Example 1.7.6]) which allows us to go back and forth between these two interpretations.

Lemma 2.1 *Let X, Y, Z be pseudo-smooth projective varieties over a field. Suppose $f : X \to Y$ and $g : Y \to Z$ are morphisms, and $\alpha \in CH^*_{\mathbb{Q}}(X \times Y)$, $\beta \in CH^*_{\mathbb{Q}}(Y \times Z)$, $\gamma \in CH^*_{\mathbb{Q}}(X \times Z)$. Then the following formulas hold:*

$$(1 \times g)_*(\alpha) = [\Gamma_g] \bullet \alpha,$$
$$(f \times 1)^*(\beta) = \beta \bullet [\Gamma_f],$$
$$(f \times 1)_*(\gamma) = \gamma \bullet [\Gamma_f^t],$$
$$(1 \times g)^*(\gamma) = [\Gamma_g^t] \bullet \gamma.$$

An important observation is that composition of correspondences is well-behaved with respect to pullback of cycles.

Lemma 2.2 *With notation as in Lemma 2.1, suppose further that g is a morphism of degree d, and $\alpha, \beta \in CH^*_{\mathbb{Q}}(Z \times Z)$. Then $(g \times g)^* \alpha \bullet (g \times g)^* \beta = d(g \times g)^*(\alpha \bullet \beta)$.*

Proof   Observe first that by the projection formula, we have:

$$[\Gamma_g] \bullet [\Gamma_g^t] = [\Delta_Z] \bullet [\Gamma_g] \bullet [\Gamma_g^t] = (g \times 1)_*(g \times 1)^*[\Delta_Z] = d[\Delta_Z].$$

Then

$$\begin{aligned}
(g \times g)^* \alpha \bullet (g \times g)^* \beta &= (1 \times g)^*(g \times 1)^* \alpha \bullet (1 \times g)^*(g \times 1)^* \beta \\
&= [\Gamma_g^t] \bullet \alpha \bullet [\Gamma_g] \bullet [\Gamma_g^t] \bullet \beta \bullet [\Gamma_g] \qquad (2.1) \\
&= d\left([\Gamma_g^t] \bullet \alpha \bullet \beta \bullet [\Gamma_g]\right) = d(g \times g)^*(\alpha \bullet \beta).
\end{aligned}$$

$\square$

The following fact about compositions of product cycles is surely well known; however, since we will be using it so frequently, we include a proof in the interest of completeness of exposition.

Lemma 2.3 *Let X be a pseudo-smooth irreducible projective variety of dimension d over some field k. Suppose $\alpha \in CH^i_{\mathbb{Q}}(X)$, $\beta \in CH^j_{\mathbb{Q}}(X)$, $\gamma \in CH^k_{\mathbb{Q}}(X)$, and $\delta \in CH^\ell_{\mathbb{Q}}(X)$. Then*

$$(\alpha \times \beta) \bullet (\gamma \times \delta) = (\gamma \times \beta) \cdot p_\emptyset^{XX*} p_{\emptyset\,*}^X(\delta \cdot \alpha).$$

In particular,

$$(\alpha \times \beta) \bullet (\gamma \times \delta) = m(\alpha, \delta)(\gamma \times \beta)$$

for some $m(\alpha, \delta) \in \mathbb{Q}$, which equals zero if $i + \ell \neq d$.

Proof

$$\begin{aligned}
(\alpha \times \beta) &\bullet (\gamma \times \delta) \\
&= p_{13}^{XXX}{}_*(p_{12}^{XXX*}(p_1^{XX*}\gamma \cdot p_2^{XX*}\delta) \cdot p_{23}^{XXX*}(p_1^{XX*}\alpha \cdot p_2^{XX*}\beta)) \\
&= p_{13}^{XXX}{}_*(p_{13}^{XXX*}p_1^{XX*}\gamma \cdot p_2^{XXX*}\delta \cdot p_2^{XXX*}\alpha \cdot p_{13}^{XXX*}p_2^{XX*}\beta) \\
&= p_{13}^{XXX}{}_*(p_{13}^{XXX*}(p_1^{XX*}\gamma \cdot p_2^{XX*}\beta) \cdot p_2^{XXX*}\delta \cdot p_2^{XXX*}\alpha) \\
&= p_1^{XX*}\gamma \cdot p_2^{XX*}\beta \cdot p_{13}^{XXX}{}_*p_2^{XXX*}(\delta \cdot \alpha) \\
&= p_1^{XX*}\gamma \cdot p_2^{XX*}\beta \cdot p_\emptyset^{XX*}p_{\emptyset\,*}^X(\delta \cdot \alpha) \\
&= (\gamma \times \beta) \cdot p_\emptyset^{XX*}p_{\emptyset\,*}^X(\delta \cdot \alpha).
\end{aligned}$$

Now let $m(\alpha, \delta) = p_\emptyset^{XX^*} p_{\emptyset*}^{X}(\delta \cdot \alpha)$. If $i + \ell \neq d$, then $p_{\emptyset*}^{X}(\delta \cdot \alpha) \in CH_\mathbb{Q}^{i+\ell-d}(\operatorname{Spec} k) = 0$. If $i + \ell = d$, then $p_\emptyset^{XX^*} p_{\emptyset*}^{X}(\delta \cdot \alpha) \in CH_\mathbb{Q}^0(X \times X) \cong \mathbb{Q}$.

□

2.3   Lefschetz algebra structure

We recall, with slight revisions, the definition of Lefschetz algebra from [Ku].

Definition 2.4 A *Lefschetz algebra* of dimension d is a triple $(R, \{\eta_i\}_{i=0}^{2d}, L, \Lambda)$ where $R = \bigoplus_{p \in \mathbb{Z}} R^p$ is a graded $\mathbb{Q}$-algebra, $L \in R^1$, $\Lambda \in R^{-1}$, and

(i) $\eta_0, \ldots, \eta_{2d}$ are elements of R^0 satisfying

$$\sum_{i=0}^{2d} \eta_i = 1 \text{ and } \eta_i \bullet \eta_j = \begin{cases} \eta_i & \text{if } i = j \\ 0 & \text{if } i \neq j. \end{cases}$$

(ii) For all i, $L \bullet \eta_i = \eta_{i+2} \bullet L$.

(iii) For all i, $\Lambda \bullet \eta_i = \eta_{i-2} \bullet \Lambda$.

(iv) $[\Lambda, L] := \Lambda \bullet L - L \bullet \Lambda = \sum_{i=0}^{2d}(d-i)\eta_i$.

The examples of primary concern to us arise when X is a pseudo-smooth projective variety of dimension d over some field, R is the ring $CH_\mathbb{Q}^{*+d}(X \times X)$, the η_i are Chow-Künneth components of $[\Delta_X]$, and $L \in CH_\mathbb{Q}^{d+1}(X \times X)$, $\Lambda \in CH_\mathbb{Q}^{d-1}(X \times X)$ are elements satisfying the identities (ii)–(iv). If $CH_\mathbb{Q}^{*+d}(X \times X)$ can be endowed with the structure of a Lefschetz algebra in this manner, we say that X is of *Lefschetz type*. Varieties of Lefschetz type are of interest largely due to the following result, which may be deduced formally from the definition.

Corollary 2.5 *[Ku, Theorem 4.1] Let R be a Lefschetz algebra as above, and define $I = \{(i, k) \in \mathbb{Z} \times \mathbb{Z} \mid \max\{0, i - d\} \leq k \leq \lfloor i/2 \rfloor\}$. Then R has a Lefschetz decomposition, i.e. there exist elements $p_{i,k} \in R^0$ satisfying:*

(i) $\sum_k q_{i,k} = \eta_i$ *for each i.*

(ii) $q_{i,k} \bullet \eta_j = \eta_j \bullet q_{i,k} = q_{i,k}$ *if $i = j$ and 0 otherwise.*

(iii) $q_{i,k} = 0$ *for $(i, k) \notin I$.*

(iv) $q_{i,k} \bullet q_{j,l} = q_{i,k}$ *if $i = j$ and $k = l$ and 0 otherwise.*

(v) $q_{i,k} \bullet L = L \bullet q_{i-2,k-1}$.

(vi) $\Lambda \bullet q_{i,k} = q_{i-2,k-1} \bullet \Lambda$.

(vii) $L \bullet \Lambda \bullet q_{i,k} = k(g - i + k + 1)q_{i,k}$.

(viii) $\Lambda \bullet L \bullet q_{i,k} = (k + 1)(g - i + k)q_{i,k}$.

Corollary 2.6 (Hard Lefschetz Theorem) *[Ku, Theorem 5.2] If X is a pseudo-smooth projective variety of dimension d over a field k, and $R = ((CH_{\mathbb{Q}}^*(X \times X), \{\eta_i\}_{i=0}^{2d}, L, \Lambda)$ is a Lefschetz algebra, then for i, $0 \le i \le d$, the correspondence L^{d-i} defines an isomorphism of motives $h^i(X) \xrightarrow{\cong} h^{2d-i}(X)(d - i)$ with inverse Λ^{d-i}, where $h^j(X)$ is the rational Chow motive $(X, \eta_j, 0)$.*

3   An explicit Chow-Künneth decompositions for blow-ups

3.1   Main construction and technical details

For the balance of the paper, we fix the following notation and hypotheses.

Assumptions.

- X is a pseudo-smooth projective variety of dimension d over some field k.

- Y is the blow-up of X along $T = \{a\}$, where a is a k-rational point of X.

- Y is pseudo-smooth and the exceptional divisor Z of the blow-up is isomorphic to $\mathbb{P}^{d-1}$.

Let

$$
\begin{array}{ccc}
E & \xrightarrow{\ j\ } & Y \\
\downarrow{\scriptstyle g} & & \downarrow{\scriptstyle f} \\
T & \xrightarrow{\ i\ } & X
\end{array}
$$

be the commutative square describing this blow-up.

The objective of this section is to describe explicitly how a CK-decomposition for X can be used to construct one on Y, and conversely. We begin by setting up the framework for our construction and proving some auxiliary results.

First, observe that $Y \times Y$ is the blow-up of $X \times X$ along the closed subscheme $S = S_1 \cup S_2$, where $S_1 = T \times X$ and $S_2 = X \times T$. The exceptional divisor of this blow-up is $E = E_1 \cup E_2$, where $E_1 = Z \times Y$ and $E_2 = Y \times Z$. Thus we have commutative diagrams:

$$
\begin{array}{ccc}
E & \xrightarrow{\tilde{j}} & Y \times Y \\
\downarrow{\scriptstyle \tilde{g}} & & \downarrow{\scriptstyle f \times f} \\
S & \xrightarrow{\tilde{i}} & X \times X
\end{array}
\qquad
\begin{array}{ccc}
E_1 & \xhookrightarrow{j \times 1} & Y \times Y \\
\downarrow{\scriptstyle g \times 1} & & \downarrow{\scriptstyle f \times 1} \\
T \times Y & \xhookrightarrow{i \times 1} & X \times Y
\end{array}
\qquad
\begin{array}{ccc}
E_2 & \xhookrightarrow{1 \times j} & Y \times Y \\
\downarrow{\scriptstyle 1 \times g} & & \downarrow{\scriptstyle 1 \times f} \\
Y \times T & \xhookrightarrow{1 \times i} & Y \times X.
\end{array}
$$

Note that even when X is smooth, S is not regularly imbedded in $X \times X$, so we cannot use the blow-up exact sequence to relate the Chow groups of $Y \times Y$ to those of $X \times X$. Instead, we use the localization sequence. Let m be an integer, $0 \leq m \leq d$, and define $U_X = X \times X - S$ and $U_Y = Y \times Y - E$. Then there is a commutative diagram with exact rows:

$$
\begin{array}{ccccccc}
CH_{\mathbb{Q}}^{m-1}(E) & \xrightarrow{\tilde{j}_*} & CH_{\mathbb{Q}}^{m}(Y \times Y) & \longrightarrow & CH_{\mathbb{Q}}^{m}(U_Y) & \longrightarrow & 0 \\
\downarrow{\scriptstyle \tilde{g}_*} & & \downarrow{\scriptstyle (f \times f)_*} & & \downarrow{\scriptstyle \cong} & & \\
CH_{\mathbb{Q}}^{m-d}(S) & \xrightarrow{\tilde{i}_*} & CH_{\mathbb{Q}}^{m}(X \times X) & \longrightarrow & CH_{\mathbb{Q}}^{m}(U_X) & \longrightarrow & 0
\end{array}
\qquad (3.1)
$$

The aim of the rest of this section is to describe a decomposition of $CH_{\mathbb{Q}}^*(Y \times Y)$ as the internal direct sum of two subgroups, A and B, and to study the multiplicative structure of $CH_{\mathbb{Q}}^*(Y \times Y)$ as a ring (under composition of correspondences) with respect to these subgroups.

To this end, define

$$
\begin{aligned}
\zeta &= (f \times f)^*[\Delta_X] \in CH_{\mathbb{Q}}^{d}(Y \times Y), \\
A &= \zeta \bullet CH_{\mathbb{Q}}^*(Y \times Y) \bullet \zeta, \text{ and} \\
A_m &= A \cap CH_{\mathbb{Q}}^{m}(Y \times Y), \; 0 \leq m \leq 2d.
\end{aligned}
\qquad (3.2)
$$

Clearly, $A = \bigoplus_{m=0}^{2d} A_m$.

Lemma 3.1 *For $\gamma \in CH_{\mathbb{Q}}^*(Y \times Y)$, $(f \times f)^*(f \times f)_*\gamma = \zeta \bullet \gamma \bullet \zeta$. In particular, if $\gamma \in (f \times f)^* CH_{\mathbb{Q}}^*(X \times X)$, then $\zeta \bullet \gamma \bullet \zeta = \gamma$. Furthermore, $A = (f \times f)^* CH_{\mathbb{Q}}^*(X \times X)$ is a ring with unit element ζ, and A_d is a subring of A.*

Proof From the projection formula and the fact that f has degree 1, it follows that $[\Gamma_f] \bullet [\Gamma_f^t] = [\Delta_X]$. Then

$$\zeta \bullet \gamma \bullet \zeta = (f \times f)^* [\Delta_X] \bullet \gamma \bullet (f \times f)^* [\Delta_X]$$
$$= [\Gamma_f^t] \bullet [\Delta_X] \bullet [\Gamma_f] \bullet \gamma \bullet [\Gamma_f^t] \bullet [\Delta_X] \bullet [\Gamma_f]$$
$$= [\Gamma_f^t] \bullet [\Gamma_f] \bullet \gamma \bullet [\Gamma_f^t] \bullet [\Gamma_f]$$
$$= (f \times f)^* (f \times f)_* \gamma.$$

The remaining assertions are clear from Lemma 2.2.

$\square$

Next, define

$$B' = (j \times 1)_* (\mathrm{Ker}\, ((f \circ j) \times 1)_*)$$

$$B'' = (1 \times j)_* (\mathrm{Ker}\, (1 \times (f \circ j))_*).$$

Let $B = B' + B''$; for $0 \le m \le 2d$, set $B'_m = B' \cap CH_{\mathbb{Q}}^m(Y \times Y)$, $B''_m = B'' \cap CH_{\mathbb{Q}}^m(Y \times Y)$, and $B_m = B'_m + B''_m$. Then there are direct sum decompositions

$$B = \bigoplus_{m=0}^{2d} B_m, \ B' = \bigoplus_{m=0}^{2d} B'_m, \ B'' = \bigoplus_{m=0}^{2d} B''_m.$$

There is a rather important orthogonality relationship between A and B.

Proposition 3.2 (Orthogonality principle) *Suppose* $\alpha \in A$, $\beta' \in B'$ *and* $\beta'' \in B''$. *Then* $\beta' \bullet \alpha = 0$ *and* $\alpha \bullet \beta'' = 0$.

Proof Write $\alpha = (f \times f)^* \delta$, $\beta' = (j \times 1)_* \varepsilon_1$ and $\beta'' = (1 \times j)_* \varepsilon_2$, where $\delta \in CH_{\mathbb{Q}}^d(X \times X)$, $\varepsilon_1 \in \mathrm{Ker}\, ((f \circ j) \times 1)_*$, and $\varepsilon_2 \in \mathrm{Ker}\, (1 \times (f \circ j))_*$. Then

$$\beta' \bullet \alpha = \varepsilon_1 \bullet [\Gamma_j^t] \bullet [\Gamma_f^t] \bullet \delta \bullet [\Gamma_f] = ((f \circ j) \times 1)_* \varepsilon_1 \bullet \delta \bullet [\Gamma_f] = 0.$$
$$\alpha \bullet \beta'' = [\Gamma_f^t] \bullet \delta \bullet [\Gamma_f] \bullet [\Gamma_j] \bullet \varepsilon_2 = [\Gamma_f^t] \bullet \delta \bullet (1 \times (f \circ j))_* \varepsilon_2 = 0. \quad \square$$

A simple chase on diagram 3.1 shows that $\sigma = [\Delta_Y] - \zeta \in B_d$. Direct calculation then shows that the formulas

$$\sigma \bullet \sigma = \sigma, \ \sigma^t = \sigma, \ \text{and} \ \sigma \bullet \zeta = \zeta \bullet \sigma = 0$$

hold. Moreover, Proposition 3.2 shows that for $\beta' \in B'$, $\beta'' \in B''$, we have $\beta' \bullet \sigma = \beta'$ and $\sigma \bullet \beta'' = \beta''$.

Given $\gamma \in CH_{\mathbb{Q}}^*(Y \times Y)$, let $\gamma' = \gamma - \zeta \bullet \gamma \bullet \zeta$. Then using Lemma 3.1 we calculate:

$$
\begin{aligned}
(f \times f)_*(\gamma') &= (f \times f)_*(\gamma - \zeta \bullet \gamma \bullet \zeta) \\
&= (f \times f)_*\gamma - (f \times f)_*(f \times f)^*(f \times f)_*\gamma \\
&= (f \times f)_*\gamma - (f \times f)_*\gamma \\
&= 0.
\end{aligned}
$$

Another chase on diagram 3.1 shows that $\gamma' \in \operatorname{Im} \tilde{j}_* \cap \operatorname{Ker} (f \times f)_*$. This shows that there are well-defined maps

$$
s : A \oplus B \to CH_{\mathbb{Q}}^*(Y \times Y) \text{ and } t : CH_{\mathbb{Q}}^*(Y \times Y) \to A \oplus B
$$

given by $s(a, b) = a + b$ and $t(\gamma) = (\zeta \bullet \gamma \bullet \zeta, \gamma - \zeta \bullet \gamma \bullet \zeta)$.

Proposition 3.3

(i) *s and t are inverse isomorphisms; thus, $CH_{\mathbb{Q}}^*(Y \times Y)$ is the internal direct sum of A and B.*

(ii) *A and B are each closed under transposition of cycles.*

Proof It is clear from the definitions that $s \circ t = 1$, so it suffices to show that s is injective and that t is surjective. To show the former, we prove $A \cap B = \{0\}$. If $\gamma \in A \cap B$, then in particular, $\gamma \bullet \zeta = \gamma = \zeta \bullet \gamma$, and also $\gamma = b' + b''$ for some $b' \in B'$ and $b'' \in B''$. Using Proposition 3.2, we calculate:

$$
\gamma = \zeta \bullet \gamma \bullet \zeta = \zeta \bullet (b' \bullet \zeta) + (\zeta \bullet b'') \bullet \zeta = 0.
$$

Now suppose $(\alpha, \beta) \in A \oplus B$. Setting $\gamma = \alpha + \beta$ and writing $\beta = \beta' + \beta''$ with $\beta' \in B'$ and $\beta'' \in B''$, we have

$$
\zeta \bullet \gamma \bullet \zeta = \zeta \bullet \alpha \bullet \zeta + \zeta \bullet \beta \bullet \zeta = \alpha + \zeta \bullet (\beta' \bullet \zeta) + (\zeta \bullet \beta'') \bullet \zeta = \alpha.
$$

Hence $t(\gamma) = (\alpha, \beta)$, and so t is surjective.

For the second statement, $\alpha \in A$ implies $\alpha = (f \times f)^*\delta$ for some $\delta \in CH_{\mathbb{Q}}^*(X \times X)$. Since $\delta^t \in CH_{\mathbb{Q}}^*(X \times X)$, obviously $\alpha^t = (f \times f)^*\delta^t \in A$. Furthermore, suppose $\beta \in B$ and write $\beta^t = \alpha' + \beta'$ for some $\alpha' \in A$, $\beta' \in B$. Then $\beta = (\beta^t)^t = \alpha'^t + \beta'^t$. Because β has a unique expression as a sum of an element of A and an element of B we must have $\alpha'^t = 0$; so $\alpha' = 0$, and thus $\beta^t \in B$. □

The following is a computational criterion convenient for testing for membership in A or B:

Corollary 3.4 *Suppose $\gamma \in CH^*_{\mathbb{Q}}(Y \times Y)$. Then $\gamma \in A$ if and only if $\zeta \bullet \gamma \bullet \zeta = \gamma$ and $\gamma \in B$ if and only if $\zeta \bullet \gamma \bullet \zeta = 0$.*

Proof If $\gamma \in A$, then $\gamma = \zeta \bullet \gamma \bullet \zeta$ by Lemma 3.1. Conversely, suppose $\gamma = \zeta \bullet \gamma \bullet \zeta$. Write $\gamma = \alpha + \beta$, where $\alpha \in A$ and $\beta \in B$ and $\beta = \beta' + \beta''$, where $\beta' \in B'$ and $\beta'' \in B$. Then, using Proposition 3.2,

$$\gamma = \zeta \bullet \gamma \bullet \zeta = \zeta \bullet \alpha \bullet \zeta + \zeta \bullet \beta' \bullet \zeta + \zeta \bullet \beta'' \bullet \zeta = \zeta \bullet \alpha \bullet \zeta = \alpha \in A.$$

The second statement is now clear.

$\square$

3.2   Blowing up

The proof of the following theorem introduces one of the main constructions used in this article. Our techniques bear some resemblance to those used by Vial in [V, Section 5]. There are, however, two main differences in approach.

  (i) We are blowing up varieties that are not necessarily smooth, but having isolated quotient singularities, so that our varieties are required to be only pseudo-smooth. This also means that we cannot make use of the blow-up exact sequence (see [F, Proposition 6.7]).

 (ii) The second difference is that we interpret $Y \times Y$ as the blow-up of $X \times X$ along the subscheme S (as explained in Section 3.1); in contrast, the arguments of Vial only involve Y as the blow-up of X along a subvariety T.

Theorem 3.5 *If X has a CK-decomposition $[\Delta_X] = \sum_{i=0}^{2d} \pi_i^X$, then Y has an explicit CK-decomposition, as defined in (3.7). If the former satisfies Poincaré duality, then so does the latter.*

Proof By Lemma 2.2, $\{(f \times f)^* \pi_i^X\}_{i=0}^{2d}$ is a set of orthogonal idempotents in $CH^d_{\mathbb{Q}}(Y \times Y)$. However, $\Sigma_{i=0}^{2d}(f \times f)^*(\pi_i^X)$ may not equal $[\Delta_Y]$, so we proceed to deal with this discrepancy. Recall $\sigma = [\Delta_Y] - \zeta = [\Delta_Y] - (f \times f)^*([\Delta_X])$. Denote by $h_1 : E_1 \hookrightarrow E$, $h_2 : E_2 \hookrightarrow E$, $k_1 : E_1 \cap E_2 \hookrightarrow E_1$, and $k_2 : E_1 \cap E_2 \hookrightarrow E_2$ the various inclusion maps. Then the sequence

$$CH^{d-2}_{\mathbb{Q}}(E_1 \cap E_2) \overset{k_{1*}+k_{2*}}{\to} CH^{d-1}_{\mathbb{Q}}(E_1) \oplus CH^{d-1}_{\mathbb{Q}}(E_2)$$

$$\overset{h_{1*}-h_{2*}}{\longrightarrow} CH^{d-1}_{\mathbb{Q}}(E) \to 0$$

is exact, so we may write $\sigma = \tilde{j}_*(h_{1*}\tau_1 + h_{2*}\tau_2)$, with $\tau_i \in CH_{\mathbb{Q}}^{d-1}(E_i)$, $i = 1, 2$. Since $\tilde{j} \circ h_1 = j \times 1$ and $\tilde{j} \circ h_2 = 1 \times j$, we have $\sigma = (j \times 1)_*\tau_1 + (1 \times j)_*\tau_2$. We stress that this is the only part of our construction which involves a choice of cycles.

By Proposition 3.3, we have also $\sigma = \sigma^t = (1 \times j)_*\tau_1^t + (j \times 1)_*\tau_2^t$; thus,

$$\sigma = \frac{1}{2}[(j \times 1)_*\tau_1 + (1 \times j)_*\tau_2] + \frac{1}{2}[(1 \times j)_*\tau_1^t + (j \times 1)_*\tau_2^t)]$$

$$= (j \times 1)_* \frac{1}{2}(\tau_1 + \tau_2^t) + (1 \times j)_* \frac{1}{2}(\tau_1 + \tau_2^t)^t.$$

This calculation shows that we may replace (τ_1, τ_2) with $(\frac{1}{2}(\tau_1 + \tau_2^t), \frac{1}{2}(\tau_1^t + \tau_2))$, and thus assume without loss of generality that $\tau_2 = \tau_1^t$.

Let $\ell \in CH_{\mathbb{Q}}^1(Z)$ be the class of a generic hyperplane, and for convenience, set $\ell_i = j_*(\ell^{i-1}) \in CH_{\mathbb{Q}}^i(Y)$ for $1 \leq i \leq d$. From the projective bundle formula, we have $\tau_1 = \sum_{i=0}^{d-1} \ell^i \times a_{d-i-1}$, where $a_{d-i-1} \in CH_{\mathbb{Q}}^{d-i-1}(Y)$. Define $a_d = 0$, $\ell_0 = 0$, $\eta_0 = 0$ and $\eta_i = (j \times 1)_*(\ell^{i-1} \times a_{d-i}) = \ell_i \times a_{d-i}$ for $1 \leq i \leq d$. If we set $\theta_i = \eta_{d-i}^t$, then η_i is a product cycle of type $(i, d-i)$ when $1 \leq i \leq d$ and θ_i is a product cycle of type $(i, d-i)$ when $0 \leq i \leq d-1$. Finally, define $\gamma_i = \eta_i + \theta_i$ for $0 \leq i \leq d$. By construction, we have $\gamma_i^t = \gamma_{d-i}$.

$\square$

Lemma 3.6 $a_0 = 0$.

Proof Since $a_0 \in CH_{\mathbb{Q}}^0(Y)$, we have $a_0 = c[Y]$ for some $c \in \mathbb{Q}$. Observe first that

$$(f \times f)_*(j \times 1)_*\tau_1 = (1 \times f)_* \sum_{m=0}^{d-1} (f \times 1)_*(j \times 1)_*(\ell^m \times a_{d-m-1})$$

$$= (1 \times f)_* \sum_{m=0}^{d-1} (i \times 1)_*(g \times 1)_*(\ell^m \times a_{d-m-1})$$

$$= (1 \times f)_* \sum_{m=0}^{d-1} i_*g_*(\ell^m) \times a_{d-m-1}.$$

For reasons of dimension, $g_*(\ell^m) = 0$ when $0 \leq m \leq d-2$ and $i_*g_*\ell^{d-1} = x \in CH_{\mathbb{Q}}^d(X)$, so $(f \times f)_*(j \times 1)_*\tau_1 = x \times a_0 = c(x \times [Y])$ and similarly $(f \times f)_*(1 \times j)_*\tau_2 = c([Y] \times x)$.

By construction, $(f \times f)_*\sigma = 0$; thus,
$(f \times f)_*(j \times 1)_*\tau_1 + (f \times f)_*(1 \times j)_*\tau_2 = c(x \times [Y] + [Y] \times x) = 0$.
Since $x \times [Y]$ is a projector which is orthogonal to $[Y] \times x$, it follows that

$(x \times [Y]) \bullet (c(x \times [Y] + [Y] \times x)) = c(x \times [Y]) = 0$. Finally, since $\Delta_Y^*(x \times Y) = x \cdot [Y] = x \neq 0$, we must have $c = 0$, and hence $a_0 = 0$.

$\square$

Returning to the proof of Theorem 3.5, we note the following important facts:

$$\theta_0 = \eta_d = 0 \text{ and } \eta_i \in B'_d, \ \theta_i \in B''_d \text{ for } 0 \leq i \leq d. \qquad (3.3)$$

One easily checks that $\sum_{j=0}^d \gamma_j = \sigma$. Since γ_i is a product cycle of type $(i, d-i)$, Lemma 2.3 implies $\gamma_i \bullet \gamma_j = 0$ when $i \neq j$. In particular, we have:

$$\sigma \bullet \gamma_i = \gamma_i \bullet \gamma_i = \gamma_i \bullet \sigma \qquad (3.4)$$

for i, $0 \leq i \leq d$. Thus, $\sigma = \sigma \bullet \sigma = \sum_{j=0}^d \gamma_j \bullet \gamma_j$, and so we have $\sum_{j=0}^d [\gamma_j \bullet \gamma_j - \gamma_j] = 0$, where the term in brackets is a product cycle of type $(j, d-j)$. Composing with γ_i on the left, we conclude:

$$\gamma_i \bullet \gamma_i \bullet \gamma_i - \gamma_i \bullet \gamma_i = 0. \qquad (3.5)$$

Now by (3.6), $\sigma \bullet (\gamma_i \bullet \gamma_i - \gamma_i) = (\sigma \bullet \gamma_i) \bullet \gamma_i - \sigma \bullet \gamma_i = \gamma_i \bullet \gamma_i \bullet \gamma_i - \gamma_i \bullet \gamma_i = 0$, and similarly $(\gamma_i \bullet \gamma_i - \gamma_i) \bullet \sigma = 0$. Thus,

$$\gamma_i \bullet \gamma_i - \gamma_i = (\zeta + \sigma) \bullet (\gamma_i \bullet \gamma_i - \gamma_i) \bullet (\zeta + \sigma) = \zeta \bullet (\gamma_i \bullet \gamma_i - \gamma_i) \bullet \zeta,$$

which by (3.4), equals

$$\zeta \bullet \sigma \bullet \gamma_i \bullet \zeta + \zeta \bullet (\eta_i + \theta_i) \bullet \zeta.$$

The first term vanishes because $\zeta \bullet \sigma = 0$ and the second vanishes because $\eta_i \in B'$ and $\theta_i \in B''$. Thus, we have $\gamma_i \bullet \gamma_i = \gamma_i$, and so γ_i is a projector satisfying

$$\sigma \bullet \gamma_i = \gamma_i = \gamma_i \bullet \sigma. \qquad (3.6)$$

Furthermore, for i and j, $0 \leq i \leq d$ and $0 \leq j \leq 2d$, we have:

$$\gamma_i \bullet (f \times f)^* \pi_j^X = (\gamma_i \bullet \sigma) \bullet (\zeta \bullet (f \times f)^* \pi_j^X \bullet \zeta) = \gamma_i \bullet (\sigma \bullet \zeta) \bullet (f \times f)^* \pi_j^X \bullet \zeta = 0$$

and similarly $(f \times f)^* \pi_j^X \bullet \gamma_i = 0$.

Finally, define for $0 \leq j \leq 2d$,

$$\delta_j = \begin{cases} \gamma_{d-\frac{j}{2}} & \text{if } j \text{ is even} \\ 0 & \text{if } j \text{ is odd} \end{cases} \quad \text{and} \quad \pi_j^Y = (f \times f)^* \pi_j^X + \delta_j. \qquad (3.7)$$

The computations above show that $[\Delta_Y] = \sum_{j=0}^{2d}(f \times f)^*(\pi_j^X) + \sigma = \sum_{j=0}^{2d} \pi_j^Y$ satisfies properties (i) and (ii) in the definition of CK-decomposition. The construction shows that $\delta_j^t = \delta_{2d-j}$ for all j; hence, the assertion about Poincaré duality follows from the definition of the π_j^Y.

It remains to show that for any Weil cohomology theory H^* and every j, $0 \leq j \leq 2d$, $cl_{Y \times Y}(\pi_j^Y)$ is the $(2d-j,j)$ Künneth component of $[\Delta_Y] \in H^{2d}(Y \times Y; \mathbb{Q})$. Using the Künneth isomorphism to make the identification $H^{2d}(Y \times Y; \mathbb{Q}) \cong \bigoplus_{i=0}^{2d} H^{2d-i}(Y; \mathbb{Q}) \otimes_{\mathbb{Q}} H^i(Y; \mathbb{Q})$, it suffices to show that $cl_{Y \times Y}(\pi_j^Y) \in H^{2d-j}(Y; \mathbb{Q}) \otimes_{\mathbb{Q}} H^j(Y; \mathbb{Q})$.

Now π_j^X is a projector in the original Chow-Künneth decomposition for X, so $cl_{X \times X}(\pi_j^X) \in H^{2d-j}(X; \mathbb{Q}) \otimes_{\mathbb{Q}} H^j(X; \mathbb{Q})$. Hence, using properties of the cycle map from the definition of Weil cohomology (see for example, [Kl, Section 3]), we have $cl_{Y \times Y}(f \times f)^* \pi_j^X = (f \times f)^* cl_{X \times X}(\pi_j^X) \in H^{2d-j}(Y; \mathbb{Q}) \otimes_{\mathbb{Q}} H^j(Y; \mathbb{Q})$. Moreover, $\delta_j = \gamma_{d-j/2}$ is a product cycle of type $(d-j/2, j/2)$; hence $\gamma_{d-j/2} = \sum_{m=0}^{r} \lambda_m \times \mu_m$, where $\lambda_m \in CH_{\mathbb{Q}}^{d-j/2}(Y)$ and $\mu_m \in CH_{\mathbb{Q}}^{j/2}(Y)$. Again using properties of the cycle map,

$$cl_{Y \times Y}(\gamma_{d-j/2}) = \sum_{m=0}^{r} cl_{Y \times Y}(\lambda_m \times \mu_m)$$

$$= \sum_{m=0}^{r} cl_Y(\lambda_m) \otimes cl_Y(\mu_m) \in H^{2d-j}(Y; \mathbb{Q}) \otimes_{\mathbb{Q}} H^j(Y; \mathbb{Q}).$$

Thus, regardless of whether j is odd or even, $cl_{Y \times Y}(\pi_j^Y) \in H^{2d-j}(Y; \mathbb{Q}) \otimes_{\mathbb{Q}} H^j(Y; \mathbb{Q})$.

$\square$

3.3   Refined projectors

In the construction of the Chow-Künneth projectors, our interest was focused on the sums $\gamma_i = \eta_i + \theta_i$ as projectors in $CH_{\mathbb{Q}}^d(Y \times Y)$. In subsequent sections, however, we will need to use the fact, established below, that η_i and θ_i are themselves mutually orthogonal projectors. Before proceeding any further, we need to make some slight modifications to our definitions in certain cases, the reasoning being that we wish to avoid the situation in which η_i and θ_i are nonzero constant multiples of each other. To this end, if $0 < i < d/2$ and η_i and θ_i happen to be nonzero constant multiples of each other, replace (η_i, θ_i) with $(\eta_i + \theta_i, 0)$ and replace $(\eta_{d-i}, \theta_{d-i})$ with $(0, \eta_{d-i} + \theta_{d-i})$. This change alters the definition of the a_i, but it does not change γ_i or γ_{d-i}, nor does it disturb the duality relation $\eta_i^t = \theta_{d-i}$. Now if d is even, $\eta_{d/2}^t = \theta_{d/2}$. In view of this, the only way for η_i and θ_i

to be nonzero constant multiples of each other (after the modification described above) is when d is even, $i = d/2$ and $\eta_{d/2} = \theta_{d/2}$ We will show in Proposition 3.8 that this situation is impossible.

Lemma 3.7 *For i, $1 \le i \le d-1$, $\ell_i \times \ell_{d-i} \ne 0$.*

Proof If $\ell_i \times \ell_{d-i} = 0$, then $\Delta_Y^*(\ell_i \times \ell_{d-i}) = \ell_i \cdot \ell_{d-i} = 0$. However, by the projection formula and the self-intersection formula,

$$\ell_i \cdot \ell_{d-i} = j_* \ell^{i-1} \cdot j_* \ell^{d-i-1} = j_*(\ell^{i-1} \cdot j^* j_* \ell^{d-i-1}) = j_*(\ell^{i-1} \cdot \ell^{d-i}) = j_*(\ell^{d-1}).$$

This is a zero cycle on Y of degree one, so it cannot be zero.   □

Proposition 3.8 *For all i, $0 \le i \le d$,*

$$\eta_i \bullet \eta_i = \eta_i, \quad \theta_i \bullet \theta_i = \theta_i, \quad \eta_i \bullet \theta_i = \theta_i \bullet \eta_i = 0.$$

Moreover, $m(\ell_i, a_{d-i}) = m(a_{d-i}, \ell_i) = m(\ell_i, \ell_{d-i}) = 1$, and $m(a_i, a_{d-i}) = 0$, where the $m(-,-)$ are the rational numbers defined in Proposition 2.3. For all i and j, $a_i \times a_j = 0$.

Proof Direct calculation using the self-intersection formula shows that

$$\begin{aligned}
m(\ell_i, \ell_{d-i}) &= m(j_* \ell^{i-1}, j_* \ell^{d-i-1}) \\
&= p_\emptyset^{YY^*} p_{\emptyset\,*}^Y (j_* \ell^{i-1} \cdot j_* \ell^{d-i-1}) \\
&= p_\emptyset^{YY^*} p_{\emptyset\,*}^Y j_* (j^* \ell^{i-1} \cdot \ell^{d-i-1}) \\
&= p_\emptyset^{YY^*} p_{\emptyset\,*}^Y j_* (\ell^i \cdot \ell^{d-i-1}) \\
&= p_\emptyset^{YY^*} p_{\emptyset\,*}^Y j_* (\ell^{d-1}) \\
&= 1
\end{aligned}$$

for i, $0 \le i \le d$. Moreover, since η_i and θ_i are product cycles on $Y \times Y$, Lemma 2.3 implies that $\eta_i \bullet \eta_i = s\eta_i$ and $\theta_i \bullet \theta_i = t\theta_i$ for some $s, t \in \mathbb{Q}$. Now $\eta_i \in B'$ and $\theta_i \in B''$ by (3.3), so by Proposition 3.2, we have $\eta_i \bullet (f \times f)^* \pi_j = 0$ and $(f \times f)^* \pi_j \bullet \theta_i = 0$ for $0 \le j \le 2d$. Using Lemma 2.3, we have

$$\eta_i \bullet \gamma_i = \eta_i \bullet \sigma = \eta_i \bullet [\Delta_Y] - \eta_i \bullet \sum_{j=0}^{2d} (f \times f)^* \pi_j = \eta_i, \text{ and similarly} \quad (3.8)$$

$$\gamma_i \bullet \theta_i = \theta_i.$$

So on one hand, $\eta_i \bullet \theta_i = \eta_i \bullet \gamma_i - \eta_i \bullet \eta_i = (1-s)\eta_i$, but also $\eta_i \bullet \theta_i = \gamma_i \bullet \theta_i - \theta_i \bullet \theta_i = (1-t)\theta_i$. Hence $(1-s)\eta_i = (1-t)\theta_i$.

First suppose η_i and θ_i are not nonzero constant multiples of each other. It must be the case that $s = 1$ or $t = 1$. If $s = 1$, then $(1-t)\theta_i = 0$, so $t = 1$ or $\theta_i = 0$; but in the latter case, we may still assume $t = 1$. If $t = 1$, we may similarly conclude that $s = 1$. Thus, $s = t = 1$, and so $\eta_i \bullet \eta_i = \eta_i$, $\theta_i \bullet \theta_i = \theta_i$ and $\eta_i \bullet \theta_i = 0$. Now $\theta_i \bullet \eta_i = (\gamma_i - \eta_i) \bullet (\gamma_i - \theta_i) = \gamma_i \bullet \gamma_i - \eta_i \bullet \gamma_i - \gamma_i \bullet \theta_i + \eta_i \bullet \theta_i = \gamma_i - \eta_i - \theta_i = 0$. The formulas $m(\ell_i, a_{d-i}) = m(a_{d-i}, \ell_i) = 1$ follow from the statement $\eta_i \bullet \eta_i = \eta_i$ and the symmetry of $m(-,-)$ in its arguments. Since $\eta_i \bullet \theta_i = 0$, we have $m(a_i, a_{d-i})(\ell_i \times \ell_{d-i}) = 0$. If $1 \leq i \leq d-1$, then by Lemma 3.7, $\ell_i \times \ell_{d-i} \neq 0$, so we must have $m(a_i, a_{d-i}) = 0$. When $i = 0$ or $i = d$, this is obvious, since $a_0 = a_d = 0$. Finally, from $\theta_i \bullet \eta_i = 0$, we have $m(\ell_i, \ell_{d-i})(a_i \times a_{d-i}) = a_i \times a_{d-i} = 0$. Then

$$a_i \times a_j = (\ell_i \times a_j) \bullet (a_i \times a_{d-i}) = 0.$$

Thus, all of asserted equations hold for $i \neq d/2$, or if $i = d/2$ and either $\eta_{d/2} = \theta_{d/2} = 0$ or $\eta_{d/2} \neq \theta_{d/2}$.

Now suppose that d is even and $\eta_{d/2} = \theta_{d/2} \neq 0$. Then $\eta_{d/2} = \theta_{d/2} = \frac{1}{2}\gamma_{d/2}$, and so $\eta_{d/2} \bullet \theta_{d/2} = \frac{1}{4}\gamma_{d/2} = \frac{1}{2}\eta_{d/2} \neq 0$. However, by Lemma 2.3,

$$\eta_{d/2} \bullet \theta_{d/2} = m(\ell_{d/2}, \ell_{d/2})a_{d/2} \times a_{d/2}$$
$$= a_{d/2} \times a_{d/2}$$
$$= (\ell_{d/2+1} \times a_{d/2}) \bullet (a_{d/2} \times a_{d/2-1})$$
$$= (\ell_{d/2+1} \times a_{d/2}) \bullet (a_{d/2+1} \times a_{d/2-1}) \bullet (a_{d/2} \times \ell_{d/2-1}).$$

Since we have already showed that $a_{d/2+1} \times a_{d/2-1} = 0$, the above calculation forces $\gamma_{d/2} = 0$, which is a contradiction. This shows that the condition $\eta_{d/2} = \theta_{d/2} \neq 0$ is impossible. □

Corollary 3.9

(i) Let $\beta_1 = \ell_i \times \lambda \in CH_\mathbb{Q}^{i+j}(Y \times Y)$. Then $\beta_1 \bullet \eta_i = \beta_1$. If $\beta_1 \in B'$ or if $a_i \times \lambda = 0$, then $\beta_1 \bullet \theta_i = 0$.

(ii) Let $\beta_2 = \lambda \times \ell_j \in CH_\mathbb{Q}^{i+j}(Y \times Y)$. Then $\theta_{d-j} \bullet \beta_2 = \beta_2$. If $\beta_2 \in B''$ or if $\lambda \times a_j = 0$, then $\eta_{d-j} \bullet \beta_2 = 0$.

Proof We prove the first statement, the second being similar. First,

$$\beta_1 \bullet \eta_i = (\ell_i \times \lambda) \bullet (\ell_i \times a_{d-i}) = m(\ell_i, a_{d-i})(\ell_i \times \lambda) = \ell_i \times \lambda = \beta_1.$$

By direct computation, $\beta_1 \bullet \theta_i = m(\ell_i, \ell_{d-i})(a_i \times \lambda) = a_i \times \lambda$; so if $a_i \times \lambda = 0$, then $\beta_1 \bullet \theta_i = 0$. If $\beta_1 \in B'$, then $\beta_1 \bullet \zeta = 0$ by Proposition 2.3 and

$\beta_1 \bullet \sigma = \beta_1 \bullet \sum_{j=0}^{d} \gamma_j = \beta_1 \bullet \gamma_i$ for reasons of dimension. Thus, $\beta_1 = \beta_1 \bullet [\Delta_Y] = \beta_1 \bullet (\zeta + \sigma) = \beta_1 \bullet \gamma_i = \beta_1 \bullet (\eta_i + \theta_i) = \beta_1 + \beta_1 \bullet \theta_i$. It follows that $\beta_1 \bullet \theta_i = 0$ in this case also.

$\square$

Corollary 3.10 *B is a two-sided ideal of $CH_{\mathbb{Q}}^*(Y \times Y)$.*

Proof We need to check that for all $\alpha \in A$, and $\beta_1, \beta_2 \in B$, the elements $\alpha \bullet \beta_1$, $\beta_2 \bullet \alpha$, and $\beta_1 \bullet \beta_2$ are in B. For the first, simply note that because $\zeta \in A$, $\zeta \bullet (\alpha \bullet \beta_1) \bullet \zeta = \zeta \bullet \alpha \bullet (\beta_1 \bullet \zeta) = 0$ by Proposition 3.2. By Corollary 3.4, $\alpha \bullet \beta_1 \in B$. The argument showing $\beta_2 \bullet \alpha \in B$ is similar.

We will show that if $\beta_1, \beta_2 \in B$, then in fact $\beta_1 \bullet \beta_2 = 0$, which is clearly in B. By linearity in each factor, we may assume that β_1, β_2 are homogeneous elements with respect to the grading on $CH_{\mathbb{Q}}^*(Y \times Y)$, i.e. $\beta_1 \in CH_{\mathbb{Q}}^k(Y \times Y)$ and $\beta_2 \in CH_{\mathbb{Q}}^l(Y \times Y)$ for some k, l. For $i = 1, 2$, write $\beta_i = \beta_i' + \beta_i''$, where $\beta_i' \in B'$ and $\beta_i'' \in B''$. Then

$$\zeta \bullet (\beta_1 \bullet \beta_2) \bullet \zeta = \zeta \bullet \beta_1' \bullet \beta_2' \bullet \zeta + \zeta \bullet \beta_1'' \bullet \beta_2' \bullet \zeta + \zeta \bullet \beta_1' \bullet \beta_2'' \bullet \zeta + \zeta \bullet \beta_1'' \bullet \beta_2'' \bullet \zeta.$$

The first, second, and fourth terms are zero by Proposition 3.2, so we may assume without loss of generality that $\beta_1 \in B'$ and $\beta_2 \in B''$. By the projective bundle formula, β_1 is a sum of elements of the form $\ell_i \times b_{k-i}$ and β_2 is a sum of elements of the form $c_{l-j} \times \ell_j$, where $1 \le i, j \le d$ and $b_m, c_n \in CH_{\mathbb{Q}}^*(Y)$. Using linearity again, we reduce to the case $\beta_1 = \ell_i \times b_{k-i}$, and $\beta_2 = c_{l-j} \times \ell_j$. Then $\beta_1 \bullet \beta_2 = (\beta_1 \bullet \eta_i) \bullet (\theta_{d-j} \bullet \beta_2)$ by Corollary 3.9. If $i = d - j$, then $\eta_i \bullet \theta_{d-j} = 0$ by Proposition 3.8. If $i \ne d - j$, then $\eta_i \bullet \theta_{d-j} = \eta_i \bullet \gamma_i \bullet \gamma_{d-j} \bullet \theta_{d-j} = 0$ where the first equality is by (3.8) and the second equality is by the orthogonality of γ_i and γ_{d-j}, as shown in Lemma 2.3. In either case, we have $\beta_1 \bullet \beta_2 = 0$.

$\square$

3.4   Blowing down

We now have the tools to prove the converse of Theorem 3.5.

By Proposition 3.3, we may write every $\delta \in CH_{\mathbb{Q}}^i(Y \times Y)$ uniquely as $(f \times f)^*(f \times f)_* \delta + b_\delta$, where $b_\delta \in B$. As we will need to use a similar argument in Section 5, we phrase the next result in somewhat general form.

Lemma 3.11 *With hypotheses as in Theorem 3.5, suppose $\delta_i \in CH_{\mathbb{Q}}^*(Y \times Y)$, $1 \le i \le 4$, satisfy $\delta_1 \bullet \delta_2 = \delta_3 \bullet \delta_4$. Then $(f \times f)_* \delta_1 \bullet (f \times f)_* \delta_2 = (f \times f)_* \delta_3 \bullet (f \times f)_* \delta_4$.*

Proof For each $i = 1, \ldots, 4$, write $\delta_i = (f \times f)^*(f \times f)_* \delta_i + b_i$, where $b_i \in B$. Substituting these expression into the assumption $\delta_1 \bullet \delta_2 = \delta_3 \bullet \delta_4$ yields

$$(f \times f)^*(f \times f)_* \delta_1 \bullet (f \times f)^*(f \times f)_* \delta_2 + (f \times f)^*(f \times f)_* \delta_1 \bullet b_2$$
$$+ b_1 \bullet (f \times f)^*(f \times f)_* \delta_2 + b_1 \bullet b_2$$
$$= (f \times f)^*(f \times f)_* \delta_3 \bullet (f \times f)^*(f \times f)_* \delta_3 + (f \times f)^*(f \times f)_* \delta_3 \bullet b_4$$
$$+ b_3 \bullet (f \times f)^*(f \times f)_* \delta_4 + b_3 \bullet b_4,$$

which by Lemma 2.2 may be rewritten

$$(f \times f)^*((f \times f)_* \delta_1 \bullet (f \times f)_* \delta_2) + (f \times f)^*(f \times f)_* \delta_1 \bullet b_2$$
$$+ b_1 \bullet (f \times f)^*(f \times f)_* \delta_2 + b_1 \bullet b_2$$
$$= (f \times f)^*((f \times f)_* \delta_3 \bullet (f \times f)_* \delta_3) + (f \times f)^*(f \times f)_* \delta_3 \bullet b_4$$
$$+ b_3 \bullet (f \times f)^*(f \times f)_* \delta_4 + b_3 \bullet b_4.$$

The first summand on each side is in A, while Corollary 3.10 shows that the other three summands are in B. By Proposition 3.3, $CH^*_{\mathbb{Q}}(Y \times Y)$ is the internal direct sum of A and B, so we must have $(f \times f)^*((f \times f)_* \delta_1 \bullet (f \times f)_* \delta_2) = (f \times f)^*((f \times f)_* \delta_3 \bullet (f \times f)_* \delta_4)$. Finally, the projection formula implies that $(f \times f)_*(f \times f)^*$ is the identity map, so $(f \times f)^*$ is injective and the assertion follows.

$\square$

By taking $\delta_4 = [\Delta_Y]$, we immediately deduce:

Corollary 3.12 *If $\delta_1, \delta_2 \in CH^*_{\mathbb{Q}}(Y \times Y)$, then $(f \times f)_*(\delta_1 \bullet \delta_2) = (f \times f)_* \delta_1 \bullet (f \times f)_* \delta_2$. In particular, if $\delta_1 \bullet \delta_2 = 0$, then $(f \times f)_* \delta_1 \bullet (f \times f)_* \delta_2 = 0$.*

As discussed in [SV, 1.4], given a surjective morphism $g : V \to W$ of projective varieties, one may identify the (Chow) motive of W with a direct summand of the Chow motive of V. In particular, there is a section $s \in CH^d_{\mathbb{Q}}(W \times V)$ such that $[\Gamma_g] \bullet s = [\Delta_W]$. If one begins with a Chow-Künneth decomposition $[\Delta_V] = \sum_{j=0}^{2d} \pi_j^V$ for V, one might attempt to construct a Chow-Künneth decomposition on W by considering the elements $[\Gamma_g] \bullet \pi_j^V \bullet s$, $0 \leq j \leq 2d$. Because the π_j^V are central modulo homological equivalence (assuming some choice of Weil cohomology theory), the cohomology classes of the elements $[\Gamma_g] \bullet \pi_j^V \bullet s, 0 \leq j \leq 2d$ actually define a *Künneth* decomposition on W, but it does not follow that these elements are idempotents when considered with respect to rational equivalence.

In the case of blowing down in our special case as considered in 3.1, however, this construction actually yields a Chow-Künneth decomposition as shown in the following corollary.

Corollary 3.13 *With hypotheses as in Section 3.1, suppose* $[\Delta_Y] = \sum_{j=0}^{2d} \pi_j^Y$
is a Chow-Künneth decomposition for Y. *Then* $[\Delta_X] = \sum_{j=0}^{2d} (f \times f)_* \pi_j^Y$
is a CK-decomposition for X. *Moreover, if the CK-decomposition for* Y
satisfies Poincaré duality, then so does the CK-decomposition for X.

Proof The formula for $[\Delta_X]$ follows by applying $(f \times f)_*$ to the expression
for $[\Delta_Y]$, noting that $f \times f$ has degree 1. Now if $i \neq j$, we have $\pi_i^Y \bullet \pi_j^Y = 0$,
so $(f \times f)_* \pi_i^Y \bullet (f \times f)_* \pi_j^Y = 0$ by Corollary 3.12. Finally,

$$(f \times f)_* \pi_i^Y \bullet (f \times f)_* \pi_i^Y = \left([\Delta_X] - \sum_{j \neq i} (f \times f)_* \pi_j^Y\right) \bullet (f \times f)_* \pi_i^Y = (f \times f)_* \pi_i^Y.$$

The remaining assertions are clear from the construction.

$\square$

4 Application to Murre's Conjectures

The goal of this section is to prove that each of Murre's Conjectures holds
for Y if and only if it holds for X. The case of Murre's Conjecture **A**
(existence of a Chow-Künneth decomposition) was completed in the previous section. In the interest of making the proofs easier to follow, we use
Greek letters for elements of $CH_{\mathbb{Q}}^*(Y \times Y)$ or $CH_{\mathbb{Q}}^*(X \times X)$ and Roman
letters for elements of $CH_{\mathbb{Q}}^*(Y)$ or $CH_{\mathbb{Q}}^*(X)$. In order to study the action
of correspondences on Chow groups, we will need some results analogous
to Lemma 2.2 and Proposition 3.2.

Lemma 4.1 *If* $\alpha \in CH_{\mathbb{Q}}^*(X \times X)$ *and* $x \in CH_{\mathbb{Q}}^*(X)$, *then*

$$f^*(\alpha \bullet x) = (f \times f)^*(\alpha) \bullet f^* x.$$

Proof Viewing x as a correspondence from $\operatorname{Spec} k$ to X, we compute using
Lemma 2.1. Keeping in mind that $[\Gamma_f] \bullet [\Gamma_f^t]$ is multiplication by $\deg f = 1$,
we have:
$$(f \times f)^* \alpha \bullet f^* x = [\Gamma_f^t] \bullet \alpha \bullet [\Gamma_f] \bullet [\Gamma_f^t] \bullet x = [\Gamma_f^t] \bullet \alpha \bullet x = f^*(\alpha \bullet x).$$
$\square$

Recall that T denotes the center of the blow-up of X and that Z is
the exceptional divisor in Y. For i, $0 \leq i \leq d$, define subgroups $C_i = f^* CH_{\mathbb{Q}}^i(X)$ and $D_i = j_*(\operatorname{Ker} g_* : CH_{\mathbb{Q}}^{i-1}(Z) \to CH_{\mathbb{Q}}^{i-d}(T))$ of $CH_{\mathbb{Q}}^i(Y)$.
Observe that $D_i = CH_{\mathbb{Q}}^{i-1}(Z) \cong \mathbb{Q}$ if $i < d$ and $D_i = 0$ if $i = d$. Then

Proposition 4.2 $CH_{\mathbb{Q}}^i(Y)$ *is the internal direct sum of C_i and D_i. Furthermore, if $\alpha \in A$ and $d_i \in D_i$, then $\alpha \bullet d_i = 0$, and if $\beta \in B$ and $c \in C_i$, then $(\beta \bullet \sigma) \bullet c_i = 0$.*

Proof Let $V_X = X - T$ and $V_Y = Y - Z$. Then $V_X \cong V_Y$, so localization gives a commutative diagram with exact rows:

$$
\begin{array}{ccccccc}
CH_{\mathbb{Q}}^{i-1}(Z) & \xrightarrow{\ j_*\ } & CH_{\mathbb{Q}}^i(Y) & \longrightarrow & CH_{\mathbb{Q}}^i(V_Y) & \longrightarrow & 0 \\
\Big\downarrow{\scriptstyle g_*} & & \Big\downarrow{\scriptstyle f_*} & & \Big\downarrow{\scriptstyle \cong} & & \\
CH_{\mathbb{Q}}^{i-d}(T) & \xrightarrow{\ i_*\ } & CH_{\mathbb{Q}}^i(X) & \longrightarrow & CH_{\mathbb{Q}}^i(V_X) & \longrightarrow & 0.
\end{array}
$$

The property $CH_{\mathbb{Q}}^i(Y) = C_i + D_i$ follows from a straightforward diagram chase and the fact that i_* is injective. Now write $\alpha = (f \times f)^* u$, $c_i = f^* v \in C_i$, and $d_i = j_* y \in D_i$. Then

$$
\begin{aligned}
\alpha \bullet d_i = (f \times f)^* u \bullet j_* y &= [\Gamma_f^t] \bullet u \bullet [\Gamma_f] \bullet [\Gamma_j] \bullet y \\
&= [\Gamma_f^t] \bullet u \bullet [\Gamma_{f \bullet j}] \bullet y \\
&= [\Gamma_f^t] \bullet u \bullet [\Gamma_{i \bullet g}] \bullet y \\
&= [\Gamma_f^t] \bullet u \bullet [\Gamma_i] \bullet g_* y \\
&= 0.
\end{aligned}
$$

Therefore, if $c_i \in C_i \cap D_i$, then (regarding c_i as an element of C_i), we have $\zeta \bullet c_i = c_i$ by Lemma 4.1, but also (regarding c_i an element of D_i), $\zeta \bullet c_i = 0$ by the above calculation. Thus, $C_i \cap D_i = \{0\}$. Finally,

$$
(\beta \bullet \sigma) \bullet c_i = \beta \bullet \sigma \bullet f^* v = \beta \bullet \sigma \bullet \zeta \bullet f^* v = \beta \bullet 0 \bullet f^* v = 0.
$$

$\square$

From this point onward, we fix identifications

$$
CH_{\mathbb{Q}}^i(Y) \cong C_i \oplus D_i \quad \text{and} \quad CH_{\mathbb{Q}}^d(Y \times Y) \cong A \oplus B
$$

and thereby justify the use of ordered pair notation for elements of $CH_{\mathbb{Q}}^*(Y)$ and $CH_{\mathbb{Q}}^*(Y \times Y)$.

Corollary 4.3 *Suppose $((f \times f)^* \alpha, \beta) \in CH_{\mathbb{Q}}^d(Y \times Y)$ and $(f^* x, y) \in CH_{\mathbb{Q}}^i(Y)$. Then*

$$
((f \times f)^* \alpha, \beta \bullet \sigma) \bullet (f^* x, y) = (f^*(\alpha \bullet x), \beta \bullet \sigma \bullet y).
$$

Proposition 4.4 *Murre's Conjecture* **B** *holds for* X *if and only if it holds for* Y, *and similarly for Conjecture* **B'**.

Proof We give the proof for Conjecture **B**. First suppose X has a Chow-Künneth decomposition $[\Delta_X] = \sum_{i=0}^{2d} \pi_i^X$ satisfying Murre's Conjecture **B**, i.e. $\pi_\ell \bullet CH_\mathbb{Q}^j(X) = 0$ when $\ell < j$ or $\ell > 2j$, and let $[\Delta_Y] = \sum_{i=0}^{2d}((f \times f)^* \pi_i^X, \delta_i)$ be the Chow-Künneth decomposition for Y as constructed in (3.7). From (3.6), the property $\delta_i = \delta_i \bullet \sigma$ holds for all i. Now fix j, $0 \le j \le d$, and consider $(f^*x, y) \in CH_\mathbb{Q}^j(Y)$. By Corollary 4.3, $((f \times f)^* \pi_\ell^X, \delta_\ell) \bullet (f^*x, y) = (f^*(\pi_\ell^X \bullet x), \delta_\ell \bullet y)$. If $\ell < j$ or $\ell > 2j$, then $\pi_\ell^X \bullet x = 0$. When ℓ is odd, clearly $\delta_\ell \bullet y = 0$; so assume ℓ is even. Then δ_ℓ is a product cycle of type $(d - \ell/2, \ell/2)$; so it suffices to show that for any $u \in CH_\mathbb{Q}^{d-\ell/2}(Y)$ and $v \in CH_\mathbb{Q}^{\ell/2}(Y)$, $(u \times v) \bullet y = 0$ when $\ell < j$ or $\ell > 2j$. Then

$$
\begin{aligned}
(u \times v) \bullet y &= p_2^{YY}{}_* (p_1^{YY*} y \cdot p_1^{YY*} u \cdot p_2^{YY*} v) \\
&= p_2^{YY}{}_* p_1^{YY*} (y \cdot u) \cdot v \\
&= p_\emptyset^{Y*} p_\emptyset^{Y}{}_* (y \cdot u) \cdot v.
\end{aligned}
$$

Note that $y \cdot u \in CH_\mathbb{Q}^{j+d-\ell/2}(Y)$. If $\ell < j$, then $j+d-\ell/2 > d$; so $y \cdot u = 0$. If $\ell > 2j$, then $j - \ell/2 < 0$; so $p_\emptyset^Y{}_* (y \cdot u) \in CH_\mathbb{Q}^{j-\ell/2}(\text{Spec } k) = 0$. Thus, this Chow-Künneth decomposition for Y satisfies Murre's Conjecture **B**.

Conversely, suppose $[\Delta_Y] = \sum_{i=0}^{2d} \pi_i^Y$ is a Chow-Künneth decomposition for Y satisfying Murre's Conjecture **B**. By Proposition 3.3, we may write $\pi_i^Y = ((f \times f)^* \pi_i^X, \delta_i)$ for some $\pi_i^X \in CH_\mathbb{Q}^d(X \times X)$. This means, in particular, that if $(f^*x, y) \in CH_\mathbb{Q}^\ell(Y)$, then $(f \times f)^* \pi_j^X \bullet f^*x = 0$ when $\ell < j$ or $\ell > 2j$. By Lemma 4.1 we have $f^*(\pi_j^X \bullet x) = 0$, and since f^* is injective, $\pi_j^X \bullet x = 0$. Corollary 3.13 then guarantees that $[\Delta_X] = \sum_{i=0}^{2d} \pi_i^X$ is a Chow-Künneth decomposition for X satisfying Murre's Conjecture **B**. $\square$

Proposition 4.5 *Murre's Conjecture* **C** *holds for* X *if and only if it holds for* Y. *Similarly, Conjecture* **D** *holds for* X *if and only if it holds for* Y.

Proof Assume first that X has a CK-decomposition $[\Delta_X] = \sum_{i=0}^{2d} \pi_i^X$ satisfying Murre's Conjecture **C**. Now let $[\Delta_Y] = \sum_{i=0}^{2d}(f \times f)^* \pi_i^X + \delta_i$ be the CK-decomposition for Y constructed in the proof of Theorem 3.5. By Proposition 4.2 and Corollary 4.3, we have $((f \times f)^* \pi_\ell^X + \delta_\ell) \bullet CH_\mathbb{Q}^i(Y) = (f \times f)^* \pi_\ell^X \bullet C_i + \delta_\ell \bullet D_i = f^*(\pi_\ell^X \bullet C_i) + \delta_\ell \bullet D_i$. In particular, this implies that the filtration induced by this CK-decomposition (as defined in Section 2.1) is described by

$$F^m CH^i_{\mathbb{Q}}(Y) = f^* F^m CH^i_{\mathbb{Q}}(X) + D_{i,m} \tag{4.1}$$

where $D_i = D_{i,0} \supseteq D_{i,1} \supseteq \ldots$ is a descending chain of subgroups. By Murre's Conjecture **C** for X, the term $F^m CH^i_{\mathbb{Q}}(X)$ is independent of the original choice of CK-decomposition for X. Also, by [Mu, Lemma 1.4.4], $F^1 CH^i_{\mathbb{Q}}(Y)$ is contained in the subgroup $CH^i_{\mathbb{Q}}(Y)_{hom} \subseteq CH^i_{\mathbb{Q}}(Y)$ of cycles homologically equivalent to zero. If $i = d$, then $D_i = 0$; so $D_{i,j} = 0$ for all j. If $i < d$, then $D_i = CH^{i-1}_{\mathbb{Q}}(Z) \cong \mathbb{Q}$ is a one-dimensional $\mathbb{Q}$-vector space, with $CH^{i-1}_{\mathbb{Q}}(Z)_{hom} = 0$. Hence $D_{i,j} = 0$ for all $j \geq 1$, showing that the filtration $F^m CH^i_{\mathbb{Q}}(Y)$ is independent of the original choice of CK-decomposition on Y. This proof also shows that if Conjecture **D** holds for X, i.e. $F^1 CH^i_{\mathbb{Q}}(X) = CH^i_{\mathbb{Q}}(X)_{hom}$, then likewise $F^1 CH^i_{\mathbb{Q}}(Y) = CH^i_{\mathbb{Q}}(Y)_{hom}$.

Conversely, suppose Y has a CK-decomposition satisfying Murre's Conjecture **C**. If $[\Delta_X] = \sum_{i=0}^{2d} \pi_i^X$ is a CK-decomposition on X, use Theorem 3.5 to construct a CK-decomposition $[\Delta_Y] = \sum_{i=0}^{2d} (f \times f)^* \pi_i^X + \delta_i$ on Y. By assumption, the filtration (4.1) defined by this CK-decomposition is independent of the original choice of CK-decomposition on X; hence $f_* F^m CH^i_{\mathbb{Q}}(Y) = F^m CH^i_{\mathbb{Q}}(X)$ is also independent on this choice, and so Conjecture **C** holds for X. The assertion concerning Conjecture **D** follows similarly.

$\square$

5 Application to Lefschetz decompositions

5.1 Lefschetz type and blowing up

The goal of this section is to prove the following result:

Theorem 5.1 *Let $f : Y \to X$ be as in Section 3.1. Then Y is of Lefschetz type if and only if X is of Lefschetz type.*

Let $d = \dim X = \dim Y$ and suppose $((CH^{*+d}_{\mathbb{Q}}(X \times X), \{\pi_i^X\}_{i=0}^{2d}, L_X, \Lambda_X)$ is a Lefschetz algebra. It follows easily from Lemma 2.2 that

$$((f \times f)^* CH^*_{\mathbb{Q}}(X \times X), \{(f \times f)^* \pi_i^X\}_{i=0}^{2d}, (f \times f)^* L_X, (f \times f)^* \Lambda_X)$$

is a Lefschetz algebra.

Next, define

$$L' = \sum_{i=1}^{d}(\ell_i \times a_{d-i+1}) + (a_i \times \ell_{d-i+1}) \in CH_{\mathbb{Q}}^{d+1}(Y \times Y)$$

and

$$\Lambda' = \sum_{j=0}^{d-1}(d-j)(j+1)\left[(\ell_j \times a_{d-j-1}) + (a_j \times \ell_{d-j-1})\right] \in CH_{\mathbb{Q}}^{d-1}(Y \times Y).$$

We will show that if we set

$$L_Y = (f \times f)^* L_X + L' \in CH_{\mathbb{Q}}^{d+1}(Y \times Y)$$

and

$$\Lambda_Y = (f \times f)^* \Lambda_X + \Lambda' \in CH_{\mathbb{Q}}^{d-1}(Y \times Y)$$

and denote by $\pi_0^Y, \ldots, \pi_{2d}^Y \in CH_{\mathbb{Q}}^d(Y \times Y)$ the projectors constructed in (3.7), then $(CH_{\mathbb{Q}}^{*+d}(Y \times Y), \{\pi_i^Y\}_{i=0}^{2d}, L_Y, \Lambda_Y)$ is a Lefschetz algebra.

Proposition 5.2 *The following identities hold for each s:*

- $L' \bullet \delta_{2s} = \delta_{2s+2} \bullet L' = \ell_{d-s} \times a_{s+1} + a_{d-s} \times \ell_{s+1}.$

- $\Lambda' \bullet \delta_{2s} = \delta_{2s-2} \bullet \Lambda' = s(d-s+1)[(\ell_{d-s} \times a_{s-1}) + (a_{d-s} \times \ell_{s-1})].$

- $\Lambda' \bullet L' - L' \bullet \Lambda' = \sum_{i=0}^{2d}(d-i)\delta_i.$

Proof

$$L' \bullet \delta_{2s} = L' \bullet \gamma_{d-s} = \sum_{i=1}^{d}[\ell_i \times a_{d-i+1} + a_i \times \ell_{d-i+1}] \bullet [\eta_{d-s} + \theta_{d-s}].$$

By Lemma 2.3, the only (possibly) nonzero term in this sum corresponds to $i = d - s$, so

$$L' \bullet \delta_{2s} = (\ell_{d-s} \times a_{s+1}) \bullet \eta_{d-s} + (\ell_{d-s} \times a_{s+1}) \bullet \theta_{d-s}$$
$$+ (a_{d-s} \times \ell_{s+1}) \bullet \eta_{d-s} + (a_{d-s} \times \ell_{s+1}) \bullet \theta_{d-s}.$$

Using Proposition 3.8 and Corollary 3.9, we see that the first term equals $\ell_{d-s} \times a_{s+1}$ and the fourth term equals $a_{d-s} \times \ell_{s+1}$, while the middle two terms vanish. Thus, $L' \bullet \delta_{2s} = \ell_{d-s} \times a_{s+1} + a_{d-s} \times \ell_{s+1}$. By similar reasoning,

$$\delta_{2s+2} \bullet L' = (\eta_{d-s-1} + \theta_{d-s-1}) \bullet [\ell_{d-s} \times a_{s+1} + a_{d-s} \times \ell_{s+1}]$$
$$= \ell_{d-s} \times a_{s+1} + a_{d-s} \times \ell_{s+1}.$$

Therefore, $L' \bullet \delta_{2s} = \delta_{2s+2} \bullet L'$, as desired. The proof of the second formula is similar.

For the third formula,

$$\Lambda' \bullet L' = \sum_{j=0}^{d-1} \sum_{i=1}^{d} (d-j)(j+1) \left[(\ell_j \times a_{d-j-1}) + (a_j \times \ell_{d-j-1}) \right]$$

$$\bullet \left[(\ell_i \times a_{d-i+1}) + (a_i \times \ell_{d-i+1}) \right].$$

By Lemma 2.3, the (i,j) term of this double sum will be zero unless $j = i - 1$; hence the expression simplifies to

$$\sum_{i=1}^{d} i(d-i+1) \left[(\ell_{i-1} \times a_{d-i}) + (a_{i-1} \times \ell_{d-i}) \right] \bullet \left[(\ell_i \times a_{d-i+1}) + (a_i \times \ell_{d-i+1}) \right]$$

$$= \sum_{i=1}^{d} i(d-i+1) [m(\ell_{i-1}, a_{d-i+1})(\ell_i \times a_{d-i}) + m(\ell_{i-1}, \ell_{d-i+1})(a_i \times a_{d-i})$$

$$+ m(u_{i-1}, a_{d-i+1})(\ell_i \times \ell_{d-i}) + m(a_{i-1}, \ell_{d-i+1})(a_i \times \ell_{d-i})]$$

$$= \sum_{i=1}^{d} i(d-i+1) [(\ell_i \times a_{d-i}) + (a_i \times \ell_{d-i})].$$

Likewise,

$$L' \bullet \Lambda' = \sum_{j=0}^{d-1} \sum_{i=1}^{d} (d-j)(j+1) \left[(\ell_i \times a_{d-i+1}) + (a_i \times \ell_{d-i+1}) \right]$$

$$\bullet \left[(\ell_j \times a_{d-j-1}) + (a_j \times \ell_{d-j-1}) \right].$$

Again, the only nonzero terms correspond to the case $j = i - 1$, so this simplifies to

$$\sum_{i=1}^{d} i(d-i+1) \left[(\ell_i \times a_{d-i+1}) + (a_i \times \ell_{d-i+1}) \right] \bullet \left[(\ell_{i-1} \times a_{d-i}) + (a_{i-1} \times \ell_{d-i}) \right]$$

$$= \sum_{i=1}^{d} i(d-i+1) [m(\ell_i, a_{d-i})(\ell_{i-1} \times a_{d-i+1}) + m(\ell_i, \ell_{d-i})(a_{i-1} \times a_{d-i+1})$$

$$+ m(a_i, a_{d-i})(\ell_{i-1} \times \ell_{d-i+1}) + m(a_i, \ell_{d-i})(a_{i-1} \times \ell_{d-i+1})]$$

$$= \sum_{i=1}^{d} i(d-i+1) [(\ell_{i-1} \times a_{d-i+1}) + (a_{i-1} \times \ell_{d-i+1})]$$

$$= \sum_{i=0}^{d-1} (i+1)(d-i) [(\ell_i \times a_{d-i}) + (a_i \times \ell_{d-i})].$$

Hence,

$$\Lambda' \bullet L' - L' \bullet \Lambda'$$

$$= -d[(\ell_0 \times a_d) + (a_0 \times \ell_d)] + \sum_{i=1}^{d-1}[i(d-i+1) - (i+1)(d-i)][(\ell_i \times a_{d-i})$$

$$+ (a_i \times \ell_{d-i})] + d[(\ell_d \times a_0) + (a_d \times \ell_0)]$$

$$= -d\gamma_0 + \sum_{i=1}^{d-1}(2i-d)\gamma_i + d\gamma_d = \sum_{i=0}^{d}(2i-d)\gamma_i$$

$$= \sum_{i=0}^{d}(2i-d)\delta_{2(d-i)} = \sum_{i=0}^{d}(d-2i)\delta_{2i}.$$

Since $\delta_j = 0$ for j odd, reindexing gives $\Lambda' \bullet L' - L' \bullet \Lambda' = \sum_{j=0}^{2d}(d-j)\delta_j$, as desired.

$\square$

Corollary 5.3

$$L' \bullet \sigma = L' = \sigma \bullet L', \quad \Lambda' \bullet \sigma = \sigma \bullet \Lambda' = \Lambda',$$

and

$$\delta_j \bullet \sigma = \sigma \bullet \delta_j = \sigma \ for \ j, \ 0 \le j \le 2d.$$

Proof By the definition of L' and Proposition 5.2,

$$L' = \sum_{i=0}^{d} L' \bullet \delta_{2(d-i)} = L' \bullet \sum_{i=0}^{d} \delta_{2(d-i)} = L' \bullet \sigma$$

and

$$L' = \sum_{i=0}^{d} \delta_{2i} \bullet L' = (\sum_{i=0}^{d} \delta_{2i}) \bullet L' = \sigma \bullet L',$$

which establishes the first set of equalities; the proof for the second set is similar. The third set of equalities is a restatement of (3.6).

$\square$

Proof (of Theorem 5.1) Suppose, as above, that

$$(CH_{\mathbb{Q}}^{*+d}(X \times X), \{\pi_i^X\}_{i=0}^{2d}, L_X, \Lambda_X)$$

is a Lefschetz algebra. Let $\{\pi_i^Y\}_{i=0}^{2d}$ denote the Chow-Künneth decomposition defined by (3.7); we will show that

$$(CH_{\mathbb{Q}}^{*+d}(Y \times Y), \{\pi_i^Y\}_{i=0}^{2d}, (f \times f)^* L_X + L', (f \times f)^* \Lambda_X + \Lambda')$$

is a Lefschetz algebra. The verification is purely formal, so we will show it only for the first formula in Definition 2.4; the rest follow by analogous reasoning. First,

$$((f \times f)^* L_X + L') \bullet ((f \times f)^* \pi_{2s}^X + \delta_{2s}) = (f \times f)^* L_X \bullet (f \times f)^* \pi_{2s}^X$$
$$+ (f \times f)^* L_X \bullet \delta_{2s} + L' \bullet (f \times f)^* \pi_{2s}^X + L' \bullet \delta_{2s}.$$

By Lemma 2.2 and the hypothesis, the first term on the right equals

$$(f \times f)^* (L_X \bullet \pi_{2s}^X) = (f \times f)^* (\pi_{2s+2}^X \bullet L_X) = (f \times f)^* (\pi_{2s+2}^X) \bullet (f \times f)^* L_X.$$

Since $(f \times f)^* L_X \in A$ and $\delta_{2s} = \sigma \bullet \delta_{2s}$ by (3.6), we may compute the second term: $(f \times f)^* L_X \bullet \delta_{2s} = (f \times f)^* L_X \bullet \zeta \bullet \sigma \bullet \delta_{2s} = 0$. Likewise, the third term is $L' \bullet (f \times f)^* \pi_{2s}^X = L' \bullet \sigma \bullet \zeta \bullet (f \times f)^* \pi_{2s}^X = 0$. The last term equals $\delta_{2s+2} \bullet L'$ by Proposition 5.2. Thus, we have:

$$((f \times f)^* L_X + L') \bullet ((f \times f)^* \pi_{2s}^X + \delta_{2s}) = (f \times f)^* (\pi_{2s+2}^X) \bullet (f \times f)^* L_X + \delta_{2s+2} \bullet L'.$$

Similarly one shows

$$(f \times f)^* (\pi_{2s+2}^X) \bullet (f \times f)^* L_X + \delta_{2s+2} \bullet L'$$
$$= ((f \times f)^* (\pi_{2s+2}^X) + \delta_{2s+2}) \bullet ((f \times f)^* L_X + L'),$$

completing the argument.

Conversely, suppose $(CH_{\mathbb{Q}}^{*+d}(Y \times Y), \{\pi_i^Y\}_{i=0}^{2d}, L_Y, \Lambda_Y)$ is a Lefschetz algebra, i.e. the identities of Definition 2.4 are satisfied. By Corollary 3.12, the analogous identities required to show that $(CH_{\mathbb{Q}}^{*+d}(X \times X), \{(f \times f)_* \pi_i^Y\}_{i=0}^{2d}, (f \times f)_* L_Y, (f \times f)_* \Lambda_Y)$ is a Lefschetz algebra also hold. $\square$

5.2 Agreement with the usual Lefschetz operator

In the proof of Theorem 5.1, we constructed the Lefschetz operator L_Y in terms of L_X and an extra term L'. In the following we show that if $d = \dim X \geq 2$ and if L_X takes the usual form of the Lefschetz operator, then the same is true for Y. Letting $\Delta_X : X \to X \times X$ and $\Delta_Y : Y \to Y \times Y$ denote the respective diagonal maps, we show that if $L_X = (\Delta_X)_*(b)$ for some divisor $b \in CH_{\mathbb{Q}}^1(X)$, then $L_Y = (\Delta_Y)_*(b')$ for some $b' \in CH_{\mathbb{Q}}^1(Y)$.

To this end, let $b \in CH^1_{\mathbb{Q}}(X)$, and write $b' = f^*b + j_*[Z] \in CH^1_{\mathbb{Q}}(Y)$. Writing $L_{b'} = (\Delta_Y)_*(b')$, we have:

$$
\begin{aligned}
(f \times f)_* L_{b'} &= (f \times f)_*(\Delta_Y)_*(b') \\
&= (\Delta_X)_* f_*(b') \\
&= (\Delta_X)_* f_* f^*(b) + (\Delta_X)_* f_* j_*[Z] \\
&= (\Delta_X)_*(b) + (\Delta_X)_* i_* g_*[Z] \\
&= L_X.
\end{aligned}
$$

By Proposition 3.3, there exists $M \in B_{d+1}$ such that $L_{b'} = \zeta \bullet L_{b'} \bullet \zeta + M$; that is,

$$
M = L_{b'} - \zeta \bullet L_{b'} \bullet \zeta = L_{b'} - (f \times f)^*(f \times f)_* L_{b'} = L_{b'} - (f \times f)^* L_X.
$$

We will show that $M = L'$; it will then follow that the Lefschetz operator L_Y constructed in Theorem 5.1 coincides with $L_{b'}$. Note also that

$$
M = L_{b'} - ([\Delta_Y] - \sigma) \bullet L_{b'} \bullet ([\Delta_Y] - \sigma) = \sigma \bullet L_{b'} + L_{b'} \bullet \sigma - \sigma \bullet L_{b'} \bullet \sigma. \quad (5.1)
$$

Observe that

$$
\begin{aligned}
\sigma \bullet L_{b'} &= \sum_{i=0}^{d}(a_i \times \ell_{d-i} + \ell_i \times a_{d-i}) \bullet (\Delta_Y)_*(b') \\
&= \sum_{i=0}^{d}(a_i \bullet (\Delta_Y)_*(b')) \times \ell_{d-i} + (\ell_i \bullet (\Delta_Y)_*(b')) \times a_{d-i}.
\end{aligned}
$$

Now for $y \in CH^i_{\mathbb{Q}}(Y)$,

$$
y \bullet (\Delta_Y)_*(b') = p_2^{YY}{}_*(p_1^{YY*} y \cdot (\Delta_Y)_*(b')) = p_2^{YY}{}_*(\Delta_Y)_*((\Delta_Y)^* p_1^{YY*} y \cdot b') = y \cdot b';
$$

thus, $\sigma \bullet L_{b'} = \sum_{i=0}^{d}(a_i \cdot b') \times \ell_{d-i} + (\ell_i \cdot b') \times a_{d-i}$. Also,

$$
\begin{aligned}
\ell_i \cdot b' &= j_* \ell^{i-1} \cdot (f^*(b) + j_*[Z]) \\
&= j_*(\ell^{i-1} \cdot (j^* f^*(b) + j^* j_*[Z])) \\
&= j_*(\ell^{i-1} \cdot g^* i^*(b) + \ell^{i-1} \cdot \ell) \\
&= j_* \ell^i = \ell_{i+1}.
\end{aligned}
$$

Thus, $\sigma \bullet L_{b'} = \sum_{i=0}^{d}(a_i \cdot b') \times \ell_{d-i} + \ell_{i+1} \times a_{d-i}$. Now each term in the above summand is a product cycle of type $(i+1, d-i)$, so we have:

$$\sigma \bullet L_{b'} = (\sigma \bullet L_{b'}) \bullet \gamma_{i+1}$$

$$= \sum_{i=0}^{d} ((a_i \cdot b') \times \ell_{d-i}) \bullet (a_{i+1} \times \ell_{d-i-1} + \ell_{i+1} \times a_{d-i-1})$$

$$\qquad + (\ell_{i+1} \times a_{d-i}) \bullet (a_{i+1} \times \ell_{d-i-1} + \ell_{i+1} \times a_{d-i-1})$$

$$= \sum_{i=0}^{d} m(\ell_{d-i-1}, a_i \cdot b')(a_{i+1} \times \ell_{d-i}) + m(a_{d-i-1}, a_i \cdot b')(\ell_{i+1} \times \ell_{d-i})$$

$$\qquad + m(\ell_{i+1}, \ell_{d-i-1})(a_{i+1} \times a_{d-i}) + m(\ell_{i+1}, a_{d-i-1})(\ell_{i+1} \times a_{d-i}).$$

By Proposition 3.8, the above simplifies to:

$$\sigma \bullet L_{b'} = \sum_{i=0}^{d} m(\ell_{d-i-1}, a_i \cdot b')(a_{i+1} \times \ell_{d-i})$$

$$\qquad\qquad + m(a_{d-i-1}, a_i \cdot b')(\ell_{i+1} \times \ell_{d-i}) + (\ell_{i+1} \times a_{d-i}).$$

Let $q : Y \to \operatorname{Spec} k$ be the structure morphism. Then we have:

$$m(\ell_{d-i-1}, a_i \cdot b') = q_*(j_* \ell^{d-i-2} \cdot a_i \cdot (f^*(b) + r j_*[Z]))$$

$$= q_*(j_*(\ell^{d-i-2} \cdot j^*(a_i) \cdot j^* f^*(b))$$

$$\qquad + q_*(j_*(\ell^{d-i-2} \cdot j^*(a_i) \cdot j^* j_*[Z])$$

$$= q_*(j_*(\ell^{d-i-2} \cdot j^*(a_i) \cdot \ell))$$

$$= q_* j_*(\ell^{d-i-1} \cdot j^*(a_i))$$

$$= q_*(j_* \ell^{d-i-1} \cdot a_i)$$

$$= m(\ell_{d-i}, a_i)$$

$$= 1.$$

Also, $m(a_{d-i-1}, a_i \cdot b') = q_*(a_{d-i-1} \cdot a_i \cdot b')$, but $a_{d-i-1} \cdot a_i = \Delta_Y^*(a_{d-i-1} \times a_i) = 0$ by Proposition 3.8, so $m(a_{d-i-1}, a_i \cdot b') = 0$.

Therefore, $\sigma \bullet L_{b'} = \sum_{i=0}^{d}(a_{i+1} \times \ell_{d-i}) + (\ell_{i+1} \times a_{d-i}) = L'$. From the definitions, one sees immediately that $\sigma^t = \sigma$, $L_{b'}^t = L_{b'}$, and the above calculation shows that $(\sigma \bullet L_{b'})^t = \sigma \bullet L_{b'}$. Hence, $L_{b'} \bullet \sigma = L_{b'}^t \bullet \sigma^t = (\sigma \bullet L_{b'})^t = \sigma \bullet L_{b'}$. By (5.1), $M = \sigma \bullet L_{b'} = L'$.

6 Applications to Kummer varieties and manifolds

Let A be an abelian variety of dimension d over an algebraically closed field k of characteristic $\neq 2$. The associated Kummer variety K_A is obtained by taking the quotient of A by the action of the group (scheme) G generated by the involution $a \mapsto -a$.

The Kummer manifold is obtained from K_A by blowing up the singular locus of A — that is, by blowing up the image of the 2-torsion points of A under the quotient map $q : A \to K_A$. As observed in [DL, p.4], K_A may be embedded in $\mathbb{P}^{2^d-1}$ using a symmetric theta divisor; thus, the image of any 2-torsion point is a singular point, étale locally isomorphic to the affine cone over the second Veronese variety of $\mathbb{P}^{d-1}$. This can be seen by observing that the negation involution of the abelian variety A acts locally by $(z_1, \cdots, z_g) \to (-z_1, \cdots, -z_g)$, because it acts so on the tangent space. The ring of invariants is generated by polynomials $z_i z_j$; hence, the exceptional divisor of the blow-up of the Kummer variety at a 2-torsion point is isomorphic to $\mathbb{P}^{d-1}$. Now if $a \in A$ is a 2-torsion point, $\overline{A}$ the blow-up of A along $\{a\}$, and $\overline{K}_A$ the blow-up of K_A along $\{q(a)\}$, then the universal property of the blow-up gives an induced map $h : \overline{A}/G \to \overline{K}_A$. Since the exceptional divisors of both blow-ups are (each) isomorphic to $\mathbb{P}^{d-1}$ and K_A is known to be normal [Sas], h is a quasi-finite proper birational map. Because $\overline{K}_A$ is also normal, h is an isomorphism by Zariski's main theorem. This proves that $\overline{K}_A$ is also a pseudo-smooth variety. A similar argument shows that the intermediate schemes obtained by successively blowing up each of the singular points on K_A also satisfy the same hypotheses. Let $f : K'_A \to K_A$ denote the composition of all these blow-up maps; K'_A is then a smooth variety, the so-called Kummer manifold associated to A.

6.1 Murre's conjectures and the Lefschetz decomposition for Kummer manifolds

Corollary 6.1 *The Kummer manifold Y has a Chow-Künneth decomposition $[\Delta_Y] = \sum_{i=0}^{2d} \pi_i^Y$ satisfying Poincaré duality, and is also of Lefschetz type. Furthermore, Y satisfies Murre's conjecture $\mathbf{B}'$, and when $d \leq 4$, Y satisfies Murre's conjecture $\mathbf{B}$.*

Proof For convenience, let $X = K_A$ and $Y = K'_A$. In [DM, Section 3], Deninger and Murre constructed a particular Chow-Künneth decomposition $[\Delta_A] = \sum_{i=0}^{2d} \pi_i^A$ for A. Using this construction, the present authors showed in [AJ1] that if we set $\pi_i^X = (q \times q)_* \pi_i^A \in CH_{\mathbb{Q}}^d(X \times X)$, then $[\Delta_X] = \sum_{i=0}^{2d} \pi_i^X$ is a Chow-Künneth decomposition for X satisfying Poincaré duality and Murre's Conjecture **B'**; when $d \leq 4$, X also satisfies Murre's Conjecture **B**. We also showed in [AJ2, Theorem 1.2] that X is of Lefschetz type. The conclusion then follows by application of Theorem 3.5, Theorem 5.1, and Proposition 4.4.

$\square$

Remarks

1. A similar result holds for any of the intermediate schemes obtained by blowing up some (but not all) of the singular points on X.

2. The referee has pointed out an alternate strategy for constructing an explicit Chow-Künneth decomposition on the Kummer manifold Y, based on a different construction of the latter. The negation map $a \mapsto -a$ on the abelian variety A defines an action of $G = \mathbb{Z}/2\mathbb{Z}$ on A in the obvious manner. If one blows up the locus of 2-torsion points on A to obtain a variety $\tilde{A}$, then the action of G on A extends in a natural way to an action of G on $\tilde{A}$. The Kummer manifold Y can then be realized as the quotient variety $\tilde{A}/G$. Therefore, if one starts with a Chow-Künneth decomposition on the Abelian variety that is stable under the action of G, one could apply the results of [V, Remark 5.5] or [SV, Proposition 2.10] to obtain an explicit Chow-Kunneth decomposition on $\tilde{A}$, which could then be descended to a Chow-Kunneth decomposition on the Kummer manifold Y as in [SV, Corollary 2.13].

6.2 Algebraic equivalence on Kummer varieties and manifolds

Continuing the notation and assumptions of the previous section, we apply our explicit construction to study powers of the relation of algebraic equivalence on X and on Y.

In [Sam], Samuel defined the notion of an *adequate equivalence relation* on algebraic cycles and proved that rational equivalence is the finest such relation. Having fixed a field k, an adequate equivalence relation E is an assignment, to every pseudo-smooth variety V over k, of a subgroup $ECH^*(V) \subseteq CH^*(V)$ which is preserved under pullback, pushforward, and intersection with arbitrary cycles. Algebraic equivalence, homological equivalence, and numerical equivalence all examples of adequate equivalence relations. Hiroshi Saito [Sai] defined the product $E * E'$ of two adequate equivalence relations E and E', and proved that $E * E'$ is itself adequate. He also proved that the operation is associative and commutative, and distributes over sums of relations (defined in the expected manner).

If A is an abelian variety of dimension d over an algebraically closed field k, there is a natural filtration on $CH^p_{\mathbb{Q}}(A)$ defined by the Deninger-Murre Chow-Künneth projectors: for $r \in \mathbb{Z}$, set $F^r CH^p_{\mathbb{Q}}(A) = \sum_{i=0}^{2p-r} \pi_i^A \bullet CH^p_{\mathbb{Q}}(A)$. A conjecture of Beauville [Be] is equivalent to the assertion that the nontrivial steps in this filtration occur only in positive degree, i.e. $F^0 CH^p_{\mathbb{Q}}(A) = CH^p_{\mathbb{Q}}(A)$. This is easily seen to be equivalent to the assertion that π_i acts as 0 on $CH^j_{\mathbb{Q}}(A) = 0$ when $i < 2j$, which is the second half of Murre's Conjecture **B**.

Now let L denote the (adequate) relation of algebraic equivalence. Its rth power L^{*r} is the so-called *r-cubical equivalence* introduced in [Sam]. In previous work of the first author, the following was proved in a slightly stronger form:

Theorem 6.2 *[A, Theorem 3.1 and Proposition 3.3] Assume Beauville's Conjecture, and let A be an abelian variety over an algebraically closed field. Then:*

(i) *For $r \geq 1$, $F^r CH^d_{\mathbb{Q}}(A) = L^{*r} CH^d_{\mathbb{Q}}(A)$.*

(ii) *For $r > d$, $L^{*r} CH^*_{\mathbb{Q}}(A) = 0$.*

The second statement is a kind of nilpotence assertion for cycles on abelian varieties. We will show that our constructions yield similar results for X and Y.

First, define filtrations on $CH^p_{\mathbb{Q}}(X)$ and $CH^p_{\mathbb{Q}}(Y)$ by

$$F^r CH^p_{\mathbb{Q}}(X) = \sum_{i=0}^{2p-r} \pi_i^X \bullet CH^p_{\mathbb{Q}}(X) \text{ and } F^r CH^p_{\mathbb{Q}}(Y) = \sum_{i=0}^{2p-r} \pi_i^Y \bullet CH^p_{\mathbb{Q}}(Y).$$

Then for i, $0 \leq i \leq 2d$ and $\alpha \in CH^*_{\mathbb{Q}}(X)$, we have:

$$(q \times q)_* \pi_i^A \bullet \alpha = q_*(\pi_i^A \bullet q^* \alpha) \tag{6.1}$$

since both sides are equal (as correspondences) to $\Gamma_q \bullet \pi_i^A \bullet \Gamma_q^t \bullet \alpha$. Note that this identity is also expressed by the formula $F^r CH^p_{\mathbb{Q}}(X) = q_* F^r CH^p_{\mathbb{Q}}(A)$.

Proposition 6.3 *Assume Beauville's conjecture. Then the conclusions of Theorem 6.2 hold when A is replaced by either X or Y.*

Proof Suppose $r \geq 1$. Then Theorem 6.2, together with adequacy of L^{*r}, implies

$$F^r CH_{\mathbb{Q}}^d(X) = q_* F^r CH_{\mathbb{Q}}^d(A) = q_* L^{*r} CH_{\mathbb{Q}}^d(A) \subseteq L^{*r} CH_{\mathbb{Q}}^d(X).$$

Likewise, since $q_* q^*$ is multiplication by $|G|$,

$$\begin{aligned} L^{*r} CH_{\mathbb{Q}}^d(X) &= q_* q^* L^{*r} CH_{\mathbb{Q}}^d(X) \subseteq q_* (L^{*r} CH_{\mathbb{Q}}^d(A)) \\ &= q_* (F^r CH_{\mathbb{Q}}^d(A)) \\ &= F^r CH_{\mathbb{Q}}^p(X). \end{aligned}$$

This proves the first statement for X. For the second statement, simply observe that for $r > d$,

$$L^{*r} CH_{\mathbb{Q}}^*(X) = q_* q^* L^{*r} CH_{\mathbb{Q}}^*(X) \subseteq q_* (L^{*r} CH_{\mathbb{Q}}^*(A)) = 0.$$

To deduce the statements for Y, apply Proposition 4.2 to write $CH_{\mathbb{Q}}^d(Y) = C_d + D_d$. Direct computation shows that $D_d = 0$, so since $\pi_i^Y = (f \times f)^* \pi_i^X + \delta_i$ and $\delta_i = \delta_i \bullet \sigma$ by Corollary 5.3, another application of Proposition 4.2 implies

$$\begin{aligned} F^r CH_{\mathbb{Q}}^d(Y) &= \sum_{i=0}^{2d-r} \pi_i^Y \bullet CH_{\mathbb{Q}}^d(Y) \\ &= \sum_{i=0}^{2d-r} (f \times f)^* \pi_i^X \bullet C_d \\ &= \sum_{i=0}^{2p-r} f^* (\pi_i^X \bullet CH_{\mathbb{Q}}^d(X)) \\ &= f^* F^r CH_{\mathbb{Q}}^d(X). \end{aligned}$$

Now, using adequacy of L^{*r}, we have, for $r \geq 1$:

$$F^r CH_{\mathbb{Q}}^d(Y) = f^* F^r CH_{\mathbb{Q}}^d(X) = f^* L^{*r} CH_{\mathbb{Q}}^d(X) \subseteq L^{*r} CH_{\mathbb{Q}}^d(Y).$$

Also, because $CH_{\mathbb{Q}}^d(Y) = f^* CH_{\mathbb{Q}}^d(X)$, we have $CH_{\mathbb{Q}}^d(Y) = f^* f_* CH_{\mathbb{Q}}^d(Y)$, so

$$\begin{aligned} L^{*r} CH_{\mathbb{Q}}^d(Y) = f^* f_* L^{*r} CH_{\mathbb{Q}}^d(Y) \subseteq f^* L^{*r} CH_{\mathbb{Q}}^d(X) &= f^* F^r CH_{\mathbb{Q}}^d(X) \\ &= F^r CH_{\mathbb{Q}}^d(Y). \end{aligned}$$

This establishes the first statement. For the second, simply note that for $r > d$,

$$L^{*r}CH_{\mathbb{Q}}^*(Y) = f^*f_*L^{*r}CH_{\mathbb{Q}}^*(Y) \subseteq f^*L^{*r}CH_{\mathbb{Q}}^*(X) = 0.$$

□

6.3 A Hard Lefschetz Theorem for Chow groups of Kummer manifolds

As an application of the explicit Lefschetz decomposition constructed in Section 5, we prove the following theorem.

Theorem 6.4 (Hard Lefschetz for Chow groups) *With notation and assumptions as in Sections 3.1 and 5, suppose further that X is of Lefschetz type and that for $2p \leq d$, the map $H_X : CH_{\mathbb{Q}}^p(X) \to CH_{\mathbb{Q}}^{d-p}(X)$ defined by $a \mapsto L_X^{d-2p} \bullet a$ is an isomorphism. Then the map $H_Y : CH_{\mathbb{Q}}^p(Y) \to CH_{\mathbb{Q}}^{d-p}(Y)$ defined by $z \mapsto L_Y^{d-2p} \bullet z$ is an isomorphism.*

Proof By the direct sum decomposition $CH_{\mathbb{Q}}^i(Y) \cong C_i \oplus D_i$ from Section 4, any $z \in CH_{\mathbb{Q}}^i(Y)$ may be written (uniquely) as $z = f^*x + y$, where $x \in CH_{\mathbb{Q}}^i(X)$ and $y \in D_i$. Then

$$L_Y \bullet z = ((f \times f)^*L_X + L') \bullet (f^*x + y)$$
$$= (f \times f)^*L_X \bullet f^*x + (f \times f)^*L_X \bullet y + L' \bullet f^*x + L' \bullet y.$$

Using Proposition 4.1 to simplify the first two terms, and the equalities $\sigma \bullet L' = L' = L' \bullet \sigma$ from Corollary 5.3, we conclude: $L_Y \bullet z = f^*(L_X \bullet x) + (f \times f)^*L_X \bullet y + L' \bullet \sigma \bullet f^*x + L' \bullet y$. By Proposition 4.2, the middle two terms are 0, so we have $L_Y \bullet z = f^*(L_X \bullet x) + L' \bullet y$. Using ordered pair notation (as in Section 4) to express the decompositions $CH_{\mathbb{Q}}^*(Y \times Y) \cong A \oplus B$ and $CH_{\mathbb{Q}}^i(Y) \cong C_i \oplus D_i$, we have:

$$L_Y \bullet z = ((f \times f)^*L_X, L') \bullet (f^*x, y) = (f^*(L_X \bullet x), L' \bullet y)$$

and hence, by induction,

$$L_Y^{d-2p} \bullet z = (f^*(L_X^{d-2p} \bullet x), L'^{d-2p} \bullet y).$$

Thus, to prove that the map $H_Y : C_p \oplus D_p \to C_{d-p} \oplus D_{d-p}$ defined above is an isomorphism, it suffices to check that the induced maps $u : C_p \to C_{d-p}$

defined by $f^*x \mapsto f^*(L_X^{d-2p} \bullet x)$ and $v : D_p \to D_{d-p}$ defined by $y \mapsto L'^{d-2p} \bullet y$ are isomorphisms. That u is an isomorphism follows formally, since H_X is an isomorphism and f^* is injective.

We will also need an unweighted version of the Λ' operator, defined by:

$$\Lambda_0' = \sum_{j=0}^{d-1} \ell_j \times a_{d-j-1} + a_j \times \ell_{d-j-1} \in CH_{\mathbb{Q}}^{d-1}(Y \times Y).$$

We claim that the map $D_{d-p} \to D_p$ defined by $y \mapsto \Lambda_0'^{d-2p} \bullet y$ is a two-sided inverse to v. Fortunately, both L' and Λ_0' are product cycles, so we can calculate their powers explicitly. By Lemma 2.3, we have:

$$L'^2 = \sum_{i=1}^{d}[(\ell_i \times a_{d-i+1}) + (a_i \times \ell_{d-i+1})] \bullet \sum_{j=1}^{d}[(\ell_j \times a_{d-j+1}) + (a_j \times \ell_{d-j+1})]$$

$$= \sum_{i=1}^{d}[(\ell_i \times a_{d-i+1}) + (a_i \times \ell_{d-i+1})] \bullet [(\ell_{i+1} \times a_{d-i}) + (a_{i+1} \times \ell_{d-i})]$$

$$= \sum_{i=1}^{d}[m(\ell_i, a_{d-i})\ell_{i+1} \times a_{d-i+1} + m(\ell_i, \ell_{d-i})a_{i+1} \times a_{d-i+1}$$

$$+ m(a_i, a_{d-i})\ell_{i+1} \times \ell_{d-i+1} + m(a_i, \ell_{d-i})a_{i+1} \times \ell_{d-i+1}.$$

By Proposition 3.8, the middle two terms vanish and the expression simplifies to $\sum_{i=1}^{d} \ell_{i+1} \times a_{d-i+1} + a_{i+1} \times \ell_{d-i+1} = \sum_{i=2}^{d} \ell_i \times a_{d-i+2} + a_i \times \ell_{d-i+2}$. Arguing inductively, we conclude

$$L'^{d-2p} = \sum_{i=d-2p}^{d} \ell_i \times a_{2d-2p-i} + a_i \times \ell_{2d-2p-i}.$$

Similarly, we compute

$$\Lambda_0'^{d-2p} = \sum_{j=0}^{2p} \ell_j \times a_{2p-j} + a_j \times \ell_{2p-j}.$$

If $z \in CH_{\mathbb{Q}}^p(Y)$, direct computation shows that for $\alpha \in CH_{\mathbb{Q}}^i(Y)$, $\beta \in CH_{\mathbb{Q}}^j(Y)$, we have $(\alpha \times \beta) \bullet z = 0$ unless $i + p = d$. Using this principle, we see $L'^{d-2p} \bullet z = [\ell_{d-p} \times a_{d-p} + a_{d-p} \times \ell_{d-p}] \bullet z$ and hence

$$\Lambda_0'^{d-2p} \bullet (L'^{d-2p} \bullet z) = [a_p \times \ell_p + \ell_p \times a_p] \bullet [\ell_{d-p} \times a_{d-p} + a_{d-p} \times \ell_{d-p}] \bullet z$$

$$= (a_{d-p} \times \ell_p + \ell_{d-p} \times a_p) \bullet z$$

$$= \gamma_{d-p} \bullet z.$$

Now suppose further that $z \in D_p \subseteq CH_{\mathbb{Q}}^p(Y)$. Then

$$z = [\Delta_Y] \bullet z = (f \times f)^*[\Delta_X] \bullet z + \sigma \bullet z.$$

The first term vanishes by Proposition 3.8, and since $\sigma = \sum_{i=0}^d \gamma_i$, where each γ_i is a product cycle of type $(i, d-i)$, we see that $\sigma \bullet z = \gamma_{d-p} \bullet z$.

Summarizing, we have

$$(\Lambda_0'^{d-2p} \bullet L'^{d-2p}) \bullet z = z.$$

A similar calculation shows that for $e \in D_{d-p} \subseteq CH_{\mathbb{Q}}^{d-p}(Y)$,

$$(L'^{d-2p} \bullet \Lambda_0'^{d-2p}) \bullet e = e.$$

This shows that the maps $z \mapsto L'^{d-2p} \bullet z$ and $e \mapsto \Lambda_0'^{d-2p} \bullet e$ are mutually inverse isomorphisms between D_p and D_{d-p}, completing the proof.

$\square$

By the results of [AJ2], the hypotheses of Theorem 6.4 are satisfied for Kummer varieties over finite fields. By taking direct limits, one easily argues that they also hold for Kummer varieties over the algebraic closure of a finite field. Thus, we have:

Corollary 6.5 *Let Y be the Kummer manifold associated to an abelian variety of dimension $d > 0$ over an algebraic closure of some finite field of characteristic different from 2. Then for $2p \leq d$, the map $CH_{\mathbb{Q}}^p(Y) \to CH_{\mathbb{Q}}^{d-p}(Y)$ defined by $z \mapsto L_Y^{d-2p} \bullet z$ is an isomorphism.*

References

[A]   R. Akhtar, *Adequate equivalence relations and Pontryagin prod-ucts*, J. Pure and Applied Algebra **196** (2005), 2–37.

[AJ1]   R. Akhtar and R. Joshua, *Künneth decompositions for quotient varieties*, Indag. Math. **17** (2006), no. 3, 319–344.

[AJ2]   R. Akhtar and R. Joshua, *Lefschetz decompositions for quotient varieties*, J. K-Theory **3** (2009), no. 3, 547–560.

[Be]   A. Beauville, *Sur l'anneau de Chow d'une variété abélienne*, Math. Ann. **273** (1986), 647-651.

[Ar]   D. Arapura, *Motivation for Hodge cycles*, Adv. Math. **207** (2006), no. 2, 762–781.

[dBN]   S. del Baño Rollin and V. Navarro Aznar, *On the motive of a quotient variety*, Collect. Math. **49** (1998), no. 2-3, 203–226.

[DM]   C. Deninger and J. Murre, *Motivic decomposition of Abelian schemes and the Fourier transform*, J. Reine. Angew. Math. **422** (1991), 201–219.

[DL]   I. Dolgachev and D. Levahi, *On isogenous principally polarized abelian surfaces, curves and abelian varieties*, 51-69, Contemp. Math., **465**, Amer. Math. Soc., (2008).

[F]   W. Fulton, *Intersection Theory*, Springer-Verlag, 1984.

[GR]   A. Grothendieck *Standard conjectures on algebraic cycles*, Algebraic Geometry (Bombay, 1968) – Tata Inst. Fund. Res. Stud. Math., **4**, Tata Inst. Fund. Res., Bombay, 1969, 139–199.

[GP]   V. Guletskii and C. Pedrini, *Finite dimensional motives and the conjectures of Beilinson and Murre*, K-Theory **30** (2003), no. 3, 243–263.

[I]   J. Iyer, *Murre's conjectures and explicit Chow-Knneth projections for varieties with a NEF tangent bundle*, Trans. Amer. Math. Soc. **361** (2009), no. 3, 1667–1681.

[KMP]   B. Kahn, J. Murre, and C. Pedrini *The transcendental part of the motive of a surface*, Algebraic cycles and motives **2**, 143–202. London Math. Soc. Lecture Note Ser. **344** (2007).

[Kl]   S. L. Kleiman, *The Standard Conjectures*, Motives, Proc. Symp. Pure Math. **55** part 1, AMS (1994), pp. 3–20.

[Ku]   K. Künnemann, *A Lefschetz decomposition for Chow motives of abelian schemes*, Invent. Math. **113** (1993), no. 1, 85–102.

[Man]   Y. Manin, *Correspondences, motives, and monoidal transformations*, Math. USSR-Sb. **6** (1968), 439–470.

[Mu]   J. Murre, *On a conjectural filtration on the Chow groups of an algebraic variety: I and II*, Indag. Math. (N.S.) **4** (1993), no. 2, 177–201.

[Sai]   H. Saito, *Generalization of Abel's theorem and some finiteness property of zero-cycles on sufraces*, Compositio Math. **84** (1992), 289–332.

[Sam]   P. Samuel, *Relation d'équivalence en géometrie algébrique*, Proceedings of the International Congress of Mathematicians, Edinburgh 1958. Cambridge U. Press, 1960, 470–487.

[Sas]   R. Sasaki, *Bounds on the degree of equations defining Kummer varieties*, J. Math. Soc. Japan **33** no. 2, (1981), 323–333.

[Sch]   A. J. Scholl, *Classical motives*, Motives, Proc. Symp. Pure Math. **55** part 1, AMS (1994), pp. 189–205.

[SV]   M. Shen and C. Vial, *The motive of the Hilbert cube $X^{[3]}$*, Forum Math. Sigma **4** (2015) [e30].

[Sh]   A. M. Shermenev, *The motive of an abelian variety,* Funct. Anal. **8** (1974), 55–61.

[V]   C. Vial, *Algebraic cycles and fibrations,* Documenta Math. **18** (2013), 1521–1533.

REZA AKHTAR, DEPARTMENT OF MATHEMATICS, MIAMI UNIVERSITY, OXFORD, OHIO, 45056, USA.
     E-mail: akhtarr@miamioh.edu
ROY JOSHUA, DEPARTMENT OF MATHEMATICS, OHIO STATE UNIVERSITY, COLUMBUS, OHIO, 43210, USA.
     E-mail: joshua@math.ohio-state.edu

K-Theory
Copyright ©2018 Tata Institute of Fundamental Research
Publisher: Hindustan Book Agency, New Delhi, India

Duality of integral étale motivic cohomology

Thomas H. Geisser

Abstract

We discuss duality pairings on integral étale motivic cohomology groups of regular and proper schemes over algebraically closed fields, local fields, finite fields, and arithmetic schemes.

1 Introduction

Let X be a smooth and projective variety of dimension d over a perfect field k. Using duality theorems for étale cohomology with finite coefficients, we show duality results on integral étale motivic cohomology groups.

If k is algebraically closed, and m is an integer prime to the characteristic of k, we construct for all n, u satisfying $w = n + u - d > 0$ a Galois invariant pairing

$$H_{\text{et}}^{2d+1-i}(X, \mathbb{Z}(u))/m \times {}_mH_{\text{et}}^i(X, \mathbb{Z}(n)) \to \mathbb{Z}/m(w) \qquad (1.1)$$

which is non-degenerate on the left. For the right kernel ${}_mH_{\text{et}}^i(X, \mathbb{Z}(n))^0$, we obtain a secondary perfect pairing

$$_mH_{\text{et}}^i(X, \mathbb{Z}(n))^0 \times {}_mH_{\text{et}}^{2d+2-i}(X, \mathbb{Z}(u))^0 \to \mathbb{Z}/m(w), \qquad (1.2)$$

and we show that for $n + u = d + 1$, $i = 2n$ and $2d + 2 - i = 2u$, this pairing becomes perfect when restricted to the divisible subgroups. Thus the pairing can be thought of as a generalization of the e_m-pairing between the Picard and Albanese abelian variety of X.

If k is finite, we construct for all m pairings

$$H_{\text{et}}^{2d+2-i}(X, \mathbb{Z}(d-n))/m \times {}_mH_{\text{et}}^i(X, \mathbb{Z}(n)) \to \mathbb{Z}/m \qquad (1.3)$$

which are non-degenerate on the left. We conjecture that the pairing is perfect, and relate this conjecture to Tate's conjecture on the surjectivity of the cycle map.

If k is local and m prime to the characteristic of k, we construct pairings

$$H_{\text{et}}^{2d+3-i}(X, \mathbb{Z}(d+1-n))/m \times {}_mH_{\text{et}}^i(X, \mathbb{Z}(n)) \to \mathbb{Z}/m$$

which are non-degenerate on the left. In case $d = 0$, the perfectness of the pairing is equivalent to the statement of local class field theory, and for $d = n = 1$ it amounts to Lichtenbaum's duality between the Picard group and the Brauer group for curves over a p-adic field [15]. In contrast, we show that the pairing can have a right kernel for curves, and we give an example showing that the pairing can have a right kernel even in the good reduction case. In particular, there is no duality "in some appropriate sense of the term" expected by Lichtenbaum [16, §6].

Notation: We denote Bloch's motivic complex by $\mathbb{Z}(n)$, a complex of étale sheaves [4, Lemma 3.1]. When we need to emphasize that $\mathbb{Z}(n)$ is considered as a complex of Zariski sheaves, we write $\mathbb{Z}(n)^{\mathrm{Zar}}$.

For an abelian group A, we denote by ${}_m A$ its m-torsion, by $A\{l\} = \mathrm{colim}_r \, {}_{l^r} A$ its subgroup of l-power torsion elements, by A^* its Pontrjagin dual $\mathrm{Hom}(A, \mathbb{Q}/\mathbb{Z})$, by $A^\wedge = \lim_m A/m$ its completion, by $A^{\wedge l} = \lim_r A/l^r$ its the l-adic completion, and by $T_l A = \lim_r \, {}_{l^r} A$ its l-adic Tate module.

2   Algebraically closed fields

Over an algebraically closed base field, Zariski and étale hypercohomology of the motivic complex agree in weights at least the dimension, i.e. if $\epsilon : X_{\mathrm{et}} \to X_{\mathrm{Zar}}$ is the change of topology map, then the adjunction map $\mathbb{Z}(n)^{\mathrm{Zar}} \to R\epsilon_* \mathbb{Z}(n)$ is a quasi-isomorphism of complexes of sheaves for the Zariski-topology for $n \geq d$. This was deduced in [6] from a theorem of Suslin away from the characteristic and from [8] at the characteristic. Since the Zariski hypercohomology admits a push-forward map for the proper map $f : X \to k$, we obtain a Gysin map for étale motivic cohomology as the composition

$$Rf_* R\epsilon_* \mathbb{Z}(w+d)_X \cong Rf_* \mathbb{Z}(w+d)_X^{\mathrm{Zar}} \to \mathbb{Z}(w)_k^{\mathrm{Zar}} \cong R\epsilon_* \mathbb{Z}(w)_k.$$

If k is a perfect field, then applying Galois cohomology $\Gamma(\mathrm{Gal}(k), -)$ to this Gysin map over the base extension to the algebraic closure, we obtain a "trace" map

$$H_{\mathrm{et}}^{2d+v}(X, \mathbb{Z}(w+d)) \to H_{\mathrm{et}}^v(k, \mathbb{Z}(w))$$

for X proper over k and any $w \geq 0$.

The cup product pairing on higher Chow groups induces a product on étale hypercohomology, hence for $w = n + u - d$ we obtain a pairing

$$H_{\mathrm{et}}^{2d+v-i}(X, \mathbb{Z}(u)) \times H_{\mathrm{et}}^i(X, \mathbb{Z}(n)) \to H_{\mathrm{et}}^{2d+v}(X, \mathbb{Z}(u+n)) \xrightarrow{tr} H_{\mathrm{et}}^v(k, \mathbb{Z}(w)).$$
$$\tag{2.1}$$

If k is algebraically closed, m invertible in k, and $w \geq 1$, then the coefficient sequence gives an isomorphism

$$\mathbb{Q}/\mathbb{Z}(w) \cong H^0_{\mathrm{et}}(k, \mathbb{Q}/\mathbb{Z}(w)) \cong \mathrm{Tor}\, H^1_{\mathrm{et}}(k, \mathbb{Z}(w)).$$

Indeed, this follows by comparing to Suslin's calculation of the K-theory of an algebraically closed field. Restricting the pairing (2.1) for $v = 1$ to the m-torsion on the right, we obtain for

$$n + u > d, \quad i + j = 2d$$

a pairing

$$H^j_{\mathrm{et}}(X, \mathbb{Z}(u))/m \times {}_m H^{i+1}_{\mathrm{et}}(X, \mathbb{Z}(n)) \to {}_m H^{2d+1}_{\mathrm{et}}(X, \mathbb{Z}(u+n)) \to \mathbb{Z}/m(w).$$

This is compatible with Poincare-duality for étale cohomology with finite coefficients

$$
\begin{array}{ccc}
H^j_{\mathrm{et}}(X, \mathbb{Z}(u))/m \times {}_m H^{i+1}_{\mathrm{et}}(X, \mathbb{Z}(n)) & \longrightarrow & {}_m H^1_{\mathrm{et}}(k, \mathbb{Z}(w)) \cong \mathbb{Z}/m(w) \\
\downarrow \quad\quad\quad \partial\uparrow & & \partial\uparrow \cong \\
H^j_{\mathrm{et}}(X, \mathbb{Z}/m(u)) \times H^i_{\mathrm{et}}(X, \mathbb{Z}/m(n)) & \longrightarrow & H^0_{\mathrm{et}}(k, \mathbb{Z}/m(w)) \cong \mathbb{Z}/m(w)
\end{array}
\tag{2.2}
$$

because $\partial(x \cup y) = x \cup \partial(y)$ if $\partial(x) = 0$. Consequently, we obtain a map of short exact sequences of finite abelian groups

$$
\begin{array}{ccccc}
H^j_{\mathrm{et}}(X, \mathbb{Z}(u))/m & \longrightarrow & H^j_{\mathrm{et}}(X, \mathbb{Z}/m(u)) & \longrightarrow & {}_m H^{j+1}_{\mathrm{et}}(X, \mathbb{Z}(u)) \\
\downarrow & & \downarrow & & \downarrow \\
{}_m H^{i+1}_{\mathrm{et}}(X, \mathbb{Z}(n))^\sharp & \longrightarrow & H^i_{\mathrm{et}}(X, \mathbb{Z}/m(n))^\sharp & \longrightarrow & (H^i(X, \mathbb{Z}(n))/m)^\sharp,
\end{array}
\tag{2.3}
$$

where for any abelian group A, $A^\sharp = \mathrm{Hom}(A, \mathbb{Q}/\mathbb{Z}(w))$ is the twist of the usual Pontrjagin dual. The middle map is an isomorphism by Grothendieck's Poincaré duality for étale cohomology, so that the snake Lemma gives an exact sequence

$$0 \to H^j_{\mathrm{et}}(X, \mathbb{Z}(u))/m \to ({}_m H^{i+1}_{\mathrm{et}}(X, \mathbb{Z}(n)))^\sharp$$

$$\xrightarrow{\delta} {}_m H^{j+1}_{\mathrm{et}}(X, \mathbb{Z}(u)) \to (H^i_{\mathrm{et}}(X, \mathbb{Z}(n))/m)^\sharp \to 0. \tag{2.4}$$

In particular, the pairings (1.1) are non-degenerate on the left, and δ induces the pairing (1.2). It is easy to see that the sequence (2.4) is compatible with varying m.

Remark 2.1 The construction of the pairing requires $u+n \geq d+1$. For example, for $X = \mathrm{Spec}\, k$ and $u = n = 0$, the diagram (2.2) does not commute. The construction also does not work for m a power of the characteristic of the base field, because then $\mathbb{Z}/m(w) = 0$ for $w > 0$.

The case $u + n = d + 1, i = 2n - 1, j = 2u - 1$

Example (Rojtman's theorem) If $n = d, u = 1, i = 2d - 1, j = 1$, then $H^j_{\mathrm{et}}(X, \mathbb{Z}(u))/m = k^\times/m = 0$. Moreover, it was shown in [7] that $H^{2d-1}_{\mathrm{et}}(X, \mathbb{Z}(d))$ modulo its divisible subgroup is isomorphic to the dual of $\mathrm{Tor}\, \mathrm{NS}\, X$. Hence the fact that $(A^\sharp/m)^\sharp = {}_m A$ for a finite group A imply that (2.4) gives a short exact sequence

$$0 \to {}_m H^{2d}_{\mathrm{et}}(X, \mathbb{Z}(d))^\sharp \to {}_m \mathrm{Pic}\, X \to {}_m \mathrm{NS}\, X \to 0.$$

From this we can deduce Rojtman's theorem away from the characteristic. Indeed, since $CH_0(X) \cong H^{2d}_{\mathrm{et}}(X, \mathbb{Z}(d))$ as explained in the beginning of this section, and the Albanese map $CH_0(X) \to \mathrm{Alb}_X(k)$ is surjective, it suffices to show that the order of the m-torsion of both sides agrees. But from the duality of the Picard and Albanese variety we know that $|_m \mathrm{Alb}_X(k)| = |_m \mathrm{Pic}^0_X(k)| = |_m \mathrm{Pic}\, X|/|_m \mathrm{NS}\, X|$.

Proposition 2.2 *Let $u + n = d + 1$ and assume that $H^{2u-1}_{\mathrm{et}}(X, \mathbb{Z}_l)$ and $H^{2n-1}_{\mathrm{et}}(X, \mathbb{Z}_l)$ are torsion free. Then we have a perfect pairing*

$$_{l^r} H^{2n}_{\mathrm{et}}(X, \mathbb{Z}(n)) \times {}_{l^r} H^{2u}_{\mathrm{et}}(X, \mathbb{Z}(u)) \to \mu_{l^r}.$$

Proof This follows because $H^{2u-1}_{\mathrm{et}}(X, \mathbb{Z}(u))$ is the extension of an l-divisible group by a finite group contained in $H^{2u-1}_{\mathrm{et}}(X, \mathbb{Z}_l)$, see [7]. It follows that under the hypothesis the outer terms in (2.4) vanish. $\qquad\square$

Note that the l-adic cohomoloy groups in question are torsion free for almost all l, and they are torsion free for all l if X is an abelian variety. To get an unconditional pairing, consider the subgroup of divisible elements

$$H^{2n}_{\mathrm{hom}}(X, \mathbb{Z}(n)) = \ker H^{2n}_{\mathrm{et}}(X, \mathbb{Z}(n)) \to \lim{}_m H^{2n}_{\mathrm{et}}(X, \mathbb{Z}(n))/m$$

$$= \ker H^{2n}_{\mathrm{et}}(X, \mathbb{Z}(n)) \to \prod_l H^{2n}_{\mathrm{et}}(X, \mathbb{Z}_l(n)).$$

By [7, Cor. 2.2], $\mathrm{Tor}\, H^{2n}_{\mathrm{et}}(X, \mathbb{Z}(n))$ is a direct summand of $H^{2n}_{\mathrm{et}}(X, \mathbb{Z}(n))$. This property is shared with its subgroup of divisible elements $H^{2n}_{\mathrm{hom}}(X, \mathbb{Z}(n))$, and from this one easily concludes that $H^{2n}_{\mathrm{hom}}(X, \mathbb{Z}(n))$ is in fact the maximal divisible subgroup of $H^{2n}_{\mathrm{et}}(X, \mathbb{Z}(n))$.

Proposition 2.3 *We have a perfect pairing*

$$_m H^{2n}_{\mathrm{hom}}(X, \mathbb{Z}(n)) \times {}_m H^{2u}_{\mathrm{hom}}(X, \mathbb{Z}(u)) \to \mu_m.$$

Proof We can assume that $m = l^r$ is a prime power. Let

$$Q = \left(H^{2n}_{\mathrm{et}}(X, \mathbb{Z}(n)) / H^{2n}_{\mathrm{hom}}(X, \mathbb{Z}(n)) \right) \otimes \mathbb{Z}_{(l)} \subseteq \lim_r H^{2n}_{\mathrm{et}}(X, \mathbb{Z}(n)) / l^r$$
$$\subseteq H^{2n}_{\mathrm{et}}(X, \mathbb{Z}_l(n))$$

be the canonical inclusion. Both inclusions are isomorphisms on torsion subgroups: The former because the cokernel of the completion map $A \to \lim_r A/l^r$ is uniquely divisible, and the latter because the Tate-module $T_l H^{2n+1}_{\mathrm{et}}(X, \mathbb{Z}(n))$ is torsion free.

Since $H^{2n}_{\mathrm{hom}}(X, \mathbb{Z}(n))$ is divisible, we obtain a short exact sequence

$$0 \to ({}_m Q)^\sharp \to ({}_m H^{2n}_{\mathrm{et}}(X, \mathbb{Z}(n)))^\sharp \to ({}_m H^{2n}_{\mathrm{hom}}(X, \mathbb{Z}(n)))^\sharp \to 0.$$

The group $H^{2u-1}_{\mathrm{et}}(X, \mathbb{Z}(u))$ is an extension of a finite group by a divisible group [7, Thm. 1.1], and the l-primary part of this finite group is $H^{2u-1}_{\mathrm{et}}(X, \mathbb{Z}_l(u))\{l\}$, which is dual to $H^{2n}_{\mathrm{et}}(X, \mathbb{Z}_l(n))\{l\}$, [7, Prop. 1.2 (1)], the l-primary part of Q by the above. We conclude that in the sequence (2.4), the image of $H^{2u-1}_{\mathrm{et}}(X, \mathbb{Z}(u))/m$ in $({}_m H^{2n}_{\mathrm{et}}(X, \mathbb{Z}(n)))^\sharp$ is exactly $({}_m Q)^\sharp$.

$\square$

Example

We have $H^2_{\mathrm{hom}}(X, \mathbb{Z}(1)) \cong \mathrm{Pic}^0(X)$, and $H^{2d}_{\mathrm{hom}}(X, \mathbb{Z}(d)) \cong CH_0(X)^0$, and obtain another proof of Rojtman's theorem.

A class is algebraically equivalent to zero if it lies in the image of some map

$$H^{2n}_{\mathrm{et}}(T \times X, \mathbb{Z}(n)) \overset{t_1^* - t_0^*}{\longrightarrow} H^{2n}_{\mathrm{et}}(X, \mathbb{Z}(n))$$

for a smooth connected scheme T (which we can assume to be a smooth curve) and closed points $t_0, t_1 \in T$. The subgroup of classes algebraically equivalent to zero is written $H^{2n}_{\mathrm{alg}}(X, \mathbb{Z}(n))$. It is a subgroup of $H^{2n}_{\mathrm{hom}}(X, \mathbb{Z}(n))$, hence we can restrict the pairing above.

Definition 2.4 A homomorphism from $H^{2n}_{\mathrm{alg}}(X, \mathbb{Z}(n))$ to the k-rational points of an abelian variety A is regular, if for every pointed smooth connected variety $t_0 \in T$ and element $\Gamma \in H^{2n}_{\mathrm{alg}}(T \times X, \mathbb{Z}(n))$, the composition with

$$T(k) \to H^{2n}_{\mathrm{alg}}(X, \mathbb{Z}(n)), \qquad t \mapsto t^* \Gamma - t_0^* \Gamma$$

is the map induced on closed points by a morphism of varieties $T \to A$.

In [19], [20], Murre studied the situation for Chow groups, and he proved that a universal homomorphism to an abelian variety exists for dimension 0, and codimensions 1 and 2.

Theorem 2.5 *There is a universal object* $\rho_n : H^{2n}_{\mathrm{alg}}(X, \mathbb{Z}(n)) \to A_n$ *for regular homomorphisms from* $H^{2n}_{\mathrm{alg}}(X, \mathbb{Z}(n))$ *to abelian varieties.*

Proof This follows by the argument of Serre-H. Saito [21] because the dimension of surjective maps to abelian varieties is bounded, see also [12].
□

Question 2.6 Is there a duality between the abelian varieties A_n and A_u of Theorem 2.5 such that the diagram below arising from the e_m-pairing is commutative?

$$
\begin{array}{ccc}
{}_m H^{2n}_{\mathrm{alg}}(X, \mathbb{Z}(n)) \times {}_m H^{2u}_{\mathrm{alg}}(X, \mathbb{Z}(u)) & \longrightarrow & \mu_m \\
\rho_n \downarrow \qquad \qquad \rho_u \downarrow & & \| \\
{}_m A_n \qquad \times \qquad {}_m A_u & \longrightarrow & \mu_m.
\end{array}
$$

3 Finite fields

Over a finite field, the pairing (2.1) for $v = 2$ and $w = 0$ becomes for

$$u + n = d, \quad i + j = 2d + 1$$

the pairing

$$H^j_{\mathrm{et}}(X, \mathbb{Z}(u)) \times H^{i+1}_{\mathrm{et}}(X, \mathbb{Z}(n)) \to H^{2d+2}_{\mathrm{et}}(X, \mathbb{Z}(d)) \xrightarrow{\mathrm{tr}} H^2_{\mathrm{et}}(k, \mathbb{Z}) \cong \mathbb{Q}/\mathbb{Z}.$$

This is compatible with Poincaré duality

$$H^i_{\mathrm{et}}(X, \mathbb{Z}/m(n)) \times H^j_{\mathrm{et}}(X, \mathbb{Z}/m(u)) \to \mathbb{Q}/\mathbb{Z}$$

for all integers m (the pairing for m a power of p is discussed in [17] using the isomorphism $\mathbb{Z}/m(n) \cong \nu_r(n)$ from [8]), i.e. the diagram analog to (2.2) commutes. We obtain as in the previous case an exact sequence

$$0 \to H^j_{\mathrm{et}}(X, \mathbb{Z}(u))/m \to ({}_m H^{i+1}_{\mathrm{et}}(X, \mathbb{Z}(n)))^*$$

$$\xrightarrow{\delta} {}_m H^{j+1}_{\mathrm{et}}(X, \mathbb{Z}(u)) \to (H^i_{\mathrm{et}}(X, \mathbb{Z}(n))/m)^* \to 0, \quad (3.1)$$

hence the resulting pairing

$$H^j_{\mathrm{et}}(X, \mathbb{Z}(u))/m \times {}_m H^{i+1}_{\mathrm{et}}(X, \mathbb{Z}(n)) \to \mathbb{Z}/m \quad (3.2)$$

is non-degenerate on the left. It is non-degenerate if and only if δ vanishes, and we show that this is equivalent to Tate's conjecture:

Theorem 3.1 *The map δ vanishes for $i \neq 2n, 2n+1$. It vanishes for $i = 2n$ if and only if $H_{et}^{2n+1}(X, \mathbb{Z}(n))$ is finite. It vanishes for $i = 2n+1$ if and only if $H_{et}^{2u+1}(X, \mathbb{Z}(u))$ is finite.*

Note that $i = 2n \Leftrightarrow j = 2u+1$ and $i = 2n+1 \Leftrightarrow j = 2u$.

Proof We want to show that the map δ_m in the following diagram is the zero map:

$$
\begin{array}{ccc}
(\operatorname{Tor} H_{et}^{i+1}(X, \mathbb{Z}(n)))^* & \xrightarrow{\delta_\infty} & TH_{et}^{j+1}(X, \mathbb{Z}(u)) \\
{\scriptstyle v}\big\downarrow & & \big\downarrow \\
(_m H_{et}^{i+1}(X, \mathbb{Z}(n)))^* & \xrightarrow{\delta_m} & {}_m H_{et}^{j+1}(X, \mathbb{Z}(u)).
\end{array}
$$

Since v is surjective, the vanishing of δ_∞ is equivalent to the vanishing of δ_m for all m. The first statement of the Proposition follows because $H_{et}^j(X, \mathbb{Q}/\mathbb{Z}(u))$ is finite for $j \neq 2u, 2u+1$ for weight reasons, and this group surjects onto $\operatorname{Tor} H_{et}^{j+1}(X, \mathbb{Z}(u))$, so that $TH_{et}^{j+1}(X, \mathbb{Z}(u)) = 0$ for $j \neq 2u, 2u+1$.

If $i = 2n$, then finiteness of $H_{et}^{2n+1}(X, \mathbb{Z}(n))$ implies that the source of δ_∞ is finite, hence cannot map non-trivially to a Tate-module. Conversely, if $\delta_m = 0$ for all m, then the duality between the two groups of cofinite type $H_{et}^{2u+1}(X, \mathbb{Z}(u))/m$ and $_m H_{et}^{2n+1}(X, \mathbb{Z}(n))$ implies that $H_{et}^{2n+1}(X, \mathbb{Z}(n))$ is finite. Reversing the roles of i and j we obtain the result for $i = 2n+1$. $\qquad\square$

The connection to Tate's conjecture is given by the following (well-known) Proposition.

Proposition 3.2 *The finiteness of $H_{et}^{2n+1}(X, \mathbb{Z}(n))$ is equivalent to Tate's conjecture on the surjectivity of the cycle map in degree n for X.*

Proof Consider the coefficient sequence

$$0 \to H_{et}^{2n}(X, \mathbb{Z}(n))^{\wedge l} \to H_{et}^{2n}(X, \mathbb{Z}_l(n)) \to T_l H_{et}^{2n+1}(X, \mathbb{Z}(n)) \to 0.$$

The middle group surjects onto $H_{et}^{2n}(\bar{X}, \mathbb{Z}_l(n))^G$ with finite kernel. On the other hand, in the composition

$$CH^n(X) \otimes \mathbb{Z}_l \to H_{et}^{2n}(X, \mathbb{Z}(n)) \otimes \mathbb{Z}_l \to H_{et}^{2n}(X, \mathbb{Z}(n))^{\wedge l}$$

the left map is an isomorphism up to torsion, and the right map is surjective (as the target is a subgroup of $H_{et}^{2n}(X, \mathbb{Z}_l(n))$, hence a finitely generated $\mathbb{Z}_l$-module). We conclude that the cycle map $CH^n(X) \otimes \mathbb{Z}_l \to H_{et}^{2n}(\bar{X}, \mathbb{Z}_l(n))^G$

is rationally surjective if and only if $T_l H_{\text{et}}^{2n+1}(X, \mathbb{Z}(n)) = 0$ if and only if (the group of cofinite type) $H_{\text{et}}^{2n+1}(X, \mathbb{Z}(n))\{l\}$ is finite. Finally, by Gabber's theorem [3], the cotorsion of $H_{\text{et}}^{2n+1}(X, \mathbb{Z}(n))\{l\}$ vanishes for almost all l.

$\square$

Example If X is a surface, we obtain $H_{\text{et}}^6(X, \mathbb{Z}(1)) = 0$, and pairings

$$k^\times \times H_{\text{et}}^5(X, \mathbb{Z}(1)) \to \mathbb{Q}/\mathbb{Z}$$
$$_m \text{Pic}(X) \times H_{\text{et}}^4(X, \mathbb{Z}(1))/m \to \mathbb{Z}/m$$
$$\text{Pic}(X)/m \times {}_m H_{\text{et}}^4(X, \mathbb{Z}(1)) \to \mathbb{Z}/m$$
$$\text{Br}(X)/m \times {}_m \text{Br}(X) \to \mathbb{Z}/m.$$

The pairings are all perfect, except for the last one, which is perfect if and only if the Brauer group is finite. In [25, Thm. 5.1], Tate defines a skew-symmetric pairing on the Brauer group whose kernel consists exactly of the divisible elements. It is easy to see from the construction that Tate's pairing is obtained by composing the above pairing with the canonical map $\text{Br}(X) \to \text{Br}(X)/m$. In the limit, this become the composition

$$\text{Br}(X) \to \text{Br}(X)^\wedge \to \text{Br}(X)^*.$$

The first map has kernel exactly the divisible elements and the second map is injective.

Remark 3.3 A small modification of étale motivic cohomology yields Weil-étale motivic cohomology groups $H_W^i(X, \mathbb{Z}(n))$ which are expected to be finitely generated for all i, n and smooth and projective X [5]. Assuming finite generation, they satisfy dualities

$$H_W^i(X, \mathbb{Z}(n))/\text{Tor} \times H_W^j(X, \mathbb{Z}(u))/\text{Tor} \to \mathbb{Z}$$
$$\text{Tor}\, H_W^i(X, \mathbb{Z}(n)) \times \text{Tor}\, H_W^{j+1}(X, \mathbb{Z}(u)) \to \mathbb{Q}/\mathbb{Z}.$$

Under the finite generation conjecture, we have

$$\text{Tor}\, H_{\text{et}}^i(X, \mathbb{Z}(n)) \cong \text{Tor}\, H_W^i(X, \mathbb{Z}(n))$$

for $i \neq 2n + 2$, and

$$\text{Tor}\, H_{\text{et}}^{2n+2}(X, \mathbb{Z}(n)) \cong H_W^{2n+2}(X, \mathbb{Z}(n)) \oplus (\mathbb{Q}/\mathbb{Z})^r,$$

where r is the rank of $H_{\text{et}}^{2n}(X, \mathbb{Z}(n))$, as one sees from the long exact sequence [5, Thm. 7.1].

Arithmetic schemes

The same discussion as for finite fields should also apply to arithmetic schemes, i.e. schemes which are regular, and proper over the spectrum B of the ring of integers of a number field or a smooth and proper curve over a finite field. See [4] for properties of Bloch's higher Chow groups on smooth schemes over a Dedekind ring. In order to get the correct 2-torsion in the presence of real embeddings, one has to consider cohomology with compact support. It is defined as the étale cohomology with compact support on B of $Rf_!\mathbb{Z}(n)$, see [11, §2]. There is an exact sequence

$$\cdots \to H_c^i(X, \mathbb{Z}(n)) \to H_{\mathrm{et}}^i(X, \mathbb{Z}(n)) \to \bigoplus_{v \in S_\infty} H_T^i(\mathbb{R}, R\Gamma_{\mathrm{et}}(X_\mathbb{C}, \mathbb{Z}(n))) \to \cdots,$$

(3.3)

where the last term is Tate-modified cohomology, a finite 2-group. To use the same argument as above, two ingredients are missing, see also the discussion in [2, §6]:

1. The cup-product

$$H_{\mathrm{et}}^j(X, \mathbb{Z}(u)) \times H_c^{i+1}(X, \mathbb{Z}(n)) \to H_c^{2d+2}(X, \mathbb{Z}(d)) \to H_c^4(B, \mathbb{Z}(1)) \cong \mathbb{Q}/\mathbb{Z}$$

   is conjectured to exist, but this is currently unknown for higher Chow groups [14]. The problem is that if two cycles are located in the same special fiber, they do not intersect in the correct codimension, so that one cycle has to be moved to lie horizontal or in another special fiber. Spitzweck [24] has announced a construction of this pairing in case that X is smooth over B.

2. Duality with finite coefficients, i.e., a perfect pairing of finite groups,

$$H_{\mathrm{et}}^i(X, \mathbb{Z}/m(n)) \times H_c^j(X, \mathbb{Z}/m(u)) \to H_c^3(B, \mathbb{Z}/m(1)) \cong \mathbb{Z}/m \quad (3.4)$$

   is not known to exist. If the fibers at all places dividing m are normal crossing schemes, then Sato proved a duality as above for $\mathbb{Z}/m(n)$ replaced by his p-adic Tate-twists $\mathbb{T}_m(n)$ [23, Thm. 1.2.1]. It is expected that $\mathbb{T}_m(n)$ and $\mathbb{Z}/m(n)$ are quasi-isomorphic, but this is only know if $\mathbb{Z}/m(n)$ is acyclic in degrees larger than n (because $\mathbb{T}_m(n)$ has this property by construction) [27]. This would follow, for example, from a Gersten resolution for $\mathbb{Z}/m(n)$, but this is only known for smooth schemes [4]. In particular, we obtain such a pairing localized away from all p where X has bad reduction at a place above p.

Conjecture 3.4 *(Lichtenbaum)* The groups $H_{\mathrm{et}}^i(X, \mathbb{Z}(n))$ are finitely generated for $i \leq 2n$, finite for $i = 2n + 1$, and of cofinite type for $i \geq 2n + 2$.

If follows from the long-exact sequence (3.3) that then the same statement holds for cohomology with compact support. On the other hand, the statement of the conjecture is wrong if one removes points from the base B.

Proposition 3.5 *Assume Conjecture 3.4 and the existence of the pairings with finite coefficients (3.4). Then we have perfect pairings*

$$H^j_{et}(X, \mathbb{Z}(u))^\wedge \times \operatorname{Tor} H^{i+1}_c(X, \mathbb{Z}(n)) \to \mathbb{Q}/\mathbb{Z},$$

$$H^j_c(X, \mathbb{Z}(u))^\wedge \times \operatorname{Tor} H^{i+1}_{et}(X, \mathbb{Z}(n)) \to \mathbb{Q}/\mathbb{Z}.$$

Proof (see also [2, Prop. 3.4]) We show the first statement, the proof of the other statement is identical. If $j \leq 2u$, then $T H^{j+1}_{et}(X, \mathbb{Z}(u)) = 0$ (as the Tate module of a finitely generated group vanishes) and $H^i_c(X, \mathbb{Z}(u)) \otimes \mathbb{Q}/\mathbb{Z} = 0$ (as the cohomology group is torsion) and we obtain

$$H^j_{et}(X, \mathbb{Z}(u))^\wedge \cong \lim H^j_{et}(X, \mathbb{Z}/m(u)) \cong H^i_c(X, \mathbb{Q}/\mathbb{Z}(n))^*$$
$$\cong (\operatorname{Tor} H^{i+1}_c(X, \mathbb{Z}(n)))^*.$$

If $j > 2u$, then $H^j_{et}(X, \mathbb{Z}(u))^\wedge$ is finite, $(H^i_c(X, \mathbb{Z}(n)) \otimes \mathbb{Q}/\mathbb{Z})^*$ is torsion free, and

$$H^j_{et}(X, \mathbb{Z}(u))^\wedge \cong \operatorname{Tor} \lim H^j_{et}(X, \mathbb{Z}/m(u)) \cong \operatorname{Tor}(H^i_c(X, \mathbb{Q}/\mathbb{Z}(n))^*)$$
$$\cong (\operatorname{Tor} H^{i+1}_c(X, \mathbb{Z}(n)))^*.$$

□

4   Local fields

Let k be a local field of characteristic zero, i.e., a complete discrete valuation field with finite residue field of characteristic p. If we take $w = 1$ and $j = 3$ in (2.1), then setting

$$n + u = d + 1, \quad i + j = 2d + 2,$$

we obtain a pairing

$$H^j_{et}(X, \mathbb{Z}(u)) \times H^{i+1}_{et}(X, \mathbb{Z}(n)) \to H^{2d+3}_{et}(X, \mathbb{Z}(d+1))$$
$$\xrightarrow{tr} H^3_{et}(k, \mathbb{Z}(1)) \cong \operatorname{Br} k \cong \mathbb{Q}/\mathbb{Z}.$$

Combining this with the duality over local fields (see [18, I Cor. 2.3] combined with Poincaré duality over algebraically closed fields),

$$H^j_{et}(X, \mathbb{Z}/m(u)) \times H^i_{et}(X, \mathbb{Z}/m(n)) \to \mathbb{Q}/\mathbb{Z},$$

we again obtain an exact sequence

$$0 \to H^j_{et}(X, \mathbb{Z}(u))/m \to (_mH^{i+1}_{et}(X, \mathbb{Z}(n)))^*$$
$$\xrightarrow{\delta_m} {}_mH^{j+1}_{et}(X, \mathbb{Z}(u)) \to (H^i_{et}(X, \mathbb{Z}(n))/m)^* \to 0, \quad (4.1)$$

and pairings which are non-degenerate on the left

$$H^j_{et}(X, \mathbb{Z}(u))/m \times {}_mH^{i+1}_{et}(X, \mathbb{Z}(n)) \to \mathbb{Q}/\mathbb{Z}.$$

If δ_m is the zero-map, then this induces in the limit a duality

$$H^j_{et}(X, \mathbb{Z}(u))^\wedge \times \mathrm{Tor}\, H^{i+1}_{et}(X, \mathbb{Z}(n)) \to \mathbb{Q}/\mathbb{Z},$$

i.e. the duality "in some appropriate sense of the term" expected by Lichtenbaum [16, §6].

Example The vanishing of δ for X the spectrum of a local field is equivalent to class field theory. Indeed, for $u = 1, j = 1$ it states that the injection

$$H^1_{et}(k, \mathbb{Z}(1))^\wedge \cong (k^\times)^\wedge \to H^2_{et}(k, \mathbb{Z})^* \cong H^1_{et}(k, \mathbb{Q}/\mathbb{Z})^* \cong \mathrm{Gal}(k)^{ab}$$

is an isomorphism. For $u = 0, j = 0$ it states that the injection

$$H^0_{et}(k, \mathbb{Z})^\wedge \cong \hat{\mathbb{Z}} \to \mathrm{Br}(k)^* \cong H^3_{et}(k, \mathbb{Z}(1))^*$$

is an isomorphism.

Example If X is a curve over a p-adic field and $n = 1$, then the statement for $i = 1, 2$ is the duality between $\mathrm{Pic}(X)$ and $\mathrm{Br}(X)$ proven by Lichtenbaum [15]. For $i = 0, 3$ it follows from

$$H^5_{et}(X, \mathbb{Z}(1)) \cong H^2(k, H^2_{et}(\bar{X}, \mathbb{Q}/\mathbb{Z}(1))) \cong H^2(k, \mathbb{Q}/\mathbb{Z}) = 0.$$

Proposition 4.1 *Assume that either $i \notin \{n, \dots, n+d+1\}$, or that X has good reduction and $i \neq 2n-1, 2n, 2n+1$. Then $\delta_m = 0$.*

Proof Again the vanishing of δ_m for all m is equivalent to the vanishing of δ_∞ in the diagram

$$
\begin{array}{ccc}
(\mathrm{Tor}\, H^{i+1}_{et}(X, \mathbb{Z}(n)))^* & \xrightarrow{\delta_\infty} & T H^{j+1}_{et}(X, \mathbb{Z}(u)) \\
\downarrow & & \downarrow \\
(_mH^{i+1}_{et}(X, \mathbb{Z}(n)))^* & \xrightarrow{\delta_m} & {}_mH^{j+1}_{et}(X, \mathbb{Z}(u))
\end{array}
$$

because the left vertical map is surjective. By [9], $H^i_{\text{et}}(X, \mathbb{Q}/\mathbb{Z}(n))$ is finite for $i \notin \{n, \ldots, n+d+1\}$ for general X, and $i \neq 2n-1, 2n, 2n+1$ for X with good reduction. This implies that Tor $H^{i+1}_{\text{et}}(X, \mathbb{Z}(n))$ is finite for $i < n$ and $i < 2n-1$, respectively, hence its dual cannot map non-trivially to the torsion free Tate-module. On the other hand, the Tate module $TH^{j+1}_{\text{et}}(X, \mathbb{Z}(u))$ vanishes for $j < u \Leftrightarrow i > n+d+1$ and $j < 2u-1 \Leftrightarrow i > 2n+1$, respectively.

$\square$

We believe that an improvement is possible:

Conjecture 4.2 *If $i \notin \{n+1, \ldots, n+d\}$ or if X has good reduction and $i \neq 2n$, then $\delta = 0$.*

We give examples for δ_∞ to be non-zero, thus giving counterexamples to duality of étale motivic cohomology over local fields. Since $H^1_{\text{et}}(X, \mathbb{Z}) = 0$, we get from the limit of (4.1) a short exact sequence

$$0 \to H^{2d+1}_{\text{et}}(X, \mathbb{Z}(d+1))^\wedge \to (\text{Tor}\, H^2_{\text{et}}(X, \mathbb{Z}))^* \to TH^{2d+2}_{\text{et}}(X, \mathbb{Z}(d+1)) \to 0.$$

For X a curve, $H^3_{\text{et}}(X, \mathbb{Z}(2)) \cong H^3_{\mathbb{M}}(X, \mathbb{Z}(2)) \cong SK_1(X)$, and it follows from S. Saito's result [22, Thm. 2.6] that the right hand side has rank equal to rank $H^1_{\text{et}}(Y, \mathbb{Z})$, where Y the special fiber of a smooth and proper model. For arbitrary dimension, Yoshida proved that its rank is the dimension of the maximal split torus of the Neron model of Alb_X [26]. Hence δ_∞ can be non-zero for a curve (with bad reduction) in weights $n = 2, u = 0$.

We now give an example, obtained with the help of S. Saito and K. Sato, showing that δ_∞ can be non-zero even for schemes with good reduction. For $n = d, i = 2d$, the limit of (4.1) gives a sequence

$$0 \to \text{Pic}(X)^{\wedge l} \to (H^{2d+1}_{\text{et}}(X, \mathbb{Z}(d))\{l\})^* \overset{\delta_\infty}{\to} T_l \,\text{Br}(X)$$
$$\to (H^{2d}_{\text{et}}(X, \mathbb{Z}(d)) \otimes \mathbb{Q}_l/\mathbb{Z}_l)^* \to 0.$$

Proposition 4.3 *Assume that X admits a smooth and proper model $\mathcal{X}$. Then we have a commutative diagram with exact rows:*

$$
\begin{array}{ccccccc}
0 & \longrightarrow & T_l \,\text{Br}\,\mathcal{X} & \longrightarrow & T_l \,\text{Br}\, X & \longrightarrow & (CH_0(X) \otimes \mathbb{Q}_l/\mathbb{Z}_l)^* \\
 & & \downarrow & & \downarrow & & \| \\
 & 0 & \longrightarrow & (H^{2d}_{\text{et}}(X, \mathbb{Z}(d)) \otimes \mathbb{Q}_l/\mathbb{Z}_l)^* & \longrightarrow & (CH_0(X) \otimes \mathbb{Q}_l/\mathbb{Z}_l)^*.
\end{array}
$$

Here the upper row is induced by the Brauer-Manin pairing and the lower row by the change of topology map.

Proof By Colliot-Thélène and Saito [1, Cor. 2.4], the kernel of the Brauer-Manin pairing is $T_l \operatorname{Br} \mathcal{X}$. The lower row is exact because the cokernel of $CH_0(X) \to H^{2d}_{\text{et}}(X, \mathbb{Z}(d))$ is torsion, hence it vanishes after tensoring with $\mathbb{Q}/\mathbb{Z}$. It remains to show that the diagram is commutative.

Lemma 4.4 *The following diagram is commutative, where the upper pairing is the Brauer-Manin pairing and the lower pairing the cup-product pairing:*

$$
\begin{array}{ccc}
CH_0(X)/m \ \times_m \operatorname{Br}(X) & \longrightarrow & \mathbb{Z}/m \\
\downarrow & \| & \| \\
H^{2d}_{\text{et}}(X, \mathbb{Z}(d))/m \times_m \operatorname{Br}(X) & \longrightarrow & \mathbb{Z}/m.
\end{array}
$$

Proof By definition of the lower pairing, it suffices to show this after adding the following commutative diagram on the bottom

$$
\begin{array}{ccc}
H^{2d}_{\text{et}}(X, \mathbb{Z}(d))/m \times \ {}_m \operatorname{Br}(X) & \longrightarrow & \mathbb{Z}/m \\
\downarrow & \uparrow & \| \\
H^{2d}_{\text{et}}(X, \mathbb{Z}/m(d)) \times H^2_{\text{et}}(X, \mathbb{Z}/m(1)) & \longrightarrow & \mathbb{Z}/m
\end{array}
$$

because the middle vertical map is surjective. The Brauer-Manin pairing is defined point by point. But if $i : \operatorname{Spec} k \to X$ is a closed point, then the commutativity follows from the projection formula, i.e. the commutativity of the following diagram

$$
\begin{array}{ccc}
H^0_{\text{et}}(k, \mathbb{Z}/m(0)) \times H^2_{\text{et}}(k, \mathbb{Z}/m(1)) & \longrightarrow & \mathbb{Z}/m \cong H^2_{\text{et}}(k, \mathbb{Z}/m(1)) \\
i_* \downarrow \quad\quad i^* \uparrow & & i_* \downarrow \sim \\
H^{2d}_{\text{et}}(X, \mathbb{Z}/m(d)) \times H^2_{\text{et}}(X, \mathbb{Z}/m(1)) & \longrightarrow & \mathbb{Z}/m \cong H^{2d+2}_{\text{et}}(X, \mathbb{Z}/m(d+1)).
\end{array}
$$

$\square$

Finally, we note that there are examples with non-vanishing $T_l \operatorname{Br} \mathcal{X}$: Let $X/\mathbb{Q}_p$ be the self product of an elliptic curve $E/\mathbb{Q}_p$ without complex multiplication and good reduction E_s. Then the graph of the Frobenius of E_s in $\operatorname{Pic}(Y)$, Y the special fiber of the proper smooth model $\mathcal{X}/\mathbb{Z}_p$, does not lift to $\operatorname{Pic}(\mathcal{X})$, because $\operatorname{End}(E)$ has rank 1 [13, p. 331]. Hence we conclude by the proper base change theorem and the following diagram

$$
\begin{array}{ccccccccc}
0 & \longrightarrow & \operatorname{Pic}(\mathcal{X})^{\wedge l} & \longrightarrow & H^2(\mathcal{X}, \mathbb{Z}_l(1)) & \longrightarrow & T_l \operatorname{Br} \mathcal{X} & \longrightarrow & 0 \\
& & \downarrow & & \| & & \downarrow & & \\
0 & \longrightarrow & \operatorname{Pic}(Y)^{\wedge l} & \longrightarrow & H^2(Y, \mathbb{Z}_l(1)) & \longrightarrow & T_l \operatorname{Br} Y & = & 0.
\end{array}
$$

Remark 4.5 We believe that there should be a better behaved duality theory for Weil-étale cohomology groups, see [10] in the case of curves. B. Morin expects that there are locally compact groups $H_W^i(X, \mathbb{Z}(n))$ and $H_W^j(X, \mathbb{R}/\mathbb{Z}(u))$, together with a trace map $H_W^{2d+2}(X, \mathbb{R}/\mathbb{Z}(d+1)) \to H_W^2(K, \mathbb{R}/\mathbb{Z}(1)) \cong \mathbb{R}/\mathbb{Z}$, such that there is a perfect Pontrjagin pairing of locally compact groups

$$H_W^i(X, \mathbb{Z}(n)) \times H_W^{2d+2-i}(X, \mathbb{R}/\mathbb{Z}(d+1-n)) \to H_W^{2d+2}(X, \mathbb{R}/\mathbb{Z}(d+1)) \to \mathbb{R}/\mathbb{Z}.$$

References

[1] J.L. Colliot-Thélène and S. Saito, *Zéro-cycles sur les variétés p-adiques et groupe de Brauer,* Internat. Math. Res. Notices 1996, no. **4**, 151–160.

[2] M. Flach and B. Morin, *Weil-étale cohomology and zeta-values of proper regular arithmetic schemes,* http://arxiv.org/abs/1605.01277

[3] O. Gabber, *Sur la torsion dans la cohomologie l-adique d'une variété,* C. R. Acad. Sci. Paris Sér. I Math. **297** (1983), no. 3, 179–182.

[4] T. Geisser, *Motivic cohomology over Dedekind rings,* Math. Z. **248** (2004), no. 4, 773–794.

[5] T. Geisser, *Weil-étale cohomology,* Math. Ann. **330** (2004), 665–692.

[6] T. Geisser, *Duality via cycle complexes,* Ann. of Math. (2) **172** (2010), no. 2, 1095–1126.

[7] T. Geisser, *On the structure of étale motivic cohomology,* J. Pure Appl. Algebra **221** (2017), no. 7, 1614–1628.

[8] T. Geisser and M. Levine, *The K-theory of fields in characteristic p,* Invent. Math. **139** (2000), no. 3, 459–493.

[9] B. Kahn, *Some finiteness results for étale cohomology,* J. Number Theory **99** (2003), no. 1, 57–73.

[10] D.A. Karpuk, *Weil-étale cohomology of curves over p-adic fields,* J. Algebra **416** (2014), 122–138.

[11] T. Geisser and A.Schmidt, *Poitou-Tate duality for arithmetic schemes,* to appear in Compositio Mathematica.

[12] T. Kohrita, *Thesis, Nagoya University*

[13] A. Langer and S. Saito, *Torsion zero-cycles on the self-product of a modular elliptic curve,* Duke Math. J. **85** (1996), no. **2**, 315–357.

[14] M. Levine, *The K-theory and motivic cohomology of schemes,* http://www.math.uiuc.edu/K-theory/336/

[15] S. Lichtenbaum, *Duality theorems for curves over p-adic fields,* Invent. Math. **7** (1969) 120–136.

[16] S. Lichtenbaum, *Values of zeta-functions at nonnegative integers,* Number theory, Noordwijkerhout 1983 (Noordwijkerhout, 1983), 127–138, Lecture Notes in Math. **1068**, Springer, Berlin, 1984.

[17] J. Milne, *Values of zeta functions of varieties over finite fields,* Amer. J. Math. **108** (1986), no. 2, 297–360.

[18] J. Milne, *Arithmetic duality theorems. Second edition,* BookSurge, LLC, Charleston, SC, 2006. viii+339 pp. ISBN: 1-4196-4274-X

[19] J. Murre, *Applications of algebraic K-theory to the theory of algebraic cycles,* Algebraic geometry, Sitges (Barcelona), 1983, 216–261, Lecture Notes in Math. **1124**, Springer, Berlin, 1985.

[20] J. Murre, *Algebraic cycles and algebraic aspects of cohomology and K-theory,* Algebraic cycles and Hodge theory (Torino, 1993), 93–152, Lecture Notes in Math. **1594**, Springer, Berlin, 1994.

[21] H. Saito, *Abelian varieties attached to cycles of intermediate dimension,* Nagoya Math. J. **75** (1979), 95–119.

[22] S. Saito, *Class field theory for curves over local fields,* J. Number Theory **21** (1985), no. 1, 44–80.

[23] K. Sato, *p-adic étale Tate twists and arithmetic duality,* Ann. Sci. École Norm. Sup. (4) **40** (2007), no. 4, 519–588.

[24] M. Spitzweck, *A commutative P^1-spectrum representing motivic cohomology over Dedekind domains,* arXiv:1207.4078.

[25] J. Tate, *On the conjectures of Birch and Swinnerton-Dyer and a geometric analog,* Sem. Bourbaki 1965/66, no. **306**.

[26] T. Yoshida, *Finiteness theorems in the class field theory of varieties over local fields,* J. Number Theory **101** (2003), no. 1, 138–150.

[27] C. Zhong, *Comparison of dualizing complexes,* J. Reine Angew. Math. **695** (2014), 1–39.

DEPARTMENT OF MATHEMATICS, RIKKYO UNIVERSITY, NISHI-IKEBU-KURO, TOSHIMAKU, TOKYO, JAPAN
E-mail: geisser@rikkyo.ac.jp

K-Theory
Copyright ©2018 Tata Institute of Fundamental Research
Publisher: Hindustan Book Agency, New Delhi, India

The Vaserstein symbol on real smooth affine threefolds

Jean Fasel

Abstract

We give a necessary and sufficient topological condition for the Vaserstein symbol to be injective on smooth affine real threefolds.

Introduction

Let R be a ring. Recall that a row $(a_1, \ldots, a_n) \in R^n$ is unimodular if the ideal generated by the a_i is R itself, i.e. $\langle a_1, \ldots, a_n \rangle = R$. It is equivalent to saying that the associated homomorphism $R^n \to R$ is surjective. The second description makes clear that the group $GL_n(R)$ acts on the set $Um_n(R)$ of unimodular rows (by precomposition) and one can consider the orbit set $Um_n(R)/GL_n(R)$. It follows that any subgroup of $GL_n(R)$ also acts on $Um_n(R)$, and in particular so does $E_n(R) \subset GL_n(R)$. It is often useful to see that set of orbits as pointed by the class of $e_1 := (1, 0, \ldots, 0)$. In [VS76, §5], the authors defined the so-called *Vaserstein symbol* $V : Um_3(R)/E_3(R) \to W_E(R)$, where the right-hand term is the elementary symplectic Witt group described in [VS76, §3] (see also [Fas11b, §2]). The Witt group $W_E(R)$ is generated by skew-symmetric invertible matrices and the symbol V can be explicitly described by

$$V(a_1, a_2, a_3) = \begin{pmatrix} 0 & -a_1 & -a_2 & -a_3 \\ a_1 & 0 & -b_3 & b_2 \\ a_2 & b_3 & 0 & -b_1 \\ a_3 & -b_2 & b_1 & 0 \end{pmatrix}$$

where b_1, b_2, b_3 are such that $\sum_{i=1}^3 a_i b_i = 1$ (the symbol is independent of the choice of such elements). Further, they prove that the map V is bijective under additional condition on R ([VS76, Theorem 5.2]) which are satisfied if R is of Krull dimension 2. Consequently, the orbit set $Um_3(R)/E_3(R)$ inherits in that case an abelian group structure which was widely used. Later, R. Rao and W. van der Kallen proved in [RvdK94, Corollary 3.5] that V

was also bijective when R is a smooth k-algebra, where k is a field which is perfect and C_1. In the same paper, they also proved that the symbol V was in general not injective for threefolds. More precisely, they proved that V was not injective when R was the coordinate ring of the real algebraic sphere of dimension 3. Later, D. R. Rao and N. Gupta showed that there was infinitely many real (singular) threefolds for which the Vaserstein symbol was not injective ([RG14]), a result later improved in [RK16] where the authors exhibited an uncountable family of smooth real threefold for which V fails to be injective. This activity culminated in a conjecture attributed to the author of the present paper ([RK16, Introduction]) which gave a necessary and sufficient topological condition making the Vaserstein symbol injective for smooth real affine threefolds.

The aim of this note is to prove the conjecture. More precisely, we prove the following theorem.

Theorem *Let $X = \operatorname{Spec}(R)$ be a smooth affine real threefold. Let $\mathcal{C}$ be the set of compact connected components of $X(\mathbb{R})$ (in the Euclidean topology). Then, the Vaserstein symbol*

$$Um_3(R)/E_3(R) \to W_E(R)$$

is injective if and only if $\mathcal{C} = \emptyset$.

Our method is as follows. In [AF14c, proof of Theorem 4.3.1], we observed that the Vaserstein symbol had an interpretation in the realm of $\mathbb{A}^1$-homotopy theory. More precisely, let k be a perfect field. The smooth affine quadric Q_5 with $k[Q_5] = k[x_1, x_2, x_3, y_1, y_2, y_3]/\langle \sum x_i y_i - 1 \rangle$ is isomorphic to the quotient of algebraic varieties SL_4/Sp_4. The latter is in turn isomorphic to the the affine scheme A'_4 representing the functor assigning to a ring R the set of invertible skew-symmetric matrices with trivial Pfaffian. The composite $Q_5 \to SL_4/Sp_4 \to A'_4$ associates to a 6-tuple $(a_1, a_2, a_3, b_1, b_2, b_3)$ such that $\sum a_i b_i = 1$ the matrix $V(a_1, a_2, a_3)$ described above. Now, there is a stabilization map $SL_4/Sp_4 \to SL_6/Sp_6$ and it turns out that this map actually determines the injectivity of the Vaserstein symbol. Indeed, let $\mathcal{H}_{\mathbb{A}^1}(k)$ be the $\mathbb{A}^1$-homotopy category defined by Morel and Voevodsky in [MV99]. We then have $\operatorname{Hom}_{\mathcal{H}_{\mathbb{A}^1}(k)}(X, Q_5) = Um_3(R)/E_3(R)$ for any smooth affine threefold $X = \operatorname{Spec} R$ and $\operatorname{Hom}_{\mathcal{H}_{\mathbb{A}^1}(k)}(X, SL_6/Sp_6) = W_E(R)$, while the stabilization map $Q_5 \to SL_6/Sp_6$ precisely induces the Vaserstein symbol. In this context, we can use the computation of the homotopy sheaves of $Q_5 \simeq \mathbb{A}^3 \setminus 0$ obtained in [AF15] to prove that the symbol V is injective if $\mathcal{C}$ is empty. To prove that this condition is also necessary, we produce explicitly a morphism $\mathbb{A}^4 \setminus 0 \to Q_5$ whose composite with $Q_5 \to SL_6/Sp_6$ is homotopy trivial and show that its real realization is the Hopf map $S^3_{\mathbb{R}} \to S^2_{\mathbb{R}}$. If

$\mathcal{C} \neq \emptyset$, this allows to produce non trivial elements in $Um_3(R)/E_3(R)$ whose image under V is trivial.

Let us now explain the organization of the paper. In Section 1, we first quickly review the notions of $\mathbb{A}^1$-homotopy theory needed to understand the proof of the main result. In particular, we explain the reinterpretation of the Vaserstein symbol sketched above, and recall the explicit models of motivic spheres obtained in [AF14c]. We also take the opportunity to provide a few cohomological computations based on a recent preprint of J. Jacobson ([Jac16]). We then proceed with the proof that the condition $\mathcal{C} = \emptyset$ is sufficient for V to be injective. In Section 3, we then show that it is necessary.

Notation and conventions   The rings considered in this paper are commutative and unital. If X is a real variety, we denote by $X_\mathbb{C}$ the scheme $X_\mathbb{C} := X \times_\mathbb{R} \mathrm{Spec}(\mathbb{C})$.

Acknowledgements   The author wishes to thank Ravi Rao for motivating him to write this paper. It is also a pleasure to thank Aravind Asok for explaining the morphism $Q_7 \to Q_4$ appearing in Section 3.

1   Preliminaries

The purpose of this section is to explain a few basic features of the motivic homotopy category $\mathcal{H}_{\mathbb{A}^1}(k)$ constructed by F. Morel and V. Voevodsky. In this part of the paper, k is an infinite perfect field.

1.1   Motivic homotopy theory

Recall from [MV99] that a space $\mathcal{X}$ is a (Nisnevich) sheaf of simplicial sets $\mathrm{Sm}_k \to \mathrm{SSets}$. Basic examples of spaces are smooth k-schemes seen as functors $\mathrm{Sm}_k \to \mathrm{Sets} \to \mathrm{SSets}$ and simplicial sets seen as constant sheaves of simplicial sets. These include the simplicial spheres S^i for any $i \geq 0$. In the category of spaces, one can perform many useful constructions, among which the smash product of (pointed) spaces. For instance, if we write $\mathbf{G}_m$ for the groups scheme representing the functor $X \mapsto O_X(X)^\times$, we can form smash products of $\mathbf{G}_m$ (pointed by 1) with itself to obtain spaces $(\mathbf{G}_m)^{\wedge j}$ for any integer $j \in \mathbb{N}$, and we can perform smash products with S^i to obtain spaces $S^{i,j} := S^i \wedge (\mathbf{G}_m)^{\wedge j}$ for any $i, j \in \mathbb{N}$ that we refer to as *motivic spheres*.

Any space $\mathcal{X}$ has stalks at the points of the Nisnevich topology, and a (pointed) morphism of spaces $f : \mathcal{X} \to \mathcal{Y}$ is said to be a (pointed) weak-equivalence if it induces a weak-equivalence of simplicial sets on stalks. One can put a model structure on Spc_k that allows to invert weak-equivalences in a good way, and the corresponding homotopy category is the *simplicial homotopy category of smooth schemes*. To get the motivic homotopy category $\mathcal{H}_{\mathbb{A}^1}(k)$, one further formally inverts the morphisms $\mathcal{X} \times \mathbb{A}^1 \to \mathcal{X}$ induced by the projection $\mathbb{A}^1 \to \mathrm{Spec}k$ via Bousefield localization. For any two spaces $\mathcal{X}$ and $\mathcal{Y}$, we denote by $[\mathcal{X}, \mathcal{Y}]_{\mathbb{A}^1}$ the set of morphisms between them in $\mathcal{H}_{\mathbb{A}^1}(k)$.

As noted above, the spaces $S^{i,j}$ are objects of $\mathcal{H}_{\mathbb{A}^1}(k)$, and it is not clear in general that they have "geometric" models in that category, i.e. that there exist smooth schemes which are isomorphic to $S^{i,j}$. While the answer is known to be negative in general, there are two important particular cases. For any $n \geq 1$, let Q_{2n-1} be the smooth affine k-scheme whose ring of global sections is $k[Q_{2n-1}] = k[x_1, \ldots, x_n, y_1, \ldots, y_n]/\langle \sum x_i y_i - 1\rangle$, and let Q_{2n} be the smooth affine scheme whose global sections are $k[Q_{2n}] = k[x_1, \ldots, x_n, y_1, \ldots, y_n, z]/\langle \sum x_i y_i - z(1-z)\rangle$. In [ADF16, Theorem 2.2.5], we proved that there are explicit isomorphisms $S^{n-1,n} \simeq Q_{2n-1} \simeq \mathbb{A}^n \setminus 0$ and $S^{n,n} \simeq Q_{2n}$ in $\mathcal{H}_{\mathbb{A}^1}(k)$ for any $n \geq 1$. Interestingly, these motivic spheres appear naturally in the context of unimodular rows. Indeed, there are natural bijections $[X, Q_{2n-1}]_{\mathbb{A}^1} = [X, \mathbb{A}^n \setminus 0]_{\mathbb{A}^1} = Um_n(R)/E_n(R)$ for any smooth affine k-scheme $X = \mathrm{Spec}R$ ([AF14a, §4.2] and [AHW15, Theorem 4.2.1]).

In $\mathcal{H}_{\mathbb{A}^1}(k)$, one can define for any $n \in \mathbb{N}$ the homotopy sheaves $\pi_n^{\mathbb{A}^1}(\mathcal{X})$ associated to a (pointed) space $\mathcal{X}$. These are Nisnevich sheaves on Sm_k which are extremely hard to compute in general in analogy with classical topology. A celebrated result of F. Morel states that the homotopy sheaves of $\mathbb{A}^n \setminus 0$ are of the following form (provided $n \geq 2$):

$$\pi_i^{\mathbb{A}^1}(\mathbb{A}^n \setminus 0) = \begin{cases} 0 & \text{if } i < n - 1. \\ \mathbf{K}_n^{\mathrm{MW}} & \text{if } i = n - 1. \end{cases}$$

The sheaf $\mathbf{K}_n^{\mathrm{MW}}$ is the *(unramified) Milnor-Witt K-theory sheaf* and is defined in [Mor12, §3]. For $i \geq n$, our knowledge is still limited (see however [AF14b] for $n = 2$ and [AF15] for $n = 3$).

To conclude this section, let us recall from [MV99, §3.3] that there is a realization functor $\mathcal{H}_{\mathbb{A}^1}(k) \to \mathcal{H}$ in case $k = \mathbb{R}$, where $\mathcal{H}$ is the usual homotopy category of topological spaces. Indeed, there is a realization functor $\mathcal{H}_{\mathbb{A}^1}(k) \to \mathcal{H}(\mathbb{Z}/2)$ where $\mathcal{H}(\mathbb{Z}/2)$ is the homotopy category of equivariant $\mathbb{Z}/2$-topological spaces. This functor associates to a smooth scheme X over $\mathbb{R}$ the topological space $X(\mathbb{C})$ together with the $\mathbb{Z}/2$-action induced by complex conjugation. Taking fixed points, we finally get a functor $\mathcal{H}_{\mathbb{A}^1}(k) \to \mathcal{H}$ associating to a smooth scheme X the topological space $X(\mathbb{R})$.

1.2 Cohomology with coefficients in $\mathbf{I}^n$

As mentioned in the previous section, the Milnor-Witt K-theory sheaf $\mathbf{K}_n^{MW}$ appears as the first nontrivial homotopy sheaf of $\mathbb{A}^n \setminus 0$. This sheaf is possibly better understood using two more classical sheaves. For the first one, assume that k is of characteristic different from 2 and consider the presheaf $X \mapsto W(X)$, where $W(X)$ is the Witt group of symmetric bilinear forms as defined by M. Knebusch ([Kne77]). For any $n \in \mathbb{N}$, one can consider its subpresheaf $I^n(X) \subset W(X)$ of powers of the fundamental ideal $I(X) \subset W(X)$ and its associated (Zariski) sheaf $\mathbf{I}^n$ on Sm_k. By construction, there is an epimorphism of sheaves $\mathbf{K}_n^{MW} \to \mathbf{I}^n$ for any $n \in \mathbb{N}$ ([MV99, §3]). For real varieties, the latter recovers very important topological information about $X(\mathbb{R})$. Indeed, for any $m \in \mathbb{N}$, let $H^m(X(\mathbb{R}), \mathbb{Z})$ be the singular cohomology group of $X(\mathbb{R})$. Then, one has the following comparison theorem which is due to J. Jacobson ([Jac16, Corollary 8.3]).

Theorem 1.1 *Let X be scheme over $\mathbb{R}$ which is separated and of finite type and let d be its dimension. Then, the signature map induces isomorphisms*

$$H^m(X, \mathbf{I}^n) \to H^m(X(\mathbb{R}), \mathbb{Z})$$

for any $n \geq d + 1$ and any integer $m \geq 0$.

Corollary 1.2 *Let X be a smooth real affine variety and let $\mathcal{C}$ be the set of compact connected components of $X(\mathbb{R})$. Let $\mathcal{C}_1 \subset \mathcal{C}$ be the subset of oriented such components and $\mathcal{C}_2 = \mathcal{C} \setminus \mathcal{C}_1$. We have then for any $j > 0$ an isomorphism*

$$H^d(X, \mathbf{I}^{d+j}) \simeq \bigoplus_{\mathcal{C}_1} \mathbb{Z} \oplus \bigoplus_{\mathcal{C}_2} \mathbb{Z}/2.$$

Proof In view of the above theorem, it suffices to show that $H^d(C, \mathbb{Z}) = \mathbb{Z}$ if $C \in \mathcal{C}_1$, $H^d(C, \mathbb{Z}) = \mathbb{Z}/2$ if $C \in \mathcal{C}_2$ and $H^d(C, \mathbb{Z}) = 0$ if C is a non compact connected component of $X(\mathbb{R})$. The first case is a direct consequence of Poincaré duality [Hat02, Theorem 3.30] and the second case follows from the universal coefficient theorem and [Hat02, Theorem 3.26, Corollary 3.28]. Finally, suppose that C is a non compact connected component of $X(\mathbb{R})$. Arguing as in [BDM06, Lemma 4.16], we can see C as a strict submanifold of a compact connected component C'. It follows easily from the two previous cases that $H^d(C, \mathbb{Z}) = 0$.

$\square$

Let now $\mathbf{K}_n^{M}$ be the unramified Milnor K-theory sheaf considered for instance in [Mor05, §2]. Its sections over a finitely generated field extension

L/k form precisely the Milnor K-theory group $K_n^M(L)$. One can consider its subsheaf $2\mathbf{K}_n^M$ and it turns out (using Voevodsky's affirmation of Milnor conjecture) that there is an exact sequence of sheaves

$$0 \to 2\mathbf{K}_n^M \to \mathbf{K}_n^{MW} \to \mathbf{I}^n \to 0.$$

Using this sequence, we now slightly adapt [Fas11a, Proposition 5.5] to get the following computation of $H^d(X, \mathbf{K}_{d+j}^{MW})$.

Proposition 1.3 *Let X be a smooth real affine variety of dimension d. Then, for any $j > 0$, we have an exact sequence*

$$H^d(X_{\mathbb{C}}, \mathbf{K}_{d+j}^{MW}) \to H^d(X, \mathbf{K}_{d+j}^{MW}) \to H^d(X, \mathbf{I}^{d+j}) \to 0.$$

Proof The finite morphism $X_{\mathbb{C}} \to X$ induces a morphism between the explicit flasque resolutions of $\mathbf{K}_{d+j}^M$ provided for instance in [Ros96, §6]. Now, [Fas11a, proof Proposition 5.4] shows that this morphism of complexes yields a surjective homomorphism $H^d(X_{\mathbb{C}}, \mathbf{K}_{d+j}^M) \to H^d(X, 2\mathbf{K}_{d+j}^M)$. We conclude using the long exact sequence induced by the exact sequence of sheaves

$$0 \to 2\mathbf{K}_{d+j}^M \to \mathbf{K}_{d+j}^{MW} \to \mathbf{I}^{d+j} \to 0.$$

$\square$

1.3   Reinterpretation of the Vaserstein symbol

We now elaborate a bit more the reinterpretation of the Vaserstein symbol in $\mathcal{H}_{\mathbb{A}^1}(k)$ hinted at in the introduction. As explained in [AF14c, proof of Theorem 4.3.1], we have an isomorphism $Q_5 \to SL_4/Sp_4$ and we can consider the composite of this isomorphism with the stabilization morphism $SL_4/Sp_4 \to SL/Sp$. This yields a map $Q_5 \to SL/Sp$ which is called *degree map* in *loc.cit*. The interesting feature of SL/Sp is that it represents the (reduced) higher Grothendieck-Witt group $GW_1^3(X)$ as defined by Schlichting (see for instance [Sch10a] and [Sch10b]). Now, this group coincides for affine varieties with the group $W_E(X)$ ([FRS12, §4]) and we get for any smooth affine variety $X = \operatorname{Spec} R$ over k an induced map

$$Um_3(R)/E_3(R) = [X, Q_5]_{\mathbb{A}^1} \to [X, SL/Sp]_{\mathbb{A}^1} = W_E(R)$$

which coincides with the (opposite of the) Vaserstein symbol ([AF14c, proof of Theorem 4.3.1]). On the other hand, it follows from [AF15, Proposition 4.2.2] that for a smooth affine threefold as above, we have $[X, SL/Sp]_{\mathbb{A}^1} = [X, SL_6/Sp_6]_{\mathbb{A}^1}$ and a fiber sequence

$$\Omega(\mathbb{A}^5 \setminus 0) \to Q_5 \to SL_6/Sp_6. \tag{1.1}$$

It follows that the term $\Omega(\mathbb{A}^5 \setminus 0)$ controls the kernel of the Vaserstein symbol.

2 A sufficient condition for the symbol to be injective

In this section, we will show that if $\mathcal{C}$ is empty, then the Vaserstein symbol

$$V : Um_3(R)/E_3(R) \to W_E(X)$$

is injective. In view of the $\mathbb{A}^1$-homotopy reinterpretation above, it suffices then to show that for a real smooth affine threefold X such that $\mathcal{C} = \emptyset$ we have a bijection $[X, \mathbb{A}^3\backslash 0]_{\mathbb{A}^1} \to [X, SL_6/Sp_6]_{\mathbb{A}^1}$. To see this, we can consider the Postnikov towers (as constructed for instance in [AF14b, §6]) associated to both $\mathfrak{X} := \mathbb{A}^3 \backslash 0$ and $\mathcal{Y} := SL_6/Sp_6$. This yields a commutative diagram of spaces

$$
\begin{array}{ccccccccc}
K(\pi_2^{\mathbb{A}^1}(\mathfrak{X}), 1) & \longrightarrow & K(\pi_3^{\mathbb{A}^1}(\mathfrak{X}), 3) & \longrightarrow & \mathfrak{X}^{(3)} & \longrightarrow & K(\pi_2^{\mathbb{A}^1}(\mathfrak{X}), 2) & \longrightarrow & K(\pi_3^{\mathbb{A}^1}(\mathfrak{X}), 4) \\
{\scriptstyle K(f,1)}\downarrow & & {\scriptstyle K(f,3)}\downarrow & & \downarrow & & {\scriptstyle K(f,2)}\downarrow & & {\scriptstyle K(f,4)}\downarrow \\
K(\pi_2^{\mathbb{A}^1}(\mathcal{Y}), 1) & \longrightarrow & K(\pi_3^{\mathbb{A}^1}(\mathcal{Y}), 3) & \longrightarrow & \mathcal{Y}^{(3)} & \longrightarrow & K(\pi_2^{\mathbb{A}^1}(\mathcal{Y}), 2) & \longrightarrow & K(\pi_3^{\mathbb{A}^1}(\mathcal{Y}), 4)
\end{array}
$$

where the vertical maps are all induced by f. Since X is a threefold, it follows from [AF14b, Proposition 6.2] that $[X, \mathfrak{X}^{(3)}]_{\mathbb{A}^1} = [X, \mathfrak{X}]_{\mathbb{A}^1}$ (respectively that $[X, \mathcal{Y}^{(3)}]_{\mathbb{A}^1} = [X, \mathcal{Y}]_{\mathbb{A}^1}$). A diagram chase shows that it suffices to prove that the maps $K(f, i)$ induce isomorphisms after applying $[X, \_]_{\mathbb{A}^1}$. Now, we know that f induces an isomorphism $\pi_2^{\mathbb{A}^1}(\mathfrak{X}) \to \pi_2^{\mathbb{A}^1}(\mathcal{Y})$ and therefore $K(f, 2)$ and $K(f, 1)$ are isomorphisms. On the other hand, $[X, K(\mathcal{F}, 4)]_{\mathbb{A}^1} = H_{Nis}^4(X, \mathcal{F}) = 0$ for any strictly $\mathbb{A}^1$-invariant sheaf $\mathcal{F}$ on Sm_k thus showing that $K(f, 4)$ also induces an isomorphism. We are then reduced to prove that $K(f, 3)$ yields an isomorphism. The fiber sequence (1.1) yields an exact sequence of sheaves

$$\pi_4^{\mathbb{A}^1}(SL_6/Sp_6) \to \mathbf{K}_5^{MW} \to \pi_3^{\mathbb{A}^1}(\mathbb{A}^3 \backslash 0) \to \pi_3^{\mathbb{A}^1}(SL_6/Sp_6) \to 0.$$

In order to prove that the morphism of sheaves $\pi_3^{\mathbb{A}^1}(\mathbb{A}^3\backslash 0) \to \pi_3^{\mathbb{A}^1}(SL_6/Sp_6)$ induces an isomorphism $H^3(X, \pi_3^{\mathbb{A}^1}(\mathbb{A}^3 \backslash 0)) \simeq H^3(X, \pi_3^{\mathbb{A}^1}(SL_6/Sp_6))$, it suffices to show that the morphism of sheaves $\pi_4^{\mathbb{A}^1}(SL_6/Sp_6) \to \mathbf{K}_5^{MW}$ induces a surjective homomorphism $H^3(X, \pi_4^{\mathbb{A}^1}(SL_6/Sp_6)) \to H^3(X, \mathbf{K}_5^{MW})$. Now, [AF15, Proposition 4.2.2] shows that $\pi_4^{\mathbb{A}^1}(SL_6/Sp_6) = \mathbf{GW}_5^3$ and that $\mathbf{GW}_5^3 = \pi_4^{\mathbb{A}^1}(SL_6/Sp_6) \to \mathbf{K}_5^{MW}$ is the morphism χ_5 considered in [AF15, proof of Theorem 4.3.1]. Consider the finite morphism $X_{\mathbb{C}} \to X$.

The transfer map for this morphism yields a commutative diagram

$$
\begin{array}{ccc}
H^3(X_{\mathbb{C}}, \mathbf{GW}_5^3) & \longrightarrow & H^3(X, \mathbf{GW}_5^3) \\
\downarrow & & \downarrow \\
H^3(X_{\mathbb{C}}, \mathbf{K}_5^{\mathrm{MW}}) & \longrightarrow & H^3(X, \mathbf{K}_5^{\mathrm{MW}})
\end{array}
$$

and Proposition 1.3 shows that the bottom horizontal map is surjective. It suffices then to prove that the left-hand vertical map is surjective to conclude. We know that the cokernel of $\mathbf{GW}_5^3 \to \mathbf{K}_5^{\mathrm{MW}}$ is isomorphic to the sheaf denoted by $\mathbf{F}_5$ in [AF15], which is a quotient of the sheaf $\mathbf{T}_5$ considered in [AF14a, §3.6]. As $\mathbf{I}^5 = 0$ on $X_{\mathbb{C}}$, The sheaf $\mathbf{T}_5$ reduces to the sheaf $\mathbf{S}_5$ which is a quotient of $\mathbf{K}_5^M/24$. We are then reduced to prove that $H^3(X_{\mathbb{C}}, \mathbf{K}_5^M/24) = 0$ which follows from the fact that $\mathbf{K}_2^M(\mathbb{C})$ is divisible.

Remark 2.1 When $X_{\mathbb{C}}$ is moreover supposed to be rational and oriented, then $H^3(X, \mathbf{K}_5^{\mathrm{MW}}) = 0$ by [Fas11a, Theorem 5.7] and the argument is much simpler.

3   A necessary condition for the symbol to be injective

We now suppose that the set $\mathcal{C}$ is non empty and derive from this that the Vaserstein symbol is not injective. Our strategy is to construct an explicit morphism $Q_7 \to \mathbb{A}^3 \backslash 0$ whose composite with $\mathbb{A}^3 \backslash 0 \to SL_6/Sp_6$ is trivial. It follows that the image of $[X, Q_7]_{\mathbb{A}^1}$ in $[X, \mathbb{A}^3 \backslash 0]_{\mathbb{A}^1}$ under this map furnishes candidates for the non injectivity of the Vaserstein symbol. We are reduced to show that under our condition, this image is non trivial.

We start with the definition of an explicit morphism $Q_7 \to Q_4$. By definition, if R is a commutative ring we have $Q_7(R) = \{a_1, \dots, a_4, b_1, \dots, b_4 \in R | \sum a_i b_i = 1\}$ and $Q_4(R) = \{x_1, x_2, y_1, y_2, z \in R | x_1 y_1 + x_2 y_2 = z(1-z)\}$. The set $Q_7(R)$ can be seen as the set $\{M, N \in M_2(R) | \det M + \det N = 1\}$ via

$$
(a_1, \dots, a_4, b_1, \dots, b_4) \mapsto \left(\begin{pmatrix} a_1 & a_2 \\ -b_2 & b_1 \end{pmatrix}, \begin{pmatrix} a_3 & a_4 \\ -b_4 & b_3 \end{pmatrix} \right)
$$

and we define the morphism $f : Q_7 \to Q_4$ via $(M, N) \mapsto (MN, \det M)$. As $\det(MN) = \det M \det N = (\det M)(1 - \det M)$, we see that the 5-tuple $(MN, \det M)$ defines an element of $Q_4(R)$. Explicitly, the morphism is given by

$$f(a_1, \ldots, a_4, b_1, \ldots, b_4) = (a_1 a_3 - a_2 b_4, a_1 a_4 + a_2 b_3,$$
$$- a_4 b_2 + b_1 b_3, a_3 b_2 + b_1 b_4, a_1 b_1 + a_2 b_2).$$

Let now $[-1] : S^0 \to \mathbf{G}_m$ be the morphism sending the non basepoint to -1. Smashing with any space $\mathcal{X}$, we get a morphism $\mathcal{X} \to \mathcal{X} \wedge \mathbf{G}_m$ that we still denote by $[-1]$. By construction, the following diagram commutes

$$\begin{array}{ccc} Q_7 & \xrightarrow{\ f\ } & Q_4 \\ {\scriptstyle [-1]}\downarrow & & \downarrow{\scriptstyle [-1]} \\ Q_7 \wedge \mathbf{G}_m & \xrightarrow[f \wedge Id]{} & Q_4 \wedge \mathbf{G}_m. \end{array}$$

We next define a morphism

$$g : Q_4 \wedge \mathbf{G}_m \to \mathbb{A}^3 \setminus 0$$

via $(x_1, x_2, y_1, y_2, z, \alpha) \mapsto (2x_1, 2x_2, (\alpha - 1)z + 1)$ and finally get a morphism

$$\mathcal{H} : Q_7 \xrightarrow{f} Q_4 \xrightarrow{[-1]} Q_4 \wedge \mathbf{G}_m \xrightarrow{g} \mathbb{A}^3 \setminus 0.$$

We know that $Q_7 \simeq \mathbb{A}^4 \setminus 0 \simeq S^{3,4}$ in $\mathcal{H}_{\mathbb{A}^1} \mathbb{R}$ and it follows from [MV99, §3.3] that $Q_7(\mathbb{R})$ is homotopy equivalent to the real sphere $S^3_{\mathbb{R}}$. On the other hand, $\mathbb{A}^3 \setminus 0 \simeq S^{2,3}$ and its real realization is thus $S^2_{\mathbb{R}}$. The morphism $\mathcal{H}$ constructed above realizes in a morphism $S^3_{\mathbb{R}} \to S^2_{\mathbb{R}}$ and is therefore a multiple of the Hopf map. The next result shows that it is indeed the Hopf map.

Proposition 3.1 *The morphism $\mathcal{H}$ realizes (up to sign) in the Hopf map $S^3_{\mathbb{R}} \to S^2_{\mathbb{R}}$.*

Proof We start by exhibiting an explicit homotopy equivalence $S^3_{\mathbb{R}} \to Q_7(\mathbb{R})$. Concretely, let $h : S^3_{\mathbb{R}} \to Q_7(\mathbb{R})$ be given by

$$(x_1, x_2, x_3, x_4) \mapsto (x_1, x_2, x_3, x_4, x_1, x_2, x_3, x_4)$$

The composite $S^3_{\mathbb{R}} \xrightarrow{h} Q_7(\mathbb{R}) \to \mathbb{R}^4 \setminus 0 \to S^3_{\mathbb{R}}$, where the second map is the projection onto the first four factors and the third map is the usual homotopy equivalence, is easily checked to be the identity. As both the second and third maps are homotopy equivalences, so is the first. Now, an easy computation yields that the composite

$$S^3_{\mathbb{R}} \xrightarrow{f_1} Q_7(\mathbb{R}) \xrightarrow{\mathcal{H}} Q_4(\mathbb{R}) \xrightarrow{[-1]} (Q_4 \wedge \mathbf{G}_m)(\mathbb{R}) \xrightarrow{g} \mathbb{R}^3 \setminus 0$$

maps (x_1, x_2, x_3, x_4) to $(2x_1x_3 - 2x_2x_4, 2x_1x_4 + 2x_2x_3, x_3^2 + x_4^2 - x_1^2 - x_2^2)$. This is the Hopf map made explicit in [RvdK94, §4.5].

$\square$

We now check that the composite

$$Q_7 \overset{\mathcal{H}}{\to} \mathbb{A}^3 \setminus 0 \to SL_6/Sp_6$$

is homotopic to the point. Using the commutative diagram

$$
\begin{array}{ccc}
Q_7 & \overset{f}{\longrightarrow} & Q_4 \\
{\scriptstyle [-1]}\downarrow & & \downarrow{\scriptstyle [-1]} \\
Q_7 \wedge \mathbf{G}_m & \underset{f \wedge Id}{\longrightarrow} & Q_4 \wedge \mathbf{G}_m
\end{array}
$$

we see that it suffices to show that any morphism $Q_7 \wedge \mathbf{G}_m \to SL_6/Sp_6$ is homotopy trivial. Now we have $[Q_7 \wedge \mathbf{G}_m, SL_6/Sp_6]_{\mathbb{A}^1} = [S^{3,5}, SL/Sp] = (\mathbf{GW}_4^3)_{-5} = \mathbf{GW}_{-1}^{-2} = 0$, proving the claim. Consequently, the composite

$$[X, \mathbb{A}^4 \setminus 0]_{\mathbb{A}^1} \to [X, \mathbb{A}^3 \setminus 0]_{\mathbb{A}^1} \to W_E(X)$$

is trivial and it remains to show that the left-hand morphism is non trivial if $\mathcal{C} \neq \emptyset$. To this end, recall that realization induces maps $[X, \mathbb{A}^4 \setminus 0]_{\mathbb{A}^1} \to [X(\mathbb{R}), S^3_{\mathbb{R}}]$ and $[X, \mathbb{A}^3 \setminus 0]_{\mathbb{A}^1} \to [X(\mathbb{R}), S^2_{\mathbb{R}}]$ and that the morphism $\mathbb{A}^4 \setminus 0 \to \mathbb{A}^3 \setminus 0$ realizes into the Hopf map. In classical topology, we have the Hopf fibration

$$S^1_{\mathbb{R}} \to S^3_{\mathbb{R}} \to S^2_{\mathbb{R}}.$$

As $\pi_1(S^3_{\mathbb{R}}) = 0$, it follows that the map $S^1_{\mathbb{R}} \to S^3_{\mathbb{R}}$ is homotopy trivial and therefore for any CW-complex Y the map $[Y, S^3_{\mathbb{R}}] \to [Y, S^2_{\mathbb{R}}]$ has trivial fiber. Suppose then that we have a class in $[X, \mathbb{A}^4 \setminus 0]_{\mathbb{A}^1}$ realizing into a non trivial element of $[X(\mathbb{R}), S^3_{\mathbb{R}}]$. It follows that its image in $[X, \mathbb{A}^3 \setminus 0]_{\mathbb{A}^1}$ is non trivial and therefore that the Vaserstein symbol is not injective. Now, $[X, \mathbb{A}^4 \setminus 0]_{\mathbb{A}^1} \simeq H^3(X, \mathbf{K}_4^{\mathrm{MW}})$ which surjects onto $H^3(X, I^4) = H^3(X(\mathbb{R}), \mathbb{Z})$ and we conclude using Corollary 1.2.

Remark 3.2 Let $X = \operatorname{Spec} R$ be a smooth real affine threefold with $\mathcal{C} \neq \emptyset$. Using the identification $[X, \mathbb{A}^n \setminus 0]_{\mathbb{A}^1} = Um_n(R)/E_n(R)$, the composite

$$[X, \mathbb{A}^4 \setminus 0]_{\mathbb{A}^1} \to [X, \mathbb{A}^3 \setminus 0]_{\mathbb{A}^1} \to W_E(X)$$

can be reinterpreted as a complex

$$Um_4(R)/E_4(R) \overset{\alpha}{\to} Um_3(R)/E_3(R) \overset{V}{\to} W_E(R).$$

where $\alpha([a_1, a_2, a_3, a_4]) = [2a_1a_3 - 2a_2a_4, 2a_1a_4 + 2a_2a_3, a_3^2 + a_4^2 - a_1^2 - a_2^2]$ for any orbit $[a_1, a_2, a_3, a_4] \in Um_4(R)/E_4(R)$. The above proof shows that any non trivial element of $Um_4(R)/E_4(R)$ (which exist in case $\mathcal{C} \neq \emptyset$) will have non trivial image in $Um_3(R)/E_3(R)$ under α providing a non trivial element in the kernel of V.

References

[ADF16] A. Asok, B. Doran, and J. Fasel, *Smooth models of motivic spheres and the clutching construction*, To appear in IMRN. Preprint available at http://arxiv.org/abs/1408.0413, (2016).

[AF14a] A. Asok and J. Fasel, *Algebraic vector bundles on spheres*, J. Topology, **7**(3), 894–926, (2014). doi:10.1112/jtopol/jtt046.

[AF14b] A. Asok and J. Fasel, *A cohomological classification of vector bundles on smooth affine threefolds*, Duke Math. J. **163**(14), 2561–2601, (2014).

[AF14c] A. Asok and J. Fasel, *An explicit KO-degree map and applications*, arXiv:1403.4588, (2014).

[AF15] A. Asok and J. Fasel, *Splitting vector bundles outside the stable range and homotopy theory of punctured affine spaces*, J. Amer. Math. Soc. **28**(4), 1031–1062, (2015). Electronically published on August 7, 2014.

[AHW15] A. Asok, M. Hoyois, and M. Wendt, *Affine representability results in $\mathbb{A}^1$-homotopy theory II: principal bundles and homogeneous spaces*, Preprint available at http://arxiv.org/abs/1507.08020, (2015).

[BDM06] S. M. Bhatwadekar, Mrinal Kanti Das, and Satya Mandal, *Projective modules over smooth real affine varieties*, Invent. Math. **166**(1), 151–184, (2006).

[Fas11a] Jean Fasel, *Some remarks on orbit sets of unimodular rows*, Comment. Math. Helv. **86**(1), 13–39, (2011).

[Fas11b] Jean Fasel, *Stably free modules over smooth affine threefolds*, Duke Math. J. **156**(1), 33–49, (2011).

[FRS12] Jean Fasel, Ravi A. Rao, and Richard G. Swan, *On stably free modules over affine algebras*, Publ. Math. Inst. Hautes Études Sci. **116**(1), 223–243, (2012).

[Hat02] A. Hatcher, *Algebraic topology*, Cambridge University Press, Cambridge, 2002.

[Jac16] J. A. Jacobson, *Real cohomology and the powers of the fundamental ideal in the Witt ring*, Preprint, 2016.

[Kne77] Manfred Knebusch, *Symmetric bilinear forms over algebraic varieties*, In Conference on Quadratic Forms—1976 (Proc. Conf., Queen's Univ., Kingston, Ont., 1976), number 46 in Queen's Papers in Pure and Appl. Math., pages 103–283. Queen's Univ., (1977).

[Mor05] F. Morel, *Milnor's conjecture on quadratic forms and mod 2 motivic complexes*, Rend. Sem. Mat. Univ. Padova, **114**, 63–101 (2006), (2005).

[Mor12] Fabien Morel, $\mathbb{A}^1$-*Algebraic Topology over a Field*, volume 2052 of Lecture Notes in Math. Springer, New York, 2012.

[MV99] Fabien Morel and Vladimir Voevodsky, $\mathbf{A}^1$-*homotopy theory of schemes*, Inst. Hautes Études Sci. Publ. Math. **90**, 45–143 (2001), (1999).

[RG14] D. R. Rao and N. Gupta, *On the non-injectivity of the Vaserstein symbol in dimension three*, J. Algebra **399**, 378–388, (2014).

[RK16] D. R. Rao and S. Kolte, *Odd orthogonal matrices and the non-injectivity of the Vaserstein symbol*, To appear in J. Algebra., 2016.

[Ros96] Markus Rost, *Chow groups with coefficients*, Doc. Math. **1**, No. 16, 319–393 (electronic), (1996).

[RvdK94] Ravi A. Rao and Wilberd van der Kallen, *Improved stability for SK_1 and WMS_d of a non-singular affine algebra*, Astérisque, **226**, 11, 411–420, (1994). K-theory (Strasbourg, 1992).

[Sch10a] Marco Schlichting, *Hermitian K-theory of exact categories*, J. K-Theory **5**(1), 105–165, (2010).

[Sch10b] Marco Schlichting, *The Mayer-Vietoris principle for Grothendieck-Witt groups of schemes*, Invent. Math. **179**(2), 349–433, (2010).

[VS76] L. N. Vaseršteĭn and A. A. Suslin, *Serre's problem on projective modules over polynomial rings, and algebraic K-theory*, Izv. Akad. Nauk SSSR Ser. Mat. **40**(5), 993–1054, 1199, (1976).

J. Fasel, Institut Fourier - UMR 5582, Université Grenoble Alpes, CS 40700, 38058 Grenoble cedex 9, France
E-mail: jean.fasel@gmail.com

K-Theory
Copyright ©2018 Tata Institute of Fundamental Research
Publisher: Hindustan Book Agency, New Delhi, India

Strong cycles and intersection products on good moduli spaces

Dan Edidin[1] and Matthew Satriano[2]

Abstract

We introduce conjectures relating the Chow ring of a smooth Artin stack $\mathcal{X}$ to the Chow groups of its possibly singular good moduli space $\mathbf{X}$. In particular, we conjecture the existence of an intersection product on a subgroup of the full Chow group $A^*(\mathbf{X})$ coming from *strong cycles* on $\mathcal{X}$.

1 Introduction

Let X be a non-singular projective variety with an action of a linearly reductive group G over an algebraically closed field of characteristic 0. Given a linearization L of the action we can define the open set, X^s, of L-stable points and the open set, X^{ss}, of L-semistable points. Mumford's GIT produces quotients X^s/G and X^{ss}/G of these open sets. The former quotient has mild (finite quotient) singularities but is not in general proper. The latter quotient is projective and contains X^s/G as an open set, but in general has worse singularities.

When the action of G has generically finite stabilizers the GIT quotient X^s/G is the *coarse moduli space* of the Deligne-Mumford (DM) quotient stack $[X^s/G]$. Likewise, the quotient X^{ss} is the *good moduli space* in the sense [Alp] of the Artin quotient stack $[X^{ss}/G]$.

There is a fully developed intersection theory on quotient stacks [EG] which assigns to any smooth quotient stack $\mathcal{X}$ a Chow ring $A^*(\mathcal{X})$. When $\mathcal{X}$ is smooth Deligne-Mumford, there is a beautiful relationship between the Chow ring of $\mathcal{X}$ and that of its coarse moduli space $\mathbf{X}$: there is a pushforward isomorphism on rational Chow groups $A^*(\mathcal{X})_\mathbb{Q} \xrightarrow{\cong} A^*(\mathbf{X})_\mathbb{Q}$. As a result, the possibly singular variety (or more generally algebraic space) $\mathbf{X}$ has an intersection product on its rational Chow groups induced from the

---

[1]Research supported in part by Simons Collaboration Fellowship 315460.
[2]Research supported in part by NSERC grant RGPIN-2015-05631.

intersection product on the Chow groups of $\mathcal{X}$. Furthermore, a fundamental result of Vistoli [Vis] states that any variety $\mathbf{X}$ with finite quotient singularities is the coarse moduli space of a smooth Deligne-Mumford stack $\mathcal{X}$, and hence by the above, $A^*(\mathbf{X})_{\mathbb{Q}}$ carries an intersection product coming from that of $A^*(\mathcal{X})_{\mathbb{Q}}$.

However, for varieties $\mathbf{X}$ with worse than finite quotient singularities, or for stacks $\mathcal{X}$ which are not Deligne-Mumford, the beautiful picture above breaks down in almost every aspect. In general, if $\mathcal{X}$ is a smooth Artin stack, the rational Chow groups $A^*(\mathcal{X})_{\mathbb{Q}}$ can be non-zero in arbitrarily high degree, so cannot be isomorphic to the rational Chow groups $A^*(\mathbf{X})_{\mathbb{Q}}$ of the good moduli space. In fact, it is not even known if the moduli map $\pi \colon \mathcal{X} \to \mathbf{X}$ induces a pushforward map $\pi_* \colon A^*(\mathcal{X})_{\mathbb{Q}} \to A^*(\mathbf{X})_{\mathbb{Q}}$. Moreover, there are good quotients by actions of reductive groups, such as the cone over a quadric hypersurface, where one can prove there is no reasonable intersection product on the Chow groups.

In this article we consider two questions about the Chow groups of Artin stacks and their good moduli spaces aimed at rectifying the above problems.

Question 1.1 Let $\mathcal{X}$ be an Artin stack with good moduli space morphism $\pi \colon \mathcal{X} \to \mathbf{X}$. Is there a geometrically meaningful pushforward map $\pi_* \colon A_*(\mathcal{X})_{\mathbb{Q}} \to A_*(\mathbf{X})_{\mathbb{Q}}$?

Since this pushforward map, if it exists, cannot be an isomorphism we are led to the following:

Question 1.2 Is there an interesting subring of the Chow ring $A^*(\mathcal{X})$ such that the restriction of π_* to this subring is injective?

Remark 1.3 (Application to Chow groups of singular varieties)
As mentioned above, there exist singular varieties $\mathbf{X}$ that are good quotients by reductive groups for which $A^*(\mathbf{X})_{\mathbb{Q}}$ carries no reasonable intersection product, e.g. the cone over a quadric hypersurface. However, if Questions 1.1 and 1.2 have affirmative answers, then we can identify an interesting *subgroup* of $A^*(\mathbf{X})_{\mathbb{Q}}$ that *does* carry an intersection product. Indeed, such a variety $\mathbf{X}$ is a good moduli space of a smooth Artin stack $\mathcal{X}$; by Question 1.1, we have a map $\pi_* \colon A^*(\mathcal{X})_{\mathbb{Q}} \to A^*(\mathbf{X})_{\mathbb{Q}}$ and our desired subgroup of $A^*(\mathbf{X})_{\mathbb{Q}}$ is the image of the subring provided by Question 1.2.

We answer Question 1.1 for the class of good moduli space morphisms that look étale locally like GIT quotients with non-empty stable loci; this is done in Section 2. We then give a conjectural answer to Question 1.2 in Section 3. We shall see that the answers to both questions are related by the concept of *strong embedding* which was the subject of the first author's TIFR Colloquium lecture.

Acknowledgements It is a pleasure to thank the organizers of the TIFR International Colloquium on K-theory for a wonderful conference and a stimulating environment.

1.1 Background on stacks and good moduli spaces

For simplicity of exposition, all stacks are assumed to be defined over an algebraically closed field of characteristic 0. We also assume that any stack is of finite type over the ground field and has affine diagonal. The following definitions are generalizations of concepts in invariant theory.

Definition 1.4 [Alp, Definition 4.1] Let $\mathcal{X}$ be an algebraic stack and let $\mathbf{X}$ be an algebraic space. We say that $\mathbf{X}$ is a *good moduli space of* $\mathcal{X}$ if there is a morphism $\pi \colon \mathcal{X} \to \mathbf{X}$ such that

1. π is *cohomologically affine* meaning that the pushforward functor π_* on the category of quasi-coherent $\mathcal{O}_{\mathcal{X}}$-modules is exact.

2. π is *Stein* meaning that the natural map $\mathcal{O}_{\mathbf{X}} \to \pi_*\mathcal{O}_{\mathcal{X}}$ is an isomorphism.

Remark 1.5 By [Alp, Theorem 6.6], a good moduli space morphism $\pi \colon \mathcal{X} \to \mathbf{X}$ is the universal morphism from $\mathcal{X}$ to an algebraic space. That is, if $\mathbf{Z}$ is an algebraic space then any morphism $\mathcal{X} \to \mathbf{Z}$ factors through a morphism $\mathbf{X} \to \mathbf{Z}$. Consequently $\mathbf{X}$ is unique up to unique isomorphism, so we will refer to $\mathbf{X}$ as *the* good moduli space of $\mathcal{X}$.

Remark 1.6 If $\mathcal{X} = [X/G]$ where G is a linearly reductive algebraic group then the statement that $\mathbf{X}$ is a good moduli space for $\mathcal{X}$ is equivalent to the statement that $\mathbf{X}$ is the good quotient of X by G.

Definition 1.7 [ER] Let $\mathcal{X}$ be an Artin stack with good moduli space $\mathbf{X}$ and let $\pi \colon \mathcal{X} \to \mathbf{X}$ be the good moduli space morphism. We say that a closed point of $\mathcal{X}$ is *stable* if $\pi^{-1}(\pi(x)) = x$ under the induced map of topological spaces $|\mathcal{X}| \to |\mathbf{X}|$. A point x of $\mathcal{X}$ is *properly stable* if it is stable and the stabilizer of x is finite.

We say $\mathcal{X}$ is stable (resp. properly stable) if there is a good moduli space $\mathcal{X} \xrightarrow{\pi} \mathbf{X}$ and the the set of stable (resp. properly stable) points is non-empty.

Remark 1.8 This definition is modeled on GIT. If G is a linearly reductive group and X^{ss} is the set of semistable points for a linearization of the action of G on a projective variety X then a (properly) stable point of $[X^{ss}/G]$ corresponds to a (properly) stable orbit in the sense of GIT. The stack $[X^{ss}/G]$ is stable if and only if $X^s \neq \emptyset$. Likewise $[X^{ss}/G]$ is properly stable if and only if $X^{ps} \neq \emptyset$. As is the case for GIT quotients, the set of stable (resp. properly stable points) is open [ER].

We denote by $\mathcal{X}^s$ (resp. $\mathcal{X}^{ps}$) the open substack of $\mathcal{X}$ of stable (resp. properly stable) points. The stack $\mathcal{X}^{ps}$ is the maximal Deligne-Mumford substack of $\mathcal{X}$ which is saturated with respect to the good moduli space morphism $\pi \colon \mathcal{X} \to \mathbf{X}$; recall that a substack $\mathcal{Z} \subseteq \mathcal{X}$ is called saturated if there is a subspace $Z \subseteq \mathbf{X}$ such that $\mathcal{Z} = Z \times_X \mathcal{X}$. In particular, a stack $\mathcal{X}$ with good moduli space $\mathbf{X}$ is properly stable if and only if it contains a non-empty *saturated* Deligne-Mumford open substack.

Example 1.9 Consider the action of $\mathbb{G}_m$ on $\mathbb{A}^1 = \operatorname{Spec} k[x]$ given by $\lambda \cdot x = \lambda x$. The quotient stack $\mathcal{X} = [\mathbb{A}^1/\mathbb{G}_m]$ has good moduli space $\operatorname{Spec} k[x]^{\mathbb{G}_m} = \operatorname{Spec} k$. Then $\mathbb{A}^1 \smallsetminus 0$, is the maximal open set on which $\mathbb{G}_m$ acts with finite (in fact trivial) stabilizers. Hence $\mathcal{X}$ has a maximal open DM substack $\mathcal{U} = [(\mathbb{A}^1 \smallsetminus 0)/\mathbb{G}_m]$, which is a point. However, this open substack is not saturated with respect to the good moduli space morphism $\mathcal{X} \to \operatorname{Spec} k$. Indeed, this action of $\mathbb{G}_m$ on $\mathbb{A}^1$ has no stable or properly stable points and $\mathcal{X}$ is not a stable stack.

1.2 Background on equivariant Chow groups and Chow groups of stacks

1.2.1 Equivariant Chow groups

If X is an equidimensional scheme or algebraic space, we use the notation $A^k(X)$ to denote the Chow group of codimension-k cycles modulo rational equivalence. The total Chow group $A^*(X)$ is the direct sum $\oplus_{k=0}^{\dim X} A^k(X)$. When X is smooth the intersection product makes $A^*(X)$ into a graded ring.

The definition of equivariant Chow groups is modeled on the Borel construction in equivariant cohomology. If a linear algebraic group G acts on X then the k-th equivariant Chow group $A^k(X)$ is defined to be $A^k(X_G)$ where where X_G is any quotient of the form $(X \times U)/G$ where U is an open set in a representation $\mathbf{V}$ of G such that G acts freely on U and $\mathbf{V} \smallsetminus U$ has codimension more than i. In [EG] it is shown that such pairs $(U, \mathbf{V})$ exist for any algebraic group and that the definition of $A_G^k(X)$ is independent of the choice of U and $\mathbf{V}$.

Note that, since representations can have arbitrarily high dimension, $A_G^k(X)$ can be non-zero in arbitrarily high degree. Thus the total equivariant Chow group $A_G^*(X)$ is the *infinite* direct sum $\oplus_{k=0}^{\infty} A_G^k(X)$. An equivariant k-cycle need not be supported on X, but only on $X \times \mathbf{V}$ where $\mathbf{V}$ is a representation of G.

Because equivariant Chow groups are defined as Chow groups of schemes (or more generally algebraic spaces) they enjoy all of the functoriality of ordinary Chow groups. In particular, if X is smooth then pullback along the diagonal defines an intersection product on the total Chow group $A_G^*(X)$.

1.2.2 Chow groups of stacks

Gillet [Gil] and Vistoli [Vis] defined Chow groups of Deligne-Mumford stacks in an analogous way to Chow groups of schemes. Namely, they considered the group generated by the classes of integral closed substacks modulo rational equivalences. With rational coefficients this theory has many desired properties such as an intersection product on the rational Chow groups of smooth stacks. Kresch [Kre] generalized their work by defining integral Chow groups for Artin stacks $\mathcal{X}$ with quasi-affine diagonal. When $\mathcal{X} = [X/G]$ is a quotient stack, Kresch's Chow groups $A^*(\mathcal{X})$ agree with the equivariant Chow groups $A_G^*(X)$.

When $\mathcal{X}$ is Deligne-Mumford with coarse space $\mathbf{X}$ the proper pushforward $\pi_* \colon A^*(\mathcal{X})_{\mathbb{Q}} \to A^*(\mathbf{X})_{\mathbb{Q}}$ is an isomorphism [Vis, EG]. The pushforward is defined on cycles by the formula $[\mathcal{Z}] \mapsto e_{\mathcal{Z}}^{-1}[\pi(\mathcal{Z})]$ where $e_{\mathcal{Z}}$ is the generic order of the stabilizer group along $\mathcal{Z}$. In particular this means that if $\mathcal{X} = [X/G]$ is a Deligne-Mumford stack then every equivariant Chow class can be represented by a G-invariant cycle on X (as opposed to $X \times \mathbf{V}$ where $\mathbf{V}$ is a representation of G). Consequently $A^k(\mathcal{X})_{\mathbb{Q}} = 0$ for $k > \dim \mathcal{X}$.

If $\mathcal{X}$ is not Deligne-Mumford then $A^k(\mathcal{X})_{\mathbb{Q}}$ will be non-zero in arbitrarily high degree, so if $\pi \colon \mathcal{X} \to \mathbf{X}$ is a good moduli space morphism we cannot expect $A^*(\mathcal{X})_{\mathbb{Q}}$ to equal $A^*(\mathbf{X})_{\mathbb{Q}}$.

1.3 Strong lci morphisms of stacks with good moduli spaces

We now introduce the key concepts of strong embeddings and strong lci morphisms.

Definition 1.10 Let $\mathcal{X}$ be an Artin stack with good moduli space $\pi \colon \mathcal{X} \to \mathbf{X}$. A closed embedding $\mathcal{Y} \to \mathcal{X}$ is *strong* if $\mathcal{Y}$ saturated with respect to the morphism π, i.e. there exists an embedding $Y \to X$ such that

$\mathcal{Y} = Y \times_X \mathcal{X}$. A regular embedding which is strong will be called a *strong regular embedding*.

Remark 1.11 In [Edi] the first author considered the notion of strong regular embeddings of tame[3] stacks.

Remark 1.12 Theorem 2.9 of [AHR] states that the good moduli space morphism $\pi\colon \mathcal{X} \to \mathbf{X}$ looks étale locally like the morphism $[\operatorname{Spec} A/G] \to \operatorname{Spec} A^G$ where G is a linearly reductive group acting on a finitely generated k-algebra A. The condition that a closed embedding $\mathcal{Y} \to \mathcal{X}$ is strong is equivalent to the assertion that the local ideals I of $\mathcal{Y}$ satisfy $I^G A = I$. This is a particularly useful criterion since it tells us that being strong embedding is local in the étale topology.

Strong regular embeddings $\mathcal{Y} \to \mathcal{X}$ are characterized by a number of equivalent properties including

1. The morphism of good moduli spaces induced by the closed embedding $\mathcal{Y} \to \mathcal{X}$ is a regular embedding and the diagram

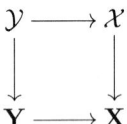

    is cartesian.

2. The stabilizer G_y of any point y of $\mathcal{Y}$ acts trivially on the fiber of the normal bundle $\mathcal{N}_{\mathcal{Y}/\mathcal{X},y}$.

These facts follows from the proof of [Edi, Theorem 2.2] for tame stacks since the proof only uses the fact a tame stack is étale locally the quotient of an affine scheme by a linearly reductive group.

We next extend the notion of strong regular embedding to that of a strong lci morphism.

Definition 1.13 A morphism of $\mathcal{Y} \to \mathcal{X}$ is a *strong lci morphism* if it factors as $\mathcal{Y} \to \mathbb{P}(\mathcal{E}) \to \mathcal{X}$ where $\mathcal{Y} \to \mathbb{P}(\mathcal{E})$ is a strong regular embedding and $\mathcal{E}$ is a vector bundle on $\mathcal{X}$ such that the stabilizer of any point x of $\mathcal{X}$ acts trivially on the fiber $\mathcal{E}_x$.

---

[3]A stack is tame if the stabilizer of every closed point is finite and linearly reductive. In characteristic 0 a stack is tame if and only if it is Deligne-Mumford.

Proposition 1.14 *If* $f: \mathcal{Y} \to \mathcal{X}$ *is a strong lci morphism and* **X** *is a good moduli space for* $\mathcal{X}$ *then* $\mathcal{Y}$ *has a good moduli space* **Y**. *Moreover, the induced morphism of good moduli spaces* **Y** → **X** *is lci and the diagram*

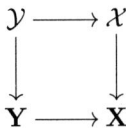

is cartesian.

Proof By assumption, the map $\mathcal{Y} \to \mathcal{X}$ has a factorization $\mathcal{Y} \to \mathbb{P}(\mathcal{E}) \to \mathcal{X}$ as in Definition 1.13. If G_x acts trivially on the fiber $\mathcal{E}_x$ for all points x of $\mathcal{X}$, then $\mathcal{E} = \pi^*\mathbf{E}$ where **E** is a vector bundle on **X**. It follows from [Alp, Proposition 4.7] that $\mathbb{P}(\mathbf{E})$ is the good moduli space of $\mathbb{P}(\mathcal{E})$. Since $\mathcal{Y}$ is strongly embedded in $\mathbb{P}(\mathbf{E})$ the proposition follows from the corresponding statement for strong regular embeddings of tame stacks [Edi, Theorem 2.2]. □

2 Pushforwards in Chow groups: an answer to Question 1.1

The goal of this section is to prove the following theorem.

Theorem 2.1 *Let* $\mathcal{X}$ *be an irreducible properly stable smooth Artin stack with good moduli space morphism* $\pi: \mathcal{X} \to \mathbf{X}$. *Then there is a pushforward map* $\pi_*: A^*(\mathcal{X})_\mathbb{Q} \to A^*(\mathbf{X})_\mathbb{Q}$ *such that*

(1) *If* x *is a closed point with finite stabilizer group* G_x, *then* $\pi_*[x] = |G_x|^{-1}[\pi(x)]$.

(2) *More generally, if* $\mathcal{Z}$ *is an irreducible closed substack of* $\mathcal{X}$ *which is contained in the maximal saturated DM substack* $\mathcal{X}^{ps}$ *then,* $\pi_*[\mathcal{Z}] = e_{\mathcal{Z}}^{-1}[\pi(\mathcal{Z})]$ *where* $e_{\mathcal{Z}}$ *is the generic order of the stabilizer group along* $\mathcal{Z}$.

(3) *The pushforward commutes with strong proper lci morphisms. Precisely, if* $f: \mathcal{Y} \to \mathcal{X}$ *is a strong lci morphism of smooth properly stable Artin stacks[4] and* $\bar{f}: \mathbf{Y} \to \mathbf{X}$ *is the induced map on good moduli spaces, then the diagram of Chow groups of stacks and good moduli*

---

[4]If $\mathcal{X}$ is properly stable then $\mathcal{Y}$ is automatically properly stable as well.

spaces commutes

$$
\begin{array}{ccc}
A^*(\mathcal{Y})_{\mathbb{Q}} & \xrightarrow{\ f_* \ } & A^*(\mathcal{X})_{\mathbb{Q}} \\
\downarrow{\scriptstyle \pi_{\mathcal{Y},*}} & & \downarrow{\scriptstyle \pi_{\mathcal{X},*}} \\
A^*(\mathbf{Y})_{\mathbb{Q}} & \xrightarrow{\ \overline{f}_* \ } & A^*(\mathbf{X})_{\mathbb{Q}}.
\end{array}
$$

2.1   The example of [EGS]

An obvious question is whether the functoriality property (3) of Theorem 2.1 holds for arbitrary lci morphisms of Artin stacks, as opposed to strong lci morphisms. In [EGS, Theorem 1], we showed that any choice of pushforward map that commutes with all regular embeddings must, in fact, be the 0 map. Thus, we cannot expect the functoriality property Theorem 2.1 (3) to extend to arbitrary lci morphisms, let alone arbitrary regular embeddings. This shows the importance of the condition that the lci morphisms be strong.

Let us briefly recall the example of [EGS]. Consider the action of $\mathbb{G}_m^3$ on $\mathbb{A}^5$ with weight matrix

$$
\begin{pmatrix}
1 & 0 & 0 & 1 & 1 \\
0 & 1 & 0 & 1 & 1 \\
0 & 0 & 1 & 0 & 1
\end{pmatrix}
$$

so that $(s,t,u) \in \mathbb{G}_m^3$ acts by $(s,t,u) \cdot (x,y,z,w,v) = (sx, ty, uz, stw, stuv)$. The quotient of the open set $X = \mathbb{A}^5 \setminus V(xyz, zw, v)$ is the projective plane $\mathrm{Proj}\, k[xyz, zw, v]$. Hence $\mathbb{P}^2$ is the good moduli space of the quotient stack $[X/\mathbb{G}_m^3]$. In [EGS, Theorem 1] we showed that any choice of pushforward map for Artin stacks with good moduli spaces that commutes with the inclusion of the smooth (and hence regularly embedded) substacks $[V(x)/\mathbb{G}_m^3]$ and $[V(y)/\mathbb{G}_m^3]$ must be the 0 map.

Note however, that neither $[V(x)/\mathbb{G}_m^3]$ nor $[V(y)/\mathbb{G}_m^3]$ is *strongly* embedded. To see this, notice that $D(v)$ is an affine open subset of X that is $\mathbb{G}_m^3$-invariant, and that $V(x) \cap D(v)$ is not defined by a $\mathbb{G}_m^3$-invariant function. Indeed the substack $[V(x)/\mathbb{G}_m^3]$ contains no stable or properly stable points, since every point with vanishing x-coordinate is in the saturation of the locus $V(x,y)$ which has positive dimensional stabilizer at all points. Similarly, $[V(y)/\mathbb{G}_m^3]$ is not strongly embedded.

By contrast, $[V(v)/\mathbb{G}_m^3]$ is a strong substack. To see this note that X is covered by the 3 affine open $\mathbb{G}_m^3$-invariant subsets $D(v)$, $D(xyz)$, and $D(zw)$. On each of these affine patches, $V(v)$ is defined by an invariant function: $V(v) \cap D(v) = \emptyset = V(1)$, $V(v) \cap D(xyz) = V(v/xyz)$, and $V(v) \cap D(zw) = V(v/zw)$.

2.2 Sketch of the proof of Theorem 2.1

The proof of Theorem 2.1 uses the following result of [ER] which is a generalization of an earlier result for toric stacks in [EM1].

Theorem 2.2 *Let $\mathcal{X}$ be an irreducible properly stable smooth Artin stack with good moduli space $\mathbf{X}$. There is a canonical sequence of birational morphisms of smooth Artin stacks $\mathcal{X}_n \to \mathcal{X}_{n-1} \to \cdots \to \mathcal{X}_0 = \mathcal{X}$ with the following properties.*

1. *The stack $\mathcal{X}_n$ is Deligne-Mumford.*

2. *Each $\mathcal{X}_k$ is smooth, irreducible, properly stable, and admits a good moduli space morphism $\mathcal{X}_k \to \mathbf{X}_k$ with $\mathbf{X}_k$ an algebraic space.*

3. *The morphism $\mathcal{X}_{k+1} \to \mathcal{X}_k$ is an isomorphism over the maximal saturated DM substack $\mathcal{X}_k^{ps} \subset \mathcal{X}$ and it is an open immersion over the complement of a proper closed substack of $\mathcal{C}_k \subset \mathcal{X}_k$.*

4. *The morphism $\mathcal{X}_{k+1} \to \mathcal{X}_k$ induces a projective birational morphism of good moduli spaces $\mathbf{X}_{k+1} \to \mathbf{X}_k$.*

5. *The maximum dimension of the stabilizers of the points of $\mathcal{X}_{k+1}$ is strictly smaller than that of $\mathcal{X}_k$.*

Remark 2.3 The birational morphisms $\mathcal{X}_{k+1} \to \mathcal{X}_k$ are called *Reichstein transformations*. They are defined as follows. Let $\mathcal{C}_k$ be the substack of $\mathcal{X}_k$ parametrizing points with maximal dimensional stabilizer. This locus is necessarily a closed smooth substack of $\mathcal{X}$. Let $\mathcal{S}_k$ be the saturation of $\mathcal{C}_k$ with respect to the good moduli space morphism $\mathcal{X}_k \to \mathbf{X}_k$. Then $\mathcal{X}_{k+1}$ is defined as the complement of the strict transform of $\mathcal{S}_k$ in the blow up of $\mathcal{X}_k$ along $\mathcal{C}_k$.

Proof (Theorem 2.1) The map $\pi_* \colon A^*(\mathcal{X})_{\mathbb{Q}} \to A^*(\mathbf{X})_{\mathbb{Q}}$ is defined as follows. The morphisms $\mathcal{X}_{k+1} \to \mathcal{X}_k$ are representable morphisms of smooth stacks. In particular they are lci so, the composite morphism $f \colon \mathcal{X}_n \to \mathcal{X}$ is as well. Hence there is a pullback of Chow groups $f^* \colon A^*(\mathcal{X}_n) \to A^*(\mathcal{X})$. On the other hand the morphisms of good moduli spaces $\mathbf{X}_{k+1} \to \mathbf{X}_k$ are birational and projective, so the composite map $\overline{f} \colon \mathbf{X}_n \to \mathbf{X}$ is also birational and projective. Thus, there is a pushforward $\overline{f}_* \colon A^*(\mathbf{X}_n) \to A^*(\mathbf{X})$. Since $\mathcal{X}_n$ is a DM stack, we also have a pushforward of Chow groups $\pi_{n*} \colon A^*(\mathcal{X}_n)_{\mathbb{Q}} \to A^*(\mathbf{X})_{\mathbb{Q}}$. We then define $\pi_* \colon A^*(\mathcal{X})_{\mathbb{Q}} \to A^*(\mathbf{X})_{\mathbb{Q}}$ as the composite $\overline{f}_* \circ \pi_{n*} \circ f^*$.

Since the maps $\mathcal{X}_{k+1} \to \mathcal{X}_k$ are isomorphisms over the properly stable locus in $\mathcal{X}$, if $\mathcal{Z} \subset \mathcal{X}$ is a closed substack contained in the properly stable locus then $\pi_*[\mathcal{Z}]$ can be identified with $\pi_{n*}[\mathcal{Z}] = e_{\mathcal{Z}}^{-1}[\pi_n(\mathcal{Z})] = e_{\mathcal{Z}}^{-1}[\pi(\mathcal{Z})]$.

This proves statement 2 of the theorem. Finally the last statement follows because the construction of [ER] is functorial for strong lci morphisms.

$\square$

Remark 2.4 A similar construction for a class of toric stacks was given in [EM2].

3   Toward a theory of strong Chow groups: a conjectural answer to Question 1.2

As discussed in Section 2.1, the example of [EGS] shows that there *does not exist* a pushforward map $\pi_* : A^*(\mathcal{X}) \to A^*(X)$ that is functorial for *all* regular embeddings (let alone all lci morphisms). So, rather than focusing on defining a pushforward from $A^*(\mathcal{X})$, we focus on the subgroup generated by strong cycles.

Definition 3.1 Let $\pi : \mathcal{X} \to \mathbf{X}$ be a good moduli space map from an irreducible properly stable Artin stack. Then the *strong relative Chow group* $A_{st}^k(\mathcal{X}/\mathbf{X})$ is the subgroup generated by $\sum c_i[\mathcal{Z}_i]$ with $\mathcal{Z}_i$ strong.

We now state a series of conjectures concerning $A_{st}^k(\mathcal{X}/\mathbf{X})$. Our first gives a conjectural answer to Question 1.1.

Conjecture 3.2 *If $\mathcal{X}$ is a properly stable smooth Artin stack and $\pi \colon \mathcal{X} \to \mathbf{X}$ its good moduli space, then the assignment*
$[\mathcal{Z}] \mapsto e_{\mathcal{Z}}^{-1}[\pi(\mathcal{Z})]$ *for strong cycles $\mathcal{Z}$ respects rational equivalence, so we obtain a pushforward map $\pi_{st,*} \colon A_{st}^k(\mathcal{X}/\mathbf{X})_{\mathbb{Q}} \to A^k(\mathbf{X})_{\mathbb{Q}}$.*

We also state a stronger form of Conjecture 3.2:

Conjecture 3.3 *If $\mathcal{X}$ is a properly stable smooth Artin stack and $\pi \colon \mathcal{X} \to \mathbf{X}$ its good moduli space, then the pushforward map $\pi_* \colon A^*(\mathcal{X})_{\mathbb{Q}} \to A^*(\mathbf{X})_{\mathbb{Q}}$ defined in Theorem 2.1 satisfies $\pi_*[\mathcal{Z}] := e_{\mathcal{Z}}^{-1}[\pi(\mathcal{Z})]$ for strong cycles $\mathcal{Z}$.*

Remark 3.4 Note that by definition, any strong 0-cycle must be contained in the stable locus so Theorem 2.1 (2) implies that Conjecture 3.3 holds for strong 0-cycles.

Remark 3.5 Conjecture 3.3 is also true for smooth strong substacks $\mathcal{Z} \subseteq \mathcal{X}$. Indeed, since $\mathcal{Z}$ is smooth, the inclusion map to $\mathcal{X}$ is a strong regular embedding and so by Theorem 2.1 (3), we reduce to the case that

$\mathcal{Z} = \mathcal{X}$ is the fundamental class. Since all maps $\mathcal{X}_{i+1} \to \mathcal{X}_i$ in Theorem 2.2 are birational, we therefore have $\pi_*[\mathcal{Z}] = e_{\mathcal{Z}}^{-1}[\mathcal{Z}]$.

We now turn to Question 1.2 where our main conjectures are as follows.

Conjecture 3.6 *If $\mathcal{X}$ is a properly stable smooth Artin stack with good moduli space $\mathbf{X}$, then $\oplus_{k=0}^{\dim \mathcal{X}} A_{st}^k(\mathcal{X}/\mathbf{X})$ is a subring of $A^*(\mathcal{X})$ under the intersection product.*

Conjecture 3.7 *Assuming Conjectures 3.2 and 3.6, then there is a subring $A_{inj}^*(\mathcal{X}/\mathbf{X})_{\mathbb{Q}}$ of $A_{st}^*(\mathcal{X}/\mathbf{X})_{\mathbb{Q}}$ which contains the subalgebra generated by $A_{st}^1(\mathcal{X}/\mathbf{X})_{\mathbb{Q}}$ and has the property that the pushforward $\pi_{st,*}$ is injective on $A_{inj}^*(\mathcal{X}/\mathbf{X})_{\mathbb{Q}}$. Moreover, if $\mathbf{X}$ has only quotient singularities then $\pi_{st,*}$ is bijective on all of $A_{st}^*(\mathcal{X}/\mathbf{X})_{\mathbb{Q}}$.*

Remark 3.8 (Conjectural answer to Question 1.2) Notice that if Conjecture 3.7 holds then it provides an answer to Question 1.2: the image of $A_{inj}^*(\mathcal{X}/\mathbf{X})$ under $\pi_{st,*}$ yields a non-trivial subgroup of $A^*(\mathbf{X})$ with an intersection product.

Remark 3.9 We will see in Example 3.12 that the assignment $[\mathcal{Z}] \mapsto e_{\mathcal{Z}}^{-1}[\mathcal{Z}]$ is not injective on *all* strong cycles. By analogy with the DM case, one might hope that it is possible to associate to every scheme $\mathbf{X}$ with reductive quotient singularities (i.e. those étale locally of the form V/G where V is a representation of a linearly reductive algebraic group) a *canonical* properly stable Artin stack $\mathcal{X}$ whose good moduli space is $\mathbf{X}$. If this were the case, then assuming Conjecture 3.7, for every such scheme $\mathbf{X}$, its Chow groups $A^*(\mathbf{X})$ are equipped with a canonical subring, namely $A_{inj}^*(\mathcal{X}/\mathbf{X})$. Moreover, since the stack $\mathcal{X}$ is canonical the full strong Chow ring $A_{st}^*(\mathcal{X}/\mathbf{X})$ is also an invariant of the scheme $\mathbf{X}$.

3.1 Examples illustrating the conjectures

Example 3.10 Let $G = \mathbb{G}_m^2$ act on $\mathbb{A}^4$ with weights $\left(\begin{smallmatrix} 1 & 0 & 1 & 1 \\ 0 & 1 & 0 & 1 \end{smallmatrix}\right)$. We let the coordinate functions on $\mathbb{A}^4$ be (x_1, x_2, x_3, z). Let $X = \mathbb{A}^4 \setminus V(x_1 x_2, x_2 x_3, z)$ and $\mathcal{X} = [X/G]$. The map $X \to \mathbb{P}^2$ given by $(x_1, x_2, x_3, z) \mapsto (x_1 x_2 : x_2 x_3 : z)$ is a good quotient.[5] Hence $\mathcal{X} \to \mathbb{P}^2$ is a good moduli space morphism. The maximal saturated DM substack is the quotient $[X^s/G]$ and $X^s = X \setminus V(x_1 x_2, x_2 x_3)$ since $V(x_1 x_2, x_2 x_3)$ is the saturation of the

---

[5]This follows because $\mathbb{A}^4 \setminus V(x_1 x_2, x_2 x_3, x_4)$ is the $(1,1)$-semi-stable locus for the action of G on $\mathbb{A}^4$ where $(1,1)$ is the character $(s,t) \mapsto st$.

locus in $V(x_1, x_2, x_3)$ where G acts with positive dimensional stabilizer. We verify all four of the above conjectures for this example.

If we denote by s, t the first Chern classes of the projection characters of $\mathbb{G}_m^2$ then

$$A^*(\mathcal{X}) = A^*_{\mathbb{G}_m^2}(X) = \mathbb{Z}[s, t] / \left(t(s + t), s^2(s + t) \right).$$

We next compute the strong Chow groups of $\mathcal{X}$. First note that since $\mathcal{X}$ is non-singular, any Weil divisor is Cartier. Now if $[D]$ is the support of a strong Cartier divisor then $D = V(f)$ where f is a function which is invariant on each G-invariant affine open in X. Such a function must necessarily be a homogeneous polynomial in the semi-invariants $(x_1 x_2, x_2 x_3, z)$ and the Chow class of such a polynomial is a multiple of $s + t$. Moreover, the divisor $V(z)$, which has Chow class $s + t$, is strong. Hence $A^1_{st}(\mathcal{X}/\mathbf{X}) \simeq \mathbb{Z}$ generated by $s + t$.

Next, $A^2_{st}(\mathcal{X}/\mathbf{X})$ is generated by $[V(x_1, z)]$ which is the class of a non-stacky closed point in $\mathcal{X}$; its Chow class is $s(s + t)$. Since $t(s + t) = 0$ in $A^*(\mathcal{X})$ we see that $[V(x_1, z)] = [V(z)]^2$. The relation $(s + t)^3 = 0$ implies $A^*_{st}(\mathcal{X}/\mathbf{X})$ is closed under multiplication, verifying Conjecture 3.6. In fact, we have shown

$$A^*_{st}(\mathcal{X}/\mathbf{X}) = \mathbb{Z}[s + t]/(s + t)^3$$

as rings.

Since $V(z)$ is contained in the stable locus, $\pi_*[V(z)] = [\pi(V(z))]$ by Theorem 2.1 (2). Now any irreducible strong divisor is of the form $V(f)$ where f is an irreducible homogeneous polynomial of degree d in the semi-invariants $(x_1 x_2, x_2 x_3, z)$. Thus $[V(f)] = d[V(z)]$ so $\pi_*[V(f)] = dh$, where h is the hyperplane class on $\mathbb{P}^2$. On the other hand, $\pi(V(f)) = V(f(A, B, C))$ where A, B, C are the projective coordinates on the quotient $\mathbb{P}^2$. Thus $[\pi(V(f))] = dh$ as well. Hence Conjecture 3.3 holds for strong divisors. Moreover, any strong 0-cycle (i.e. an element of $A^2_{st}(\mathcal{X}/\mathbf{X})$) is contained in the stable locus so Conjecture 3.3 holds for all strong cycles, and Conjecture 3.2 follows as well.

Finally, as shown above, $A^*_{st}(\mathcal{X}/\mathbf{X}) = \mathbb{Z}[s + t]/(s + t)^3$ and moreover, $\pi_*\colon A^*_{st}(\mathcal{X}/\mathbf{X}) \to A^*(\mathbb{P}^2) = \mathbb{Z}[h]/h^3$ sends $s + t$ to h. Thus, π_* is an isomorphism on $A^*_{st}(\mathcal{X}/\mathbf{X})$, verifying Conjecture 3.7.

Remark 3.11 (Generically strong cycles) Note that $[V(x_1)] + [V(x_2)] = s + t$. Since the image of $V(x_2)$ is the point $(0 : 0 : 1)$ one might expect that $\pi_*[V(x_2)] = 0$ and thus $\pi_*[V(x_1)] = h$ is the hyperplane class.

In this example, the cycle $V(x_1)$ is not strong, but the "extra" component in its saturation, $V(x_2)$, does not dominate $\pi(V(x_1))$. We call a cycle $\mathcal{Z} \subset \mathcal{X}$ *generically strong* if any extra components of the saturation of $\mathcal{Z}$ do not dominate $\pi(\mathcal{Z})$. We conjecture that $\pi_*([\mathcal{Z}]) = \pi(\mathcal{Z})$ for all such cycles, which would strengthen Theorem 2.1 (2).

Example 3.12 (Good quotient with worse than quotient singularities) Consider the action of $\mathbb{G}_m^2$ on $\mathbb{A}^5$ with weight matrix

$$\begin{pmatrix} 1 & 0 & 1 & 0 & 1 \\ 0 & 1 & 0 & 1 & 1 \end{pmatrix}.$$

We denote the coordinates as (x_1, x_2, x_3, x_4, z) and let

$$X = \mathbb{A}^5 \smallsetminus V(x_1 x_2, x_1 x_4, x_2 x_3, x_3 x_4, v).$$

The good quotient X/G is Proj $k[x_1 x_2, x_1 x_4, x_2 x_3, x_3 x_4, v]$. This is the projective closure in $\mathbb{P}^4$ of the cone over the quadric hypersurface in $\mathbb{P}^3$; its singularity is not a quotient singularity. We again verify all four of the conjectures for this example modulo an assumption about the structure of $A_{st}^2(\mathcal{X}/\mathbf{X})$.

If we denote by s, t the first Chern classes of the projection characters of $\mathbb{G}_m^2$ then

$$A^*(\mathcal{X}) = A_{\mathbb{G}_m^2}^*(X) = \mathbb{Z}[s, t] / \left(s^2(s+t), t^2(s+t) \right).$$

We now (conjecturally) compute the strong relative Chow groups of $\mathcal{X}$. First note that since $\mathcal{X}$ is non-singular, any Weil divisor is Cartier. Now if $[D]$ is the support of a strong Cartier divisor then $D = V(f)$ where f is a function which is invariant on each G-invariant affine open in X. Such a function must necessarily be a homogeneous polynomial in the semi-invariants $(x_1 x_2, x_1 x_4, x_2 x_3, x_3 x_4, v)$ and the Chow class of any such polynomial is a multiple of $s + t$. Moreover, the divisor $V(v)$, which has Chow class $s + t$, is strong. Hence $A_{st}^1(\mathcal{X}/\mathbf{X}) \simeq \mathbb{Z}$.

At the other extreme, $A_{st}^3(\mathcal{X}/\mathbf{X})$ is generated by $[V(x_1, x_2, v)]$ which is the class of a non-stacky closed point in $\mathcal{X}$. Its Chow class is $st(s+t)$.

For $A_{st}^2(\mathcal{X}/\mathbf{X})$ we only have a conjectural description. If we assume that the strong Chow group is generated by classes of substacks which are regularly embedded then $A_{st}^2(\mathcal{X}/\mathbf{X})$ is generated by the equivariant classes of $V(x_1, z)$ and $V(x_2, z)$. (It is easy to check that these are both strong cycles.) Since $[V(x_1, z)] = s(s+t)$ and $[V(x_2, z)] = t(s+t)$, we have $A_{st}^2(\mathcal{X}/\mathbf{X}) = \mathbb{Z}[V(x_1, z)] + \mathbb{Z}[V(x_2, z)]$.

If we let α, β, and γ be classes of $s+t$, $t(s+t)$, and $s(s+t)$ respectively, then $A_{st}^*(\mathcal{X}/\mathbf{X})$ is closed under multiplication in $A^*(\mathcal{X})$ and is equal to the ring

$$\mathbb{Z}[\alpha, \beta, \gamma] / \left(\alpha^2 - (\beta + \gamma), \alpha\beta - \alpha\gamma, \beta\gamma, \beta^2 \right).$$

Thus, modulo our assumption that $A^2_{st}(\mathcal{X}/\mathbf{X})$ is generated by regularly embedded substacks we see that Conjecture 3.6 holds.

We next verify Conjecture 3.3, and hence 3.2. We have already shown that every strong divisor is rationally equivalent to an integer multiple of $V(v)$. Since $V(v)$ misses the semistable locus, $\pi_*[V(v)] = [\pi(V(v))]$, so Conjecture 3.3 holds for all strong divisors. By Remark 3.4, the conjecture also holds for strong 0-cycles, so we need only consider strong curves. This follows from the observation that if we identify X/G as the projective variety Proj $k[A, B, C, D, V]/(AD - BC)$, then $\pi_*[V(x_1, v)]$ is the class of $A = B = V = 0$ and $\pi_*[V(x_2, v)]$ is the class of $A = C = V = 0$.

Lastly, we turn to Conjecture 3.7. Interestingly, π_* is *not* injective on strong cycles since the pushforwards of $\beta = [V(x_1, v)]$ and $\gamma = [V(x_2, v)]$ are both equal to the class of a line through the vertex of the cone of $\mathbf{X}$. This shows that in this example, we cannot take $A^*_{inj}(\mathcal{X}/\mathbf{X})$ equal to $A^*_{st}(\mathcal{X}/\mathbf{X})$, however, we can take it to be the subring generated by α. Then π_* maps this ring injectively to the subring of $A^*(\mathbf{X})$ generated by powers of the hyperplane class.

Remark 3.13 This example shows (*cf.* Remark 3.9) that the strong relative Chow groups $A^*_{st}(\mathcal{X}/\mathbf{X})$, are a more refined geometric invariant of the good moduli space $\mathbf{X}$ since the rulings of the quadric embedded in $\mathbf{X} = \text{Proj} k[A, B, C, D, V]/(AD - BC)$ can be distinguished in $A^*_{st}(\mathcal{X}/\mathbf{X})$, as β and γ, but not in $A^*_{st}(\mathbf{X})$.

Example 3.14 (Verifying the conjectures for [EGS] example) We use the notation of Section 2.1. A look at the weight matrix shows that the action has generically one-dimensional stabilizer along the linear subspace $V(x, y)$ and two-dimensional stabilizer along the subspace $V(x, y, z, w)$. Thus the stable locus for the action of $\mathbb{G}^3_m$ is the complement of the saturation of $V(x, y)$ which is the union of the 3 coordinate planes $V(x), V(y), V(z)$; i.e. the quotient stack $[X^s/\mathbb{G}^3_m]$ is the maximal saturated DM substack of $\mathcal{X}$. Since the action of $\mathbb{G}^3_m$ on X^s is free, the good moduli space morphism restricts to an isomorphism on this open set. If A, B, C are coordinates on $\mathbb{P}^2$ corresponding to the semi-invariant functions xyz, zw, v respectively then $[X^s/\mathbb{G}^3_m]$ can be identified with the open set $\mathbb{A}^2 = \mathbb{P}^2 \smallsetminus V(A)$.

We have

$$A^*(\mathcal{X}) = \mathbb{Z}[s, t, u]/ (u(s + t + u), s(s + t)(s + t + u), t(s + t)(s + t + u))$$

where s, t, u denote the first Chern classes of the 3 projection characters $\mathbb{G}^3_m \to \mathbb{G}_m$. With this notation the coordinate hyperplanes x, y, z, w, v have Chow classes $s, t, u, s + t, s + t + u$ respectively corresponding to the weight of the action on each coordinate.

We next calculate the strong Chow groups. Any strong divisor is given by $V(f)$ where f is a homogeneous polynomial in the semi-invariant coordinates (xyz, zw, v). This implies that the class of such a divisor is a multiple of $s + t + u$. Since $[V(v)] = s + t + u$, we see $A^1_{st}(\mathcal{X}/\mathbf{X}) \simeq \mathbb{Z}$ generated by this class. Next, $A^2_{st}(\mathcal{X}/\mathbf{X})$ is generated by the class of a non-stacky point $[V(w, v)] = (s + t)(s + t + u)$. Since $u(s + t + u) = 0$ we see that $[V(w, v)] = [V(v)]^2$ in $A^2(\mathcal{X})$. It is easy to show $(s + t + u)^3 = 0$ so $[V(v)]^3 = 0$. It follows that $A^*_{st}(\mathcal{X}/\mathbf{X})$ is closed under multiplication and equals the ring $\mathbb{Z}[s + t + u]/(s + t + u)^3$, verifying Conjecture 3.6.

Furthermore, since $V(v)$ is smooth and strong, Remark 3.5 shows $\pi_*[V(v)] = [\pi(V(v))] = h$, the class of a hyperplane on $\mathbb{P}^2$. Combined with Remark 3.4, this shows Conjecture 3.3 holds for all strong cycles, and hence Conjecture 3.2 holds as well. Finally, since $\pi_*(s + t + u) = h$, notice that $\pi_* : A^*_{st}(\mathcal{X}/\mathbf{X}) \to A^*(\mathbb{P}^2) = \mathbb{Z}[h]/h^3$ is an isomorphism, verifying Conjecture 3.7.

Remark 3.15 In Example 3.14, we were able to verify all four of the conjectures without ever calculating $\pi_*(s)$ and $\pi_*(t)$. We show how one may calculate these quantities, modulo the assumption that $\pi_*[V(z)] = 0$: this seems like a reasonable assumption since $V(z)$ is in the saturation of the locus $(0 : 0 : 1)$, but it *does not* follow from any of our conjectures. Note that $[V(v)] = [V(x)] + [V(y)] + [V(z)]$ and the automorphism $\mathcal{X} \to \mathcal{X}$ which exchanges x and y is a strong regular embedding, so $\pi_*[V(x)] = \pi_*[V(y)]$. On the other hand, $\pi(V(x)) = \pi(V(y)) = V(A)$. Since $\pi_*([V(x)] + [V(y)]) = [V(A)]$ we have $\pi_*[V(x)] = \pi_*[V(y)] = 1/2[\pi(V(x))] = h/2$. Note that $V(x), V(y), V(z)$ are all contained in the complement of the stable locus.

In codimension 2 we have a similar calculation. The locus $V(w, v)$ consists of a single closed orbit whose image is the point $(1 : 0 : 0)$ in $\mathbb{P}^2$, so we know that $\pi_*([V(w, v)]) = [\pi(V(w, v))] = h^2$ which is the class of a point in $\mathbb{P}^2$. Since $[V(w, v)] = [V(x, v)] + [V(y, v)]$ we know that $\pi_*[V(x, w)] = \pi_*[V(y, w)] = h^2/2$.

References

[Alp] Jarod Alper, *Good moduli spaces for Artin stacks*, Ann. Inst. Fourier (Grenoble) **63** (2013), no. 6, 2349–2402.

[AHR] Jarod Alper, Jack Hall, and David Rydh, *A Luna étale slice theorem for algebraic stacks*, arXiv:1504.06467 (2015).

[Edi] Dan Edidin, *Strong regular embeddings of Deligne-Mumford stacks and hypertoric geometry*, Michigan Math. J. **65** (2016), no. 2, 389–412.

[EGS]   Dan Edidin, Anton Geraschenko, and Matthew Satriano, *There is no degree map for 0-cycles on Artin stacks*, Transform. Groups **18** (2013), no. 2, 385–389.

[EG]   Dan Edidin and William Graham, *Equivariant intersection theory*, Invent. Math. **131** (1998), no. 3, 595–634.

[EM1]   Dan Edidin and Yogesh Moore, *Partial desingularizations of good moduli spaces of Artin toric stacks*, Michigan Math. **61** (2012), no. 3, 451–474.

[EM2]   Dan Edidin and Yogesh More, *Integration on Artin toric stacks and Euler characteristics*, Proc. Amer. Math. Soc. **141** (2013), no. 11, 3689–3699.

[ER]   Dan Edidin and David Rydh, *Canonical reduction of stabilizers of Artin stacks*, arXiv:1710.03220.

[Gil]   Henri Gillet, *Intersection theory on algebraic stacks and Q-varieties*, Proceedings of the Luminy conference on algebraic K-theory (Luminy, 1983), vol. 34, 1984, pp. 193–240.

[Kre]   Andrew Kresch, *Cycle groups for Artin stacks*, Invent. Math. **138** (1999), no. 3, 495–536.

[Vis]   Angelo Vistoli, *Intersection theory on algebraic stacks and on their moduli spaces*, Invent. Math. **97** (1989), no. 3, 613–670.

DAN EDIDIN, DEPARTMENT OF MATHEMATICS, UNIVERSITY OF MISSOURI, COLUMBIA MO 65211
   E-mail: edidind@missouri.edu

MATTHEW SATRIANO, PURE MATHEMATICS, UNIVERSITY OF WATERLOO, 200 UNIVERSITY AVENUE WEST, WATERLOO, ONTARIO, CANADA N2L 3G1
   E-mail: msatriano@uwaterloo.ca

K-Theory

Orientation Theory in Arithmetic Geometry

Frédéric Déglise*

Abstract

This work is devoted to study orientation theory in arithmetic geometry within the motivic homotopy theory of Morel and Voevodsky. The main tool is a formulation of the absolute purity property for an *arithmetic cohomology theory*, either represented by a cartesian section of the stable homotopy category or satisfying suitable axioms. We give many examples, formulate conjectures and prove a useful property of analytical invariance. Within this axiomatic, we thoroughly develop the theory of characteristic and fundamental classes, Gysin and residue morphisms. This is used to prove Riemann-Roch formulas, in Grothendieck style for arbitrary natural transformations of cohomologies, and a new one for residue morphisms. They are applied to rational motivic cohomology and étale rational ℓ-adic cohomology, as expected by Grothendieck in [BGI71, XIV, 6.1].

Introduction

History. One of the most striking intuition of Riemann is that the natural domain of definition of abelian integrals are (branched) surfaces rather than the complex plane. Retrospectively, one is amazed that this single idea contained in seeds the modern development of both analytical and algebraic geometry, whose varieties are now studied through their sheaf of functions. In this long and deep evolution, the Riemann-Roch formula played a catalytic role, or rather that of a lighthouse.

In his 1857 masterpiece, [Rie57], Riemann studied his new complex functions, defined in modern terms on a compact Riemann surface Σ. Notably, he described the general form of these functions once we prescribed m given (simple) poles in Σ. The striking new idea is the appearance of a geometrical invariant of S that Riemann had discovered before, that we now know as the *genus p* of Σ. Riemann established that complex functions on S

---

*Partially supported by the ANR (grant No. ANR-12-BS01-0002)

with m given poles depend upon (at least) $m - p + 1$ constants (see *loc. cit.*, §5). A formula that we now read as:

$$l(D) \geq \deg(D) - p + 1, \tag{R}$$

where D is the divisor on S made by the formal sum of the m given points, with degree $\deg(D)$ equal to m, and $l(D)$ is the dimension of the space of functions f on Σ whose associated divisor is D — meaning it admits simple poles exactly at the m points of the support of D. A few years later, Roch in [Roc65] interpreted analytically the difference of the two members of (R) as the "number of linearly disjoint integrands which can vanish at the m given poles" (see [Gra98, second par. p. 802]). In modern terminology, this becomes the *Riemann-Roch formula*, which we write today as:

$$l(D) - l(K - D) = \deg(D) - p + 1 \tag{RR}$$

where K is the canonical divisor on Σ.

Looking through the glass of a century of research, one is amazed by the exceptional role that took up this simple formula. This is particularly visible in the algebraic reformulation of Riemann's ideas by Clebsch, and then Brill-Noether (Max), where one of the driving motivation was to define the genus of an algebraic curve (complex plane projective) in order to prove the Riemann-Roch formula. The same problem is addressed slightly later in the development of algebraic surfaces by M. Noether, and then by the Italian school (Castelnuovo, Enriques, Severi, ...), whose guide was to formulate the correct extension of the Riemann inequality in dimension 2 and in particular to find the good notion of genus.

But the most historically surprising application of the formula came almost eighty years after its introduction by Riemann when F.K. Schmidt extended it to the case of function fields K over a finite field and use it to prove the rationality of the Zeta function associated with K (1931). The formula followed the path opened up by the influential 1882 work of Dedekind-Weber, who developed the birational point of view initiated by Riemann by transporting his work to the purely arithmetical world of function fields over the complex numbers. The impact of Schmidt's proof on the modern formulation of algebraic geometry is of primary importance, as it lead Weil to his work on abelian varieties and most of all to the formulation of his conjectures on Zeta functions.

As said by Dieudonné in his history of algebraic geometry, the followers of Riemann split into several branches without much interactions (see [Die74], beginning of chap. VI). So while the notion of cohomology in arithmetic geometry was slowly revealing itself, the algebraic geometers working on the Riemann-Roch problem for surfaces were discovering the theory of *canonical classes*: M. Noether for surfaces (1886), Severi (1932), Segre and Todd in higher dimensions. Meanwhile, Poincaré introduced singular homology and topologists started to study *characteristic classes* of vector bundles (Stiefel and Whitney 1935, Chern 1946) without any connections with the theory of canonical classes. The unifying tool was to be the theory of sheaves invented by Leray during World War II. Only a few years after its introduction, this theory was fully developed first by Cartan and Serre for analytical varieties and secondly by Kodaira and Spencer for Kahlerian varieties. The problem of extending the Riemann-Roch formula in higher dimensions was crystallized in those years of boiling development around the notion of sheaves, in particular through the attempts of Kodaira and Serre. The first one had already solved the extension problem for Kahlerian varieties of dimension 2 (1951) and dimension 3 (1952) and linked the problem with computations of the canonical classes of Todd, while Serre had remarked its link with duality, used Thom cobordism theory to treat a special case[1] and conjectured a general form for the extended Riemann-Roch formula. Shortly after these advances, it belonged to Hirzebruch (1954, see also [Hir66]) to prove (and make precise[2]) the formula conjectured by Serre, formula that we now call the Hirzebruch-Riemann-Roch formula:

$$\chi(X, E) = \deg \left(\operatorname{ch}(E). \operatorname{Td}(E) \right) \qquad \text{(HRR)}$$

where E is a vector bundle on an analytical variety X, $\chi(X, E)$ (resp. $\operatorname{ch}(E)$, $\operatorname{Td}(E)$) its Euler characteristic (resp. Chern character, Todd class) with values in singular cohomology. The proof of Hirzebruch, rather technical, uses (and developed) the theory of characteristic classes (Chern, Todd classes,...) and makes use again of Thom cobordism theory.[3]

But the final revolution of Riemann's original problem was imagined by a single man whose ideas were to change completely our conception of it, Grothendieck. Shortly after the proof of Hirzebruch (see [BS58]), Grothendieck gave a new and meaningful interpretation of the formula, whose first practical interest was the simplicity of its proof and its validity for algebraic varieties over an arbitrary base field. The two main ideas introduced there, which had never been anticipated before, was first a relative formulation (*i.e.* for a morphism rather than a single algebraic variety) and secondly a purely cohomological interpretation of the (RR)-formula by introducing a generalized cohomology theory that would soon become famous as K-theory. From a conceptual point of view, the Grothendieck-Riemann-

---

[1]This work is unpublished but see the account of [Die74, VIII. 12.].

[2]In a letter to Kodaira and Spencer, Serre conjectured that the Euler characteristic should be expressed by some polynomial expression on Chern classes.

[3]To anticipate the content of this paper, one remarks that the universality of cobordism theory was already fully playing its role here.

Roch formula expresses the defect of functoriality of a natural transformation of cohomology theories with respect to the exceptional covariant
functoriality; in the assumptions of Borel-Serre, given a proper morphism
$f : Y \to X$ of non singular quasi-projective varieties over any field k, T_X
(resp. T_Y) being the tangent bundle of X (resp. Y), one has for any element
y, in the K-group $K(Y)$ of *virtual vector bundles* over Y,

$$\mathrm{ch}(f_*(y)). \, \mathrm{Td}(T_X) = f_*\big(\mathrm{ch}(y). \, \mathrm{Td}(T_Y)\big) \qquad \text{(GRR)}$$

where ch denotes the Chern character from K-theory to rational Chow
groups, and Td is the Todd class of a vector bundle.

During the period just described, history tells us that interactions between topology, geometry and algebraic geometry were very strong[4]. Therefore, soon after the appearance of the (GRR)-formula, Atiyah and Hirzebruch introduced topological K-theory, that they immediately understood
as a generalized cohomology theory[5] and proved the topological formulation
of the (GRR)-formula. It was soon realized that the covariant functoriality
involved in the formula should be a consequence of Poincaré duality, on
the model of the covariant functoriality discovered by Gysin in his study
of sphere bundles (1942). Consequently, a very general (GRR)-formula,
in which one considers an arbitrary natural transformation of generalized
cohomology theories, each equipped with a *complex orientation* to get the
usual theory of characteristic classes, was written by Dyer (see [Dye62]) —
and stated as a folklore theorem, only 4 years after the original formulation
of Grothendieck !

History must stop at some point. We will end it by two cornerstones
of which our work is a direct continuation. The first one is Quillen dis-
covery of the universality of Thom complex cobordism theory in terms of
formal group laws and oriented cohomology theories: [Qui69]. The second
one is Grothendieck's final extension of his Riemann-Roch formula to the
arithmetic setting in [BGI71].

Motivic stable homotopy and cohomology theories. The purpose
of this work is to extend the arithmetic formulation of the Grothendieck-
Riemann-Roch formula of [BGI71] in the same way that by Dyer (again
[Dye62]) extends the topological formulation of the Hirzebruch-Riemann-
Roch formula following Atiyah-Hirzebruch. To that end, the natural frame-
work is Morel-Voevodsky's stable motivic homotopy theory, as it is defined

---

[4]It was after a seminar in Princeton which gathered most of the main characters
discussed here that Hirzebruch found his proof.

[5]*i.e.* a cohomology theory that satisfies all the axiom of Eilenberg-Steenrod except
the dimension axiom;

by a clear analogy with the ordinary stable homotopy category of topological spaces used by Dyer.[6]

The objects of the stable homotopy category, both classical and motivic, are meant to represent cohomology theories. Called spectra, they form a triangulated category whose distinguished triangles correspond to universal long exact sequences in cohomology. Similarly, all structures or properties of the stable homotopy category are reflected in the cohomologies representable by spectra. Probably the most important example of such a structure is the existence of a (symmetric) tensor product, called the *smash product*: a (commutative) monoid[7] on a spectra induces a product structure on its cohomology. These monoids are of primary importance in (motivic) stable homotopy; they are called *(motivic) ring spectra*.

In the motivic setting, we work over a base scheme[8] S; the motivic stable homotopy category of Morel-Voevodsky is denoted by $S\mathscr{H}(S)$. The starting point of this paper is that the representability of a cohomology theory has many interesting consequences. Let us first describe the obvious ones, for a given spectrum $\mathbb{E}$ (see Prop. 1.2.10 for details): the cohomology represented by $\mathbb{E}$ is a contravariant functor $\mathbb{E}^{**}$ from smooth S-schemes to bigraded abelian groups (the first index is the degree and the second one is called the *twist*). It satisfies the homotopy invariance property with respect to the affine line $\mathbb{A}^1$ (the affine line $\mathbb{A}^1$ is contractible), stability property with respect to the projective line $\mathbb{P}^1$ (seen as the analogue of the circle in topology). Moreover, it can be extended to a *cohomology theory with support*: given a smooth S-scheme X and a closed subscheme $Z \subset X$, one can define a bigraded abelian group $\mathbb{E}_Z^{**}(X)$ of cohomology classes with support in Z, in an appropriate functorial way and such that $\mathbb{E}_X^{**}(X) = \mathbb{E}^{**}(X)$ (see Def. 1.2.5 for details). This theory with support satisfies the (Nisnevich) excision property (analogue of the excision property of the Eilenberg-Steenrod axioms in topology, see Sec. 1.4, property (Nis) for details).

In this paper, we will use another important property of $S\mathscr{H}(S)$, its basic functoriality in S: it is a fibered category over the category of schemes. A cartesian section[9] with respect to this fibered structure, eventually restricted to a subcategory $\mathscr{S}$ of the category of schemes, will be called an

---

[6]When we work over the field of complex numbers, the stable motivic homotopy category can be realized in the ordinary stable homotopy category so that any motivic construction or statement has a realization in the topological world.

[7]In this introduction and in the whole paper, all monoid structures will be assumed to be commutative.

[8]In this introduction and in the whole paper, all schemes will be assumed to be Noetherian of finite dimension.

[9]Explicitly: the data of a ring spectrum $\mathbb{E}_X$ for each scheme X, with a given transition isomorphism $f^*(\mathbb{E}_X) \simeq \mathbb{E}_Y$ for any morphism $f : Y \to X$.

absolute spectrum (see Def. 1.2.1 and Ex. 1.2.3 for examples). Such an absolute spectrum represents a cohomology which is defined over the whole category $\mathscr{S}$, allowing to avoid the restriction to smooth S-schemes.[10] It still satisfies all the properties enumerated above. But moreover, under the presence of a ring structure on the absolute spectrum, we get an important product on cohomology with support which is not commonly used (but see [Del77, IV]). We call it the *refined product*: given closed subschemes $T \subset Z \subset X$, it has the form

$$\mathbb{E}_T^{**}(Z) \otimes \mathbb{E}_Z^{**}(X) \to \mathbb{E}_T^{**}(X) \tag{$*$}$$

and will be an essential technical tool to our study of fundamental classes (see Par. 1.2.8).

Note that all the basic properties of the cohomologies representable by an absolute spectrum are gathered in Proposition 1.2.10.

Absolute purity. The absolute purity conjecture of Grothendieck, formulated in [Gro77, I 3.1.4], has been a major problem in étale cohomology, because of its consequences on finiteness and duality as pointed out by Grothendieck (see *loc. cit.*). It was solved by Thomason in [Tho84] under some assumptions (on the coefficients) and in full generality by Gabber in [Fuj02]. Roughly, the conjecture says that the cohomology of a regular scheme X with support in a closed regular scheme Z is isomorphic to the cohomology of Z. The case of smooth schemes over a field (or even over some base) can be treated easily (see [AGV73, XVI, 3.9]). The main problem in this conjecture is to treat the case of regular schemes of unequal characteristics, which are the objects of the so-called *arithmetic geometry*.

This problem was confined in the étale setting until D.C. Cisinski and the author discovered that a similar statement could be formulated and proved in the newly defined setting of rational mixed motives (see [CD12b, 14.4.1])[11]. This naturally raises the question of extending the problem of absolute purity to any representable cohomology theories.

So we introduce here (Def. 1.3.2) the property of *absolute purity* for an absolute spectrum and any closed subscheme $Z \subset X$ whose immersion is regular. The formulation of this property is the new technical ingredient introduced by this work. As in the case of étale cohomology, for smooth schemes over a field (or over some base), the property is always fulfilled according to a fundamental result of Morel-Voevodsky. Thus the interest of this property lies in the case of arithmetic geometry. Fortunately, there

---

[10]Using a terminology introduced by Beilinson, this is an absolute cohomology; this justifies the terminology *absolute spectrum*.

[11]As in the étale setting, this result has important consequences for rational mixed motives. Note the importance of absolute purity was anticipated by Ayoub in [Ayo07a].

are several cohomology theories which satisfies absolute purity in the arithmetic case. The matrix case is integral algebraic K-theory according to the localization theorem of Quillen. The cases of rational motivic cohomology and rational cobordism theory follows (see Ex. 1.3.4).

We think that the problematic of absolute purity is an important question for homotopy theory. A new aspect of our definition of this property is that it is intrinsic and do not depend on the choice of a purity isomorphism.[12] Moreover, it is formulated for any cohomology theory without assuming the existence of an orientation (see below). In particular, we conjecture that this property holds integrally not only for the algebraic cobordism spectrum but also for the sphere spectrum (see conjectures A and B p. 266). This result would have several interesting consequences: see Remark 1.3.5.

An important ingredient in the proof of absolute purity by Gabber is the so-called analytical invariance of étale cohomology with support. While we do not attack the previous conjectures on absolute purity, we nevertheless prove the *analytical invariance property* for any cohomology representable by an absolute spectrum (see Thm. 1.4.6), extending a result already obtained by Wildeshaus [Wil06]. Note in particular that the result can be applied to the absolute spectrum representing *rigid cohomology* over a field of characteristic $p > 0$ — according to Ex. 1.3.4(1). Thus our proposition contains in particular Theorem 1.1 of [Ouw14].

Orientation theory: characteristic and fundamental classes. At this point, we connect motivic homotopy theory with the two fundamental notions that were developed has a natural evolution of the Riemann-Roch problem: characteristic classes and fundamental classes. Recall the first ones have first been studied in algebraic topology, and evolved naturally into the theory of orientation while the second one was the domain of algebraic geometry and evolved into intersection theory.

The idea of transporting orientation theory from algebraic topology to algebraic geometry is comparatively quite recent as it takes its origin in the first proof of the Milnor conjecture by Voevodsky (see [Voe96]). The theory has grown out of an unpublished work of Morel which was developed by several authors.

Let us recall in this introduction the basics of this theory. An orientation c of a ring spectrum $\mathbb{E}$ over a base scheme S is a cohomological class — in degree $(2, 1)$ — of the infinite projective space with coefficients in $\mathbb{E}$ (Def. 2.1.2). Giving such a class allows to derive *canonically* a lot of interesting cohomological invariants and structures. First, we can define Chern classes associated with vector bundles — the first Chern class follows directly from

---

[12]This is due to the use of the deformation space.

the orientation, and the other ones follow from the computation of the cohomology of any projective bundle according to the classical construction of Grothendieck: see Section 2.1.

The connection with the work of Quillen appears at this point. As in topology, Chern classes of an arbitrary oriented spectrum need not be additive: the first Chern class of a tensor product of line bundles is not the sum of the Chern classes of each bundle. Instead, it is described according to a *formal group law* that is canonically associated with the chosen orientation (see Par. 2.1.19 for more details). This is connected with the theory of *Thom classes* that one derives from Chern classes.[13] Indeed, these classes uniquely define a canonical structure of **MGL**-algebra on $\mathbb{E}$ where **MGL** is the algebraic cobordism spectrum (analogue of the complex cobordism spectrum **MU** in topology). Orientations on $\mathbb{E}$ are in one to one correspondence with structures of **MGL**-algebra and it is widely believed that the formal group law of **MGL** is the universal formal group law.[14]

Secondly, the orientation determines fundamental classes. In fact, assuming the absolute purity property for a regular closed immersion $Z \subset X$, the Thom class of the normal bundle of Z in X gives us directly the *refined* fundamental class of Z in X, as a cohomology class of X with support Z. Using the refined product by this fundamental class — $(*)$ with $T = Z$ — gives us the usual purity isomorphism:

$$\mathbb{E}^{**}(Z) \xrightarrow{\sim} \mathbb{E}^{**}_Z(X).$$

With our formalism, all this follows easily. But note however that this is the first appearance of this form of the purity isomorphism in motivic homotopy theory — and even in étale cohomology.[15]

These generalized fundamental classes are the trace of classical intersection theory, though they can be defined in very general theories such as algebraic K-theory or algebraic cobordism. We illustrate this concretely by proving some of the classical formulas known for Chow groups: the excess intersection formula (Cor. 2.4.4) and associativity (compatibility with composition, Th. 2.4.9). Note that the excess intersection formula allows to get back the classical link between fundamental classes and characteristic classes: when Z can be parametrized by a section of a vector bundle over X, the fundamental class of Z in X equals the top Chern class of the vector bundle (see Cor. 2.4.6 for details). Let us also recall that the compatibility

---

[13]The Thom class of a vector bundle E, as a class in the projective completion of E, equals the top Chern class of the universal quotient bundle (see Ex. 2.2.4).

[14]According to works respectively of Levine and of Hoyois, given a field k of exponential characteristic p, this is true for **MGL**$[1/p]$ if one restricts to k-schemes.

[15]We were especially inspired by the formulation of Poincaré duality of Bloch and Ogus in [BO74].

with composition of fundamental classes is a key technical problem in orientation theory. The geometrical tool (the double deformation space) used here is not new but the use of the refined product ($*$) allows both a finer expression of this result and an easier proof.

Residues and Gysin morphisms. A formal consequence of the purity isomorphism for a regular closed immersion $i : Z \to X$ is the existence of a localization long exact sequence in cohomology, so that the theory described just above for an oriented absolute ring spectrum canonically leads to such a sequence. It is made of two interesting morphisms. The first one is a morphism that we have called the *residue map associated with* (X, Z). Our interest to that kind of maps come from our comparison between cycle modules and homotopy sheaves with transfers ([Dég11]) as our residues on sheaves corresponds to Rost residues on cycle modules. One illustrates this phenomenon by the computation of our residue maps when X is a trait, Z its closed point: for motivic cohomology in bidegree (n, n), our residue map coincides with Milnor residue symbol. More generally, the residue map on symbols in classical cohomologies always agree with Milnor residue map (see Par. 5.4.1 for details). More interestingly, we prove here that the residue we have defined by purely geometrical means agrees with *Tate residue* on De Rham cohomology (Th. 5.4.5) — this is an application of the analytical invariance of our residue map. This implies in particular, that for divisors, the residue map we have defined here — by deformation to the normal bundle and orientation theory — agrees with that of Leray. These classical constructions gets also extended to rigid cohomology (see Ex. 5.4.2).

The second interesting map is the so-called *Gysin morphism*[16] associated with the immersion i; in other words, the covariant functoriality of cohomology. According to our formalism, it is simply equal to the multiplication by the refined fundamental class (which gives a class with support in Z) followed by the obvious map which forgets the support (see (3.1.2.a) for details). Thus the good properties of (refined) fundamental classes give all the basic expected properties of these particular Gysin maps.

Once all this ground work is in place, one can introduce the main construction of orientation theory, that of *Gysin morphisms* for certain projective morphisms. In our work, it comes mainly in two settings: the geometric one, for schemes smooth over some fixed base, and the arithmetic one, for any regular schemes. The way we phrased and use absolute purity allows us to treat these two cases in a single turn, thus building Gysin morphisms associated with any projective morphism between one of these two kind of

---

[16]This is the usual terminology in algebraic geometry. It is named that way after the pioneering work of Gysin that we have described in the historical part.

schemes. In fact, the process to build these Gysin maps is very classical: you need to consider the case of closed immersions and of projections of a projective bundle, and then find the sufficient condition so that these two cases can be glued. However, because we deal with general oriented spectra, the case of the projection of a projective bundle is not trivial. As in our previous work on the subject ([Dég08a]), to treat that case, we use a duality argument that can be summarized as follows: because of the projective bundle formula, the cohomology of a projective bundle P/X is a finite free module over the cohomology of the base: thus it is dualizable. Then, one shows that the Gysin morphism associated with the diagonal embedding of P induces a duality pairing on $\mathbb{E}^{**}(P)$. Finally one defines the Gysin morphism of the projection $P \to X$ as the transpose of the pullback with respect to this duality (see Def. 3.2.2). The condition for gluing then follows, using a computation of characteristic classes, as well as the main properties of Gysin morphisms for closed immersions: compatibility with composition, projection formula, compatibility with transversal pullbacks, excess of intersection formula.

One of the basic examples of a representable cohomology theory is *Beilinson motivic cohomology* which is representable by absolutely oriented ring spectrum because of the ground work [CD12b]. Our construction gives Gysin morphisms for this cohomology with respect to any projective morphism between regular schemes. This is an improvement of the constructions of [Sou85, Th. 9]. For more general cohomology theories, such as algebraic K-theory or algebraic cobordism, Gysin morphisms reveal a third kind of characteristic classes, the *cobordism classes* (see Def. 3.2.13). Note these classes are interesting only when the formal group law associated with the considered oriented ring spectrum is non additive — see in particular formula (3.2.14.a). The non triviality of these classes in the case of non additive formal group law explains why the definition of Gysin morphisms in our context is far more difficult than in the ordinary case.

As a prelude to the Riemann-Roch formula, note that, inspired by a result of Panin, we give a uniqueness statement for Gysin morphisms by simple axioms (see Th. 3.3.1). This allows us to compare the Gysin morphisms defined here with more classical constructions (see the examples in 3.3.4).

Riemann-Roch formulas. The beauty of the Grothendieck-Riemann-Roch formula lies in its generality and the simplicity of its proof. The same phenomena happens here and the reader will see that the main technical work was to define Gysin morphisms.

The topological interpretation of the Riemann-Roch formula can be summarized as an answer to the following question: what happens if we *change the choice of orientation* on an absolutely pure oriented spectrum $\mathbb{E}$?

The answer is that all the data defined through the orientation theory, as described above, change and in particular the Gysin morphism: the change of the later is exactly measured by the Riemann-Roch formula (see Section 5.2 for that point of view).

The setting of our general Riemann-Roch formula is to consider two absolute oriented ring spectra $(\mathbb{E}, c)$, $(\mathbb{F}, d)$ and an arbitrary morphism of ring spectra $\varphi : \mathbb{E} \to \mathbb{F}$. Then one realizes that $\varphi(c)$ is an orientation of $\mathbb{F}$, which does not necessarily coincide with the given orientation d. Thus, we come back to the question of changing the orientation on a given spectrum and understanding its effect on Gysin morphisms. In a word, this change of orientation is measured by the Todd class Td_φ associated with the morphism φ (see Prop. 4.1.2). With this definition in hands, we prove the analogue of the classical Grothendieck-Riemann-Roch formula: for any local complete intersection, any projective morphism $f : Y \to X$ in $\mathscr{S}$ with virtual tangent bundle τ_f, one has for a cohomology class $y \in \mathbb{E}^{**}(Y)$ (see also Th. 4.3.2):

$$\varphi_X(f_*(y)) = f_*\big(\mathrm{Td}_\varphi(\tau_f)\cup\varphi_Y(y)\big). \tag{GRR$_\varphi$}$$

At this point, the proof is quite easy and follows the initial proof of Borel and Serre: one reduces to the case of closed regular embeddings and projective smooth morphisms. In our situation, the last case follows by duality from the first one — here the proof differs slightly from that of Borel and Serre. The case of closed embedding is by reduction to divisors in which case it is tautological.

Inspecting this last case, we fall onto the surprising result that the Riemann-Roch formula for closed regular embedding $i : Z \to X$ has a companion formula which was never observed till now and involves the residue map $\partial_{X,Z}$. We call it the residual Riemann-Roch formula. Using the Todd class of the normal bundle $N_Z X$ of Z in X, it reads as follows for a cohomology class $u \in \mathbb{E}^{**}(X - Z)$ (see also Th. 4.2.3):

$$\partial_{X,Z}\big(\varphi_U(u)\big) = \mathrm{Td}_\varphi(-N_Z X)\cup\varphi_Z\big(\partial_{X,Z}(u)\big). \tag{∂RR}$$

Applications and comparison with the work of Gillet and Soulé.
Before going to the applications, let us explain with more details the crucial situation of changing between two orientations c and d on an absolute ring spectrum $\mathbb{E}$. As explained above, the orientations c and d correspond over a base scheme X to formal group laws F_c and F_d with coefficients in the ring $\mathbb{E}^{**}(X)$. Moreover, it follows automatically that these formal group laws are isomorphic: say from F_d to F_c, the isomorphism corresponds to a

power series $\Psi(t)$ with coefficients in $\mathbb{E}^{**}(X)$ of the form $\Psi(t) = (t + \cdots)$. Then the Todd class which corresponds to changing the orientation from c to d is uniquely defined in terms of $\Psi(t)$ (see section 5.2 for a detailed discussion).

This understanding of Todd classes allows to enlighten the classical case. When $\mathscr{S}$ is the category of regular schemes, Quillen algebraic K-theory (resp. Beilinson motivic cohomology) is representable by an oriented absolute ring spectrum $\mathbf{KGL}$ (resp. $\mathbf{H}_B$). Then the classical (higher) Chern character as defined by Gillet corresponds to an isomorphism of absolute ring spectra:

$$\mathrm{ch} : \mathbf{KGL} \to \oplus_{i \in \mathbb{Z}} \mathbf{H}_B(i)[2i]$$

that was first given in these terms by Riou (in a slightly different form, see [Rio10, 6.2.3.9]). The formal group law associated with $\mathbf{KGL}$ (resp. $\mathbf{H}_B$) is the multiplicative (resp. additive) formal group law — see Example 2.1.21. It follows from the theory of formal group laws that there is only one isomorphism of the form $\Psi(t)$ from the additive to the multiplicative formal group law: this is the exponential power series and one recovers the classical definition of the Todd power series, with a conceptual explanation why it has this precise form. The Riemann-Roch theorem that we get for ch is a cohomological version of the formulas obtained by Gillet, [Gil81, Th. 4.1]: in fact, one will recognize most of the principles of motivic homotopy theory in the work of Gillet. The main difference is that we have built the Gysin morphisms appearing in the formula while in *loc. cit.* they are part of the axioms (of a "duality theory with support"). Therefore, the cohomological formulation of our Riemann-Roch formula for ch is new, valid for any projective morphism between regular schemes. It extends the classical Grothendieck-Riemann-Roch formula for K_0 to higher degrees. Note this formula was proved by Riou in [Rio10] in the particular case of smooth projective morphisms and was obtained in full generality in [AH10] but the construction of Gysin morphisms in *op. cit.* is merely a reference to [Dég08a] — see Example 2.4 in *op. cit.* Note finally we have given a special attention to the residual Riemann-Roch formula in the case of the Chern character resulting in some purely algebraic results (see Section 5.5).

Our other examples are more concerned with pure orientation theory. First the analysis of the change of orientations shows that on motivic cohomology, as well as on mixed Weil cohomologies (they are representable according to [CD12a]), there is *only one* possible orientation (see 5.1.10 for the general statement). Therefore, on these cohomologies, there is only one possible theory of Chern classes and Gysin morphisms (see also Th. 3.3.1 and Ex 3.3.4). This settles the point of unnecessary questions about signs, uniqueness, that frequently occur in this kind of situation. We also give a new proof, as well as a conceptual explanation, of a formula of Quillen for computing the cobordism class of a non necessarily trivial projective bundle, but only with rational coefficients (see Th. 5.2.4 and Example 5.2.7).[17] The idea of the proof is to look at the Riemann-Roch formula that one gets by changing the formal group law of algebraic cobordism (conjecturally the universal formal group law) to the additive one — this is why we need rational coefficients.

Comparison with the work of Panin and étale cohomology. The last setting to which our work can be compared is the fundamental work of Panin on oriented cohomology theories. In [Pan04], Panin proves our general Riemann-Roch formula for cohomology theories satisfying suitable axioms but defined over the category of smooth k-schemes. So our work should be viewed as an extension of the axiomatic of oriented cohomology theory to the arithmetic case. This is what we prove in the last section of this paper: we have extracted a list of all the properties of representable cohomology theories that we have used in this work, dubbed here *arithmetic cohomology* for short (see Def. 6.1.1).[18] This axiomatic indeed generalizes the one of Panin and the proofs of this paper show that our results still apply to it, yielding Gysin morphisms, residues, characteristic classes and their formulas such as the Riemann-Roch ones.

This point of view is in fact useful because it applies especially to the étale l-adic cohomology of $\mathbb{Z}[1/l]$-schemes. In fact, we do not know if this cohomology is representable by an *absolute* ring spectrum.[19] But however, it satisfies all the axioms of an arithmetic cohomology — in particular because of Thomason result about absolute purity. Thus, the constructions of this paper apply to that cohomology, and its rational version, and in particular give Gysin morphisms. We deduce the classical Riemann-Roch formula for rational ℓ-adic cohomology of regular schemes, as expected by Grothendieck in [BGI71, XIV, 6.1] and even the higher Riemann-Roch formula as well as the residual Riemann-Roch formula (∂RR) (see Cor. 6.2.4). Note finally that one can also apply the recent work of [CD14] on h-motives, to get that étale motivic cohomology with coefficients in any ring R is an arithmetic cohomology.[20] The Gysin morphisms that one get extend to arbitrary coefficients, in the regular case, the recent construction of Gabber-

---

[17]This is enough to get the integral formula as well as the case of schemes of characteristic 0: see Remark 5.2.6(2).

[18]A longer but more precise terminology is also introduced: *absolutely pure oriented ringed cohomology with support*.

[19]According to the results of [CD14], it is representable over any scheme by a ring spectrum but we do not know these ring spectra form a cartesian section of $S\mathscr{H}$ — unless one restricts to schemes over a field.

[20]Recall that in the case where R is a torsion ring of characteristic exponent invertible on the schemes considered, this later cohomology agrees with the usual étale cohomology with coefficients in R.

Riou ([Rio14], see Remark 6.2.5 for the comparison).

Outline of the work

The paper is organized as follows. In the first Section, we recall the basics
on the motivic stable homotopy category give the basic properties of representable cohomologies, and states the absolute purity property. We end-up
the section with a discussion about analytical invariance. In Section 2, we
recall orientation theory in motivic homotopy theory, gives the construction of (refined) fundamental classes and their properties. In Section 3, we
define the residues and Gysin morphisms and study their properties. Section 4 is centered around the general Riemann-Roch formula and its proof.
In Section 5, we treat examples and give applications of our formulas. Finally, Section 6 discusses the axiomatic of arithmetic cohomologies and the
case of étale cohomology, motivic étale cohomology and continuous étale
cohomology with various coefficients.

Notations and conventions

All schemes in this paper are assumed to be noetherian of finite dimension.
We will say that an S-scheme X, or equivalently its structure morphism, is
projective if it admits an S-embedding into $\mathbb{P}^n_S$ for a suitable integer n.[21]
    In the whole text, unless stated otherwise, $\mathscr{S}$ stands for a sub-category
of the category of such schemes. We will assume that $\mathscr{S}$ is stable by blow-
up and contains any open subscheme of (resp. projective bundle over) a
scheme in $\mathscr{S}$. The category $\mathscr{S}$ can be the category of all schemes, especially
in the examples and definitions which do not deal with absolute purity. On
the contrary, when dealing with the absolute purity property, the relevant
examples for applications area:

- the category $\mathscr{R}eg$ of all regular schemes (*i.e.* its local rings are regular.)

- the category $\mathscr{S}m_S$ of smooth S-schemes, for an arbitrary base
  scheme S.

    By convention, unless explicitly stated, when we speak of the rank of a
vector bundle, the dimension of a morphism, the codimension of a closed

---

[21]For example, if one works with quasi-projective schemes over a noetherian affine
scheme (or more generally a noetherian scheme which admits an ample line bundle),
then a morphism is proper if and only if it is projective with our convention — use
[Gro61, Cor. 5.3.3].

subscheme (or a closed immersion), it will always be assumed to be *constant.*[22]

Given a projective bundle P/X associated with a vector bundle E/X, we will call *canonical line bundle* on P/X the tautological line bundle λ on P characterized by the property $\lambda \subset P \times_X E$ (it corresponds with the notation $\lambda = \mathcal{O}(-1)$).

The letter $\mathbb{N}$ denotes the set of non negative integers.

1 Absolute cohomology and purity

1.1 Functoriality in stable homotopy

1.1.1 Recall that the stable homotopy category of schemes defines a 2-functor from the category of schemes to the category of symmetric monoidal closed triangulated categories. This means that for any morphism of schemes $f : T \to S$, we have a pullback functor

$$f^* : S\mathscr{H}(S) \to S\mathscr{H}(T)$$

which is symmetric monoidal and such that for any composable morphisms of schemes f, g, we have the relation: $f^* g^* = (gf)^*$.

We will use the following properties:

(A1) For any morphism (resp. smooth morphism) f, the functor f^* admits a right (resp. left) adjoint denoted by f_* (resp. $f_\sharp$).

(A2) For any cartesian square:

$$\begin{array}{ccc} Y & \xrightarrow{q} & X \\ g \downarrow & & \downarrow f \\ T & \xrightarrow{p} & S \end{array}$$

such that f is a smooth morphism, the base change map

$$p^* f_\sharp \to g_\sharp q^*$$

is an isomorphism.

(A3) For any smooth morphism $f : Y \to X$ and any spectrum $\mathbb{E}$ over Y (resp. $\mathbb{F}$ over X), the canonical transformation:

$$f_\sharp(\mathbb{E} \wedge f^*(\mathbb{F})) \to f_\sharp(\mathbb{E}) \wedge \mathbb{F}$$

is an isomorphism.

---

[22]If one wants to avoid this convention, see Remarks 2.2.3 and 2.3.2.

(A4) For any closed immersion $i : Z \to X$ with complementary open immersion j, there exists a unique natural transformation $\partial_i : i_* i^* \to j_\sharp j^*[1]$ which fits in a distinguished triangle of the form:

$$j_\sharp j^* \xrightarrow{ad'} Id \xrightarrow{ad} i_* i^* \xrightarrow{\partial_i} j_\sharp j^*[1]$$

where ad (resp. ad') is the unit (resp. counit) map of the adjunction (i^*, i_*) (resp $(j_\sharp, j^*)$).

Except for the last property, these are easy consequences of the construction of $S\mathcal{H}$ — see [Ayo07b]. Property (A4) is a consequence of [MV99, §3, th. 2.21] — see [Ayo07b, 4.5.47] for details.

Remark 1.1.2 One can deduce from (A4) that i_* is fully faithful. Moreover, when i is a nil-immersion, i^* is fully faithful, so that (i^*, i_*) is an equivalence of categories.

Note also that we can derive from properties (A1) and (A4) the following ones:

(A5) For any closed immersion i, the functor i_* admits a right adjoint denoted by $i^!$.

(A6) For any cartesian square:

$$\begin{array}{ccc} T & \xrightarrow{k} & Y \\ {\scriptstyle g}\downarrow & & \downarrow{\scriptstyle f} \\ Z & \xrightarrow{i} & X \end{array} \qquad\qquad (1.1.2.a)$$

such that i is a closed immersion, the base change morphism

$$f^* i_* \to k_* g^*$$

is an isomorphism.

(A7) For any closed immersion $i : Z \to X$ and any spectrum $\mathbb{E}$ over Z (resp. $\mathbb{F}$ over X), the canonical transformation:

$$i_*(\mathbb{E} \wedge i^*(\mathbb{F})) \to i_*(\mathbb{E}) \wedge \mathbb{F}$$

is an isomorphism.

For these three last properties, we refer the reader to [Ayo07a] or for a more compact reference to [CD12b], 2.3.3, 2.3.8 and 2.3.15 respectively.

1.1.3 For any smooth S-scheme X, we denote by $\Sigma_S^\infty X$ the infinite suspension spectrum associated with the sheaf represented by X with a base point added. Recall that $\Sigma_S^\infty X = f_\sharp(\mathbb{1}_X)$.

Consider again the notations of axiom (A4). We simply denote by X/U the cokernel of the map induced by j in the category of Nisnevich sheaves of sets over $\mathscr{S}m_S$. Note it is pointed by identifying U with the base point. As j is a monomorphism, we deduce a homotopy cofiber sequence

$$U \to X \to X/U$$

in the $\mathbb{A}^1$-local model category of simplicial sheaves over S ([MV99, Sec. 3.2, p. 105]). According to (A4) and Remark 1.1.2, the canonical map $\mathbb{1}_Z \to i^*\Sigma^\infty(X/U)$ is an isomorphism.

To simplify notations, we will still denote by X/U the infinite suspension spectrum associated with the sheaf X/U. Given any object K of $S\mathscr{H}(X)$, we get a canonical isomorphism: $i^*\big((X/U) \wedge K\big) = i^*(K)$ which, by adjunction, induces a map

$$(X/U) \wedge K \to i_*i^*(K). \tag{1.1.3.a}$$

In fact, the localization axiom (A4) for i is equivalent to the fact that this map is an isomorphism and i_* is fully faithful — see [CD12b, 2.3.15].

Remark 1.1.4 In what follows, we will consider the isomorphisms listed in the above properties as identities unless it involves a non trivial commutativity statement.

1.2 Absolute cohomology

As usual, a ring spectrum over a scheme S will be a commutative monoid of the symmetric monoidal category $S\mathscr{H}(S)$.

Definition 1.2.1 An $\mathscr{S}$-*absolute spectrum* (resp. *ring spectrum*) $\mathbb{E}$ is a collection of spectra (resp. ring spectra) $\mathbb{E}_X$ over X for a scheme X in $\mathscr{S}$ and the data for any morphism $f : Y \to X$ in $\mathscr{S}$ of an isomorphism of spectra (resp. ring spectra) $\epsilon_f : f^*\mathbb{E}_X \to \mathbb{E}_Y$ satisfying the usual cocycle condition.[23]

A *morphism* $\varphi : \mathbb{E} \to \mathbb{F}$ of $\mathscr{S}$-absolute spectra (resp. ring spectra) is a collection of morphisms $\varphi_X : \mathbb{E}_X \to \mathbb{F}_X$ of spectra (resp. ring spectra)

---

[23] In other words, this is a cartesian section of the $\mathscr{S}$-fibered category of spectra (resp. ring spectra).

indexed by schemes X in $\mathscr{S}$ such that for any morphism $f : Y \to X$ in $\mathscr{S}$, the following diagram commutes:

$$
\begin{array}{ccc}
f^*\mathbb{E}_X & \xrightarrow{\ f^*\varphi_X\ } & f^*\mathbb{F}_X \\
{\scriptstyle \epsilon_f^{\mathbb{E}}}\downarrow & & \downarrow{\scriptstyle \epsilon_f^{\mathbb{F}}} \\
\mathbb{E}_Y & \xrightarrow{\ \varphi_Y\ } & \mathbb{F}_Y.
\end{array}
$$

As the category $\mathscr{S}$ is fixed in the entire paper we will abusively say *absolute* for $\mathscr{S}$-absolute. However, when the category $\mathscr{S}$ is the category of all S-schemes, we will say S-absolute for $\mathscr{S}$-absolute.

Remark 1.2.2 In the whole paper, we will be interested only in $\mathscr{S}$-absolute spectra. However, it is also convenient to consider the sections $(\mathbb{E}_X)_{X \in \mathscr{S}}$ over $\mathscr{S}$ of the stable homotopy category which are not necessarily cartesian (*i.e.* the given transition morphisms $\epsilon_f : f^*(\mathbb{E}_X) \to \mathbb{E}_Y$ are not isomorphisms). We will call these objects *weak $\mathscr{S}$-absolute spectra*. As in the above definition, they can also be equiped with a ring structure.

Example 1.2.3 (1) Let S be a fixed scheme.

Then, up to the choice of the isomorphisms ϵ_f in the above definition, an S-absolute ring spectrum $\mathbb{E}$ is determined by its value on S. Reciprocally, given any ring spectrum $\mathbb{E}_S$ over S, we get a canonical S-absolute spectrum $\mathbb{E}$ by putting $\mathbb{E}_X = f^*(\mathbb{E}_S)$ for any $f : X \to S$. In the S-absolute case, we will frequently identify $\mathbb{E}_S$ and $\mathbb{E}$ to simplify notations.

A basic example of this kind of situation is given by the concept of mixed Weil theory. Let us recall the setting. We let k be a field (non necessarily perfect) and K be a field of characteristic 0. A K-linear mixed Weil theory E over k is a presheaf of commutative differential graded K-algebras over the category of smooth affine k-algebras satisfying axioms: homotopy invariance, Nisnevich excision, dimension, stability and Künneth formula (see [CD12a]). To such a theory one canonically associates a ring spectrum $\mathbb{E}$ over k.

We then get an absolute ring spectrum over the category of k-schemes by the preceding procedure (cf. [CD12b, 17.2.5]). The original cohomology defined by E for smooth affine k-schemes gets extended to any k-schemes. This extension is uniquely characterized by the h-descent property (cf. [CD12b, 17.2.6]) and by the commutation with limit property (cf. Lemma 1.2.13 below).

Examples are given by the classical Weil theories: Betti, De Rham, geometric étale and rigid cohomologies.

(2) The 0-sphere spectrum S^0 — unit for the smash product — is obviously an absolute ring spectrum.

(3) For any scheme S, one can consider one of the following ring spectra:

- The Beilinson motivic cohomology spectrum $\mathbf{H}_{\mathrm{B},S}$ (see [Rio10, CD12b]).
- The homotopy invariant K-theory ring spectrum $\mathbf{KGL}_S$ (see [Voe98, Rio10]).[24]
- The cobordism ring spectrum $\mathbf{MGL}_S$ (see [PPR08, 2.1]).

Then each of these examples defines an absolute ring spectrum denoted respectively by $\mathbf{H}_{\mathrm{B}}$, $\mathbf{KGL}$, $\mathbf{MGL}$ (see the respective reference given above for this assertion).

(4) In [Voe98, §6], Voevodsky defined the *motivic Eilenberg-Mac Lane spectrum* over any smooth k-scheme S where k is a perfect field, representing motivic cohomology with integral coefficients. This construction was generalized to any base scheme S in the work of [DRØ03, Ex. 3.4].

In [CD12b, 11.2.17], D.C. Cisinski and the author gave a general construction of this spectrum relying on other ideas of Voevodsky. Let Λ be a localization of the ring $\mathbb{Z}$ and S be any scheme. Then one defines a ring spectrum $\mathbf{H}\Lambda_S$ such that the abelian group

$$\mathrm{Hom}_{S\mathscr{H}(S)}(S^0, \mathbf{H}\Lambda_S(m)[n]) = H_{\mathcal{M}}^{n,m}(S, \Lambda)$$

is Voevodsky's motivic cohomology of S with coefficients in Λ.[25] According to [CD12b, 11.2.21], for any morphism $f : S' \to S$ of schemes, there exists a canonical morphism of ring spectra:

$$\tau_f : f^*(\mathbf{H}\Lambda_S) \to \mathbf{H}\Lambda_{S'}.$$

In other word, we have a weak absolute ring spectrum (over the category of all schemes) that we will denote $\mathbf{H}\Lambda^w$. A fundamental conjecture of Voevodsky (cf [Voe02, Conj. 17] for the case of integral coefficients) can be reformulated saying that $\mathbf{H}\Lambda^w$ is in fact a (strong) $\mathscr{S}$-absolute ring spectrum over the category of all schemes.

---

[24]Recall it represents *homotopy invariant K-theory* according to [Cis13].

[25]Note that, in the current state of the theory, the consideration of the ring Λ for coefficients is crucial as, given a localization Λ' of Λ, we do not always have:

$$\mathbf{H}(\Lambda')_S \simeq \mathbf{H}\Lambda_S \otimes_\Lambda \Lambda'.$$

See however Proposition 11.2.19 of *loc. cit.* for cases where this identification holds.

Unfortunately, this conjecture is still not known. In [CD12b, 16.1.7], we prove it when $\Lambda = \mathbb{Q}$ if one restricts to the case of geometrically unibranch schemes.[26] On the other hand, by construction, the above map τ_f is obviously an isomorphism when f is smooth.[27] Thus, if $\mathscr{S}$ is the category of smooth schemes over an arbitrary base scheme S, then the collection $\mathbf{H}\Lambda_X$ defines an S-absolute ring spectrum simply denoted by $\mathbf{H}\Lambda_{/S}$. Moreover, from the perspective of this paper, we can safely extend $\mathbf{H}\Lambda_{/S}$ as an S-absolute ring spectrum taking the various pullbacks of $\mathbf{H}\Lambda_S$ over any S-scheme as in the first Example above. This is particularly relevant if S is the spectrum of a prime field (we refer the reader to [CD14] for more details).

(5) Recall finally a very interesting construction of Markus Spitzweck. Let Λ be a localization of $\mathbb{Z}$. In [Spi13], Spitzweck defines a ring spectrum $\mathbf{M}\Lambda_{\mathbb{Z}}$ over $\mathbb{Z}$ (*loc. cit.*, Def. 4.27) whose pullbacks to any field $\operatorname{Spec}(k) \to \operatorname{Spec}(\mathbb{Z})$ is Voevodsky's Eilenberg-Mac Lane spectrum (*loc. cit.* 6.7, 9.16 and 9.17). Thus, the absolute ring spectrum $\mathbf{M}\Lambda$ obtained by considering for any scheme $f : X \to \operatorname{Spec}(\mathbb{Z})$ the ring spectrum $f^*(\mathbf{M}\Lambda_{\mathbb{Z}})$ is an interesting candidate for motivic cohomology in general. Note in particular that according to *loc. cit.*, 7.19, for any smooth $\mathbb{Z}$-scheme X,

$$\mathbf{M}\mathbb{Z}^{n,m}(X) = CH_{2m-n}(X, m)$$

where the right hand side is Bloch's higher Chow group as defined by Levine.

1.2.4 In the remainder of this section, we consider an absolute ring spectrum $\mathbb{E}$ (which can be weak in the sense or Remark 1.2.2 until Paragraph 1.2.8) and define structures on its associated cohomology theory.

As usual, we call closed (resp. open) pair any couple (X, Z) (resp. (X/U)) such that X is a scheme and Z (resp. U) is a closed (resp. open) subscheme of X.

Definition 1.2.5 Given a closed pair (X, Z), corresponding to a closed immersion i, and a couple $(n, m) \in \mathbb{Z}^2$, we define the relative cohomology of (X, Z) in bi-degree (n, m) with coefficients in $\mathbb{E}$ by one of the following

---

[26]In fact, we proved that the ring spectrum $\mathbf{H}\mathbb{Q}_S$ coincides with $\mathbf{H}_{\mathrm{B},S}$ when S is geometrically unibranch.

[27]Using the notations of [CD12b, Par. 11.2.21], when f is smooth, the exchange morphism $f^*\varphi_* \to \varphi_* f^*$ used to construct τ_f is an isomorphism.

equivalent formulas:

$$\mathbb{E}_Z^{n,m}(X) := \mathrm{Hom}_X(X/X - Z, \mathbb{E}_X(m)[n])$$
$$= \mathrm{Hom}_X(i_*(\mathbb{1}_Z), \mathbb{E}_X(m)[n]) = \mathrm{Hom}_Z(\mathbb{1}_Z, i^! \mathbb{E}_X(m)[n]).$$

When it clarifies the notation, we will also put: $\mathbb{E}^{n,m}(X, Z) := \mathbb{E}_Z^{n,m}(X)$. When $X = Z$ we simply put as usual: $\mathbb{E}^{n,m}(X) := \mathbb{E}_X^{n,m}(X)$.

A morphism $\varphi : \mathbb{E} \to \mathbb{F}$ of absolute ring spectra obviously induces for any closed pair (X, Z) a morphism

$$\varphi_* : \mathbb{E}_Z^{n,m}(X) \to \mathbb{F}_Z^{n,m}(X). \tag{1.2.5.a}$$

1.2.6 *Contravariant functoriality*: Consider a closed pair (X, Z) and a morphism $f : Y \to X$ in $\mathscr{S}$. Let $T = Y \times_X Z$ be the pullback in the category of schemes, considered as a closed subscheme of Y. Then we get a morphism of abelian groups:

$$f^* : \mathbb{E}_Z^{n,m}(X) \to \mathbb{E}_T^{n,m}(Y)$$

by one of the following equivalent definitions:

- According to Remark 1.1.3, $f^*(X/X - Z) = Y/Y - T$. Thus, for a cohomology class $\rho : (X/X - Z) \to \mathbb{E}_X(m)[n]$, the pullback map $f^*(\rho)$ gives the desired map:

$$(Y/Y - T) = f^*(X/X - Z) \to f^* \mathbb{E}_X(m)[n] \xrightarrow{\epsilon_f} \mathbb{E}_Y(m)[n].$$

- Consider the pullback square (1.1.2.a) with f as above and i the immersion of Z in X. Then property (A6) gives a canonical identification: $f^* i_*(\mathbb{1}_Z) = k_* g^*(\mathbb{1}_Z) = k_*(\mathbb{1}_T)$. Thus, taking a cohomology class $\rho : i_*(\mathbb{1}_Z) \to \mathbb{E}_X(m)[n]$, the pullback map $f^*(\rho)$ gives the desired map:

$$k_*(\mathbb{1}_T) = f^* i_*(\mathbb{1}_Z) \to f^* \mathbb{E}_X(m)[n] \xrightarrow{\epsilon_f} \mathbb{E}_Y(m)[n].$$

1.2.7 *Covariant functoriality*: Consider closed immersions $T \xrightarrow{\nu} Z \xrightarrow{i} X$ and put $k = i \circ \nu$.
We define a pushforward in cohomology with support

$$\nu_! : \mathbb{E}_T^{n,m}(X) \to \mathbb{E}_Z^{n,m}(X)$$

by one of the following equivalent definitions:

- By functoriality of homotopy colimits, the immersion $(X - Z) \to (X - T)$ induces a canonical map $\bar{\nu} : (X/X - Z) \to (X/X - T)$ in $S\mathscr{H}(X)$. Then we associate to a cohomology class $\rho : (X/X - T) \to \mathbb{E}_X(m)[n]$ the map $\nu_!(\rho) := \rho \circ \bar{\nu}$.
- The unit map of the adjunction (ν^*, ν_*) gives a morphism

$$ad_\nu : i_*(\mathbb{1}_Z) \to i_* \nu_* \nu^*(\mathbb{1}_Z) = k_*(\mathbb{1}_T)$$

in $S\mathscr{H}(X)$. Then for any cohomology class $\rho : k_*(\mathbb{1}_T) \to \mathbb{E}_X(m)[n]$, we put: $\nu_!(\rho) := \rho \circ ad_\nu$.

1.2.8 *Products*: Consider the assumption of the preceding paragraph, except that now we ask $\mathbb{E}$ is an absolute ring spectrum. We define a *refined product* in cohomology with supports

$$\mathbb{E}_T^{n,m}(Z) \otimes \mathbb{E}_Z^{s,t}(X) \to \mathbb{E}_T^{n+s,m+t}(X), (\lambda, \rho) \mapsto \lambda \cdot \rho \qquad (1.2.8.\mathrm{a})$$

as follows. Consider cohomology classes:

$$\lambda : \nu_*(\mathbb{1}_T) \to \mathbb{E}_Z(t)[s], \quad \rho : i_*(\mathbb{1}_Z) \to \mathbb{E}_X(m)[n].$$

Applying i_* to λ, we get a map:

$$k_*(\mathbb{1}_T) = i_* \nu_*(\mathbb{1}_T) \to i_*(\mathbb{E}_Z(m)[n])$$

$$\xrightarrow{\epsilon_i^{-1}} i_* i^*(\mathbb{E}_X(m)[n]) \simeq \mathbb{E}_X \wedge i_*(\mathbb{1}_Z)(m)[n]$$

where the last identification uses (A7). Let us simply denote by $i_*(\lambda)$ this composite map. We define the product $\lambda \cdot \rho$ as the following composite morphism:

$$k_*(\mathbb{1}_T) \xrightarrow{i_*(\lambda)} \mathbb{E}_X \wedge i_*(\mathbb{1}_Z)(m)[n] \xrightarrow{Id \wedge \rho}$$

$$\mathbb{E}_X \wedge \mathbb{E}_X(m+s)[n+t] \xrightarrow{\mu_X} \mathbb{E}_X(m+s)[n+t].$$

One can deduce from this product the usual product of cohomology with support as follows. Assume we are given a cartesian square of closed immersions:

$$\begin{array}{ccc} T & \xrightarrow{i'} & Z' \\ \nu \downarrow & & \downarrow \iota \\ Z & \xrightarrow{i} & X. \end{array}$$

Then one defines the *cup-products* by the following formula:

$$\mathbb{E}_Z^{n,m}(X) \otimes \mathbb{E}_{Z'}^{s,t}(X) \to \mathbb{E}_T^{n+s,m+t}(X), (\alpha, \beta) \mapsto \alpha \cup \beta = \iota^*(\alpha) \cdot \beta. \quad (1.2.8.\mathrm{b})$$

One can also describe this product using the following identification in $S\mathcal{H}(X)$:
$$(X/X - Z) \wedge (X/X - Z') = (X/X - T).$$

This can be obtained by a direct computation of homotopy colimits or by applying formula (1.1.3.a) and (A6) as follows:

$$(X/X - Z') \wedge (X/X - Z) = \iota_*(\mathbb{1}_{Z'}) \wedge i_*(\mathbb{1}_Z) =$$
$$\iota_*(\iota^* i_*(\mathbb{1}_Z)) = \iota_* i'_* \nu^*(\mathbb{1}_Z) = (X/X - T).$$

Then given $\alpha : (X/X - Z) \to \mathbb{E}(m)[n]$ and $\beta : (X/X - Z') \to \mathbb{E}_X(t)[s]$, one checks that $\iota^*(\alpha) \cdot \beta$ is equal to the following composite map:

$$(X/X - T) = (X/X - Z) \wedge (X/X - Z') \xrightarrow{\alpha \wedge \beta} \mathbb{E}_X \wedge \mathbb{E}_X(m+s)[n+s]$$
$$\xrightarrow{\mu_X} \mathbb{E}_X(m+s)[n+r].$$

Note that when $Z = Z' = X$ the product (1.2.8.b) describes the usual cup-product in $\mathbb{E}$-cohomology.

Remark 1.2.9 The need of the refined product just defined is the only reason for us to use absolute ring spectra. Indeed, we have not only used the existence of the structural map ϵ_f but also the fact it is an isomorphism.

We have gathered the basic properties of these operations in the following proposition:

Proposition 1.2.10 *Given an absolute ring spectrum $\mathbb{E}$, the following properties hold:*

(E1) $f^* g^* = (gf)^*$, $\nu'_! \nu_! = (\nu' \nu)_!$ *whenever defined.*

(E2) *When ν is a closed nil-immersion, $\nu_!$ is an isomorphism.*

(E3) *Consider the following cartesian squares:*

$$\begin{array}{ccc}
T' & \xrightarrow{\nu'} Z' \longrightarrow & X' \\
\downarrow & \downarrow{\scriptstyle g} & \downarrow{\scriptstyle f} \\
T & \xrightarrow{\nu} Z \longrightarrow & X
\end{array}$$

where horizontal maps are closed immersions. Then for any cohomology class $\rho \in \mathbb{E}_T^{**}(X)$, $f^* \nu_!(\rho) = \nu'_! f^*(\rho)$.

(E4) *Consider closed immersions: $W \xrightarrow{\iota} T \xrightarrow{\nu} Z \xrightarrow{i} X$. Then for any triple $(\lambda, \alpha, \beta) \in \mathbb{E}_W^{**}(T) \times \mathbb{E}_T^{**}(Z) \times \mathbb{E}_Z^{**}(X)$, one has:*
$$\lambda \cdot (\alpha \cdot \beta) = (\lambda \cdot \alpha) \cdot \beta.$$

(E5) *Under the assumption of (E3), for any couple* $(\lambda, \rho) \in \mathbb{E}_T^{**}(Z) \times \mathbb{E}_Z^{**}(X)$, *one has:*
$$f^*(\lambda \cdot \rho) = g^*(\lambda) \cdot f^*(\rho).$$

(E6) *Under the assumption of (E4), for any couple* $(\lambda, \rho) \in \mathbb{E}_T^{**}(Z) \times \mathbb{E}_Z^{**}(X)$, *one has:*
$$\nu_!(\lambda \cdot \rho) = \nu_!(\lambda) \cdot \rho.$$

(E7) *Consider the following diagram:*

$$
\begin{array}{ccc}
T' & \xrightarrow{\nu'} & Z' \\
{\scriptstyle h}\downarrow & & \downarrow{\scriptstyle g} \\
T & \xrightarrow{\nu} & Z \xrightarrow{i} X
\end{array}
$$

made of closed immersions and such that the square is cartesian. Then for any couple $(\lambda, \rho) \in \mathbb{E}_T^{**}(Z) \times \mathbb{E}_{Z'}^{**}(X)$, *one has:* $h_!(g^*(\lambda) \cdot \rho) = \lambda \cdot g_!(\rho).$

Proof (E1) is clear and (E2) follows from Remark 1.1.2.

Consider (E3). Recall that $\nu_!$ (resp. $\nu_!'$) is induced by (pre)composition with the canonical map $\epsilon_X : (X/X - Z) \to (X/X - T)$ (resp. $\epsilon_Y : (Y/Y - Z') \to (Y/Y - T'))$. Then (E3) simply follows from the fact $f^*(\epsilon_X) = \epsilon_Y$.

Point (E4) follows easily using the associativity of the product on the ring spectrum $\mathbb{E}_X$. Point (E5) follows easily using (A6) and point (E6) is clear.

Point (E7) is the most difficult one. Let us denote by i' (resp. k, k') the obvious embedding of Z' (resp. T, T') in X. The cohomology classes of the statement to be proved can be written as follows:

$$\lambda : \nu_*(\mathbb{1}_T) \to \mathbb{E}_Z, \quad \rho : i'_*(\mathbb{1}_{Z'}) \to \mathbb{E}_X.$$

We let $ad_g : i_*(\mathbb{1}_Z) \to i'_*(\mathbb{1}_{Z'})$ (resp. $ad_h : k_*(\mathbb{1}_T) \to k'_*(\mathbb{1}_{T'})$) be the map induced by the unit of the adjunction (g_*, g^*) (resp. (h^*, h_*)). Then the left hand side of the relation to be proved is equal to:

$$k_*(\mathbb{1}_T) \xrightarrow{ad_h} k'_*(\mathbb{1}_{T'}) = i'_*g^*\nu_*(\mathbb{1}_Z) \xrightarrow{i'_*g^*(\lambda)} i'_*g^*(\mathbb{E}_Z) = i'_*(\mathbb{E}_{Z'}) =$$

$$\mathbb{E}_X \wedge i'_*(\mathbb{1}_{Z'}) \xrightarrow{Id \wedge \rho} \mathbb{E}_X \wedge \mathbb{E}_X \xrightarrow{\mu} \mathbb{E}_X,$$

while the right hand side is:

$$k_*(\mathbb{1}_T) = i_*\nu_*(\mathbb{1}_Z) \xrightarrow{i_*(\lambda)} i_*(\mathbb{E}_Z) = \mathbb{E}_X \otimes i_*(\mathbb{1}_Z) \xrightarrow{Id \wedge ad_g} \mathbb{E}_X \wedge i'_*(\mathbb{1}_{Z'})$$

$$\xrightarrow{Id \wedge \rho} \mathbb{E}_X \wedge \mathbb{E}_X \xrightarrow{\mu} \mathbb{E}_X.$$

To check the identity, we prove the commutativity of the following diagram:

$$
\begin{array}{ccccccccc}
k_*h_*h^*(\mathbf{1}_T) & = & i_*g_*\nu'_*h^*(\mathbf{1}_T) & \xrightarrow{b^{-1}} & i_*g_*g^*\nu_*(\mathbf{1}_T) & \xrightarrow{\lambda} & i_*g_*g^*(\mathbb{E}_Z) & = & \mathbb{E}_X \wedge i_*g_*g^*(\mathbf{1}_Z) \\
{\scriptstyle ad_h}\uparrow & & {\scriptstyle (1)} & & {\scriptstyle ad_g}\uparrow & & {\scriptstyle ad_g}\uparrow & & {\scriptstyle Id \wedge ad_g}\uparrow \\
k_*(\mathbf{1}_T) & = & = & i_*\nu_*(\mathbf{1}_T) & \xrightarrow{\lambda} & i_*(\mathbb{E}_Z) & = & \mathbb{E}_X \wedge i_*(\mathbf{1}_Z)
\end{array}
$$

where the maps with label ad_g (resp. ad_h) are induced by the unit of the adjunction (g^*, g_*) (resp. (h^*, h_*)) and the map b, whose inverse appears in the above diagram, stands for the base change isomorphism obtained from (A6). Only the commutativity of part (1) is non trivial. It follows from the description of the base change map b as the composite:

$$
g^*\nu_* \xrightarrow{ad_h} g^*\nu_*h_*h^* = g^*g_*\nu'_*h^* \xrightarrow{ad'_g} \nu'_*h^*
$$

and the relation between the unit and the counit ad'_g of the adjunction (g^*, g_*).

$\hfill\square$

1.2.11 It is usually convenient to introduce the following notions. We define a *morphism of closed pairs* $\Delta : (Y, T) \to (X, Z)$ as being a commutative diagram

$$
\begin{array}{ccc}
T & \to & Y \\
{\scriptstyle g}\downarrow & & \downarrow{\scriptstyle f} \\
Z & \to & X
\end{array}
$$

such that the induced map $T \xrightarrow{\nu} Y \times_X Z$ is a nil-immersion.[28] We say that the morphism Δ is *cartesian* when the above square is cartesian in the category of schemes — i.e. ν is an isomorphism. We also use the notation (f, g) for Δ when we want to refer to the morphisms in the above square. The composition of morphisms of closed pairs is given in categorical terms by the vertical composition of squares.

Using (E2), we associate to Δ the following composite morphism of abelian groups:

$$
\Delta^* : \mathbb{E}_Z^{n,m}(X) \xrightarrow{f^*} \mathbb{E}_{Z \times_X Y}^{n,m}(Y) \xrightarrow{\nu_!^{-1}} \mathbb{E}_T^{n,m}(Y).
$$

The relation $\Delta^*\Theta^* = (\Theta\Delta)^*$ is clear from (E1) and (E3).

Remark 1.2.12 Given a transformation $\varphi : \mathbb{E} \to \mathbb{F}$, it is clear that the associated morphism (1.2.5.a) is natural with respect to contravariant and covariant functorialities. Moreover, it is compatible with all the products of Paragraph 1.2.8.

---

[28] In other words, the diagram Δ is topologically cartesian.

The following property is not essential to our purpose so that we do not list it among the fundamental axioms of an absolute cohomology.[29]

Lemma 1.2.13 *Let $(X_\alpha, Z_\alpha)_{\alpha \in A}$ be an essentially affine projective system of closed pairs of $\mathscr{S}$ whose projective limit[30] (X, Z) is still a closed pair in $\mathscr{S}$.*

Then, given any $\mathscr{S}$-absolute ring spectrum $\mathbb{E}$, the canonical map:

$$\varinjlim_{\alpha \in A^{op}} \left(\mathbb{E}_{Z_\alpha}^{n,i}(X_\alpha) \right) \to \mathbb{E}_Z^{n,i}(X)$$

is an isomorphism.

Proof According to [CD12b, 4.3.6], the category $S\mathscr{H}$ is a continuous motivic category in the sense of *loc. cit.*, 4.3.2. Thus the lemma follows by applying *lo. cit.*, 4.3.4 to the projective systems

$$(X_\alpha/X_\alpha - Z_\alpha)_{\alpha \in A}, (\mathbb{E}_{X_\alpha})_{\alpha \in A}$$

given that for any index $\alpha \in A$, the object $(X_\alpha/X_\alpha - Z_\alpha)$ of $S\mathscr{H}(X_\alpha)$ is constructible.

$\square$

1.3 Absolute purity

1.3.1 Let (X, Z) be a closed pair in $\mathscr{S}$. We say that (X, Z) is *regular* if the inclusion $i : Z \to X$ is a regular embedding.

Assume (X, Z) is regular. We let $N_Z X$ (resp. $B_Z X$) be the normal cone (resp. blow-up) of Z in X. Recall the definition of the *deformation space* of (X, Z) as:

$$D_Z X = B_{0 \times Z}(\mathbb{A}_X^1) - B_Z X$$

(see [Ros96], or [Dég08a, §4.1] for this presentation).[31] This is a flat scheme over $\mathbb{A}^1$ whose fiber over 1 (resp. 0) is X (resp. $N_Z X$). Note also that $D_Z Z = \mathbb{A}_Z^1$ is a closed subscheme of $D_Z X$ so that we finally get a *deformation diagram* of closed pairs:

$$(X, Z) \xrightarrow{\sigma_1} (D_Z X, \mathbb{A}_Z^1) \xleftarrow{\sigma_0} (N_Z X, Z), \qquad (1.3.1.a)$$

made of cartesian morphisms. Note this diagram is natural with respect to cartesian morphisms of closed pairs.

---

[29]It will only be applied in Example 1.3.4 in the case of k-absolute ring spectra.
[30]in the category of pairs of schemes,
[31]Note that according to our convention on $\mathscr{S}$, this is a scheme in $\mathscr{S}$.

Definition 1.3.2 Let $\mathbb{E}$ be a weak $\mathscr{S}$-absolute (ring) spectrum (Remark 1.2.2).

For any closed pair (X, Z), we say that (X, Z) is $\mathbb{E}$-pure if (X, Z) is regular and the morphisms

$$\mathbb{E}^{**}(X, Z) \xleftarrow{\sigma_1^*} \mathbb{E}^{**}(D_Z X, \mathbb{A}_Z^1) \xrightarrow{\sigma_0^*} \mathbb{E}^{**}(N_Z X, Z)$$

induced by the above deformation diagram (see notation in Definition 1.2.5) are isomorphisms of bigraded abelian groups.

We say that $\mathbb{E}$ is *absolutely pure* if any regular closed pair in $\mathscr{S}$ is $\mathbb{E}$-pure.

Note the following trivial stability properties of spectra satisfying the absolute purity property with respect to a given closed pair:

Proposition 1.3.3 *Let* (X, Z) *be a regular closed pair in* $\mathscr{S}$.

Then the category of absolute spectra (resp. weak absolute spectra) $\mathbb{E}$ *such that* (X, Z) *is* $\mathbb{E}$-pure is stable by suspension and twists, direct factors, infinite direct sums and distinguished triangles.*[32]

Example 1.3.4 We refer the reader to Example 1.2.3 for the absolute ring spectra appearing below:

(1) Let $\mathbb{E}$ be any absolute ring spectrum and S be a scheme. We will say that a closed S-pair (X, Z) is smooth if X and Z are smooth over S.

According to [MV99, §3.2, th. 2.23, p. 115], any smooth closed S-pair (X, Z) is $\mathbb{E}$-pure. Thus, if $\mathscr{S}$ is the category of smooth S-schemes, any $\mathscr{S}$-absolute ring spectrum is absolutely pure. Note in particular this is the case for the $\mathscr{S}m_S$-absolute ring spectrum $\mathbf{H}_\Lambda$, representing motivic cohomology with Λ-coefficients.

(2) Let k be a perfect field.

According to Popescu theorem, any regular closed pair (X, Z) over k can be written as a projective limit of smooth closed pairs over k provided X is regular. Thus, according to the computation of the cohomology with supports of a projective limit (cf. Lemma 1.2.13), we deduce from the previous example that any k-absolute ring spectrum is absolutely pure. This is in particular the case for the k-absolute motivic Eilenberg-MacLane spectrum $\mathbf{H}_\Lambda$.

(3) According to [CD12b], respectively Theorems 14.4.1 and 13.6.3, the absolute ring spectra $\mathbf{H}_{\mathbb{B}}$ and $\mathbf{KGL}$ are absolutely pure.

---

[32] A distinguished triangle of (weak) absolute spectra is the datum of distinguished triangles over each schemes which are compatible with the structural base change maps.

(4) Recall from [NSØ09b, 10.5] that there is an isomorphism of absolute ring spectra:
$$\mathbf{MGL} \otimes_{\mathbb{Z}} \mathbb{Q} = \mathbf{H}_{\mathbb{B}}[b_1, b_2, \ldots]$$
where b_i is a generator of degree $(2i, i)$. Then the preceding example implies $\mathbf{MGL} \otimes \mathbb{Q}$ is absolutely pure. We deduce from that example that any Landweber spectrum (cf. [NSØ09b, th. 7.3]) with rational coefficients is absolutely pure.

The only integral example of an absolutely pure ring spectrum is given by homotopy invariant K-theory. However, in view of the previous examples, we think it is reasonable to conjecture the following:

Conjecture A   *The absolute ring spectrum* $\mathbf{MGL}$ *is absolutely pure.*

Conjecture B   *The absolute ring spectrum* S^0 *is absolutely pure.*

Remark 1.3.5   (1) Recall the notion of cellular spectra, first introduced in [DI05]. Over a scheme S, one defines the category of cellular spectra as the smallest thick triangulated subcategory of $S\mathcal{H}(S)$ which contains the spheres that we denote here $S^0(m)[n]$ for any integers $(n, m) \in \mathbb{Z}$. According to the previous proposition, Conjecture B implies that any absolute spectrum $\mathbb{E}$ which is cellular over $\mathrm{Spec}\,(\mathbb{Z})$ is absolutely pure.

In particular, Conjecture B implies Conjecture A because $\mathbf{MGL}_S$ is cellular (see [DI05, 6.4]) — and the former conjecture would reprove the absolute purity for $\mathbf{KGL}$ (see [DI05, 6.4]).

(2) Let S be any scheme and Λ be the localization of $\mathbb{Z}$ at the primes which are not invertible on S. Let $\mathbf{M}\Lambda$ be the absolute ring spectrum of Spitzweck (see Ex. 1.2.3(5)). According to [Spi13, Cor. 11.4], $\mathbf{M}\Lambda_S$ is cellular. Thus, according point (1) above, Conjecture B implies that $\mathbf{M}\Lambda$ is absolutely pure over the category of S-schemes.

(3) Supporting these conjectures:

- $\mathbf{MGL}$ and S^0 are absolutely pure for closed pairs which are smooth over some base.
- The conjecture for $\mathbf{MGL} \otimes \mathbb{Q}$ is true according to point (3) of the preceding example.
- The conjecture for $S^0 \otimes \mathbb{Q}$ is true if one restricts to base schemes over which -1 is a sum of squares. In fact, under this assumption, according to Morel theorem one has $S^0 \otimes \mathbb{Q} \simeq \mathbf{H}_{\mathbb{B}}$ (see [CD12b, Cor. 16.2.14]) and we are reduced to point (3) of the above example.

(4) Let us consider the Eilenberg-MacLane motivic ring spectrum, as a weak absolute ring spectrum, $\mathbf{H}\mathbb{Z}^w$ — Example 1.2.3(4). Independently of the conjecture of Voevodsky which ask whether $\mathbf{H}\mathbb{Z}^w$ is an absolute ring spectrum, it is interesting to ask if $\mathbf{H}\mathbb{Z}^w$ is absolutely pure. Note that if this was true, then using the coniveau spectral sequence we will get for any regular scheme S an isomorphism

$$H_{\mathcal{M}}^{2n,n}(S,\mathbb{Z}) \simeq CH^n(S)$$

where the right hand side is the group of n-codimensional cycles in S modulo rational equivalence (see [Ful98, Gil05]).[33] The existence of this isomorphism is particularly interesting as there is a well defined product of Voevodsky's motivic cohomology while it is still an open question to define a product on the classical Chow (see [BGI71, XIV, §8]).

(5) We have separated the case of $\mathbf{MGL}$ with that of S^0 because $\mathbf{MGL}$ is oriented (see next section) and S^0 is not. Note however that in our formulation of absolute purity, we do not need any orientation. In the particular, one can see that the Conjecture B is equivalent to ask the following:

> For any closed immersion $i : Z \to X$ between regular schemes, with normal bundle $N_Z X$, there exists a canonical isomorphism in $S\mathcal{H}(Z)$:
>
> $$i^{!}(S^0) \simeq \mathrm{Th}(-N_Z X)$$
>
> — see Par. 2.2.1 for recall on the definition of the right hand side.

The resulting map in $S\mathcal{H}(S)$

$$i_*(\mathrm{Th}(-N_Z X)) \to S^0$$

would be called the (unoriented) fundamental class of i (relative to S^0). This is the unoriented version of Definition 2.3.1. Moreover, one can see that this class would be universal among the fundamental classes constructed in this paper (using that any ring spectrum is an algebra over S^0).

---

[33]The absolute purity property allows to compute the E_1-term as the $\mathbf{H}\mathbb{Z}^w$-cohomology of the residue fields for $\mathbf{H}\mathbb{Z}^w$, as S is regular. As the motivic cohomology of fields is known, one gets the mentioned computation after identifying one of the differential with the classical divisor class map.

1.4   Analytical invariance

By construction of the stable homotopy category, an absolute cohomology theory satisfies cohomological descent for the Nisnevich topology. A convenient way to express this property uses the so called *excision property*. Let us start by an elementary geometric fact which will link excision with analytical invariance.

Proposition 1.4.1 *Let $f : Y \to X$ be a morphism locally of finite type, Z a closed subscheme of X, $T = f^{-1}(Z)$. Then the following assertions are equivalent:*

(i) *f is étale at all points of the scheme T and the induced morphism $f|_T : T_{red} \to Z_{red}$ is an isomorphism.*

(ii) *the induced morphism $\hat{f} : \hat{Y}_T \to \hat{X}_Z$ between the respective formal completions of Y at T and X at Z is an isomorphism.*

Proof   Note first that given any point $x \in Z$, the completion of the local ring of $\hat{X}_Z$ at x coincides with the completion of the local ring of X at x and the corresponding isomorphism is natural in (X, Z, x).

Thus the equivalence of the assertions follow from [GD67, 17.6.3] which asserts that f is étale at a point y of T, $x = f(y)$, if and only if the induced morphism $\hat{\mathcal{O}}_{X,x} \to \hat{\mathcal{O}}_{Y,y}$ between the respective completed local rings is an isomorphism.

$\square$

Definition 1.4.2 Let $\Delta = (f, g) : (Y, T) \to (X, Z)$ be a morphism of closed pairs. One says that Δ is *excisive* if the morphism f and the closed scheme $Z \subset X$ satisfy the equivalent assertions of the preceding proposition.

Thus, generalizing slightly [MV99, Lem. 1.6, p. 98], we get the following property of the absolute cohomology represented by $\mathbb{E}$:

(Nis) For any excisive morphism $\Delta : (Y, T) \to (X, Z)$, the associated pull-back $\Delta^* : \mathbb{E}_Z^{**}(X) \to \mathbb{E}_T^{**}(Y)$ is an isomorphism.

Remark 1.4.3 In fact, it is well known since the work of Morel-Voevodsky that this property characterizes Nisnevich descent, though one only needs excisive morphisms which are globally étale (see for example [CD12a, Prop. 1.1.10] for a precise statement).

An interesting corollary of the absolute purity property is the following stronger statement (see the end of this section for a stronger result):

Proposition 1.4.4 *Let $\mathbb{E}$ be an absolutely pure $\mathscr{R}$eg-spectrum. Let X be a regular local scheme with closed point x, $\hat{X}$ be its completion at the point x and consider the canonical map $f : (X, x) \to (\hat{X}, x)$.*

Then the induced morphism of cohomology with support:

$$f^* : \mathbb{E}_x^{**}(\hat{X}) \to \mathbb{E}_x^{**}(X)$$

is an isomorphism.

Remark 1.4.5 This result will be used to compute residues in Proposition 5.4.5.

Using the Artin approximation property, one can give a much stronger result than the preceding proposition according to an initial idea of Wildeshaus (cf. [Wil06, §5], in the case of motives over a field).

Theorem 1.4.6 *Let S be an excellent scheme, (X, Z) and (Y, T) be closed pairs made of S-schemes essentially of finite type.*

Let $\mathbb{E}$ be an S-absolute spectrum.

Assume there exists an isomorphism $\mathfrak{f} : \hat{Y}_T \to \hat{X}_Z$ between the respective formal completions. Then there exists an isomorphism:

$$\mathfrak{f}^* : \mathbb{E}_Z^{**}(X) \to \mathbb{E}_T^{**}(Y)$$

which depends only on $\mathfrak{f}$.

Proof The following proof follows that of [Wil06, 5.5] using the more advanced theory we now have at our will.

We can assume $Z = T$, seen as a reduced scheme. Let us fix a point $z \in Z$, and let $X_{(z)}$, $Y_{(z)}$ be the respective local schemes of X, Y at z. According to [Swa98, 2.4], the henselisation of $\mathcal{O}_{X,z}$ satisfies the Artin approximation property. Thus according to [Art69, 2.6], there exists a common Nisnevich neighborhood $W_{(z)}$ of $(X_{(z)}, z)$ and $(Y_{(z)}, z)$. Moreover, one can lift the situation in a neighborhood of z, both in X and Y: there exists an S-scheme W essentially of finite type, which lifts $W_{(z)}$ and fits into the following commutative diagram:

$$
\begin{array}{ccccc}
 & & W & & \\
 & {}^{f}\swarrow & \uparrow & \searrow^{g} & \\
X' & \leftarrow & Z' & \rightarrow & Y' \\
\updownarrow & & \updownarrow & & \updownarrow \\
X & \leftarrow & Z & \rightarrow & Y
\end{array}
$$

where the squares are cartesian, X', Y', Z' are open neighborhood of z in X, Y, Z respectively, and f (resp. g) is a Nisnevich neighborhood of Z' in X (resp. Y).

Using this construction, one can further find Zariski hypercoverings $\mathcal{X}$, $\mathcal{Y}$, $\mathcal{Z}$ of X, Y, Z and a simplicial scheme $\mathcal{W}$ which fits into the following commutative diagram:

$$
\begin{array}{ccc}
 & \mathcal{W} & \\
{}^{f}\swarrow \; {}^{\uparrow} \; {}^{g}\searrow & & \\
\mathcal{X} \longleftarrow \mathcal{Z} \longrightarrow \mathcal{Y} & & \\
{}^{p}\downarrow \quad \downarrow \quad \downarrow {}^{q} & & \\
X \longleftarrow Z \longrightarrow Y & &
\end{array}
\tag{1.4.6.a}
$$

such that the squares are cartesian and for each integer $n \geq 0$, $\mathcal{W}_n$ is a Nisnevich neighborhood of $\mathcal{Z}_n$ in $\mathcal{X}_n$ (resp. $\mathcal{Y}_n$).

To finish the proof, one has to use the fact that the stable homotopy category $S\mathcal{H}(S)$ can be extended to simplicial schemes according to [Ayo07b, Chap. 4] or [CD12b, 3.1]. This implies that the cohomology $\mathbb{E}^{**}$ can be extended to simplicial S-schemes and simplicial Z-pairs. Then the following sequence gives us almost the desired isomorphism

$$
\mathfrak{f}_{\mathcal{W}}^{*} : \mathbb{E}_{Z}^{**}(X) \xrightarrow{p^{*}} \mathbb{E}_{\mathcal{Z}}^{**}(\mathcal{X}) \xrightarrow{f^{*}} \mathbb{E}_{\mathcal{Z}}^{**}(\mathcal{W}) \xrightarrow{(g^{*})^{-1}} \mathbb{E}_{\mathcal{Z}}^{**}(\mathcal{Y}) \xrightarrow{(q^{*})^{-1}} \mathbb{E}_{Z}^{**}(Y).
$$

Indeed, according to the Zariski descent property of $S\mathcal{H}$ as formulated in [CD12b, 3.2.7, 3.3.5], the maps p^{*} and q^{*} are isomorphisms. Moreover, we can derive from property (Nis) the fact the maps f^{*} and g^{*} are isomorphisms (either we apply the Zariski descent spectral sequence or we argue directly in $S\mathcal{H}$).

To finally get $\mathfrak{f}^{*}$, one has to take the limit of the isomorphisms $\mathfrak{f}_{\mathcal{W}}^{*}$ over the filtering category of diagrams of the form (1.4.6.a).

$\square$

Remark 1.4.7 (1) The fact $\mathfrak{f}^{*}$ depends only on $\mathfrak{f}$ can also be supplemented by the following cocycle condition: given composable isomorphisms $\mathfrak{f}$ and $\mathfrak{g}$ of certain formal completions, one gets: $(\mathfrak{f} \circ \mathfrak{g})^{*} = \mathfrak{g}^{*} \circ \mathfrak{f}^{*}$.

(2) Under the presence of a ring structure on $\mathbb{E}$, the isomorphism $\mathfrak{f}^{*}$ is compatible with products as defined in Paragraph 1.2.8.

As a corollary, we get the following reinforcement of Prop. 1.4.4:

Corollary 1.4.8 *Let $\mathbb{E}$ be an absolute spectrum. Let X be a local scheme with closed point x, $\hat{X}$ be its completion at the point x and consider the* *canonical map* $f : (X, x) \to (\hat{X}, x)$. *Then the induced morphism of coho-mology with support:*

$$f^* : \mathbb{E}_x^{**}(\hat{X}) \to \mathbb{E}_x^{**}(X)$$

is an isomorphism.

Example 1.4.9 The preceding theorem and its corollary can be applied in particular to any of the absolute ring spectrum of Example 1.2.3. In particular we get another proof of analytical invariance for De Rham co-homology and a proof in the case of rigid cohomology. The later case was also proved independently by Ouwehand in [Ouw14].

Note that it also holds for K-theory and algebraic cobordism. Even in the first case, this seems to be new.

Remark 1.4.10 The preceding corollary is especially useful in dealing with absolute purity. In order to prove it for a given absolute ring spec-trum, one easily reduces to the case of a closed pair (X, x) where X is a local regular scheme with closed point x. According to the preceding corol-lary, one derives that we can further assume that X is the spectrum of a complete local scheme.

2 Orientation and characteristic classes

2.1 Orientation theory and Chern classes

The considerations of this section are well known in motivic homotopy theory (see for example [Bor03, Vez01] — in the case of a base field). They were also studied, with a slightly different formalism, in our paper [Dég08a].

2.1.1 Let S be a scheme. We will assume that the scheme $\mathbb{P}_S^n$ is pointed by the infinite point (of homogeneous coordinates $[0 : \cdots : 0 : 1]$ to fix ideas). Then we get a tower of pointed S-schemes

$$\mathbb{P}_S^1 \to \mathbb{P}_S^2 \to \cdots \to \mathbb{P}_S^n \xrightarrow{\iota_n} \mathbb{P}_S^{n+1} \to \cdots$$

where ι_n denotes the embedding of the last n-th coordinates. The colimit of this tower in the category of pointed sheaves defines an object $\mathbb{P}_S^\infty$ of the pointed homotopy category $\mathscr{H}_\bullet(S)$ — see [MV99]. We still denote by $\iota_1 : \mathbb{P}_S^1 \to \mathbb{P}_S^\infty$ the induced map in the homotopy category.

The following definition is now basic in motivic homotopy theory:

Definition 2.1.2 Let $\mathbb{E}$ be an absolute ring spectrum with unit $\eta_S : S^0 \to \mathbb{E}_S$ over a scheme S in $\mathscr{S}$. We can see η_S as a class in the reduced cohomology $\tilde{\mathbb{E}}^{2,1}(\mathbb{P}^1_S)$.[34]

An *orientation* of $\mathbb{E}$ over S is a class c_S in the reduced cohomology $\tilde{\mathbb{E}}^{2,1}(\mathbb{P}^\infty_S)$ such that $\iota_1^*(c_S) = \eta_S$.

An (absolute) *orientation* of $\mathbb{E}$ is a family of classes $c = (c_S)$ for any scheme S in $\mathscr{S}$ such that for any morphism $f : T \to S$, $f^*(c_S) = c_T$. In this situation, we also say that $(\mathbb{E}, c)$ is an absolute oriented ring spectrum.

Remark 2.1.3 To give an orientation of an S-absolute ring spectrum $\mathbb{E}$ it is enough and sufficient to give an orientation of $\mathbb{E}_S$.

Example 2.1.4 Let us review the absolute ring spectra of Example 1.2.3:

(1) According to [CD12a, 2.2.8], the k-absolute ring spectrum associated with a Mixed Weil theory is canonically oriented.

(2) Voevodsky's Einlenberg-Mac Lane motivic ring spectrum $\mathbf{H}_\Lambda$, considered as an S-absolute ring spectrum, is oriented according to [CD12b, Sec. 11.3].

(3) The cobordism ring spectrum $\mathbf{MGL}$ is canonically oriented. This follows from theoretical reasons (see recall in Prop. 2.2.6) or can be directly seen from the construction (see Par. 2.2.5).

(4) The K-theory ring spectrum $\mathbf{KGL}$ is oriented. Let S be a regular scheme ($S = \mathrm{Spec}\,(\mathbb{Z})$ would be enough). By construction, we have a canonical isomorphism:
$$K_0(S) \simeq \mathbf{KGL}^{0,0}(S).$$

In particular, any line bundle L/S defines an element $[L] \in \mathbf{KGL}^{0,0}(S)$. Moreover, Bott periodicity theorem implies the existence of the following isomorphism:
$$\mathbf{KGL}^{0,0}(S) \simeq K_0(S) \simeq \tilde{K}_0(\mathbb{P}^1_S, \infty) \simeq \widetilde{\mathbf{KGL}}^{0,0}(\mathbb{P}^1_S, \infty).$$

The image of 1 by this isomorphism is an element β in $\mathbf{KGL}^{-2,-1}(S)$ called the *Bott element*. Let λ be the canonical line bundle on $\mathbb{P}^\infty_S$. Then we define the orientation of $\mathbf{KGL}$ as follows:[35]
$$c^{\mathbf{KGL}} := \beta^{-1}.(1 - [\lambda^\vee]) \in \mathbf{KGL}^{2,1}(\mathbb{P}^\infty_S).$$

---

[34] By definition of the Tate twist, $\mathbb{E}^{2,1}(\mathbb{P}^1_S) = \mathbb{E}^{2,1}(S) \oplus \mathbb{E}^{0,0}(S)$.

[35] This choice of orientation coincides with the one of [LM07, Ex. 1.1.5]. In the literature however, one can find different choices of orientations of the ring spectrum $\mathbf{KGL}$. The present choice is justified by Example 3.2.9 as well as the correct form of the Todd class appearing in the Riemann-Roch theorem 5.3.4.

(5) Beilinson motivic cohomology spectrum $\mathbf{H}_B$ is canonically oriented (see [CD12b, 14.1.5], [NSØ09b]).

It is worth to point out that we will show in Cor. 5.1.10 and Ex. 5.1.11 that the orientations involved in points (1), (2) and (5) are unique.

Remark 2.1.5 The presence of the Bott element in formulas involving the spectrum $\mathbf{KGL}$ can be explained as follows. By construction (see [Rio10]), for any integer n and any regular scheme S, one has a canonical contravariantly functorial isomorphism:

$$\varphi_S^n : \mathbf{KGL}^{n,0}(S) \to K_{-n}(S)$$

where the right hand side is the n-th Quillen K-theory of S. According to the definition above, multiplication by β on $\mathbf{KGL}^{**}(S)$ induces an isomorphism. Thus, for any couple of integers (r, n), we get a canonical isomorphism:

$$\varphi_S^{n,i} : \mathbf{KGL}^{n,i}(S) \to \mathbf{KGL}^{n-2i,0}(S) \to K_{2i-n}(S), x \mapsto \varphi_S^{n-2i}(\beta^i.x).$$
$$(2.1.5.a)$$

2.1.6 Recall from [MV99, §4, 3.7] there exists a canonical isomorphism in $\mathscr{H}_\bullet(S)$:

$$B\mathbb{G}_m \simeq \mathbb{P}_S^\infty$$

where $B\mathbb{G}_m$ is the (Nisnevich) classifying space of $\mathbb{G}_m$. This immediately gives an application

$$\mathrm{Pic}(S) \overset{(*)}{=} \mathrm{Hom}_{\mathscr{H}_\bullet^s(S)}(S_+, B\mathbb{G}_m) \to \mathrm{Hom}_{\mathscr{H}_\bullet(S)}(S_+, B\mathbb{G}_m)$$
$$\simeq \mathrm{Hom}_{\mathscr{H}_\bullet(S)}(S_+, \mathbb{P}_S^\infty) \quad (2.1.6.a)$$

where $\mathscr{H}_\bullet^s(S)$ denotes the simplicial homotopy category and the first map is induced by the projection functor $\mathscr{H}_\bullet^s(S) \to \mathscr{H}_\bullet(S)$ — the target category being the $\mathbb{A}^1$-localization of the source category. We have used [MV99, §4, 1.15] for the identification $(*)$.[36]

Remark 2.1.7 For any integer n, we let λ_n be the canonical line bundle on $\mathbb{P}_S^n$ (see Notations and conventions, p. 252). Then the family $(\lambda_n)_{n \in \mathbb{N}}$ defines an element λ of $\mathrm{Pic}(\mathbb{P}_S^\infty)$ — which generates this group as a ring of formal power series over $\mathbb{Z}$ as soon as S is local and regular. The map (2.1.6.a) is characterized by the fact it sends λ to the canonical projection $\mathbb{P}_{S+}^\infty \to \mathbb{P}_S^\infty$.

---

[36] Note also Morel and Voevodsky proved the map (2.1.6.a) is an isomorphism whenever S is regular ; *op. cit.* Prop. 3.8.

Definition 2.1.8 Let $(\mathbb{E}, c)$ be an absolute oriented ring spectrum.

For any scheme S, we associate to the class c a canonical morphism of sets:

$$c_1 : \mathrm{Pic}(S) \xrightarrow{(2.1.6.a)} \mathrm{Hom}_{\mathscr{H}_\bullet(S)}(S_+, \mathbb{P}^\infty_S) \xrightarrow{\Sigma^\infty} \mathrm{Hom}_{S\mathscr{H}(S)}(\Sigma^\infty S_+, \Sigma^\infty \mathbb{P}^\infty_S)$$

$$\xrightarrow{(c_S)_*} \mathrm{Hom}_{S\mathscr{H}(S)}(\Sigma^\infty S_+, \mathbb{E}_S(1)[2]) = \mathbb{E}^{2,1}(S),$$

called the *first Chern class*.

Example 2.1.9 In the case of the orientation of **KGL** defined in Ex. 2.1.4, one gets for any line bundle over a regular scheme S:

$$c_1^{\mathbf{KGL}}(L) = \beta^{-1}.(1 - [L^\vee]).$$

2.1.10 According to this definition and the preceding remark, we get the following properties:

(a) For any morphism of schemes $f : T \to S$ and any line bundle λ on S, $f^* c_1(\lambda) = c_1(f^{-1}\lambda)$.

(b) Let $n \geq 0$ be an integer, λ_n be the line bundle over $\mathbb{P}^n_S$ considered in the above remark and $\nu_n : \mathbb{P}^n_S \to \mathbb{P}^\infty_S$ be the obvious morphism.

Then $c_1(\lambda_n) = \nu_n^*(c_S)$ as classes in $\mathbb{E}^{2,1}(\mathbb{P}^n_S)$, according to the above remark.

Remark 2.1.11 One must be careful that the relation $c_1(\lambda \otimes \lambda') = c_1(\lambda) + c_1(\lambda')$ does not necessarily hold. This is due to the fact that the second of the three maps considered in the definition of c_1 is not a morphism of abelian groups. This remark will be made more precise latter (see 2.1.22).

2.1.12 Next we recall the projective bundle theorem for an absolute oriented ring spectrum $(\mathbb{E}, c)$.

Let $p : P \to S$ be a projective bundle of rank n with canonical line bundle λ. We define the following morphism:

$$\epsilon_P : \oplus_{i=0}^n \mathbb{E}^{**}(X) \to \mathbb{E}^{**}(P), (x_0, \dots, x_n) \mapsto \sum_i p^*(x_i).c_1(\lambda)^i.$$

Theorem 2.1.13 *With the above assumptions and notations, the morphism ϵ_P is an isomorphism.*

In other words, $\mathbb{E}^{**}(P)$ is a free graded $\mathbb{E}^{**}(X)$-module with basis $(1, c_1(\lambda), \dots, c_1(\lambda)^n)$ (as usual).

Proof The proof, essentially due to Morel, is the same as the proof of Th. 3.1 in [Dég08a]. We recall the main steps for the convenience of the reader.

Using the Mayer-Vietoris long exact sequence associated to an open cover by two open subsets, we reduce to the case where P is trivializable and then to the case $P = \mathbb{P}^n_S$ — this uses only property 2.1.10(a).

Then the proof goes on by induction on n ; the case $n = 0$ is trivial and the case $n = 1$ is an immediate consequence of the definition of the orientation c_S.

The principle of the induction is to use the following facts:

- Let $\mathbb{P}^n_S/\mathbb{P}^{n-1}_S$ be the cokernel of the embedding ι_{n-1} in the category of pointed sheaves. Then the sequence

$$\mathbb{P}^{n-1}_S \xrightarrow{\iota_n} \mathbb{P}^n_S \xrightarrow{\pi_n} \mathbb{P}^n_S/\mathbb{P}^{n-1}_S$$

 is homotopy exact ; in particular, it induces a long exact sequence:

$$\cdots \to \mathbb{E}^{**}(\mathbb{P}^n_S/\mathbb{P}^{n-1}_S) \xrightarrow{\pi^*_n} \mathbb{E}^{**}(\mathbb{P}^n_S) \xrightarrow{\iota^*_n} \mathbb{E}^{**}(\mathbb{P}^{n-1}_S) \to \cdots$$

- There exists a canonical isomorphisms in $\mathscr{H}_\bullet(S)$:

$$\tau_n : \mathbb{P}^n_S/\mathbb{P}^{n-1}_S \to (\mathbb{P}^1_S)^{\wedge,n}.$$

Then we are reduce to prove the following relations:

- $c_1(\lambda_{n-1})^n = 0.$
- Using the isomorphism τ_n, we get an isomorphism $\tau^*_n : \mathbb{E}^{2n,n}(\mathbb{P}^n_S/\mathbb{P}^{n-1}_S) \simeq \mathbb{E}^{0,0}(S)$. Then $\tau^*_n(\overline{c_1(\lambda_n)^n}) = \eta_S$, the unit of the ring spectrum $\mathbb{E}_S$ — where we have put: $\pi^*_n(\overline{c_1(\lambda_n)^n}) = c_1(\lambda_n)^n$.

These relations can easily be deduced from the following lemma:

Lemma 2.1.14 (Morel) *Let* $\delta_n : \mathbb{P}^n_S \to (\mathbb{P}^n_S)^{\wedge,n}$ *be the n-th diagonal of the pointed scheme* $\mathbb{P}^n_S/S$. *Then the following square commutes in* $\mathscr{H}_\bullet(S)$:

$$
\begin{array}{ccc}
\mathbb{P}^n_S & \xrightarrow{\delta_n} & (\mathbb{P}^n_S)^{\wedge,n} \\
{\scriptstyle \pi_n}\downarrow & & \uparrow{\scriptstyle (\iota_1)^{\wedge,n}} \\
\mathbb{P}^n_S/\mathbb{P}^{n-1}_S & \xrightarrow{\tau_n} & (\mathbb{P}^1_S)^{\wedge,n}.
\end{array}
$$

For this lemma, we refer the reader to the proof of [Dég08a, lem. 3.3].

$\square$

As first remarked by Morel, the projective bundle theorem admits the following corollary — whose proof can be easily adapted from [Dég08a, Cor. 3.6].

Corollary 2.1.15 *Let $\mathbb{E}$ be an orientable absolute ring spectrum. Then for any scheme X in $\mathscr{S}$ and any closed subschemes Z, Z' of X, one has the following property:*

$$\forall(x,y) \in \mathbb{E}_Z^{n,p}(X) \times \mathbb{E}_{Z'}^{m,q}(X),\ x \cup y = (-1)^{nm} y \cup x.$$

Following the method of Grothendieck, we can now introduce the following definition.

Definition 2.1.16 Let $(\mathbb{E}, c)$ be an absolute oriented ring spectrum.

Let E/S be a vector bundle of rank n. We let $P = \mathbb{P}(E)$ be the associated projective bundle, with projection p and canonical line bundle λ.

Using the previous theorem, we define the *Chern classes of E/S* with coefficients in $(\mathbb{E}, c)$ as the elements $c_i(E)$ of $\mathbb{E}^{2i,i}(S)$ for an integer $i \in [0, n]$ such that

$$\sum_{i=0}^{n} p^*(c_i(E)).\big(-c_1(\lambda)\big)^{n-i} = 0 \qquad\qquad (2.1.16.\mathrm{a})$$

and $c_0(E) = 1$. We put $c_i(E) = 0$ for $i \notin [0, n]$.

Indeed, the above theorem guarantees the existence and uniqueness of the Chern classes $c_i(E)$.

2.1.17 We deduce from this definition the following (usual) properties of Chern classes:

(a) For any vector bundle E/S, and any morphism $f : T \to S$, $f^* c_i(E) = c_i(f^{-1}E)$.

(b) For any scheme S and any isomorphism of vector bundles $E \simeq E'$ over S, $c_i(E) = c_i(E')$.

(c) For any scheme S, any exact sequence of vector bundles:

$$0 \to E' \to E \to E'' \to 0$$

and integer $k \geq 0$, $c_k(E) = \sum_{i+j=k} c_i(E').c_j(E'')$.

For details on the proof of formula (c) — *Whitney sum formula* — we refer the reader to [Dég08a, 3.13].

Remark 2.1.18 Let $(\mathbb{E}, c)$ be an absolute oriented ring spectrum.

Let Gr_S be the infinite Grassmannian, seen as a Nisnevich simplicial sheaf of sets over $\mathscr{S}m_S$. According to [MV99, 4.3.7], it is isomorphic in $\mathscr{H}_\bullet(S)$ to the classifying space BGL_S of the infinite general linear group over S. Moreover, when S is regular and for any integer $n \geq 0$, one gets a canonical isomorphism:

$$\mathrm{Hom}_{\mathscr{H}_\bullet(S)}(S^n, \mathbb{Z} \times Gr) \simeq K_n(S),$$

according to [MV99, 4.3.13], where $\mathbb{Z} \times Gr$ is the product of $\mathbb{Z}$-copies of Gr_S. This map is compatible with pullbacks of regular schemes.

Using Chern classes, it is well known how to compute $\mathbb{E}^{**}(Gr_S)$ (see [NSØ09a, 6.2]):

$$\mathbb{E}^{**}(Gr_S) \simeq \mathbb{E}^{**}(S)[[\mathfrak{c}_1, \ldots, \mathfrak{c}_n, \ldots]]$$

where $\mathfrak{c}_i$ is a cohomology class of bidegree $(2i, i)$. According to the above isomorphism, it corresponds to a map in $\mathscr{H}_\bullet(S)$:

$$\mathfrak{c}_i : \mathbb{Z} \times Gr_S \to Gr_S \to \Omega^\infty(\mathbb{E}_S(i)[2i]).$$

Moreover, using the techniques of Riou (cf. [Rio10], Th. 1.1.6 as in the proof of Th. 6.2.1.2), one gets an isomorphism:

$$\mathrm{Hom}_{\mathscr{H}_\bullet(S)}(\mathbb{Z} \times Gr, \Omega^\infty(\mathbb{E}(i)[2i]) \to \mathrm{Hom}(K_0(-), \mathbb{E}^{2i,i}(-))$$

where the right hand side stands for the morphisms of presheaves of sets on $\mathscr{S}m_S$. Under this isomorphism, the map $\mathfrak{c}_i$ corresponds to the natural transformation c_i that we have just defined.

As in [Gil81], this allows to automatically extends Chern classes to higher Chern classes with support in a closed subscheme $Z \subset S$ as follows:

$$c_i^Z : K_n^Z(S) \simeq [S^n \wedge (S/S - Z), \mathbb{Z} \times Gr]$$
$$\xrightarrow{(\mathfrak{c}_i)_*} [S^n \wedge (S/S - Z), \Omega^\infty(\mathbb{E}(i)[2i])] \simeq \mathbb{E}_Z^{2i-n,i}(S).$$

2.1.19 *The associated formal group law.–* The usual Segre embeddings

$$\sigma_{nm} : \mathbb{P}_X^n \times_X \mathbb{P}_X^m \to \mathbb{P}_X^{n+m+nm},$$

indexed by a pair of positive integers (n, m), induce a multiplication map

$$\sigma : \mathbb{P}_X^\infty \times_X \mathbb{P}_X^\infty \to \mathbb{P}_X^\infty. \tag{2.1.19.a}$$

This gives a structure of an H-group to the object $\mathbb{P}_X^\infty$ of $\mathcal{H}_\bullet(X)$ which in turn induces a group structure on the target of the map (2.1.6.a). We deduce from the previous paragraph and the equality

$$\sigma_{nm}^{-1}(\lambda_{n+m+nm}) = \lambda_n \times_X \lambda_m \qquad (2.1.19.b)$$

that (2.1.6.a) is a morphism of abelian groups.

Suppose $(\mathbb{E}, c)$ is an absolute oriented ring spectrum. According to Theorem 2.1.13, the pullback along σ corresponds to a (comultiplication) morphism:

$$\mathbb{E}^{**}(X)[[c]] \xrightarrow{\sigma^*} \mathbb{E}^{**}(X)[[x, y]]$$

where x and y stands for the Chern classes of the two canonical line bundles $p_1^{-1}(\lambda)$ and $p_2^{-1}(\lambda)$. It follows that σ^* is determined by $\sigma^*(c)$ which is a power series of the form:

$$F_X(x, y) = \sum_{i,j \geq 0} a_{ij}^X . x^i y^j. \qquad (2.1.19.c)$$

As σ^* is a comultiplication, one deduces that F_X is a commutative formal group law with coefficients in $\mathbb{E}^{**}(X)$ — see [Dég08a, §3.7]. In fact, the classes a_{ij}^S enjoy the following properties:

- a_{ij}^S has bidegree $(2 - 2i - 2j, 1 - i - j)$ in $\mathbb{E}^{**}(S)$,
- $a_{0,1}^S = 1$, $a_{0,i}^S = 0$ if $i \neq 1$,
- for every couple (i, j), $a_{ij}^S = a_{ji}^S$,
- for any morphism of schemes $f : T \to S$, $f^*(a_{ij}^S) = a_{ij}^T$.

Definition 2.1.20 Given the notations above, we will say that F_S is the *formal group law* associated with the oriented ring spectrum $(\mathbb{E}, c)$ above S.

We will say that $(\mathbb{E}, c)$ (or just c) is *additive* (resp. *multiplicative with parameter u*) if for any scheme S, $F_S(x, y) = x + y$ (resp. $F_S(x, y) = x + y + u.x.y$).

Example 2.1.21 Consider the absolute oriented ring spectra of Example 2.1.4:

(1) The k-absolute oriented ring spectrum associated with a Mixed Weil theory is additive (cf. [CD12a, 2.2.10]).

The absolute oriented ring spectrum $\mathbf{H}_\mathbb{B}$ is also additive. This last fact follows from the definition. To be more precise, given any regular scheme S, the canonical bijection:

$$\mathbf{H}_\mathbb{B}^{2,1}(S) \simeq Gr_\gamma^1 K_0(S)_\mathbb{Q} \simeq Pic(S)_\mathbb{Q}$$

is in fact an isomorphism of abelian groups. Note also that, restricting to the category of smooth S-schemes, for an arbitrary base S, the absolute oriented ring spectrum $\mathbf{H}_\Lambda$ is additive according to [CD12b, 11.3.5].

(2) The absolute oriented ring spectrum $\mathbf{KGL}$ is multiplicative with parameter $(-\beta)$ as follows from the easy computation, with $l = [L^\vee]$ and $l' = [L'^\vee]$:

$$c_1^{\mathbf{KGL}}(L \otimes L') = \beta^{-1}.(1 - [(L \otimes L')^\vee]) = \beta^{-1}.(1 - ll')$$
$$= \beta^{-1}((1-l) + (1-l') - -(1-l).(1-l'))$$
$$= c_1^{\mathbf{KGL}}(L) + c_1^{\mathbf{KGL}}(L') - \beta.c_1^{\mathbf{KGL}}(L).c_1^{\mathbf{KGL}}(L').$$

(3) Let k be a field of exponential characteristic p. Let $\mathbf{MGL}[1/p]$ be the absolute ring spectrum obtained from $\mathbf{MGL}$ by inverting p. Then, as a consequence of the Theorem of Hopkins-Morel-Hoyois (cf. [Hoy15, Th. 7.12]), we know that the formal group law of the absolute oriented ring spectrum $\mathbf{MGL}[1/p]$, considered over the category of all k-schemes, is isomorphic to the universal formal group law.

More precisely, if (L, F_{univ}) denotes the Lazard ring equipped with its canonical formal group law, according to *loc. cit.*, Prop. 8.2, there exists an isomorphism of formal group laws:

$$(L[1/p], F_{univ}) \to (\mathbf{MGL}_{(2,1)*}(k)[1/p], F_{\mathbf{MGL}}).$$

Proposition 2.1.22 *Consider the notations of the previous definition.*

(1) *For any vector bundle E/X and any integer $i > 0$, the class $c_i(E)$ is nilpotent in $\mathbb{E}^{**}(X)$.*

(2) *For any line bundles L_1, L_2 over X,*

$$c_1(L_1 \otimes L_2) = F_X\big(c_1(L_1), c_1(L_2)\big) \in \mathbb{E}^{2,1}(X).$$

Proof Point (1) follows from the hypothesis that X is noetherian according to the proof of [Dég08a, 3.8(1)]. Point (2) is tautological by definition of the first Chern class c_1 and of the formal group law F_X. $\qquad\square$

Remark 2.1.23 Note that unlike in [Dég08a, Prop. 3.8], to prove Point (2), we do not need that X admits an ample line bundle. This is because we consider cohomology theories that are representable in $S\mathcal{H}$: in fact, any line bundle L/X can be represented by a map $S^0 \to \mathbb{P}_X^\infty$ in $\mathcal{H}_\bullet(X)$ according to the theorem of Morel-Voevodsky. (If L/X is not generated by its sections, this map cannot be lifted in the category of schemes.)

2.2   Thom classes and MGL-modules

2.2.1 Let $(\mathbb{E}, c)$ be an absolute oriented ring spectrum (see Def. 2.1.2).

Let E/X be a vector bundle of rank n, $\mathbb{P}(E)$ (resp. $\bar{E}$, $E^{\times}$) be the associated projective bundle (resp. projective completion, complement of the zero section). Recall from [MV99, §3, 2.16] that one defines the Thom space of E/X as the pointed sheaf

$$\mathrm{Th}(E) = E/E^{\times}.$$

According to *loc. cit.*, Prop. 2.17, we get a canonical $\mathbb{A}^1$-equivalence of simplicial sheaves $\mathrm{Th}(E) \simeq \bar{E}/\mathbb{P}(E)$, the right hand side being the cokernel of the canonical embedding $\nu : \mathbb{P}(E) \to \bar{E}$ in the category of sheaves, equipped with its obvious base point. Thus we get a homotopy cofiber sequence for the $\mathbb{A}^1$-local model structure:

$$\mathbb{P}(E) \xrightarrow{\nu} \bar{E} \xrightarrow{\pi} \mathrm{Th}(E)$$

which induces a long exact sequence

$$\cdots \to \mathbb{E}^{**}(\mathrm{Th}(E)) \xrightarrow{\pi^*} \mathbb{E}^{**}(\bar{E}) \xrightarrow{\nu^*} \mathbb{E}^{**}(\mathbb{P}(E)) \to \cdots$$

According to Theorem 2.1.13, ν^* is a split epimorphism of free $\mathbb{E}^{**}(X)$-modules of respective ranks n and $n-1$. Thus $\mathbb{E}^{**}(\mathrm{Th}(X))$ is a free $\mathbb{E}^{**}(X)$-module of rank 1, isomorphic to $\ker(\nu^*)$.

Definition 2.2.2 Consider the notations and assumptions above.

We define the *Thom class* of E/X as the following element of $\mathbb{E}^{2n,n}(\bar{E})$:

$$\mathfrak{t}(E) = \sum_{i=0}^{n} p^*(c_i(E)).\big(-c_1(\lambda)\big)^{n-i}$$

where p is the canonical projection. We define the *refined Thom class* $\bar{\mathfrak{t}}(E)$ of E as the unique element of $\mathbb{E}^{2n,n}(\mathrm{Th}(E))$ such that

$$\pi^*\big(\bar{\mathfrak{t}}(E)\big) = \mathfrak{t}(E).$$

When the base of the vector bundle E is not clear, we indicate it as follows: $\mathbb{P}_X(E)$, $\mathrm{Th}_X(E)$, $\mathfrak{t}(E/X)$, $\bar{\mathfrak{t}}(E/X)$.

Note that $\mathbb{E}^{**}(\mathrm{Th}(E)) = \mathbb{E}_X^{**}(E)$ (see Definition 1.2.5). According to the preceding paragraph, the canonical map:

$$\mathbb{E}^{**}(X) \to \mathbb{E}_X^{**}(E), \lambda \mapsto \lambda.\bar{\mathfrak{t}}(E) \qquad\qquad (2.2.2.\mathrm{a})$$

defined in Paragraph 1.2.8 is an isomorphism called the *Thom isomorphism*.

We deduce from formulas (a), (b) of 2.1.17 that Thom classes are compatible with base change and invariant under isomorphisms of vector bundles.

Remark 2.2.3 Recall that in general the rank of a vector bundle E/X is Zariski locally constant on X. In other words, it is a function $r : \pi_0(X) \to \mathbb{N}$. On the other hand, $\mathbb{E}$-cohomology of X is additive. Moreover, we can give sense to the formula defining the refined Thom class of $\mathbb{E}$ without requiring E/X is of (constant) rank equal to n. Then $\bar{\mathfrak{t}}(E)$ is an element of $\mathbb{E}^{**}(\mathrm{Th}(E))$, which still induces a Thom isomorphism as above. Then, we can define the bidegree of this class as the function $\pi_0(X) \to \mathbb{Z}^2$ which to a connected component X_i of X associates the couple $(2r(X_i), r(X_i))$. The Thom isomorphism is then Zariski locally homogeneous on X with the same bidegree.

Example 2.2.4 Recall the universal quotient bundle ξ on $\bar{E}$ is defined by the exact sequence

$$0 \to \lambda \to p^{-1}(E \oplus 1) \to \xi \to 0.$$

Thus the Whitney sum formula 2.1.17(c) gives the following formula:

$$\mathfrak{t}(E) = c_n(\xi). \tag{2.2.4.a}$$

2.2.5 The cobordism spectrum **MGL** is given by the sequence of Thom spaces $\mathrm{Th}(\gamma_n/\mathrm{BGl}_n)$ for $n > 0$ where γ_n is the tautological rank n vector bundle over the classifying space of GL_n. As $\mathrm{Th}(\gamma_1) = B\mathbb{G}_m = \mathbb{P}^\infty$, $\mathbf{MGL}_S$ is canonically oriented. We denote by $c^{\mathbf{MGL}}$ this orientation.

Given an absolute oriented ring spectrum $(\mathbb{E}, c)$, the Thom class defined previously allows to define a morphism $\varphi_c : \mathbf{MGL} \to \mathbb{E}$. This is the key observation of the following proposition (see [Vez01, 4.3]):

Proposition 2.2.6 *Let $\mathbb{E}$ be an absolute ring spectrum. Then the following sets correspond bijectively:*

  (i) *orientations c of $\mathbb{E}$;*
  (ii) *maps of absolute ring spectra $\varphi : \mathbf{MGL} \to \mathbb{E}$.*

by the following maps:

$$(ii) \to (i), \varphi \mapsto \varphi_*(c^{\mathbf{MGL}}), \ \varphi_* \ \textit{induced map in cohomology},$$
$$(i) \to (ii), c \mapsto \varphi_c.$$

2.2.7 A module over the ring spectrum $\mathbf{MGL}_S$ is an $\mathbf{MGL}_S$-module in $S\mathscr{H}(S)$ in the classical sense: a spectrum $\mathbb{E}$ over S equipped with a multiplication map $\gamma_{\mathbb{E}} : \mathbf{MGL}_S \wedge \mathbb{E} \to \mathbb{E}$ satisfying the usual identities — see [ML98].

Given two $\mathbf{MGL}_S$-modules $\mathbb{E}$ and $\mathbb{F}$, a morphism of $\mathbf{MGL}_S$-modules is a morphism $f : \mathbb{E} \to \mathbb{F}$ in $S\mathscr{H}(S)$ such that the following diagram is commutative:

$$
\begin{array}{ccc}
\mathbf{MGL}_S \wedge \mathbb{E} & \xrightarrow{\ 1 \wedge f\ } & \mathbf{MGL}_S \wedge \mathbb{F} \\
{\scriptstyle \gamma_{\mathbb{E}}}\downarrow & & \downarrow{\scriptstyle \gamma_{\mathbb{F}}} \\
\mathbb{E} & \xrightarrow{\quad\quad f \quad\quad} & \mathbb{F}.
\end{array}
$$

We will denote by $\mathbf{MGL}-mod^w{}_S$ the additive category of $\mathbf{MGL}_S$-modules.

Given any spectrum $\mathbb{E}$, $\mathbf{MGL}_S \wedge \mathbb{E}$ has an obvious structure of an $\mathbf{MGL}_S$-module. The assignment $L^w_{\mathbf{MGL}} : \mathbb{E} \mapsto \mathbf{MGL}_S \wedge \mathbb{E}$ defines a functor left adjoint to the inclusion functor $\mathcal{O}^w_{\mathbf{MGL}}$ and we get an adjunction of categories:

$$
L^w_{\mathbf{MGL}} : S\mathscr{H}(S) \leftrightarrows \mathbf{MGL}-mod^w{}_S : \mathcal{O}^w_{\mathbf{MGL}}. \tag{2.2.7.a}
$$

The category $\mathbf{MGL}-mod^w{}_S$ is not well behaved: it has no triangulated monoidal structure. In [Dég08a, Ex. 2.12(2)] and [Dég13, Sec. 2.2], we introduced the homotopy category of *strict* $\mathbf{MGL}_S$-modules. We will denote it by $\mathbf{MGL}-mod_S$ and call it the *homotopy category of $\mathbf{MGL}_S$-modules*. This is an enrichment of the category of $\mathbf{MGL}_S$-modules: one defines a canonical commutative diagram of functors:

$$
\begin{array}{ccc}
 & S\mathscr{H}(S) & \\
{\scriptstyle L_{\mathbf{MGL}}}\swarrow & & \searrow{\scriptstyle L^w_{\mathbf{MGL}}} \\
\mathbf{MGL}-mod_S & \xrightarrow{\ \ \mathcal{O}'_{\mathbf{MGL}}\ \ } & \mathbf{MGL}-mod^w{}_S
\end{array} \tag{2.2.7.b}
$$

where $L_{\mathbf{MGL}}$ is a triangulated functor and $\mathcal{O}'_{\mathbf{MGL}}$ is a conservative functor. Therefore, using the previous commutative diagram and Proposition 2.2.6, we get the following result.

Proposition 2.2.8 *Let $(\mathbb{E}, c)$ be an absolute oriented ring spectrum.*

Then for any scheme S, the functor $\varphi_{\mathbb{E}} = \mathrm{Hom}_{S\mathscr{H}(S)}(-, \mathbb{E})$ induces a canonical functor $\tilde{\varphi}_{\mathbb{E}}$ which fits in the following commutative diagram:

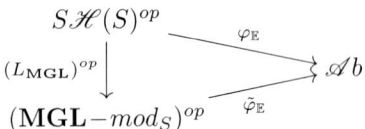

2.2.9 Let S be a scheme. Recall the considerations of [Dég08a, §2.3.2]. For any cartesian square of smooth S-schemes

$$\begin{array}{ccc} W & \longrightarrow & V \\ \downarrow & \Delta & \downarrow \\ U & \longrightarrow & X \end{array}$$

made of immersions, we denote by $\frac{X/U}{V/W}$ the colimit of Δ in the category of sheaves over S.[37]

We denote by $\mathbf{MGL}_S\left(\frac{X/U}{V/W}\right)$ the image of the infinite suspension spectrum $\Sigma^\infty\left(\frac{X/U}{V/W}\right)$ by the functor $L_{\mathbf{MGL}}$ defined above. When $V = W = \varnothing$ (resp. $U = V = W = \varnothing$), we simply denote this object by $\mathbf{MGL}_S(X/U)$ (resp. $\mathbf{MGL}_S(X)$) — according to the notation of Par. 1.1.3.

As explained in [Dég08a, 2.3.2, Ex. 2.12(2)], the category $\mathbf{MGL}-mod_S$ together with the canonical orientation $c^{\mathbf{MGL}}$ of $\mathbf{MGL}_S$ satisfies all the axioms of [Dég08a, §2.1]. Therefore we can apply the results of *loc. cit.* to that category.

Then, according to the previous proposition, for any smooth closed S-pair (X, Z), we get:

$$\mathbb{E}_Z^{n,m}(X) = \mathrm{Hom}_{S\mathscr{H}(S)}(X/X - Z, \mathbb{E}(m)[n])$$
$$= \tilde{\varphi}_{\mathbb{E}}\left(\mathbf{MGL}_S\left(X/X - Z\right)(-m)[-n]\right).$$

The important thing for us is that, given smooth closed S-pairs (X, Z) and (Y, T) and a couple of integers (n, m), any morphism of $\mathbf{MGL}_S$-modules of the form:

$$\mathbf{MGL}_S(X/X - Z) \to \mathbf{MGL}_S(Y/Y - T)(m)[n]$$

induces a canonical homogeneous morphism of bigraded abelian groups

$$\mathbb{E}_T^{**}(Y) \to \mathbb{E}_Z^{*+n,*+m}(X).$$

Obviously, this association is compatible with composition.

Example 2.2.10 Let $f : Y \to X$ be a projective S-morphism between smooth S-schemes. Assume f has dimension d. Then, applying [Dég08a, 5.12], we get the Gysin morphism

$$f^* : \mathbf{MGL}_S(X) \to \mathbf{MGL}_S(Y)(-d)[-2d]$$

---

[37]As all maps in this diagram are cofibrations, this is the homotopy colimit of the diagram in the $\mathbb{A}^1$-local model category of simplicial sheaves. It measures the obstruction for the square Δ to be homotopy cocartesian.

which in turn induces a push-forward in cohomology:

$$f_* : \mathbb{E}^{*,*}(Y) \to \mathbb{E}^{*-d,*-2d}(X).$$

The purpose of the next sections is to generalize this pushforward in the case of regular schemes.

Remark 2.2.11 In fact, all the orientation theory exposed here for ring spectra can be done without ring structure by replacing an orientation c by a structure of a module over $\mathbf{MGL}_S$, on a given spectrum $\mathbb{E}$. This is the content of *loc. cit.* in the case $\mathscr{S} = \mathscr{S}m_S$.[38]

Then, instead of having Chern classes in $\mathbb{E}$ cohomology, we get an *action* of Chern classes, through the $\mathbf{MGL}^{**}$-module structure of $\mathbb{E}^{**}$ — see also Example 5.1.3. Similarly, to anticipate what follows, we do not get fundamental classes in $\mathbb{E}$-cohomology but an action of fundamental classes with coefficients in $\mathbf{MGL}$. Note this is enough to get the purity isomorphism (2.3.1.b) and therefore, Gysin long exact sequence (as in Def. 3.1.2), Gysin morphisms for projective lci morphisms (as in 3.2.7).

However, the general case $\mathscr{S} = \mathscr{R}\mathrm{eg}$ is not covered by [Dég08a]. In fact, in this case, Conjecture A (of section 1.3) seems very natural.

2.3   Fundamental classes

In this section and the following one, we fix an absolute oriented ring spectrum $(\mathbb{E}, c)$.

Definition 2.3.1 Let (X, Z) be a regular closed pair of codimension n and normal bundle $N_Z X$. Let $i : Z \to X$ be the corresponding embedding.

When (X, Z) is $\mathbb{E}$-pure we define the *refined fundamental class* of Z in X (with coefficients in $\mathbb{E}$) as the image of the Thom class $\bar{\mathfrak{t}}(N_Z X)$ by the isomorphisms (see Definition 1.3.2):

$$\mathbb{E}^{2n,n}(\mathrm{Th}(N_Z X)) \xrightarrow{(\sigma_0^*)^{-1}} \mathbb{E}_{\mathbb{A}_Z^1}^{2n,n}(D_Z X) \xrightarrow{\sigma_1^*} \mathbb{E}_Z^{2n,n}(X). \qquad (2.3.1.\mathrm{a})$$

We denote it by $\bar{\eta}_X(Z)$.

We also define the *fundamental class* of Z in X as the following class of $\mathbb{E}^{2n,n}(X)$:

$$\eta_X(Z) := i_!\left(\bar{\eta}_X(Z)\right).$$

---

[38] The case of weak $\mathbf{MGL}$-modules can be deduced from the case of strict $\mathbf{MGL}$-modules which is treated in *loc. cit.* using the commutative diagram (2.2.7.b).

Because σ_0^*, σ_1^* are isomorphisms of $\mathbb{E}^{**}(Z)$-modules, and according to the Thom isomorphism (2.2.2.a), the class $\bar{\eta}_X(Z)$ is in fact a base of the $\mathbb{E}^{**}(Z)$-bigraded module $\mathbb{E}_Z^{**}(X)$. In other words, we get an isomorphism

$$\mathfrak{p}_{(X,Z)} : \mathbb{E}^{**}(Z) \to \mathbb{E}_Z^{**}(X), z \mapsto z.\bar{\eta}_X(Z) \qquad (2.3.1.b)$$

called the *purity isomorphism.*

Remark 2.3.2 For the reader who does not want to restrict to pure codimension pairs, let us indicate that in general, the codimension of a regular closed pair (X, Z) is Zariski local in Z, as well as the rank of its normal bundle $N_Z X$. Then the convention of 2.2.3 will apply both to the refined fundamental class and the purity isomorphism associated with (X, Z).

Example 2.3.3 Let E/X be a vector bundle. Then the closed X-pair (E, X) corresponding to the zero section is $\mathbb{E}$-pure — see Example 1.3.4(1). Then we get from the above definition the equality of classes in $\mathbb{E}^{**}(\mathrm{Th}(E/X))$:

$$\bar{\eta}_E(X) = \mathfrak{t}(E/X). \qquad (2.3.3.a)$$

Indeed, one can identify the deformation space $D_X E$ (resp. normal bundle $N_X E$) with the affine line $\mathbb{A}_E^1$ (resp. the X-vector bundle E itself) and the maps σ_1, σ_0 of (1.3.1.a) with respectively the unit and zero section of $\mathbb{A}_E^1$ (see [Dég08a, Rem. 4.2]) so that the above identification follows from the previous definition. Moreover, if $\bar{E}/X$ denotes the projective completion of E/X, so that $E \subset \bar{E}$ is an open subscheme, one gets by Nisnevich excision an identification $\mathbb{E}_X^{**}(\bar{E}) = \mathbb{E}_X^{**}(E)$ through which $\bar{\eta}_{\bar{E}}(X)$ corresponds to the class $\bar{\eta}_E(X) = \mathfrak{t}(E/X)$. In particular, one gets the equality of (non refined) fundamental classes:

$$\eta_{\bar{E}}(X) = \mathfrak{t}(E/X). \qquad (2.3.3.b)$$

Remark 2.3.4 Consider a smooth closed S-pair (X, Z) of codimension n. Recall from Example 1.3.4(1) that (X, Z) is $\mathbb{E}$-pure. We have defined in [Dég08a, 4.6] a purity isomorphism in the homotopy category of $\mathbf{MGL}_S$-modules:

$$\mathbf{MGL}_S(X/X - Z) \xrightarrow{\sim} \mathbf{MGL}_S(Z)(n)[2n].$$

According to the above example and [Dég08a, 4.3, 4.4], we get that the induced morphism in $\mathbb{E}$-cohomology (see 2.2.9) coincides exactly with the purity isomorphism $\mathfrak{p}_{(X,Z)}$ introduced above. In particular, the class $\bar{\eta}_X(Z)$ in $\mathbb{E}_Z^{**}(X)$ introduced here coincides with that obtained from [Dég08a, 4.14] by considering $\mathbf{MGL}_S$-modules.[39]

---

[39]The terminology here differs slightly from that of *loc. cit.*: we use the adjective "refined" instead of "localized" which seems more classical.

2.4 Intersection theory

Definition 2.4.1 Consider a morphism $\Delta : (Y, T) \to (X, Z)$ of closed pairs (Par. 1.2.11) such that (X, Z) and (Y, T) are $\mathbb{E}$-pure.

Then, according to Definition 2.3.1, there exists a unique class e_Δ in $\mathbb{E}^{**}(T)$ such that

$$\Delta^*(\bar{\eta}_X(Z)) = e_\Delta \cdot \bar{\eta}_Y(T).$$

We call e_Δ the *defect* of the morphism Δ.[40]

The following result generalizes [Dég08a, Prop. 4.16]:

Theorem 2.4.2 *With the notations of the above definition, assume* $\Delta = (f, g)$ *is cartesian. Then it induces a monomorphism* $\nu : N_T Y \to g^{-1}(N_Z X)$ *of vector bundles over* T*. We put*

$$\xi = g^{-1}(N_Z X)/N_T Y \qquad\qquad (2.4.2.a)$$

and denote by e *the rank of this vector bundle.*
    Then $e_\Delta = c_e(\xi)$.

Remark 2.4.3 One can apply this theorem in two cases:

- Given a base scheme S, Δ is a morphism of smooth closed S-pairs. This case is already known from [Dég08a, 4.16].
- $\mathbb{E}$ is $\mathscr{R}eg$-absolutely pure and the schemes X, Z, Y, $T = f^{-1}(Z)$ are all regular.

Proof Recall the deformation diagram (1.3.1.a) is functorial with respect to the morphism Δ, because it is assumed to be cartesian. Thus, going back to the definition of fundamental classes (2.3.1), using the deformation diagrams respectively for the closed pairs (X, Z) and (Y, T) we are reduced to compare Thom classes for the corresponding normal bundles, as summarized in the following cartesian square:

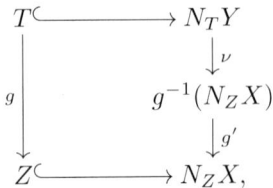

---

[40]One could call this class the *defect of transversality*. In fact, there are two kinds of possible defect: excess of intersection, ramification.

where ν is a monomorphism of vector bundles over T. This now follows from [Dég08a, Lem. 4.18] (using the considerations of Paragraph 2.2.9).

$\square$

In terms of fundamental classes, we get:

Corollary 2.4.4 *Let $(f, g) : (Y, T) \to (X, Z)$ be a cartesian morphism of* $\mathbb{E}$-*pure closed pairs. Let $f^* : \mathbb{E}_Z^{**}(X) \to \mathbb{E}_T^{**}(Y)$ be the morphism defined in Paragraph 1.2.6.*
Then, using Definition (2.4.2.a), we get the following formula in $\mathbb{E}_T^{**}(Y)$:

$$f^*(\bar{\eta}_X(Z)) = c_e(\xi). \, \bar{\eta}_Y(T).$$

In particular when f is transversal to Z,

$$f^*(\bar{\eta}_X(Z)) = \bar{\eta}_Y(T). \tag{2.4.4.a}$$

Remark 2.4.5 According to formulas (E3) and (E7) of Proposition 1.2.10, we get in the assumptions of the above corollary the even more usual formula in $\mathbb{E}^{**}(Y)$:

$$f^*(\eta_X(Z)) = c_e(\xi). \, \eta_Y(T).$$

Applying this formula in the case where f is the zero section $s : X \to E$ of a vector bundle of rank n, we get the relation

$$s^*(\eta_E(X)) = c_n(E).^{41}$$

These two relations give the following classical trick, which will be used later, to compute fundamental classes.

Corollary 2.4.6 *Let (X, Z) be an $\mathbb{E}$-pure closed pair of codimension n corresponding to an immersion $i : Z \to X$. Assume there exists a vector bundle E/X with a section s transversal to the zero section s_0 and such that $s_0^{-1}(s) = i$. Then*

$$\eta_X(Z) = c_n(E).$$

An important point of intersection theory is the associativity of the intersection product. In our setting, this can be expressed nicely using the refined product of Paragraph 1.2.8. We begin with a particular case of Theorem 2.4.9 which will be the crucial step.

---

[41] In other words, given a vector bundle E/X the fundamental class of its zero section coincides up to homotopy with its *Euler class*. This fact justifies our choice of Thom class in Definition 2.2.2.

Proposition 2.4.7 *Consider an exact sequence* (σ) *of vector bundles over a scheme* X:

$$0 \to E' \xrightarrow{\nu} E \to E'' \to 0.$$

Then the following relation holds in $\mathbb{E}^{**}(\mathrm{Th}(E))$:

$$\bar{\mathfrak{t}}(E/X) = \bar{\mathfrak{t}}(E'/X).\,\bar{\mathfrak{t}}(E/E')$$

using the pairing $\mathbb{E}_X^{**}(E') \otimes \mathbb{E}_{E'}^{**}(E) \to \mathbb{E}_X^{**}(E)$ *(Paragraph 1.2.8).*

Proof Note that according to formula (2.3.3.a), we have to prove the equality:

$$\bar{\eta}_E(X) = \bar{\eta}_{E'}(X).\,\bar{\eta}_E(E'). \tag{2.4.7.a}$$

As in the proof of [Rio10, 4.1.1], we can find a torsor T over the X-vector bundle $\underline{\mathrm{Hom}}(E'', E')$ such that the sequence (σ) splits over T. The morphism $T \to X$ is an $\mathbb{A}^1$-weak equivalence ; thus, by compatibility of the Thom class with base change, we can assume the sequence (σ) splits.

Let us consider P' (resp. P'') the projective completion of E'/X (resp. E''/X), $P = P' \times_X P''$, and the following commutative diagram made of cartesian squares:

$$
\begin{array}{ccccc}
X & \longrightarrow & E' & \longrightarrow & E \\
\| & & \downarrow{\scriptstyle j'} & & \downarrow{\scriptstyle j} \\
X & \xrightarrow{s'} & P' & \xrightarrow{s''} & P \\
& & \underset{s}{\searrow} & &
\end{array}
$$

where j (resp. s, s', s'') stands for the natural open immersion (resp. canonical sections coming from the obvious zero sections). Remark that if we apply the functor $j^* : \mathbb{E}_X^{**}(P) \to \mathbb{E}_X^{**}(E)$ to the following equality:

$$\bar{\eta}_P(X) = \bar{\eta}_{P'}(X).\,\bar{\eta}_P(P'), \tag{2.4.7.b}$$

we get (2.4.7.a) — use formula (E5) of Prop. 1.2.10 and formula (2.4.4.a). Thus we are reduced to prove that equality (2.4.7.b) holds in $\mathbb{E}_X^{**}(P)$.

According to the projective bundle formula, the morphism $s_! : \mathbb{E}_X^{**}(P) \to \mathbb{E}^{**}(P)$ is a split monomorphism (see the end of Par. 2.2.1 for more details). Thus it is sufficient to prove that (2.4.7.b) holds after applying $s_!$. That means we have to prove $\eta_P(X)$ is equal to:

$$s_!\big(\bar{\eta}_{P'}(X).\,\bar{\eta}_P(P')\big) = s_!'' s_!'\big(\bar{\eta}_{P'}(X).\,\bar{\eta}_P(P')\big)$$

$$\overset{(E6)}{=} s_!''\big(\eta_{P'}(X).\,\bar{\eta}_P(P')\big) = s_!''\,\big(s''^*[\eta_P(P'')].\,\bar{\eta}_P(P')\big)$$

$$\overset{(E7)}{=} \eta_P(P'').\,\eta_P(P')$$

using once again the properties of the refined product enumerated in Proposition 1.2.10. This last check follows from a direct computation — see [Dég08a, Lem. 4.25].

$\square$

Remark 2.4.8 Consider a scheme X. Given vector bundles E' and E'' over X, we get using the product defined in Paragraph 1.2.8 a canonical map of Künneth type:

$$\mathbb{E}^{**}(\mathrm{Th}(E')) \otimes_{\mathbb{E}^{**}(X)} \mathbb{E}^{**}(\mathrm{Th}(E'')) \to \mathbb{E}^{**}(\mathrm{Th}(E' \oplus E'')).$$

According to the above proposition, $\bar{\mathfrak{t}}(E' \oplus E'') = \bar{\mathfrak{t}}(E').\bar{\mathfrak{t}}(E'')$. In other words, the above map is an isomorphism (of bigraded $\mathbb{E}^{**}(X)$-modules).

Now, given an exact sequence (σ) of vector bundles as in the above proposition, Riou in the proof of [Rio10, 4.1.1] shows there exists a canonical isomorphism in $S\mathscr{H}(X)$

$$\Sigma^\infty \mathrm{Th}(E') \wedge \Sigma^\infty \mathrm{Th}(E'') \xrightarrow{\epsilon_\sigma} \Sigma^\infty \mathrm{Th}(E).$$

Using this isomorphism, one defines a canonical product of bigraded $\mathbb{E}^{**}(X)$-modules:

$$\mathbb{E}^{**}(\mathrm{Th}(E'))\otimes_{\mathbb{E}^{**}(X)}\mathbb{E}^{**}(\mathrm{Th}(E'')) \longrightarrow \mathbb{E}^{**}(\mathrm{Th}(E')\wedge\mathrm{Th}'(E'')) \xrightarrow{(\epsilon^*_\sigma)^{-1}} \mathbb{E}^{**}(\mathrm{Th}(E)).$$
$$a\otimes b \longmapsto \langle a,b\rangle_\sigma$$

Then the relation obtained in the previous proposition can be rewritten as follows:

$$\bar{\mathfrak{t}}(E) = \langle \bar{\mathfrak{t}}(E'), \bar{\mathfrak{t}}(E'')\rangle_\sigma.$$

Thus, the map $\langle -, -\rangle_\sigma$ is an isomorphism.

Recall that Deligne defines in [Del87, 4.12] the category of *virtual vector bundle* denoted by $\underline{K}(X)$. As remarked by Riou, the isomorphisms of type ϵ_σ show that the functor $\Sigma^\infty \mathrm{Th}_X$ induces a canonical functor

$$\mathrm{Th}_X : \underline{K}(X) \to S\mathscr{H}(X).$$

In fact the preceding considerations show that, for any virtual vector bundle ξ in $\underline{K}(X)$, the bigraded $\mathbb{E}^{**}(X)$-module $\mathbb{E}^{**}(\mathrm{Th}(\xi))$ is free of rank 1. Moreover, it admits a canonical trivialization which we will denote $\bar{\mathfrak{t}}(\xi)$: if $\xi = [E] - [E']$ for two vector bundles E and E', one puts:

$$\bar{\mathfrak{t}}(\xi) := \bar{\mathfrak{t}}(E).\bar{\mathfrak{t}}(E')^*.$$

A conceptual way of stating this result: the canonical functor $\mathbb{E}^{**} \circ \tilde{\mathrm{Th}}_X$ induces a functor of Picard categories:

$$\left(\underline{K}(X)\right)^{op} \to \mathrm{Pic}\left(\mathbb{E}^{**}(X)\right)$$

where the right hand side category is the Picard category of bigraded virtual line bundles over $\mathbb{E}^{**}(X)$.

Theorem 2.4.9 *Consider regular closed immersions $Z \xrightarrow{k} Y \xrightarrow{i} X$ in $\mathscr{S}$ and assume $\mathbb{E}$ is absolutely pure.*
Then we get the following equality in $\mathbb{E}_Z^{**}(X)$:

$$\bar{\eta}_X(Z) = \bar{\eta}_Z(Y).\bar{\eta}_X(Y),$$

using the pairing $\mathbb{E}_Z^{**}(Y) \otimes \mathbb{E}_Y^{**}(X) \to \mathbb{E}_Z^{**}(X)$ (Paragraph 1.2.8) for the right hand side.

Proof Recall from Paragraph 1.3.1 the deformation space $D_Z X$ associated with the closed pair (X, Z). Following [Ros96, §10], we define the double deformation space by the following formula:

$$D(X, Y, Z) = D(D_Z(X), D_Z(X)|_Y).$$

This scheme is flat over $\mathbb{A}^2$. Its fiber over $(1, 1)$ is X while its fiber over $(0, 0)$ is the normal bundle $E := N\left(N_Z X, N_Z Y\right)$. Let us put $D' = D(Y, Y, Z)$ and $E' = N_Z Y$. Then we get the following diagram made of cartesian squares:

$$
\begin{array}{ccccc}
Z & \xrightarrow{s'} & E' & \xrightarrow{\nu} & E \\
{\scriptstyle s_1}\downarrow & & \downarrow & & \downarrow{\scriptstyle d_0} \\
\mathbb{A}_Z^2 & \longrightarrow & D' & \longrightarrow & D \\
{\scriptstyle s_0}\uparrow & & \uparrow & & \uparrow{\scriptstyle d_1} \\
Z & \xrightarrow{k} & Y & \xrightarrow{i} & X
\end{array}
$$

where the first (resp. third) row is seen as the $(0, 0)$-fiber (resp. $(1, 1)$-fiber) of the second row, with respect to its canonical projection to $\mathbb{A}^2$. Because this projection is flat, all the squares in this diagram are transversal.

Thus, one can apply Corollary 2.4.4 to the morphism of closed pairs $(X, Z) \xrightarrow{d_1} (D, \mathbb{A}_Z^2)$ (resp. $(E, Z) \xrightarrow{d_0} (D, \mathbb{A}_Z^2)$): because s_1 (resp. s_0) is a strong $\mathbb{A}^1$-homotopy equivalence, and using the absolute purity property, the pullback morphism $d_1^* : \mathbb{E}_{\mathbb{A}_Z^2}^{**}(D) \to \mathbb{E}_Z^{**}(X)$ (resp. $d_0^* : \mathbb{E}_{\mathbb{A}_Z^2}^{**}(D) \to \mathbb{E}_Z^{**}(E)$) is an isomorphism.

Applying again Corollary 2.4.4, we deduce that to prove the theorem, it is enough to prove the relation

$$\bar{\eta}_E(Z) = \bar{\eta}_{E'}(Z).\bar{\eta}_E(E').$$

In view of formula (2.3.3.a), this is precisely Proposition 2.4.7.

$\square$

Remark 2.4.10 In the case where $\mathscr{S}$ is the category of smooth S-schemes, for a base scheme S, the above theorem is very close to [Dég08a, 4.30]. The proof given here is considerably simpler, as we use the refined product of 1.2.8 — it uses the localization property of $S\mathscr{H}$, a very strong result.

3 Gysin morphisms and localization long exact sequence

In this section, we fix an absolute oriented ring spectrum $(\mathbb{E}, c)$ — recall Def. 2.1.2.

3.1 Residues and the case of closed immersions

3.1.1 Consider a closed immersion $i : Z \to X$ with complementary open immersion $j : U \to X$. Property (A4) of Paragraph 1.1.1 implies the existence of the *localization long exact sequence*:

$$\cdots \mathbb{E}_Z^{n,m}(X) \xrightarrow{i_!} \mathbb{E}^{n,m}(X) \xrightarrow{j^*} \mathbb{E}^{n,m}(U) \xrightarrow{\delta_{X,Z}} \mathbb{E}_Z^{n+1,m}(X) \cdots \qquad (3.1.1.a)$$

If we assume that the corresponding closed pair (X, Z) is regular of codimension c and $\mathbb{E}$-pure, the associated purity isomorphism (2.3.1.b) induces a long exact sequence in cohomology from the preceding one:

$$\cdots \mathbb{E}^{n-2c,m-c}(Z) \xrightarrow{i_*} \mathbb{E}^{n,m}(X) \xrightarrow{j^*} \mathbb{E}^{n,m}(U) \xrightarrow{\partial_{X,Z}} \mathbb{E}^{n-2c+1,m-c}(Z) \cdots$$
$$(3.1.1.b)$$

Definition 3.1.2 Under the notations above, assuming (X, Z) is $\mathbb{E}$-pure of codimension c, we call (3.1.1.b) (resp. i_*, $\partial_{X,Z}$) the Gysin long exact sequence (resp. Gysin morphism, residue morphism) associated with the closed pair (X, Z) (or with the immersion i).

Note that in terms of the refined fundamental class (Def. 2.3.1), the Gysin morphism can be described by the following formula for a cohomology class $z \in \mathbb{E}^{**}(Z)$:

$$i_*(z) := i_!\big(z.\bar{\eta}_X(Z)\big). \qquad (3.1.2.a)$$

Moreover, the residue of a cohomology class $u \in \mathbb{E}^{**}(U)$ is uniquely determined by the following property:

$$\delta_{X,Z}(u) = \partial_{X,Z}(u).\bar{\eta}_X(Z). \qquad (3.1.2.b)$$

Remark 3.1.3 Note that the Gysin morphism i_* defined above do depend on the orientation c on the absolute ring spectrum $\mathbb{E}$, fixed at the beginning of the section — this dependence is responsible for the Riemann-Roch theorem 4.2.3. Usually, there will be no possible confusion about the chosen orientation, but in case there might, we will denote by i_*^c the corresponding Gysin morphism (see in particular 5.0.3).

Proposition 3.1.4   (1) *Assume $\mathbb{E}$ is absolutely pure. Then for any $\mathbb{E}$-pure closed immersions*

$$T \xrightarrow{k} Z \xrightarrow{i} X,$$

the following equality holds: $i_*k_* = (ik)_*$.

(2) *Consider an $\mathbb{E}$-pure closed immersion $i : Z \to X$. For any couple (x, z) in $\mathbb{E}^{**}(X) \times \mathbb{E}^{**}(Z)$,*

$$i_*\big(i^*(x).z\big) = x.i_*(z).$$

(3) *Consider a cartesian square*

$$\begin{array}{ccc} T & \xrightarrow{k} & Y \\ {\scriptstyle g}\downarrow & & \downarrow{\scriptstyle f} \\ Z & \xrightarrow{i} & X \end{array}$$

in $\mathscr{S}$ such that i and k are $\mathbb{E}$-pure closed immersions. Let $h : (Y - T) \to (X - Z)$ be the morphism induced by f. Let ξ be the vector bundle over T defined by formula (2.4.2.a) with respect to the morphism of closed pairs (f, g). Let e be the rank of ξ.

Then the following formulas hold:

$$f^*i_*(z) = k_*\big(c_e(\xi).g^*(z)\big). \qquad\qquad (3.1.4.\text{a})$$

$$\partial_{Y,T}h^*(u) = c_e(\xi).g^*\partial_{X,Z}(u). \qquad\qquad (3.1.4.\text{b})$$

Proof Taking care of formulas (3.1.2.a) and (3.1.2.b), points (1), (2), (3) are respectively consequences of Th. 2.4.9, Proposition 1.2.10(E4)+(E7), Corollary 2.4.4.

$\square$

Remark 3.1.5 Let $i : Z \to X$ be a closed immersion between smooth S-schemes. According to point (1) of Example 1.3.4, the closed pair (X, Z) is $\mathbb{E}$-pure. Then both Example 2.2.10 and the preceding definition gives a pushforward of the form

$$i_* : \mathbb{E}^{**}(Z) \to \mathbb{E}^{**}(X).$$

Remark 2.3.4 shows that these two constructions coincide.

As we will see in the next examples, residues are connected to the *tame residue symbols* of Milnor in K-theory. This connection can be reduced to the following interesting property:

Proposition 3.1.6 *Let X be a regular scheme and Z be a principal divisor of X parametrized by a regular function $\pi : X \to \mathbb{A}_k^1$. Let us denote by $i : Z \to X$ ($j : U \to X$) be the corresponding closed immersion (resp. complementary open immersion). Let us consider the following composite map:*

$$\gamma_\pi : \mathbb{E}^{n,m}(U) \xrightarrow{(1)} \mathbb{E}^{n,m}(U) \oplus \mathbb{E}^{n+1,m+1}(U) \simeq \mathbb{E}^{n+1,m+1}(\mathbb{G}_m \times U)$$

$$\xrightarrow{(2)} \mathbb{E}^{n+1,m+1}(U)$$

where (1) is the obvious inclusion and (2) is the pullback by the graph of the induced map $\pi|_U : U \to \mathbb{G}_m$.
  Then if (X, Z) is $\mathbb{E}$-pure, the following relation holds:

$$\partial_{X,Z} \circ \gamma_\pi \circ j^* = i^*.$$

Proof The proof is essentially the same as the one explained in [Dég08b, Prop. 2.6.5]. Let us indicate the main steps.
  According to the definition of the purity isomorphism through the deformation diagram (1.3.1.a), and the fact the canonical morphism $\sigma_1 : X \to D_Z X$ is a split monomorphism (cf *loc. cit.*), we restrict to the case of $(N_Z X, Z)$ where $N_Z X$ is the normal bundle of Z in X, while π can be considered as a global trivialization of this line bundle.
  Up to isomorphism, we thus are reduced to the case of $(\mathbb{A}_Z^1, Z)$ and the canonical parametrization t of the affine line (given by the identity). Then, the problem can be unfolded to a trivial identity given at the level of schemes (cf. the end of the proof of *loc. cit.*).

$\square$

Example 3.1.7 Let R be a discrete valuation ring with valuation v, fraction field K and residue field k.
  Recall first that the *tame residue symbol* on Milnor K-theory:

$$\partial_v : K_n^M(K) \to K_{n-1}^M(k) \tag{3.1.7.a}$$

is uniquely characterized by the following properties, where π is a given uniformizing parameter:

(1) $\partial_v(\{\pi\}) = 1$;

(2) for units $u_1, \ldots, u_r \in R^\times$, $\partial_v(\{\pi, u_1, \ldots, u_r\}) = \{\bar{u}_1, \ldots, \bar{u}_r\}$;

(3) for units $u_1, \ldots, u_r \in R^\times$, $\partial_v(\{u_1, \ldots, u_r\}) = 0$.

If we assume that $(\mathrm{Spec}\,(R), \mathrm{Spec}\,(k))$ is $\mathbb{E}$-pure, we get an abstract residue map with coefficients in $\mathbb{E}$:

$$\partial_v^{\mathbb{E}} : \mathbb{E}^{n,m}(K) \to \mathbb{E}^{n-1,m-1}(k).$$

When R is of equal characteristics, we can apply this construction to the case of Voevodsky's Einlenberg-MacLane motivic ring spectrum $\mathbf{H}_{\mathbb{Z}}$ — S is the spectrum of the prime field of R. Then, the induced map

$$\partial_v : K_n^M(K) \simeq \mathbf{H}_{\mathcal{M}}^{n,n}(K) \to \mathbf{H}_{\mathcal{M}}^{n-1,n-1}(k) \simeq K_{n-1}^M(k)$$

coincides with (3.1.7.a). In fact, relations (1) and (2) follows from the previous proposition while relation (3) follows from the fact $\partial_v \circ j^* = 0$ according to the Gysin long exact sequence (3.1.1.b).

In the unequal characteristics case, we get the same result with rational coefficients by considering Beilinson motivic cohomology ring spectrum $\mathbf{H}_{\mathrm{B}}$.

Remark 3.1.8 Other applications of the previous proposition will be given in Section 5.5.

3.2   Projective lci morphisms

3.2.1 Let X be a scheme, put $P = \mathbb{P}_X^n$ and consider the canonical projection $p : P \to X$. Recall we have introduced in Example 2.2.10 the Gysin morphism: $p_* : \mathbb{E}^{*,*}(\mathbb{P}_X^n) \to \mathbb{E}^{*-2n,*-n}(X)$ based on the construction of [Dég08a]. We give an alternative construction which uses the point of view considered in the present setting. It is based on the following facts which follow directly from the projective bundle theorem (2.1.13). Consider the bigraded ring $A = \mathbb{E}^{**}(X)$. In the paragraph and definition that follow, we work in the category of bigraded A-modules and refer to them simply as A-modules:

- The bigraded group $\mathbb{E}^{**}(P)$ is a free A-module of finite rank.
- The Künneth map

$$\mathbb{E}^{**}(P) \otimes_A \mathbb{E}^{**}(P) \to \mathbb{E}^{**}(P \times_X P), (x, y) \mapsto p_1^*(x) \cup p_2^*(y)$$

  is an isomorphism of A-modules.

From the first point, we deduce that the A-module $\mathbb{E}^{**}(P)$ is dualizable. Let $\mathbb{E}^{**}(P)^\vee$ be its A-dual and

$$ev : \mathbb{E}^{**}(P)^\vee \otimes_A \mathbb{E}^{**}(P) \to A$$

be the evaluation map. Let $\delta : P \to P \times_X P$ be the diagonal immersion of P/X. Using the Gysin morphism and the second point, we get a co-pairing of A-modules:

$$\epsilon_P : A = \mathbb{E}^{**}(X) \xrightarrow{p^*} \mathbb{E}^{**}(P) \xrightarrow{\delta_*} \mathbb{E}^{**}(P \times_X P) \simeq \mathbb{E}^{**}(P) \otimes_A \mathbb{E}^{**}(P). \tag{3.2.1.a}$$

We claim this is a duality co-pairing in the sense that the induced map

$$\mathcal{D}_P : \mathbb{E}^{**}(P)^\vee \xrightarrow{1 \otimes \epsilon_P} \mathbb{E}^{**}(P)^\vee \otimes_A \mathbb{E}^{**}(P) \otimes_A \mathbb{E}^{**}(P) \xrightarrow{ev \otimes 1} \mathbb{E}^{**}(P) \tag{3.2.1.b}$$

is an isomorphism. Indeed, the computation of the matrix of $\mathcal{D}_P$ in the base given by the projective bundle theorem is precisely the same as that of [Dég08a, Lem. 5.5]: it is a triangular lower matrix with the identity diagonal. Note that $\mathcal{D}_P$ is homogeneous of degree $(2n, n)$.

Definition 3.2.2 Using the above notations, one defines the Gysin morphism associated with p as the following composite morphism:

$$p_* : \mathbb{E}^{**}(P) \xrightarrow{\mathcal{D}_P^{-1}} \mathbb{E}^{**}(P)^\vee \xrightarrow{(p^*)^\vee} \mathbb{E}^{**}(X)^\vee = \mathbb{E}^{**}(X),$$

homogeneous of degree $(-2n, -n)$.

In other words, p_* is the transpose of p^* with respect to the duality given by the co-pairing (3.2.1.a). In view of [Dég08a, Prop. 5.26], the above definition coincides with that of Example 2.2.10.

Remark 3.2.3 Here is a practical way to determine the Gysin map p_* constructed in the preceding definition. Let us again denote by A the cohomology ring $\mathbb{E}^{**}(X)$.

First, one determines the map ϵ_P — see (3.2.1.a). According to the projective bundle theorem, the $\mathbb{E}$-cohomology of $P \times_X P$ is a free A-module with basis $c^i . d^j$, $0 \le i, j \le n$, where c (resp. d) is the first Chern class of the canonical line bundle on the first (resp. second) factor of $P \times_X P$ (see Theorem 2.1.13). The A-linear map ϵ_P is uniquely determined by its value on the multiplicative identity element 1 of A, which can be a priori written as:

$$\epsilon_P(1) = \delta_*(1) = \sum_{0 \le i, j \le n} \pi^*\left(\eta_{ij}^{(n)}\right) . c^i d^j.$$

where π is the projection map of $P \times_X P/X$.

Second, the map $\mathcal{D}_P$ — see (3.2.1.b) — is an A-linear map between rank $n + 1$ free A-modules. Then the matrix M of $\mathcal{D}_P$ with respect to the basis on the source and targets coming from the projective bundle theorem applied to P/X can be easily computed with the previous notation as:

$$M := \left(\eta_{ij}^{(n)} \right)_{\{0 \leq i,j \leq n\}}.$$

From what was said in the paragraph preceding the above definition, the matrix M is invertible — it is in fact symmetric, lower triangular with 1 on the anti-diagonal; see Example 3.2.14 for a computation of this matrix in terms of the formal group law associated with $(\mathbb{E}, c)$.

According to the preceding definition, the Gysin map p_* is then completely determined by the inverse matrix M^{-1}. In fact, here is a way to compute the image of an element $x \in \mathbb{E}^{**}(P)$ under the Gysin map p_*:

- We write:
$$x = \sum_{0 \leq i \leq n} p^*(x_i).c^i$$

  where c is the canonical line bundle on P. Let X be the element of $A^{\{0,\dots,n\}}$ whose value at the i-th place is x_i.

- The map $(p^*)^\vee$ just extract the coefficient of the element $(c^0)^\vee$ with respect to the decomposition of elements of $\mathbb{E}^{**}(P)$ in the dual base $((c^i)^\vee)$. Therefore, we get that $p_*(x)$ is the 0-th coefficient of $M^{-1}.X$, seen as an element of A^n.

Note that, given the cofactors formula to compute the inverse of a matrix, we get in particular the following formula:

$$p_*(1) = \frac{1}{\det(M)} \cdot \det \left(\eta_{ij}^{(n)} \right)_{\{1 \leq i,j \leq n\}}. \qquad (3.2.3.a)$$

This will be made precise in Example 3.2.14.

3.2.4 The Gysin morphism introduced above satisfies the following properties:

1. For any integers $n, m \geq 0$, considering the canonical projections

$$
\begin{array}{ccc}
\mathbb{P}_X^n \times_X \mathbb{P}_X^m & \xrightarrow{q'} & \mathbb{P}_X^n \\
p' \downarrow & & \downarrow p \\
\mathbb{P}_X^m & \xrightarrow{q} & X
\end{array}
$$

one has: $p_* q'_* = q_* p'_*$.

2. For any cartesian square

$$
\begin{array}{ccc}
\mathbb{P}^n_Z & \xrightarrow{\;l\;} & \mathbb{P}^n_X \\
q\downarrow & & \downarrow p \\
Z & \xrightarrow{\;i\;} & X
\end{array}
$$

where i is an $\mathbb{E}$-pure closed immersion and p is the canonical projection: $i_* q_* = p_* l_*$.

3. For any integer $n \geq 0$ and any section s of the projection $p : \mathbb{P}^n_X \to X$: $p_* s_* = 1$.

For the proof of these properties, we use Remark 3.1.5 and refer the reader to [Dég08a], respectively Lemmas 5.8, 5.9 and 5.10. Then the following lemma follows formally from these properties — see [Dég08a, Lem. 5.11]:

Lemma 3.2.5 *Assume $\mathbb{E}$ is absolutely pure and consider a commutative diagram :*

$$
\begin{array}{ccc}
 & \mathbb{P}^n_X & \\
Y \overset{k}{\nearrow} & & \overset{p}{\searrow} X \\
\searrow_i & \mathbb{P}^m_X & \nearrow_q
\end{array}
$$

where i (resp. k) is a closed immersion and p (resp. q) is the canonical projection. Then, using the above definitions, $p_* k_* = q_* i_*$.

3.2.6 Recall that an X-scheme Y is said to be a *local complete intersection* if it admits locally an immersion into an affine space $\mathbb{A}^n_X$ — see [BGI71, VIII, 1.1]. We will say abusively *lci* instead of local complete intersection. Given a projective lci morphism $f : Y \to X$ in $\mathscr{S}$, it admits a factorization of the form

$$
Y \xrightarrow{\;i\;} \mathbb{P}^n_X \xrightarrow{\;p\;} X \tag{3.2.6.a}
$$

where i is a regular closed immersion and p is the canonical projection. This follows from our convention on projective morphisms and [BGI71, 1.2].

Now, assume that $\mathbb{E}$ is absolutely pure. From the preceding lemma, the composite morphism:

$$
\mathbb{E}^{**}(Y) \xrightarrow{\;i_*\;} \mathbb{E}^{**}(\mathbb{P}^n_X) \xrightarrow{\;p_*\;} \mathbb{E}^{**}(X)
$$

obtained using Definition 3.1.2 and Paragraph 3.2.1 is independent of the choice of the factorization of f.

Definition 3.2.7 Consider the above assumptions and notations. We put $f_* = p_* i_*$ and call it the Gysin morphism associated with f.

Note that if f has dimension d, then f_* is homogeneous of bidegree $(-2d, -d)$.

Remark 3.2.8 The warning of Remark 3.1.3 can be equally applied to the preceding Gysin morphism. In fact, it is proved in Theorem 3.3.1 that the preceding Gysin morphisms are uniquely determined by the chosen orientation c on $\mathbb{E}$.

Example 3.2.9 (1) *Motivic cohomology and Chow groups.–* Let k be a perfect field and consider the k-absolute ring spectrum $\mathbf{H}_{\mathbb{Z}}$ of Example 2.1.4. It is well known that for any integer $n \geq 0$ and any smooth k-scheme:
$$\mathbf{H}_{\mathcal{M}}^{2n,n}(X, \mathbb{Z}) \simeq CH^n(X).$$
In fact, we know that the Nisnevich sheaf $H^n(\mathbb{Z}(n))$ is the unramified Milnor K-theory $\mathcal{K}_n^M$, and from the hypercohomology spectral sequence associated with the homotopy t-structure along with some known vanishing[42], one gets:
$$\mathbf{H}_{\mathcal{M}}^{2n,n}(X, \mathbb{Z}) \simeq H_{\mathrm{Zar}}^n(X, \mathcal{K}_*^M) \simeq CH^n(X).$$
Then, from [Dég11, Prop. 3.16], we get that the Gysin morphism associated with $\mathbf{H}_{\mathbb{Z}}$ and a projective morphism in the previous definition coincides in degree $(2n, n)$ with the usual pushforward on Chow groups.[43] (See also Example 3.3.4(2)).

(2) *The spectrum* $\mathbf{KGL}$ *and algebraic K-theory.–* the Gysin morphism on $\mathbf{KGL}$ associated with a projective morphism of regular schemes agrees with the usual functoriality: see Example 3.3.4(3).

Proposition 3.2.10 *Assume $\mathbb{E}$ is absolutely pure.*

(1) *For any composable projective lci morphisms f and g: $f_* g_* = (fg)_*$.*
(2) *For any projective lci morphism $f : Y \to X$ and any couple (x, y) in $\mathbb{E}^{**}(X) \times \mathbb{E}^{**}(Y)$,*
$$f_*\big(f^*(x).y\big) = x.f_*(y).$$

Proof Point (1) follows formally from the preceding facts (for details, see the proof of [Dég08a, Prop. 5.14], using: relations (1), (2) of Paragraph 3.2.4, point (1) of Proposition 3.1.4 and the preceding lemma).

---

[42]namely: $H^q(\mathbb{Z}(n)) = 0$ if $q > n$ and $H^p(X, H^q(\mathbb{Z}(n))) = 0$ if $p > n$, which follows from the Gersten resolution satisfied by the sheaf $H^q(\mathbb{Z}(n))$.

[43]Let us indicate also that one can go from the case of smooth k-schemes to arbitrary regular schemes of equal characteristics using Popescu's desingularization theorem together with Lemma 1.2.13.

Point (2) follows from relation (2) of Proposition 3.1.4 for the case of a closed immersion together with [Dég08a, Cor. 5.18] for the case of the projection of $\mathbb{P}_X^n/X$.

□

3.2.11 Consider a cartesian square

$$
\begin{array}{ccc}
Y' & \xrightarrow{q} & X' \\
g\downarrow & \Delta & \downarrow f \\
Y & \xrightarrow{p} & X
\end{array}
$$

in $\mathscr{S}$ such that p is a projective lci morphism.

Choose an X-embedding ν of Y into a suitable projective bundle $\mathbb{P}_X^n$. The preceding square induces a cartesian morphism of closed pairs Θ : $(\mathbb{P}_{X'}^n, Y') \to (\mathbb{P}_X^n, Y)$ and we denote by ξ the vector bundle over Y' defined by formula (2.4.2.a) with respect to Θ. According to [Ful98, Prop. 6.6(c)] the vector bundle ξ is independent up to isomorphism of the choice of ν and we call it the *excess intersection bundle* associated with the square Δ.

Proposition 3.2.12 *Assume $\mathbb{E}$ is absolutely pure and consider the above notations. Let e be the rank of ξ over Y'.*

Then for any $y \in \mathbb{E}^{**}(Y)$, one has: $f^*p_*(y) = q_*\big(c_e(\xi).g^*(y)\big).$

This follows easily from the definition of ξ and point (3) of Proposition 3.1.4.

As usual, one says the square Δ is transversal if $e = 0$ — *i.e.* for any point a in Y', the dimension of q at a is equal to the dimension of p at $g(a)$. In this case, the preceding formula reads simply: $f^*p_* = q_*g^*$.

As a companion of the Gysin morphism, one gets another kind of characteristic classes, the cobordism classes. The definition obtained here is valid in a slightly better generality than that of [Dég08a, §5.27].

Definition 3.2.13 Assume $\mathbb{E}$ is an absolutely pure oriented ring spectrum.

Given any base scheme S in $\mathscr{S}$ and any projective S-scheme X in $\mathscr{S}$ with structural morphism p, we define the *cobordism class* of X/S as the following element of $\mathbb{E}^{**}(S)$:

$$[X/S] := p_*(1).$$

Note that if X/S has constant relative dimension d, the class $[X/S]$ has cohomological degree $(-2d, -d)$.

Example 3.2.14 Let us now recall the explicit computation of the cobordism class of the linear projective n-plane obtained in [Dég08a, 5.31]. In fact, we will actually correct a sign mistake *loc. cit.* that was indicated to us by the referee.

According to Remark 3.2.3, the key point is to compute the *fundamental class of the diagonal* δ of the X-scheme $P = \mathbb{P}_X^n$; in other words, the element $\delta_*(1)$, with the notation of Definition 3.1.2.

The main point used to compute that class is the classical fact that the diagonal subscheme $\delta(P)$ of $P \times_X P$ is the zero locus of a canonical section of the vector bundle

$$\lambda_1^\vee \otimes \xi_2 = \underline{\mathrm{Hom}}(\lambda_1, \xi_2)$$

where λ_1 (resp. ξ_2) is the canonical line (resp. quotient) bundle on the first (resp. second) factor of $P \times_X P$, seen as a vector bundle over $P \times_X P$ (see [Dég08a, Ex. 5.29]). Therefore according to Corollary 2.4.6, one gets:

$$\delta_*(1) = c_n\left(\lambda_1^\vee \otimes \xi_2\right).$$

Then, if $(a_{ij})_{i,j \in \mathbb{N}}$ denotes the coefficients of the formal group law associated with the orientation c of $\mathbb{E}$, one gets the following formula[44]:

$$\delta_*(1) = \sum_{0 \le i,j \le n} \pi^*(a_{1,i+j-n}).c^i.d^j.$$

Therefore, with the notations of Remark 3.2.3, the matrix M corresponding to $\delta_*(1)$ is:

$$\begin{pmatrix} 0 & \cdots & 0 & 1 \\ & & & a_{1,1} \\ 0 & & & \\ 1 & a_{1,1} & \cdots & a_{1,n} \end{pmatrix}.$$

Note a basic computation gives that the determinant of this matrix is $(-1)^{\lfloor(n+1)/2\rfloor}$. Therefore, using formula (3.2.3.a) together with the preceding computation, we get the following expression of the cobordism class of the projective n-plane:

$$[\mathbb{P}^n/S] = (-1)^{\lfloor(n+1)/2\rfloor}. \begin{vmatrix} 0 & \cdots & 0 & 1 & a_{1,1} \\ & & & & a_{1,2} \\ 0 & & & & \\ 1 & & & & \\ a_{1,1} & a_{1,2} & \cdots & & a_{1,n} \end{vmatrix}. \qquad (3.2.14.\mathrm{a})$$

---

[44] We refer the reader to [Dég08a, proof of 5.30] for a proof.

Beware that the sign written here corrects the sign in [Dég08a, Cor. 5.31].
In fact, this formula coincides with its analogue in topology given by the
classical Myschenko formula.

Given an arbitrary projective S-scheme X, we get the following method
to compute the cobordism class of X/S: choose an embedding $i : X \to \mathbb{P}_S^N$;
compute the fundamental class of i in terms of the canonical basis $(c^i, 0 \leq i \leq N)$ of $\mathbb{E}^{**}(\mathbb{P}_S^n)$ as a free $\mathbb{E}^{**}(S)$-module, where c is the first Chern class
of dual canonical invertible bundle:

$$\eta_{\mathbb{P}_S^n}(X) = \sum_{i=0}^{N} x_i . c^i, \, x_i \in \mathbb{E}^{**}(S).$$

Then the projection formula yields the following expression:

$$[X/S] = \sum_{i=0}^{N} x_i . [\mathbb{P}^{N-i}/S].$$

Remark 3.2.15 Using the Riemann-Roch formula, we will reprove Quillen
formula which computes the cobordism class of an arbitrary projective bundle (see Ex. 5.2.7)

3.3 Uniqueness

In our setting, we can extend the uniqueness statement of [Pan09, 4.1.4],
to obtain the following characterization of the Gysin morphisms we have
introduced.

Theorem 3.3.1 *Assume $\mathbb{E}$ is an absolutely pure ring spectrum over $\mathscr{S}$.
For any scheme S, we will denote by η_S the class in $\mathbb{E}^{2,1}(\mathbb{P}_S^1)$ obtained by
$\mathbb{P}^1$-suspension of the unit $S^0 \to \mathbb{E}$ of the ring spectrum structure.*

*Suppose given for any projective lci morphism $f : Y \to X$ a morphism
of bigraded abelian groups (non necessarily homogeneous):*

$$f_\star : \mathbb{E}^{**}(Y) \to \mathbb{E}^{**}(X)$$

such that the following properties hold:

(1) $(fg)_\star = f_\star g_\star$;

(2) $f_\star$ *is $\mathbb{E}^{**}(X)$-linear:* $f_\star(f^*(x).y) = x.f_\star(y)$;

(3) *for any square Δ as in 3.2.11, which is in addition transversal,* $f^* p_\star = q_\star g^*$;

(4) *if i is a closed immersion with complementary open immersion i',*
$\mathrm{Im}(i_\star) = \mathrm{Ker}(i'^*)$.

For any integer $n > 0$, let λ_n be the canonical line bundle on $\mathbb{P}^n_S$ and s_n be its zero section. Let us put:

$$c_{n,S} = s_n^* s_{n\star}(1).$$

Then the following conditions are equivalent:

- (i) the sequences $c_S = (c_{n,S})_{n>0}$ indexed by a scheme S form an absolute orientation of $\mathbb{E}$;
- (ii) for any $n > 0$, $c_{n,S}$ has bidegree $(2,1)$ and $c_{1,S} = \eta_S$ in $\mathbb{E}^{2,1}(\mathbb{P}^1_S)$.

When these equivalent conditions are fulfilled, we get in addition for any scheme S which admits an ample family of line bundles:

- (5) for any section s of a line bundle L/S, $s^* s_\star(1) = c_1(L)$ where the right hand side is the first Chern class associated with the orientation c of $\mathbb{E}$ given by the above condition (i).

Finally, if we assume that condition (5) holds for any scheme S in $\mathscr{S}$, we get:

- for any projective lci morphism f, one has: $f_\star = f_*$, the last morphism being the Gysin morphism defined in 3.2.7 with respect to $(\mathbb{E}, c)$.

Remark 3.3.2 To summarize: the Gysin morphism of an absolutely pure oriented ring spectrum $(\mathbb{E}, c)$ is uniquely characterized by properties (1)–(5).

Proof The equivalence between (i) and (ii) is obvious from Definition 2.1.2 and the fact that Point (3) implies: $\iota_n^*(c_{n+1}^S) = c_n^S$.

Let us assume these equivalent conditions are satisfied. To prove (5), using homotopy invariance, we can assume s is the zero section of L/S. Because S admits an ample family of line bundles, we can use Jouanolou's trick which reduces us to the case where S is affine. Then L is generated by a finite number of its sections and one can find an immersion $i : S \to \mathbb{P}^n_S$ such that $L = i^{-1}(\lambda_n)$. Using (3), one reduces to the case where S is $\mathbb{P}^n_S$, $L = \lambda_n$ which holds by definition.

Let us now prove the final point. Given any projective lci morphism f, we put $f_\flat = f_* - f_\star$. We have proved previously that the Gysin morphisms $f_\star$ satisfies properties (1)–(4) stated above. In particular, $f_\flat$ satisfies properties (1)–(3) and in the situation of (4), we get:

(4') $\mathrm{Im}(i_\flat) \subset \mathrm{Ker}(i'^*)$.

Case of closed immersions: assuming $f = i : Z \to S$ is a closed immersion, we show $i_\flat = 0$. Using (3) for $?_\flat$, the deformation diagram (1.3.1.a) induces a commutative diagram:

$$
\begin{array}{ccccc}
\mathbb{E}^{**}(Z) & \xleftarrow{\;s_1^*\;} & \mathbb{E}^{**}(\mathbb{A}_Z^1) & \xrightarrow{\;s_0^*\;} & \mathbb{E}^{**}(Z) \\
\downarrow{\scriptstyle i_\flat} & & \downarrow{\scriptstyle k_\flat} & & \downarrow{\scriptstyle s_\flat} \\
\mathbb{E}^{**}(X) & \xleftarrow{\;\sigma_1^*\;} & \mathbb{E}^{**}(D_Z X) & \xrightarrow{\;\sigma_0^*\;} & \mathbb{E}^{**}(N_Z X).
\end{array}
$$

We use the following lemma:

Lemma 3.3.3 *Consider the notations above and let $k' : (D_Z X - \mathbb{A}_Z^1) \to D_Z X$ be the complementary open immersion to k. Then the morphism: (k'^*, σ_0^*) is injective.*

Indeed, let $x \in \mathbb{E}^{**}(D_Z X)$ being an element of the kernel. Because of property (4), we get a cohomology class y such that $x = k_\star(y)$. On the other hand, because of (3), $\sigma_0^* k_\star = s_\star s_0^*$. But $s_\star$ is a split monomorphism (because of (1)), and s_0^* is an isomorphism. Thus we deduce $y = 0$ and this concludes.

With the help of this lemma, and the fact $k'^* k_\flat = 0$ from (4') above, we see that it is sufficient to prove $\sigma_0^* k_\flat = 0$. Thus, we are reduced to show that $s_\flat = 0$, where again s is the zero section of the vector bundle $E = N_Z X$ over Z.

Using the splitting principle applied to the vector bundle E/Z and property (1), we reduce to the case where s is the zero section of a line bundle $p : L \to Z$. According to (2), $s_\flat(x) = s_\flat(1).p^*(x)$. The fact $s_\flat(1) = 0$ is nothing else than assumption (5).

Case of a projective bundle: let $p : \mathbb{P}_S^n \to S$ be the canonical projection, and let us show $p_\flat = 0$. Let δ be the diagonal immersion of $\mathbb{P}_S^n/S$. Let us consider the bigraded ring $A = \mathbb{E}^{**}(S)$. Below any A-module is assumed to be a bigraded A-module, but we do not require morphisms of bigraded A-modules are homogeneous. The symbol $\otimes_A$ means the tensor product of bigraded A-modules. Then, according to the projective bundle theorem, $M = \mathbb{E}^{**}(\mathbb{P}_S^n)$ is a free A-module of finite rank. Thus it is a rigid object of the category of A-modules. On the other hand, one gets using (2) the following morphisms of A-modules:

$$
p_\star \delta^* : M \otimes_A M \longrightarrow A
$$
$$
\delta_\star p^* : A \longrightarrow M \otimes_A M.
$$

Using (1), (2) and (3), we get that the following composite is the identity morphism:

$$M \xrightarrow{M \otimes_A \delta_* p^*} M \otimes_A M \otimes_A M \xrightarrow{p_* \delta^* \otimes_A M} M.$$

The same is true when replacing $\star$ by $*$. Thus, as we have already proved that $\delta_\star = \delta_*$, we deduce that:

$$(p_\flat \delta^* \otimes_A M) \circ (M \otimes_A \delta_* p^*) = 0.$$

As δ_* and p^* are split monomorphisms, we deduce that $p_\flat \delta^* \otimes_A M = 0$. This allows to conclude because M is faithfully flat over A and δ^* is a split epimorphism.

□

Example 3.3.4 As said before, the case where $\mathscr{S}$ is the category of smooth k-schemes was already obtained by Panin and Smirnov, later published by Panin in [Pan09, 4.1.4].[45]

(1) It is worthwhile to mention that this result applies in particular to any Mixed Weil cohomology theory: example 1.2.3(1), where $\mathscr{S}$ is the category of regular k-schemes. Thus, there is only one way to define pushforwards on any such cohomology satisfying conditions (1)–(5) — and in particular compatible with a well behaved first Chern class.

(2) It can also be applied to the usual Chow groups: in [Dég08a], we have shown that the category of stable motivic complexes $DM(k)$ is endowed with a t-structure, called the homotopy t-structure[46] and whose heart is the category of cycle modules defined by Rost. In particular, the cycle module corresponding to Milnor K-theory defines an object in $DM(k)$ which is nothing else than the 0-th cohomology object of the unit in $DM(k)$ (see *loc. cit.*, Th. 5.11). Using the forgetful functor

$$DM(k) \to SH(k)$$

we can see this object as a ring spectrum in $SH(k)$: by definition, it corresponds to the unramified Milnor-K-theory sheaf $\mathcal{K}_*^M$, seen over the Nisnevich site of smooth k-schemes, and represents the Chow group in $SH(k)$, with its product structure. Note also this is a direct factor of the motivic Eilenberg-MacLane spectrum $\mathbf{H}_{\mathcal{M},k}^{\mathbb{Z}}$ representing motivic cohomology (cf. 1.2.3): it corresponds to cut-out the groups of bidegree $(2*, *)$. As usual, one extends $\mathcal{K}_k$ to a k-absolute ring spectrum $\mathcal{K}$ by taking pullbacks within the fibred category SH.

---

[45] See [PS00] for the original work of Panin and Smirnov. I thank the referee for attracting my attention on the paper of Smirnov [Smi06a].

[46] extending the homotopy t-structure of Voevodsky that he defined on $DM_-^{eff}(k)$.

Then the preceding theorem applies to the latter ring spectrum $\mathcal{K}$, when $\mathscr{S}$ is the category of regular k-schemes (because of Lemma 1.2.13 and the fact that usual Chow groups commute with limits of regular k-schemes). Thus, it says that prescribing the orientation on CH^*, there exists a unique way of defining pushforwards satisfying properties (1)–(5).

(3) More interestingly, the proposition applies to **KGL**, the absolute ring spectrum representing K-theory when $\mathscr{S}$ is the category of regular schemes (Example 1.2.3(3)). Recall from [Qui73] the corresponding cohomology has well defined pushforward that satisfies conditions (1)–(4) of the proposition. Moreover, with the choice of orientation of Example 2.1.4(4), condition (5) of the above Theorem has been proved by Thomason in [Tho93, Th. 3.1]. Thus, the isomorphism (2.1.5.a) is covariantly functorial with respect to projective morphisms between regular schemes, where on the source we consider the pushforward defined above with respect to the absolute oriented ring spectrum $(\mathbf{KGL}, c^{\mathbf{KGL}})$ and on the aim we consider Quillen pushforward.

Note also that we need Thomason excess intersection formula only in the case of line bundles: the machinery developed here gives a proof in the general case. Note also that if we make the "hypothèse paresseuse" (referred to by Thomason in the end of the introduction of *loc. cit.*) that schemes, in addition to being regular, admit an ample line bundle, we even get a new proof of the excess intersection formula for higher K-theory — because (5) is then automatically satisfied by our choice of orientation and we can compare our Gysin morphism with the one defined by Quillen.

4 Riemann-Roch formulas

4.0.5 In this section, we will consider two absolute oriented ring spectra $(\mathbb{E}, c)$, $(\mathbb{F}, d)$ and a morphism of absolute ring spectra (cf Definition 1.2.1):

$$\varphi : \mathbb{E} \to \mathbb{F}.$$

When considering the constructions of orientation theory as described previously for $\mathbb{E}$ (resp. $\mathbb{F}$), we will will put an index $\mathbb{E}$ (resp. $\mathbb{F}$) in the notation. However, when no confusion is possible, we drop this index.

In the particular case $\mathbb{E} = \mathbb{F}$, we will also put an upper-index c (resp. d) in the notation of orientation theory (Chern classes, Gysin morphisms, ...) with respect to the orientation c (resp. d).

4.1 Todd classes

4.1.1 Let S be a scheme (in $\mathscr{S}$). We deduce from φ a morphism of graded rings:

$$\mathbb{E}^{**}(\mathbb{P}_S^\infty) \xrightarrow{\varphi_{\mathbb{P}_S^\infty}} \mathbb{F}^{**}(\mathbb{P}_S^\infty)$$

using the notations of Paragraph 2.1.1. According to the projective bundle theorem (2.1.13), this corresponds to a morphism of ring:

$$\mathbb{E}^{**}(S)[[u]] \to \mathbb{F}^{**}(S)[[t]]$$

and we denote by $\Psi_S(t)$ the image of u by this map. In other words, this formal power series is characterized by the relation:

$$\varphi_{\mathbb{P}_S^\infty}(c) = \Psi_S(d). \tag{4.1.1.a}$$

Note that the restriction of $\varphi_{\mathbb{P}_S^\infty}(c)$ to $\mathbb{P}_S^0$ (resp. $\mathbb{P}_S^1$) is 0 (resp. the multiplicative identity element of the ring $\mathbb{F}^{**}(\mathbb{P}_S^1)$) because c is an orientation and φ is a morphism of ring spectra. Thus, as d is also an orientation of $\mathbb{F}$, we can write $\Psi_S(t)$ as:

$$\Psi_S(t) = t + \sum_{i>1} \alpha_i^S.t^i$$

where $\alpha_i^S \in \mathbb{F}^{2-2i,1-i}(S)$. Obviously, the power series $\Psi_S(t)/t$ is invertible.

We will also consider the commutative monoid $\mathcal{M}(S)$ generated by the isomorphism classes of vector bundles over S modulo the relations $[E] = [E'] + [E'']$ coming from exact sequences

$$0 \to E' \to E \to E'' \to 0.$$

Then $\mathcal{M}$ is a presheaf of monoids on $\mathscr{S}$ whose presheaf of abelian groups is the functor K_0.

Note that $\mathbb{F}^{00}(S)$, equipped with cup-product, is a commutative monoid. We will denote by $\mathbb{F}^{00\times}(S)$ the group made by its invertible elements.

Proposition 4.1.2 *There exists a unique natural transformation of presheaves of monoids over $\mathscr{S}$*

$$\mathrm{Td}_\varphi : \mathcal{M} \to \mathbb{F}^{00}$$

such that for any line bundle L over a scheme X,

$$\mathrm{Td}_\varphi(L) = \frac{t}{\Psi_S(t)}(t = d_1(L)). \tag{4.1.2.a}$$

Moreover, it induces a natural transformation of presheaves of abelian groups:

$$\mathrm{Td}_\varphi : K_0 \to \mathbb{F}^{00\times}.$$

Proof The proof is very classical: the uniqueness statement follows from the *splitting principle* while the existence statement follows from the use of *Chern roots*. Note also that the relation (4.1.2.a) is well defined because $c_1(L)$ is nilpotent (see Proposition 2.1.22). The final assertion follows from the fact $t/\Psi_S(t)$ is an invertible formal power series.

$\square$

Remark 4.1.3 According to the construction of the first Chern classes for the oriented ring spectra $(\mathbb{E}, c)$ and $(\mathbb{F}, d)$ together with Relations (4.1.1.a) and (4.1.2.a), we get for any line bundle L/S the following identity in $\mathbb{F}^{2,1}(S)$:

$$\varphi_S\big(c_1(L)\big) = \mathrm{Td}_\varphi(-L) \cup d_1(L). \qquad (4.1.3.\mathrm{a})$$

Definition 4.1.4 Consider the context and notations of the previous proposition.

Given any virtual vector bundle e over X, the element $\mathrm{Td}_\varphi(e) \in \mathbb{F}^{00}(X)$ is called the *Todd class* of e over X associated with the morphism of ring spectra φ.

4.2 The case of closed immersions

4.2.1 Consider a regular closed immersion $i : Z \to X$, $U = X - Z$. As by assumption $\mathbb{E}$ (resp. $\mathbb{F}$) is absolutely pure, we can consider the associated refined fundamental class $\bar\eta_X^{\mathbb{E}}(Z)$ in $\mathbb{E}_Z^{**}(X)$ (resp. $\bar\eta_X^{\mathbb{F}}(Z)$ in $\mathbb{F}_Z^{**}(X)$) — Definition 2.3.1. The morphism φ induces a map in relative cohomology:

$$\mathbb{E}_Z^{**}(X) \xrightarrow{\varphi_{X,Z}} \mathbb{F}_Z^{**}(X).$$

According to the definition of the purity isomorphism (2.3.1.b), we deduce there exists a unique class $\tau_\varphi(X, Z) \in \mathbb{F}^{00}(Z)$ such that

$$\varphi_{X,Z}\big(\bar\eta_X^{\mathbb{E}}(Z)\big) = \tau_\varphi(X, Z) \cdot \bar\eta_X^{\mathbb{F}}(Z). \qquad (4.2.1.\mathrm{a})$$

This relation together with the definition of the localization long exact sequence (3.1.1.b) immediately gives the following commutative diagram:

$$
\begin{array}{ccccc}
\mathbb{E}^{**}(U) & \xrightarrow{\partial_{X,Z}} & \mathbb{E}^{**}(Z) & \xrightarrow{i_*} & \mathbb{E}^{**}(X) \\
\varphi_U \downarrow & & \tau_\varphi(X,Z)\cdot\varphi_Z \downarrow & & \downarrow \varphi_X \\
\mathbb{F}^{**}(U) & \xrightarrow{\partial_{X,Z}} & \mathbb{F}^{**}(Z) & \xrightarrow{i_*} & \mathbb{F}^{**}(X).
\end{array}
\qquad (4.2.1.\mathrm{b})
$$

The commutativity of the square on the right hand side is a tautological Riemann-Roch formula. Our proof of the actual Riemann-Roch theorem in fact lies in the computation of the class involved in this formula:

Lemma 4.2.2 *In the above assumptions, the following relation holds:*

$$\tau_\varphi(X, Z) = \mathrm{Td}_\varphi(-N_Z X).$$

Proof As Relation (4.2.1.a) characterizes uniquely the class $\tau_\varphi(X, Z)$, whatever the regular closed pair (X, Z) is, the deformation diagram (2.3.1.a) gives the relation: $\tau_\varphi(X, Z) = \tau_\varphi(N_Z X, Z)$.

Thus, we are reduced to prove that for any scheme X and any vector bundle E/X,
$$\mathrm{Td}_\varphi(-E) = \tau_\varphi(E, X).$$

Using again the characterizing relation (4.2.1.a), which involves refined Thom classes according to (2.3.3.a), we deduce:

- for any morphism $f : Y \to X$ and any vector bundle E/X, $f^* \tau_\varphi(E, X) = \tau_\varphi(f^{-1}(E), Y)$.
- for any scheme X and any exact sequence of vector bundles over X,

$$0 \to E' \to E \to E'' \to 0,$$

one has: $\tau_\varphi(E, X) = \tau_\varphi(E', X)\tau_\varphi(E'', X)$.

More precisely: for the first relation, one uses the compatibility of the refined Thom class with pullback and for the second, one applies Proposition 2.4.7.

In other words, one gets a morphism of monoids

$$M(X) \to F^{00}(X), [E] \mapsto \tau_\varphi(E, X)$$

which is contravariantly natural in X. According to the uniqueness statement of Proposition 4.1.2, we are reduced to the case of line bundles.

Let L/X be a line bundle and P be its projective completion. Let ξ be the universal quotient bundle on P. According to Relation (4.1.3.a), one obtains:

$$\varphi_P\big(c(\xi)\big) = \mathrm{Td}_\varphi(-\xi) \cup d(\xi).$$

Let s be the section of P/X induced by the zero section of L/X. Using the commutativity of the right square of (4.2.1.b) when the closed pair (X, Z) is (P, X), we obtain:

$$\varphi_P\big(s_*(1)\big) = s_*(\tau_\varphi(P, X)) = s_*(\tau_\varphi(L, X))$$

According to (3.1.2.a), (2.3.3.b) and (2.2.4.a): $s_*(1) = c(\xi)$. Thus we obtain:

$$\mathrm{Td}_\varphi(-\xi) \cup s_*(1) = s_*(\tau_\varphi(L, X)).$$

Let $p : P \to X$ be the canonical projection. Of course, $p \circ s = Id$. Thus, it is sufficient to apply p_* to the preceding relation to conclude:

$$p_*\big(\mathrm{Td}_\varphi(-\xi)\cup s_*(1)\big) \overset{(*)}{=} p_* s_*\big(s^* \mathrm{Td}_\varphi(-\xi)\big) = s^*(\mathrm{Td}_\varphi(-\xi)) = \mathrm{Td}_\varphi(-L),$$

where $(*)$ is the projection formula 3.1.4(2).

$\square$

Finally the following Riemann-Roch formulas are consequences of the commutative diagram (4.2.1.b) in conjunction with the preceding lemma.

Theorem 4.2.3 *Under the assumptions of the previous paragraph, the following formulas hold:*

$$\varphi_X(i_*(z)) = i_*\big(\mathrm{Td}_\varphi(-N_Z X)\cup\varphi_Z(z)\big),$$
$$\partial_{X,Z}\big(\varphi_U(u)\big) = \mathrm{Td}_\varphi(-N_Z X)\cup\varphi_Z\big(\partial_{X,Z}(u)\big).$$

4.3 The general case

4.3.1 Consider a projective lci morphism $f : Y \to X$ (Par. 3.2.6). Given a factorization (3.2.6.a) of f, we define the *virtual tangent bundle* of f as the element of $K_0(Y)$:

$$\tau_f := [i^{-1}T_p] - [N_i]$$

where T_p is the tangent space of p and N_i is the normal bundle of i. Standard considerations show this definition is independent of the chosen factorization (see [BGI71, VIII, 2.2]).

Theorem 4.3.2 *Consider the notations above. Then for any element $y \in \mathbb{E}^{**}(Y)$, the following formula holds:*

$$\varphi_X(f_*(y)) = f_*\big(\mathrm{Td}_\varphi(\tau_f)\cup\varphi_Y(y)\big).$$

Proof According to the multiplicativity of the Todd class and the compatibility of the Gysin morphism with respect to composition, this formula can be divided in two cases according to a factorization (3.2.6.a) of f. The case of a closed immersion was treated above (4.2.3) and it remains to consider the case of the canonical projection p of a projective space $P = \mathbb{P}^n_X/X$.

We first treat the case where $\varphi_{\mathbb{P}^\infty_X}(c) = d$. This implies that $\mathrm{Td}_\varphi = 1$, so we have to prove that for $y \in \mathbb{E}^{**}(P)$, one has:

$$\varphi_X(p_*(y)) = p_*\big(\varphi_Y(y)\big). \tag{4.3.2.a}$$

Consider the notations of Paragraph 3.2.1 with respect to both $\mathbb{E}$ and $\mathbb{F}$. The Riemann-Roch formula for the diagonal embedding δ gives the formula:

$$\varphi_{P\times_X P}\delta_* = \delta_*\varphi_P.$$

This is sufficient to conclude in view of Definition 3.2.2. Let us give more details with the considerations of Remark 3.2.3. The preceding formula implies that the map $\varphi_X : \mathbb{E}^{**}(X) \to \mathbb{F}^{**}(X)$ sends each of the coefficients of the matrix $M^{\mathbb{E}}$ of $\delta_*(1) \in \mathbb{E}^{**}(P \times_X P)$ to the corresponding coefficient of the matrix $M^{\mathbb{F}}$ of $\delta_*(1) \in \mathbb{F}^{**}(P\times_X P)$. Therefore, the same result holds for the inverse matrices of $M^{\mathbb{E}}$ and $M^{\mathbb{F}}$. Thus finally, Formula (4.3.2.a) comes from the algorithm to compute p_* given at the end of Remark 3.2.3.

It remains now to consider the case where $\mathbb{E} = \mathbb{F}$, $\varphi = 1_{\mathbb{E}}$, for which we adopt the convention of the last paragraph of 4.0.5. Thus, we have to prove

$$\forall \alpha \in \mathbb{E}^{**}(P), \; p_*^c(\alpha) = p_*^d(\mathrm{Td}_\varphi(T_p).\alpha). \qquad (4.3.2.b)$$

We consider again the notations of Paragraph 3.2.1 with respect to both $(\mathbb{E}, c)$ and $(\mathbb{E}, d)$. Let $\pi : P \times_X P \to P$ be the projection to the second factor. The Riemann-Roch formula for δ can be read in this case as:

$$\delta_*^c(\alpha) = \delta_*^d\big(\mathrm{Td}_\varphi(-N_\delta)\cup\alpha\big) = \delta_*^d\big(\delta^*\pi^*(\mathrm{Td}_\varphi(-N_\delta))\cup\alpha\big)$$
$$= \pi^*(\mathrm{Td}_\varphi(-N_\delta))\cup\delta_*^d(\alpha)$$

— we use Prop. 3.2.10 for the last equality. Recall from Paragraph 3.2.1 we have a Künneth isomorphism $\mathbb{E}^{**}(P\times_X P) = \mathbb{E}^{**}(P)\otimes_A \mathbb{E}^{**}(P)$. Through this identification, and the fact $N_\delta = T_p$, the class $\pi^*(\mathrm{Td}_\varphi(-N_\delta))$ corresponds to $1 \otimes_A \mathrm{Td}_\varphi(-T_p)$. Thus the preceding equality leads to the commutativity of the following diagrams:

$$
\begin{array}{ccc}
A & \xrightarrow{\epsilon_P^c} & \mathbb{E}^{**}(P) \otimes_A \mathbb{E}^{**}(P) \\
\| & & \downarrow{\scriptstyle 1\otimes_A \mathrm{Td}_\varphi(T_p)} \\
A & \xrightarrow{\epsilon_P^d} & \mathbb{E}^{**}(P) \otimes_A \mathbb{E}^{**}(P)
\end{array}
\qquad \Rightarrow \qquad
\begin{array}{ccc}
\mathbb{E}^{**}(P)^\vee & \xrightarrow{\mathcal{D}_P^c} & \mathbb{E}^{**}(P) \\
\| & & \downarrow{\scriptstyle \mathrm{Td}_\varphi(T_p)} \\
\mathbb{E}^{**}(P)^\vee & \xrightarrow{\mathcal{D}_P^d} & \mathbb{E}^{**}(P).
\end{array}
$$

Then formula (4.3.2.b) finally follows from Definition 3.2.2.

$\square$

5   Examples and applications

5.0.3 In this section, we will adopt the following notations. Orientations of an absolute ring spectrum (Def. 2.1.2) will be denoted by letters c (or

c', d, ...). Then the corresponding Chern classes (Def. 2.1.16) with be denoted by the same letter with a lower index: c_1, c_2,... The corresponding Gysin morphisms (resp. fundamental classes, residue morphisms) will be denoted by p_*^c (resp.$\bar{\eta}_X^c(Z)$, $\partial_{X,Z}^c$).

All spectra in this section will be $\mathscr{S}$-absolutely pure ring spectra, so we simply say spectra for $\mathscr{S}$-absolutely pure ring spectra. Recall that an orientation c of such a spectrum $\mathbb{E}$ is a family of orientation c_S of $\mathbb{E}_S$ indexed by schemes S of $\mathscr{S}$, and stable by pullbacks (Def. 2.1.2). As seen above, such a collection of orientations gives rise to a collection F of formal group laws F_S on the ring $\mathbb{E}^{**}(S)$, which will simply be called the formal group law associated with $(\mathbb{E}, c)$ — or just c when $\mathbb{E}$ is clear.

5.1 Principle of computation

Definition 5.1.1 Let $(\mathbb{E}, c)$ and $(\mathbb{F}, d)$ be oriented spectra.

A *pseudo-morphism* of oriented spectra $\varphi : (\mathbb{E}, c) \to (\mathbb{F}, d)$ is simply a morphism of ring spectra. We will say φ is a *morphism* of oriented spectra if for any scheme S, one has:

$$\varphi_{\mathbb{P}_S^\infty}(c) = d$$

in $\tilde{\mathbb{F}}^{2,1}(\mathbb{P}_S^\infty)$, the reduced cohomology of the pointed scheme $\mathbb{P}_S^\infty$ pointed by the infinite point of $\mathbb{P}_S^1$.

We will also say that a pseudo-morphism φ is *identical* if, as a morphism of ring spectra, it is the identity.

Note that one immediately deduces from the construction of Chern classes and from the Riemann-Roch formulas 4.2.3 and 4.3.2 the following result:

Proposition 5.1.2 *Let $\varphi : (\mathbb{E}, c) \to (\mathbb{F}, d)$ be a morphism of oriented spectra.*

(1) *For any vector bundle E/S and any integer $n \geq 0$, one has: $\varphi_S(c_n(E)) = d_n(E)$.*

(2) *For any projective lci morphism $f : Y \to X$ one has: $\varphi_X \circ f_*^c = f_*^d \circ \varphi_Y$.*

(3) *For any regular closed immersion $i : Z \to X$, $U = X - Z$, one has: $\varphi_Z \circ \partial_{X,Z}^c = \partial_{X,Z}^d \circ \varphi_U$.*

Example 5.1.3 Let $(\mathbb{E}, c)$ be an oriented spectrum. According to Proposition 2.2.6, the choice of c uniquely corresponds to a morphism of oriented spectra:

$$\varphi : \left(\mathbf{MGL}, c^{\mathbf{MGL}}\right) \to (\mathbb{E}, c)$$

where $c^{\mathbf{MGL}}$ is the canonical orientation of $\mathbf{MGL}$.

Then, according to the preceding proposition, the morphism φ is compatible with all the constructions of orientation theory given in this paper.

5.1.4 Given any pseudo-morphism as in the above definition, we obviously get that $c' = \varphi_{\mathbb{P}_S^\infty}(c)$ is an orientation of $\mathbb{F}$. In particular, the pseudo-morphism φ admits a canonical factorization:

$$(\mathbb{E}, c) \xrightarrow{\tilde{\varphi}} (\mathbb{F}, c') \xrightarrow{\psi} (\mathbb{F}, c) \qquad\qquad (5.1.4.a)$$

such that $\tilde{\varphi}$ (resp. ψ) is a morphism (resp. identical pseudo-morphism) of oriented spectra.

According to Definition 4.1.4, the Todd class of φ is equal to the Todd class of ψ. Thus the computation of the Todd class of a pseudo-morphism can always be reduced to the case of an identical pseudo-morphism of the aim, which corresponds to the effect of *changing the orientation*. We will describe this case in more details below (see in particular 5.2.1).

Remark 5.1.5 According to Proposition 2.2.6, oriented (ring) spectra correspond to $\mathbf{MGL}$-algebras. In the light of this analogy, pseudo-morphisms (resp. morphisms) of oriented spectra corresponds to morphisms of rings (resp. $\mathbf{MGL}$-algebras) between $\mathbf{MGL}$-algebras. In light of this analogy, the preceding factorization can be understood as follows: $\tilde{\varphi}$ corresponds to a morphism of $\mathbf{MGL}$-algebras while ψ corresponds to a change of the $\mathbf{MGL}$-algebra structure on $\mathbb{F}$.

5.1.6 Recall that given a ring R and formal group laws F, G with coefficients in R, a morphism $\Phi : (R, F) \to (R, G)$ of formal group laws is a power series $\Phi(t) \in R[[t]]$ of positive valuation such that

$$\Phi(F(x, y)) = G(\Phi(x), \Phi(y))$$

in $R[[x, y]]$. Such a morphism is an isomorphism if and only if Φ admits a composition inverse — equivalently, $\Phi'(0)$ is invertible in R. It is called a *strict* isomorphism if $\Phi'(0) = 1$.

According to the conventions of this section, a morphism of formal group laws arising from orientations of an (absolute) ring spectrum will be a family of morphisms indexed by schemes of $\mathscr{S}$ and stable by pullbacks.

Proposition 5.1.7 *Let $\mathbb{E}$ be a ring spectrum and c be an orientation of $\mathbb{E}$ with associated formal group law F_c. Consider the following sets:*

(1) *the orientations d of $\mathbb{E}$ (equivalently: the identical pseudo-morphisms of ring spectra with source $(\mathbb{E}, c)$);*

(2) *the strict isomorphisms* Φ *of formal group laws with source* F_c *such that for any scheme* S, $\Phi_S(t)$ *can be written as a power series of the form:*

$$t + \sum_{i>1} \alpha_i^S . t^i$$

where α_i^S *is an element of* $\mathbb{E}^{2-2i,1-i}(S)$.

Then the map

$$(2) \xrightarrow{(*)} (1), \Phi \mapsto \Phi(c)$$

is a well defined bijection.

Proof We prove the map $(*)$ is well defined. First note that, because of the condition on the degree of the coefficients of Φ_S, the cohomology class $\Phi_S(c)$ is of degree $(2,1)$. The fact it is an orientation of $\mathbb{E}_S$ simply follows from the form of $\Phi_S(t)$ which implies: $\Phi_S(0) - 0$ and $\Phi'_S(0) = 1$.

To prove that $(*)$ is a bijection, we construct its inverse. Let d be an orientation of $\mathbb{E}$ and S be a base scheme. Then d_S is an element of the bigraded algebra

$$\mathbb{E}^{**}(\mathbb{P}_S^\infty) \simeq \mathbb{E}^{**}(S)[[c]].$$

This means there is a unique power series

$$\Phi_S(t) = \sum_{i\geq 0} \alpha_i^S . t^i$$

with coefficients in $\mathbb{E}^{**}(S)$ such that $d_S = \Phi_S(c_S)$. Because d_S has bidegree $(2,1)$, we deduce that necessarily, α_i^S has bidegree $(2-2i, 1-i)$. Moreover, because d_S is an orientation of $\mathbb{E}_S$, we deduce that $\alpha_0^S = 0$ and $\alpha_1^S = 1$. By uniqueness of Φ_S, we deduce that the family $\Phi = (\Phi_S)_{S\in\mathscr{S}}$ is stable by pullbacks.

Let us consider the Segre embedding (Par. 2.1.19):

$$\sigma : \mathbb{P}_S^\infty \to \mathbb{P}_S^\infty \times_S \mathbb{P}_S^\infty.$$

By definition of the formal group law F_c (resp. F_d) associated with c (resp. d) we obtain (dropping the reference to the base S):

$$\sigma^*(c) = F_c(c', c''), \text{ resp. } \sigma^*(d) = F_d(d', d'').$$

where, on the left hand side c', c'' (resp. d', d'') corresponds to the first Chern class associated with the orientation c (resp. d) of the canonical line bundle on the first (resp. second) factor of $\mathbb{P}_S^\infty \times_S \mathbb{P}_S^\infty$. Thus, we obtain that Φ_S is a strict isomorphism from F_c to F_d:

$$\Phi(F_c(c', c'')) = \Phi(\sigma^*(c)) = \sigma^*(\Phi(c)) = \sigma^*(d) = F_d(d', d'') = F_d(\Phi(c'), \Phi(c'')).$$

Therefore $\Phi_S(t)$ is a strict isomorphism $(\mathbb{E}^{**}(S), F) \to (\mathbb{E}^{**}(S), F_d)$ of formal group laws.

The uniqueness of Φ_S shows that we have indeed constructed an inverse map to $(*)$.

$\square$

Remark 5.1.8 This proposition is essentially an elaboration of the construction done in Paragraph 4.1.1. Note to be more precise that, given an identical pseudo-morphism $\varphi : (\mathbb{E}, c) \to (\mathbb{E}, d)$, the power series $\Phi(t)$ obtained by applying the previous proposition is the composition inverse of the power series $\Psi(t)$ obtained in 4.1.1. Thus, in fact, we get reciprocal strict isomorphisms of formal group laws on $\mathbb{E}^{**}$:

$$\Phi : F_c \leftrightarrows F_d : \Psi \tag{5.1.8.a}$$

From a categorical perspective, having fixed an oriented spectrum $(\mathbb{E}, c)$, the proposition defines a covariant functor $(*)$ from the category of orientations on $\mathbb{E}$ with morphisms the identical pseudo-morphisms to the category of ($\mathscr{S}$-families) of formals group laws on the bigraded ring $\mathbb{E}^{**}$.

The functor $(*)$ induces a bijection of groupoids when one restricts morphisms on the target category to strict isomorphisms satisfying the condition on the degrees of point (2) above.

Remark 5.1.9 According to this proposition, two orientations on a given (orientable ring) spectrum $\mathbb{E}$ necessarily yields (strictly) isomorphic formal group laws. Thus, there is a uniquely defined isomorphism class of formal group law associated with an orientable ring spectrum $\mathbb{E}$ and we can safely qualify such a ring spectrum as being additive, multiplicative, etc. Note that it is not known whether any formal group law can be realized as the formal group law associated with an orientable ring spectrum. However, interesting new examples are provided in [LYZ13, Th. A].

Corollary 5.1.10 *Let $\mathbb{E}$ be an absolute ring spectrum satisfying the following property:*

(Ann) *For any integer $i < 0$, and any scheme S, $\mathbb{E}^{2i,i}(S) = 0$.*

Then the following assertions hold:

   (i) *If an orientation exists on $\mathbb{E}$ over a scheme S, it is unique. Moreover, if $\mathbb{E}$ is rational,[47] then the formal group law associated with an orientation on $\mathbb{E}$ is necessarily additive.*

---

[47] *i.e. $\mathbb{E}$ is isomorphic to its rationalization.*

(ii) *Assume* $\mathbb{E}$ *is oriented with orientation* c *and* $(\mathbb{F}, d)$ *is an oriented spectrum. Then, any morphism of ring spectra* $\varphi : \mathbb{F} \to \mathbb{E}$ *automatically satisfies* $\varphi(d) = c$.

Example 5.1.11 Assumption (Ann) is most common in algebraic geometry. It is fulfilled in any of the examples (1), (2), (5) of 2.1.4.

Over an algebraically closed field $S = \mathrm{Spec}\,(k)$, or more generally any scheme S such that -1 is a sum of squares in all residue fields, we get from a theorem of Morel (cf [CD12b, 16.2.14]) that any S-absolute rational ring spectrum satisfying assumption (Ann) is uniquely oriented and the corresponding orientation is additive.

5.2 Change of orientation

5.2.1 Let $\mathbb{E}$ be a ring spectrum equipped with two orientations c, d whose respective associated formal group laws are F_c and F_d. This is pictured by the following identical pseudo-morphism of oriented ring spectra:

$$\varphi : (\mathbb{E}, c) \to (\mathbb{E}, d).$$

Let us consider the reciprocal isomorphism of formal group laws (5.1.8.a) uniquely associated with φ in Proposition 5.1.7. Recall the convention: $d = \Phi(c)$, $c = \Psi(d)$.

Then, from Proposition 4.1.2, we get two expressions of the Todd class of a line bundle L/S associated with φ:

$$\mathrm{Td}_\varphi(L) = \left.\frac{t}{\Psi_S(t)}\right|_{t=d_1(L)} = \left.\frac{\Phi_S(t)}{t}\right|_{t=c_1(L)}.$$

Another way of saying this is the relation:

$$d_1(L) = \mathrm{Td}_\varphi(L).c_1(L).$$

5.2.2 In general, it is not easy to compute the strict isomorphism associated with a change of orientations on a ring spectrum. However, let us recall that for any $\mathbb{Q}$-algebra A, given any formal group law F with coefficients in A, there exists a unique strict isomorphism between F and the additive formal group law, called the *logarithm of* F:

$$\log_F : F \to F_{add}.$$

There exists a well known formula for the logarithm. First, one attach to F a formal differential form

$$\omega_F(x) = \left(\left.\frac{\partial F(x,y)}{y}\right|_{y=0}\right)^{-1} .dx$$

and then one defines the logarithm as the primitive of this differential form:

$$\log_F(t) = \int \omega_F(x). \tag{5.2.2.a}$$

The composition inverse for this power series is usually called the *exponential of F* and denoted by:

$$\exp_F : F_{add} \to F.$$

As a consequence, we get the well-known fact that any two formal group laws with coefficients in A are uniquely strictly isomorphic.

This gives the following formula for computing the Todd class associated with a change of orientations. One determines the formal group laws F_c and F_d associated with c and d. Then the strict isomorphisms $\Phi : F_c \leftrightarrows F_d : \Psi$ are given by the power series:

$$\Phi(t) = \exp_{F_d} \circ \log_{F_c}(t), \Psi(t) = \exp_{F_c} \circ \log_{F_d}(t).$$

Example 5.2.3 It is easy to compute the Todd class using the splitting principle. Let us consider the abstract case of a rational oriented ring spectrum $(\mathbb{E}, c)$ with an abstract formal group law

$$F(x, y) = \sum_{i,j} a_{ij} x^i y^j,$$

and assume $\bar{c} = \log_F(c)$ is the canonical orientation corresponding to the additive formal group law (cf. the preceding examples).

It is convenient to denote by $p_i = [\mathbb{P}^i]$ the cobordism class of the projective space of dimension i — whose expression in terms of the coefficients $a_{1,i}$ can be found in Example 3.2.14. Indeed, according to the Myschenko formula, one gets:

$$\log_F(t) = \sum_{i=0}^{\infty} \frac{p_i}{i+1} . t^{i+1}.$$

Then, associated with the identical pseudo-morphism $(\mathbb{E}, c) \to (\mathbb{E}, \bar{c})$, the Todd class of an arbitrary vector bundle E/S of rank n can be expressed as a polynomial of the p_i and of the Chern classes $\bar{c}_i := \bar{c}_i(E)$ for $1 \leq i \leq n$.

The first few terms are as follows:

$$\mathrm{Td}(E) = 1 + \left(\frac{1}{2}p_1\right)\bar{c}_1 + \left(\frac{3}{4}p_1^2 - \frac{2}{3}p_2\right)\bar{c}_2$$

$$+ \left(-\frac{1}{4}p_1^2 + \frac{1}{3}p_2\right)\bar{c}_1^2 + \left(-\frac{7}{8}p_1^3 + \frac{5}{3}p_1p_2 - \frac{3}{4}p_3\right)\bar{c}_1\bar{c}_2$$

$$+ \left(-\frac{13}{16}p_1^4 + 2p_1^2p_2 - \frac{5}{4}p_1p_3 - \frac{1}{3}p_2^2 + \frac{2}{5}p_4\right)\bar{c}_2^2 + \left(\frac{1}{4}p_1^3 - \frac{1}{2}p_1p_2 + \frac{1}{4}p_3\right)\bar{c}_1^3$$

$$+ \left(\frac{11}{8}p_1^4 - \frac{43}{12}p_1^2p_2 + \frac{17}{8}p_1p_3 + \frac{8}{9}p_2^2 - \frac{4}{5}p_4\right)\bar{c}_1^2\bar{c}_2 + \cdots$$

Note we can also give an expression of the Todd class in function of the original Chern classes $c_i = c_i(E)$:

$$\mathrm{Td}(E) = 1 + \left(\frac{1}{2}p_1\right)c_1 + \left(\frac{1}{4}p_1^2 - \frac{2}{3}p_2\right)c_2$$

$$+ \left(\frac{1}{3}p_2\right)c_1^2 + \left(\frac{1}{6}p_1p_2 - \frac{3}{4}p_3\right)c_1c_2 + \left(-\frac{1}{4}p_1p_3 + \frac{1}{9}p_2^2 + \frac{2}{5}p_4\right)c_2^2$$

$$+ \left(\frac{1}{4}p_3\right)c_1^3 + \left(\frac{1}{8}p_1p_3 - \frac{4}{5}p_4\right)c_1^2c_2 + \left(-\frac{3}{10}p_1p_4 + \frac{1}{12}p_2p_3\right)c_1c_2^2$$

$$+ \left(-\frac{2}{15}p_2p_4 + \frac{1}{16}p_3^2\right)c_2^3 + \cdots$$

As an illustration of the Riemann-Roch formula in the case of a change of orientations, we give the following simple proof of a formula due to Quillen (in complex cobordism, [Qui69]):

Theorem 5.2.4 *Let E be a vector bundle of rank $n + 1$ over a scheme S, $P = \mathbb{P}(E)$ be the associated projective bundle, $p : P \to S$ be the canonical projection and $\lambda = \mathcal{O}(1)$ be the canonical dual line bundle on P.*

Put $A = \mathbf{MGL}^{**}(S)$ and let c be the canonical orientation of $\mathbf{MGL}$ over S. Let $F(x, y) \in A[[x, y]]$ be the formal group law associated with $(\mathbf{MGL}, c)$.

Then for any polynomial $\phi(t) \in A[t]$, the following formula holds in $A \otimes_{\mathbb{Z}} \mathbb{Q}$:

$$p_*\left(\phi(c_1(\lambda))\right) = \mathrm{Res}_t\left(\frac{\phi(t).\omega_F(t)}{\prod_i F(t, l_i)}\right)$$

where l_i are the Chern roots of E with respect to the orientation c.

Proof [D., Levine, Vishik][48] Let $\log_F(t)$ (resp. $\exp_F(t)$) be the logarithm (resp. exponential) associated with the formal group law F — see above.

---

[48] I thank a lot M. Levine and A. Vishik for helping me to finish this proof.

Then the class $\bar{c} = \log_F(c)$ is an orientation of $\mathbf{MGL}_{\mathbb{Q}}$ whose formal group law is additive. Let us denote by $\bar{p}_*$ the Gysin morphism associated with p with respect to $\bar{c}$.

Then the Riemann-Roch formula for the identical pseudo-morphism

$$(\mathbf{MGL}_{\mathbb{Q}}, c) \to (\mathbf{MGL}_{\mathbb{Q}}, \bar{c})$$

and the morphism p reads:

$$p_*\big(\phi(d)\big) = \bar{p}_*\big(P(d).\,\mathrm{Td}(T_p)\big) \tag{5.2.4.a}$$

where we have put $d = c_1(\lambda)$ and T_p denotes the tangent bundle of P/S. The following lemma is a reformulation of a well known formula in the (classical) theory of Chern classes:

Lemma 5.2.5 *Let $(\mathbb{E}, \bar{c})$ be an additive oriented ring spectrum over S. Consider the total Chern class of E:*

$$\bar{c}_t(E) = \sum_{i \geq 0} \bar{c}_i(E).\,t^i.$$

as an invertible power series with coefficients in $\mathbb{E}^{**}(S)$. Put $\bar{d} = \bar{c}_1(\lambda)$.

Then for any power-series $\psi(t)$ with coefficients in $\mathbb{E}^{**}(S)$, one has the following equality in $\mathbb{E}^{**}(S)$:

$$\bar{p}_*(\psi(\bar{d})) = \mathrm{Res}_t\left(\frac{\psi(t).\,dt}{t^{n+1}.\bar{c}_{t^{-1}}(E)}\right)$$

where Res_t stands for the residue of the indicated Laurent power-series.[49]

Let us write $\psi(t) = \sum_i \alpha_i.\,t^i$ and $\bar{c}_t(E)^{-1} = \sum_{i \geq 0} \bar{c}_i(-E).\,t^i$. Recall that for any element $\alpha \in \mathbb{E}^{**}(P)$, $\bar{p}_*(\alpha)$ is the coefficient of $\bar{d}^n$ in the decomposition of α within the A-basis $(\bar{d}^i)_{0 \leq i \leq n}$ of $\mathbb{E}^{**}(P)$. A classical computation according to the defining relation of Chern classes (2.1.16.a) in the additive case gives us:

$$\bar{p}_*(\bar{d}^i) = \bar{c}_{i-n}(-E).$$

In particular, one obtains:

$$\bar{p}_*\left(\sum_i \alpha_i.\bar{d}^i\right) = \sum_i \alpha_i.\bar{c}_{i-n}(-E).$$

---

[49]The formula inside Res_t is a Laurent power series because the Chern classes of E are nilpotent (2.1.22).

To end the proof of the lemma, one has only to realize that the right hand side is the coefficient of t^n in the following Laurent power-series:

$$\left(\sum_i \alpha_i.t^i\right).c_{t^{-1}}(-E).$$

Let us now compute the right hand side of (5.2.4.a). From the exact sequence of vector bundles over P:

$$0 \to \mathcal{O}_P \to \lambda \otimes p^{-1}(E) \to T_p \to 0$$

we get $\mathrm{Td}(T_p) = \mathrm{Td}(\lambda \otimes p^{-1}(E))$. By assumption, the classes $\bar{l}_i = \log_F(l_i)$ are the Chern roots of $p^{-1}(E)$ with respect to the additive orientation $\bar{c}$. Thus, by definition of the Todd class (see also the formula of 5.2.1 with $\Psi_S = \exp_F$), one gets:

$$\mathrm{Td}(\lambda \otimes p^{-1}(E)) = \prod_{i \in I} \frac{t}{\exp_F(t)} \ (t = \bar{d} + \bar{l}_i).$$

Note that because $\bar{l}_i$ are Chern roots of E with respect to $\bar{c}$, one has:

$$\prod_{i \in I}(\bar{d} + \bar{l}_i) = \sum_i \bar{c}_{n+1-i}(E).\bar{d}^i = \left(t^{n+1}.c_{t^{-1}}(E)\right)\big|_{t=\bar{d}}.$$

Because $F(x,y) = \exp_F(\log_F(x) + \log_F(y))$, one also has:

$$\exp_F(\bar{d} + \bar{l}_i) = F(d, l_i) = F\left(\exp_F(\bar{d}), l_i\right).$$

Thus to compute the right hand side of (5.2.4.a) using the formula of the preceding lemma, we are led to introduce the following power series:

$$\psi(t) = \frac{\phi(\exp_F(t)).t^{n+1}.c_{t^{-1}}(E)}{\prod_{i \in I} F\left(\exp_F(t), l_i\right)}.$$

Applying the preceding lemma:

$$\bar{p}_*\left(\phi(d).\mathrm{Td}(T_p)\right) = \bar{p}_*(\psi(\bar{d})) = \mathrm{Res}_t\left(\frac{\phi(\exp_F(t)).dt}{\prod_{i \in I} F\left(\exp_F(t), l_i\right)}\right).$$

Computing this residue with the change of variables $x = \exp_F(t)$, one gets the desired result: by (5.2.2.a), $d\log_F(x) = \omega_F(x)$.

$\square$

Remark 5.2.6 (1) In view of Remark 5.1.3, the preceding formula is universal: It is valid without any change for any S-absolute oriented ring spectrum $\mathbb{E}$ with associated formal group law F. Anyway, it is clear in the above proof that one can faithfully replace $\mathbf{MGL}$ by $\mathbb{E}$.

(2) The above proof is particularly simple but it works only with rational coefficients whereas Quillen formula is stated in [Qui69] for complex cobordism with integral coefficients. However, one can at least deduce from this proof the formula with integral coefficients in characteristic 0: indeed, in this case, one reduces to $S = \mathrm{Spec}\,(\mathbb{Q})$ and we know from [Lev09] that $A = \mathbf{MGL}^{**}(\mathbb{Q})$ is the Lazard ring and thus has no torsion. One gets back the usual Quillen formula by using the complex realization functor.

(3) As a particular case of the preceding formula, one gets the classical Myschenko formula computing the cobordism class of $\mathbb{P}_S^n$ for any integer n: take $E = \mathbb{A}_S^{n+1}$ and $P(t) = 1$. May be it is worth to summarize the proof in this case:

$$p_*(1) = \bar{p}_*(\mathrm{Td}(T_{\mathbb{P}^n})) = \bar{p}_*(\mathrm{Td}(\lambda \otimes \mathbb{A}^{n+1}))$$

$$= \bar{p}_*\left(\left(\frac{\bar{d}}{\exp \bar{d}}\right)^{n+1}\right) = \mathrm{Res}_t\left(\frac{dt}{\exp_F(t)^{n+1}}\right) = \mathrm{Res}_x\left(\frac{\omega_F(x)}{x^{n+1}}\right).$$

Note also that this computation, including the change of variable $x = \exp_F(t)$, was used by Borel and Serre in [BS58] to prove the classical Grothendieck-Riemann-Roch formula for $p : \mathbb{P}_S^n \to S$ — replacing $\mathbf{MGL}$ by $\mathbf{KGL}$: as explained below (5.3.3), the Chern character corresponds to changing the natural orientation on $\mathbf{KGL}$ to the additive orientation.

(4) The general formula of Quillen has been proved integrally in the context of oriented cohomology theories in [Shi07] and [Vis07, App. B]. The proofs given here are equally valid in our more general context.

Example 5.2.7 An interesting particular case of Quillen formula is the following computation of the cobordism class of a projective bundle P/S associated with a vector bundle E/S:

$$[P/S] = \mathrm{Res}_t\left(\frac{\omega_F(t)}{\prod_i F(t, l_i)}\right)$$

where l_i are the Chern roots of E.

Example 5.2.8 We end-up this series of illustration of the (generalized) Riemann-Roch formula by explaining how one can also compare the Chern classes arising from different orientations.

Let us consider again the general setting of Paragraph 5.2.1. In general, given a vector bundle E/S of rank n, we can use the splitting principle to compute the two different type of Chern classes of E associated respectively with the orientations d and c. For any integer $0 \le r \le n$, one gets:

$$d_r(E) = d_r(\oplus_{i=1}^n L_i) = \sum_\alpha d_1(L_{\alpha_1}) \dots d_1(L_{\alpha_r})$$

$$= \sum_\alpha \Phi(c_1(L_{\alpha_1})) \dots \Phi(c_1(L_{\alpha_r})),$$

where α runs over the r-uple of integers $(\alpha_1, \dots, \alpha_r)$ such that

$$1 \le \alpha_1 < \alpha_2 < \cdots < \alpha_r \le n.$$

Once again, one has to express the right hand side in terms of the elementary symmetric functions in the $c_1(L_i)$ to obtain an expression in terms of $c_i(E)$.

In general, the formulas are pretty complicated. However, one can remark there is a simple formula when $r = n$. Then, one simply obtains:

$$d_n(E) = \mathrm{Td}(E).c_n(E),$$

generalizing the case of a line bundle (Par. 5.2.1).

Another case that can be computed is the case $r = 1$. Assume the strict isomorphism of formal group law Φ such that $d = \Phi(c)$ has been written as:

$$\Phi(x) = x + \sum_{i>1} \alpha_i.x^i.$$

Recall the following classical relation (determinantal form of the Newton's identity) between the power sum symmetric polynomials p_i and the elementary symmetric polynomials e_j:

$$p_i = \begin{vmatrix} e_1 & 1 & 0 & \cdots & 0 \\ 2e_2 & e_1 & & & \\ & e_2 & & & 0 \\ & & & & 1 \\ i.e_i & e_{i-1} & \cdots & e_2 & e_1 \end{vmatrix}.$$

Using the splitting principle together with this relation, one obtains the following relation:

$$d_1(E) = c_1(E) + \sum_{i>1} \alpha_i \cdot \begin{vmatrix} c_1(E) & 1 & 0 & \cdots & 0 \\ 2c_2(E) & c_1(E) & & & \\ & c_2(E) & & & 0 \\ & & & & 1 \\ i.c_i(E) & c_{i-1}(E) & \cdots & c_2(E) & c_1(E) \end{vmatrix}.$$

Note the sum above is finite because Chern classes are always nilpotent (cf. Prop. 2.1.22(1)).

Example 5.2.9 As a last illustration of the change of orientation principle, let us consider an arbitrary oriented ring spectrum $(\mathbb{E}, c)$. One defines a new orientation on E by putting for any line bundle L over a scheme S:

$$c_1'(L) = -c_1(L^\vee),$$

where $L^\vee$ denotes the dual of L/S. Indeed, if one denotes by $\mu(t)$ the formal inverse associated with the formal group law of $(\mathbb{E}, c)$, then the strict isomorphism corresponding to this change of orientation is the composition inverse of the power series $-\mu(t)$.

Using the splitting principle, one easily checks that for any vector bundle E/S, the following relation holds:

$$c_i'(E) = (-1)^i . c_i(E^\vee). \tag{5.2.9.a}$$

Let λ be the canonical line bundle on $\mathbb{P}(E)$. Then, from this formula and the relation (2.1.16.a) defining Chern classes with respect to the orientation c', one gets the following formula[50] expressing Chern classes in terms of the dual canonical line bundle $\lambda^\vee$:

$$\sum_{i=0}^{n} (-1)^i p^* \left(c_i(E^\vee) \right) . c_1\left(\lambda^\vee \right)^{n-i} = 0. \tag{5.2.9.b}$$

5.3 Universal formulas and the Chern character

In the next proposition, we work over a fixed base scheme S. Recall we have seen in Example 5.1.11 that there exists a unique orientation $c^{\mathbf{H}_\mathbb{B}}$ on the Beilinson motivic cohomology ring spectrum $\mathbf{H}_{\mathbb{B},S}$ whose formal group law is necessarily additive. Using a fundamental result of [CD12a], we get slightly more:

---

[50]This formula was suggested to me by Alberto Navarro.

Proposition 5.3.1 *Let* $(\mathbb{E}, c)$ *be an additive oriented ring spectrum with rational coefficients over S.*

Then there exists a unique morphism of absolute ring spectra

$$\sigma : \mathbf{H}_{\mathbb{E},S} \to \mathbb{E}$$

which is moreover a morphism of oriented ring spectra: $\sigma(c^{\mathbf{H}_{\mathbb{E}}}) = c$.

In particular, the morphism σ is compatible with Gysin morphisms, residues, Chern classes and fundamental classes as constructed in the preceding sections.

Proof The existence and the uniqueness of σ follow from [CD12b, 14.2.16]. The fact σ is necessarily a morphism of oriented ring spectra follows from Proposition 5.1.7 and the fact that the unique strict automorphism of a formal group law on a rational ring is the identity (fact recalled in Paragraph 5.2.2). The last assertion is then an application of the Riemann-Roch formula (or its extension for Chern classes and fundamental classes) to σ. □

Remark 5.3.2 Another way of stating the previous proposition is that $\mathbf{H}_{\mathbb{E}}$ is the universal absolute *orientable* rational ring spectrum, — whereas **MGL** is the universal absolute *oriented* ring spectrum.

Thus, the proposition (as well as its integral counterpart stated in a few paragraphs) answers a *desideratum* raised in the introduction (cf. second paragraph) of Beilinson's fundamental work [Beĭ84].

The morphism σ could be called the *higher cycle class* morphism: in degree $(2n, n)$, it gives a morphism from the Chow group of n-codimensional cycles.[51] Over a field and in the integral case (see below), this fits well with the fact motivic cohomology (of regular schemes) is given by Bloch's higher Chow groups.

5.3.3 Recall that the decomposition of algebraic K-theory according to the eigenvalues of the Adams operation has been lifted by J. Riou to the stable homotopy category of schemes resulting in a canonical isomorphism of ring spectra:

$$\mathrm{ch}_t : \mathbf{KGL}_{\mathbb{Q}} \to \oplus_{i \in \mathbb{Z}} \mathbf{H}_{\mathbb{E}}(i)[2i].$$

This is essentially the results of [Rio10] as explained in [CD12b], Lemma 14.1.4 for the existence of this isomorphism and Corollary 14.2.17 for the

---

[51]Note that, even when $\mathbb{E}$ is the spectrum representing Deligne-Beilinson cohomology (say over $\mathbb{Q}$, see [AH10]), the map σ is not exactly the *regulator map* as described in [Sou86, §3.3].

fact it is a morphism of ring spectra. By definition, for any regular scheme X and any integer n, this isomorphism induces the canonical decomposition

$$K_n(X)_{\mathbb{Q}} \to \oplus_{i \geq 0} \mathrm{Gr}^i_{\gamma} K_n(X)_{\mathbb{Q}}$$

for the γ-filtration on Quillen K-theory with rational coefficients (cf. [Sou85]). According to [Sou85, §7], the projector on the i-th graded part of this decomposition is given by the i-th Chern character. Thus the morphism of ring spectra ch_t lifts to the category of spectra the usual Chern character in higher K-theory with values in Beilinson motivic cohomology (see again [Sou85]): for any pair $(n, r) \in \mathbb{N}^2$ and any regular scheme, one gets the usual (higher) Chern character:

$$\mathrm{ch}_{r,n} : K_r(X) \to \mathbf{H}_{\mathcal{B}}^{2n-r,n}(X).$$

For $r = 0$, this coincide withs the Chern character of [BGI71]. In particular, it is uniquely characterized by its value on the class of a line bundle L/X:

$$\mathrm{ch}_{0,n}\left([L]\right) = \frac{1}{n!} c_1(L)^n.$$

Using the principle explained in paragraphs 5.1.4 and 5.2.1, we can easily determine the Todd class associated with the morphism of spectra ch_t. Indeed, the formal group law associated with $c^{\mathbf{KGL}}$ is $F_{\mathbf{KGL}}(x, y) = x + y - \beta.xy$ (Ex. 2.1.21). By definition of β, we get that $\mathrm{ch}_t(\beta) = 1$. We deduce that the formal group law associated with the orientation $\mathrm{ch}_t(c^{\mathbf{KGL}})$ on the aim is the following multiplicative formal group law $F_{mult}(x, y) = x + y - xy$. On the other hand $c^{\mathbf{H}}_{\mathcal{B}}$ is additive, thus the pair of reciprocal strict isomorphisms associated with ch_t is the logarithm/exponential of the formal group law F_{mult}. Formula (5.2.2.a) easily yields:

$$\Phi(t) = \log_{F_{mult}}(t) = -\ln(1 - t),$$
$$\Psi(t) = \exp_{F_{mult}}(t) = 1 - e^{-t}.$$

Thus, from 5.2.1, we obtain that the Todd class associated with ch_t is defined by the power series $\frac{t}{1-e^{-t}}$ (Prop. 4.1.2). In particular, it coincides exactly with the usual Todd class

$$\mathrm{Td} : K_0 \to \mathrm{Gr}^*_{\gamma} K_0(X)_{\mathbb{Q}} \simeq \mathbf{H}_{\mathcal{B}}^{2*,*}(X)$$

of [BGI71].

In the end, taking care about the isomorphisms (2.1.5.a), Theorem 4.3.2 yields the following higher arithmetic Grothendieck-Riemann-Roch formula:

Proposition 5.3.4 *Consider the preceding notations. Let $f : Y \to X$ be a projective morphism between regular schemes and $\tau_f \in K_0(Y)$ be its virtual tangent bundle (see 4.3.1).*

Then for any integer $r \geq 0$, the following diagram is commutative:

$$
\begin{array}{ccc}
K_r(Y)_{\mathbb{Q}} & \xrightarrow{\quad f_* \quad} & K_r(X)_{\mathbb{Q}} \\
{\scriptstyle \mathrm{Td}(\tau_f).\,\mathrm{ch}_t} \downarrow & & \downarrow {\scriptstyle \mathrm{ch}_t} \\
\oplus_{n \geq 0} \mathbf{H}_{\mathrm{B}}^{2n-r,n}(Y) & \xrightarrow{\quad f_* \quad} & \oplus_{n \geq 0} \mathbf{H}_{\mathrm{B}}^{2n-r,n}(X).
\end{array}
$$

In other words, for any integer $n \geq 0$ and any element $y \in K_r(Y)$, one has:

$$
\mathrm{ch}_{r,n} \left(f_*(y) \right) = f_* \left(\sum_{i+j=n} \mathrm{Td}_i(\tau_f).\,\mathrm{ch}_{r,j}(y) \right).
$$

Remark 5.3.5 Taking into account Proposition 5.3.1, the preceding formula is universal. Indeed, given any additive oriented rational ring spectrum $\mathbb{E}$, the Chern character and the Todd class with values in $\mathbb{E}$ are induced by those with values in Beilinson motivic cohomology. Then we get the following commutative diagram:

$$
\begin{array}{ccccc}
K_r(Y)_{\mathbb{Q}} & \xrightarrow{\mathrm{Td}(\tau_f).\,\mathrm{ch}_t} & \oplus_{n \geq 0} \mathbf{H}_{\mathrm{B}}^{2n-r,n}(Y) & \xrightarrow{\sum \sigma_Y^{2n-r,n}} & \oplus_{n \geq 0} \mathbb{E}^{2n-r,n}(Y) \\
{\scriptstyle f_*} \downarrow & & \downarrow {\scriptstyle f_*} & & \downarrow {\scriptstyle f_*} \\
K_r(X)_{\mathbb{Q}} & \xrightarrow{\mathrm{ch}_t} & \oplus_{n \geq 0} \mathbf{H}_{\mathrm{B}}^{2n-r,n}(X) & \xrightarrow{\sum \sigma_X^{2n-r,n}} & \oplus_{n \geq 0} \mathbb{E}^{2n-r,n}(X),
\end{array}
$$

Note that the map ch_t exists over any scheme. On the other hand, it is known from [Cis13] that the spectrum $\mathbf{KGL}_{\mathbb{Q}}$ represents Weibel homotopy invariant K-theory KH_*. Therefore, because $\mathbf{KGL}_{\mathbb{Q}}$ and $\mathbf{H}_{\mathrm{B}}$ are $\mathscr{S}m_S$-absolutely pure for any scheme S, the preceding proposition admits the following version, which is finer when S is singular:

Proposition 5.3.6 *Let S be any scheme and $f : Y \to X$ be a projective morphism between smooth S-schemes with virtual tangent bundle $\tau_f \in K_0(Y)$.*

Then for any integer $r \geq 0$, the following diagram is commutative:

$$
\begin{array}{ccc}
KH_r(Y)_{\mathbb{Q}} & \xrightarrow{\quad f_* \quad} & KH_r(X)_{\mathbb{Q}} \\
{\scriptstyle \mathrm{Td}(\tau_f).\,\mathrm{ch}_t} \downarrow & & \downarrow {\scriptstyle \mathrm{ch}_t} \\
\oplus_{n \geq 0} \mathbf{H}_{\mathrm{B}}^{2n-r,n}(Y) & \xrightarrow{\quad f_* \quad} & \oplus_{n \geq 0} \mathbf{H}_{\mathrm{B}}^{2n-r,n}(X).
\end{array}
$$

Remark 5.3.7 The morphism f_* in the above proposition stands *a priori* for the Gysin morphisms we have defined. However, recall from [TT90, 3.16.4], that K-theory is covariant with respect to proper maps of finite Tor-dimension, thus a fortiori with respect to the morphism f in the above proposition. The definition of KH_* makes it clear that this covariant functoriality extends to KH_* in such a way that the following diagram commutes:

$$
\begin{array}{ccc}
K_r(Y) & \xrightarrow{\ f_*\ } & K_r(X) \\
\downarrow & & \downarrow \\
KH_r(Y) & \xrightarrow{\ f_*\ } & KH_r(X).
\end{array}
$$

Applying Theorem 3.3.1 as in Example 3.3.4(3), we get that this last morphism coincides with our Gysin morphism. Therefore, because of the preceding commutative square, the above Riemann-Roch formula is even true when replacing KH_* by Thomason-Trobaugh algebraic K-theory.[52]

5.3.8 Finally, we show how to get an almost integral (weaker) version of Proposition 5.3.1 using the recent Hopkins-Morel-Hoyois theorem (cf. [Hoy15]). Let us restate this theorem for the sake of notations. Consider a field k of exponential characteristic p. The canonical orientation on **MGL** induces a formal group law on its ring of coefficients, say over k. By universality of the Lazard ring, we get a canonical morphism of ring:

$$
L \longrightarrow \mathbf{MGL}_{**} := \mathbf{MGL}^{-*,-*}(\mathrm{Spec}\,(k))
$$

which in fact induces the following isomorphism (cf. [Hoy15, 8.2]) of graded rings:

$$
L[1/p] \longrightarrow \mathbf{MGL}_{(2,1)*}[1/p] \tag{5.3.8.a}
$$

where $\mathbf{MGL}_{(2,1)*}$ denotes the elements of $\mathbf{MGL}_{**}$ of bidegree $(2n, n)$ for an integer $n \in \mathbb{Z}$. The Lazard ring is in fact a graded polynomial algebra $L = \mathbb{Z}[b_1, b_2, \ldots]$ where b_i has degree i. In particular, any element b_i corresponds to an element of $\mathbf{MGL}_{2i,i}$. Following [Hoy15], paragraph after Corollary 6.9, one can define the quotient spectrum

$$
\mathbf{MGL}/\{b_i, i > 0\}
$$

by killing the elements b_i. In fact, this amounts to kill the coefficients a_{ij} of the formal group law of **MGL** — which, after inverting p, corresponds to the universal formal group law on the Lazard ring through the isomorphism (5.3.8.a).

---

[52]Recall it coincides with Quillen's definition whenever X admits an ample line bundle.

As the formal group law associated with the canonical orientation on the motivic Eilenberg Mac Lane spectrum $\mathbf{H}_{\mathbb{Z}}$ is additive, one gets a canonical map of spectra:

$$(\mathbf{MGL}/\{b_i, i > 0\}) \to \mathbf{H}_{\mathbb{Z}} \qquad (5.3.8.b)$$

which becomes an isomorphism after inverting p (cf. [Hoy15, 7.12]). As a corollary, we get:

Proposition 5.3.9 *Let k be a field of exponential characteristic p and $\mathbb{E}$ be a $\mathbb{Z}[1/p]$-linear k-absolute ring spectrum.*
   Then the following conditions are equivalent:

- $\mathbb{E}$ *admits an additive orientation c.*
- *There exists a morphism of absolute ring k-spectra:*

$$\sigma : \mathbf{H}_{\mathbb{Z}} \to \mathbb{E}.$$

Moreover, when these conditions are fulfilled, the additive orientation c on $\mathbb{E}$ is unique and is the image under σ of the canonical orientation on $\mathbf{H}_{\mathbb{Z}}$. In particular, σ is compatible with Gysin morphisms, residues, Chern classes and fundamental classes as constructed in the preceding sections.

Proof The fact (i) implies (ii) is obvious: the orientation of $\mathbf{H}_{\mathbb{Z}}$ induces an orientation of $\mathbb{E}$, which is necessarily additive.
   Reciprocally. According to Prop. 2.2.6, the orientation c of $\mathbb{E}$ corresponds to a morphism of ring spectra over $\mathrm{Spec}\,(k)$

$$\varphi : \mathbf{MGL} \to \mathbb{E}.$$

This map induces a morphism of formal group law. Moreover, according to the commutative diagram:

$$\begin{array}{ccc} \mathbf{MGL}^{**}(\mathbb{P}^\infty) & \longrightarrow & \mathbf{MGL}^{**}(\mathbb{P}^\infty \times \mathbb{P}^\infty) \\ \varphi_* \downarrow & & \downarrow \varphi_* \\ \mathbb{E}^{**}(\mathbb{P}^\infty) & \longrightarrow & \mathbb{E}^{**}(\mathbb{P}^\infty \times \mathbb{P}^\infty), \end{array}$$

the definition of the formal group law associated with an oriented ring spectrum and the fact $(\mathbb{E}, c)$ is additive, we obtain that $\varphi_* : \mathbf{MGL}_{**}(k) \to \mathbb{E}_{**}(k)$ sends all the elements a_{ij} to 0 as soon as $(i, j) \neq (1, 0), (0, 1)$. Thus, φ induces the morphism of spectra σ using the isomorphism induced by (5.3.8.b) and the assumption that $\mathbb{E}$ is $\mathbb{Z}[1/p]$-linear.
   Having fixed an additive orientation c of $\mathbb{E}$, from [Vez01, Th. 4.3], the orientations d on $\mathbb{E}$ are in one-to-one correspondence with the isomorphisms

of formal group laws on $\mathbb{E}_{(2,1)*}(k)$ from the additive formal group to the formal group law F_d associated with d. As there is only one automorphism of the additive formal group law, the uniqueness statement follows. The remaining assertion is clear from what we have seen.

$\square$

Remark 5.3.10 Another way of stating the preceding proposition is that given a representable $\mathbb{Z}[|1/p]$-linear cohomology theory $\mathbb{E}^{**}(-)$ over (smooth) k-schemes, there is only one possible way to define Chern classes which are additive. Moreover the other structures on the cohomology, that is residues, Gysin morphisms and fundamental classes, are unique. Similarly, the cycle class map, from classical Chow groups, is unique.

5.4   Residues and symbols

5.4.1 Let us consider one of the following two cases:

- $S = \mathrm{Spec}\,(\mathbb{Z})$, $\Lambda = \mathbb{Q}$;
- $S = \mathrm{Spec}\,(k)$, k a field of exponential characteristic p, $\Lambda = \mathbb{Z}[1/p]$.

Let us denote simply by $\mathbf{H}_\Lambda$ either the Beilinson motivic cohomology spectrum in the first case or the Eilenberg-Mac Lane motivic ring spectrum with coefficients in Λ in the second case.

According to the preceding section, any S-absolute oriented ring spectrum $\mathbb{E}$ with additive formal group law has a structure of $\mathbf{H}_\Lambda$-algebra with structural morphism σ.

In particular, given any field K over S, we get a canonical symbol map:

$$K_n^M(K) \otimes_\mathbb{Z} \Lambda \simeq \mathbf{H}_\Lambda^{n,n}(K) \xrightarrow{\sigma_*} \mathbb{E}^{n,n}(K), \{f_1,\ldots,f_n\} \mapsto \{f_1,\ldots,f_n\}_\mathbb{E}.$$

If K admits a discrete valuation v such that its ring of integers is an S-scheme, with residue field k, we have obtained (Ex. 3.1.7) a residue map with coefficients in $\mathbb{E}$:

$$\partial_v^\mathbb{E} : \mathbb{E}^{n,m}(K) \to \mathbb{E}^{n-1,m-1}(k).$$

As an easy application of the Riemann-Roch residual theorem, we get the following computation of this residue: given any elements $f_1,\ldots,f_n$ in $K^\times$,

$$\partial_v^\mathbb{E}(\{f_1,\ldots,f_n\}_\mathbb{E}) = \sigma_*\big(\partial_v(\{f_1,\ldots,f_n\})\big)$$

where ∂_v is Milnor tame residue symbols (cf Example 3.1.7). In fact, this follows from Theorem 4.2.3 applied to σ: the Todd class is equal to 1 in this case as σ is a morphism of oriented ring spectra.[53]

---

[53]One could also obtained this computation directly as in Example 3.1.7 using Proposition 3.1.6.

Example 5.4.2   (1) *De Rham cohomology.*– Assume $S = \text{Spec}(k)$ where k is a field of characteristic 0 and consider $\mathbb{E}_{dR}$ the k-absolute ring spectrum representing De Rham cohomology (cf. 1.2.3(1)). Recall the twist on that cohomology is just given by the tensor product with the 1-dimensional k-vector space $k(1) := H^1_{dR}(\mathbb{G}_m)^\vee$.

Note that $H^1_{dR}(\mathbb{G}_m) = k.d\log(t)$, where $\mathbb{G}_m = \text{Spec}(k[t, t^{-1}])$. The choice of the generator $d\log(t)$ determines an isomorphism:

$$\mathbb{E}^{n,i}_{dR}(X) \simeq H^n_{dR}(X/k)$$

functorial in any smooth k-scheme X. As already mentioned, the fact $\mathbb{E}_{dR}$ is a k-absolute ring spectrum extends De Rham cohomology to any k-scheme. In the particular case of an extension field K/k, the choice of $d\log(t)$ gives a canonical isomorphism:

$$\mathbb{E}^{n,i}_{dR}(K) \simeq H^n_{dR}(K/k).$$

Through this isomorphism (case $i = n$), the *de Rham symbol* associated with a family of units $f_1, \ldots, f_n$ in an extension field K/k is given by the classical formula:

$$\{f_1, \ldots, f_n\}_{dR} = d\log(f_1) \wedge \cdots \wedge d\log(f_n) \in H^n_{dR}(K/k).$$

Given now a discrete valuation ring (K, v) over k, according to the previous paragraph one gets a residue map:

$$\partial^{dR}_v : H^1_{dR}(K/k) \simeq \mathbb{E}^{1,1}_{dR}(K) \to \mathbb{E}^{0,0}_{dR}(\kappa(v)) \simeq k.$$

According to the previous computation of residues on symbols, one gets:

$$\partial^{dR}_v(d\log(f)) = v(f). \tag{5.4.2.a}$$

(2) *Rigid cohomology.*– Let V be a complete discrete valuation ring with fraction field E and residue field k. Let $\mathbb{E}_{rig}$ be the k-absolute ring spectrum representing rigid cohomology $H_{rig}(-/E)$.

The situation is analogous to the previous one though the constructions are less concrete. By definition, $H^1_{rig}(\mathbb{G}_m/E)$ is the rational part of the first cohomology group of the weakly complete De Rham complex associated with the weakly complete V-algebra $V\{t, t^{-1}\}$. In particular, it is generated by the differential form $d\log(t)$ of $V\{t, t^{-1}\}$. For smooth (or even singular) k-scheme X, the choice of this differential gives a canonical isomorphism

$$\mathbb{E}^{n,i}_{rig}(X) \simeq H^n_{rig}(X/E).$$

Given a unit $f : X \to \mathbb{G}_m$, this isomorphism in the case $n = i = 1$ sends the symbol $\{f\}_{rig}$ to the element $d\log(f) := f^*(d\log(t))$. Similarly, $\{f_1, \ldots, f_n\}_{rig}$ corresponds to $d\log(f_1) \wedge \cdots \wedge d\log(f_n)$.

Given an extension field K/k, one gets:

$$\mathbb{E}_{rig}^{n,i}(K) = \varinjlim_{A \subset K} H_{rig}^n(\operatorname{Spec}(A)/E) =: H_{rig}^n(K/E)$$

where A runs over the sub-rings of K which are smooth of finite type over the inseparable closure of k in K. And for any discrete valuation v on K, we get a tame residue symbol:

$$\partial_v^{rig} : H_{rig}^1(K/E) \to H_{rig}^0(K/E) \simeq E$$

satisfying the expected property on symbols.

Remark 5.4.3 Symbols in differential calculus have a beautiful history starting from van der Kallen formula ([vdK71]) and going to the Bloch-Kato conjecture modulo p ([BK86, 2.1]).

5.4.4 Consider again the notations of point (1) of the preceding example. Let C be a proper connected regular curve over k with function field K. Let us fix a closed point of C, in other words a discrete valuation v on K trivial on k. In [Tat68], Tate gives a purely algebraic definition of a residue map: $\partial_v^{Tate} : H_{dR}^1(K/k) \to k$. In view of the preceding example, the reader should not be surprised by the following comparison result:

Proposition 5.4.5 *Using the above notations, $\partial_v^{dR} = \partial_v^{Tate}$.*

Proof Both definitions are invariant under completion with respect to the valuation v (see Prop. 1.4.4 for ∂_v^{dR}). Thus we can replace K by $\hat{K} \simeq k((t))$ and we are reduced to identify the two residues on differential forms of the form $\omega = f(t)dt$ when $f(t)$ is a power series with coefficients in k. By continuity and additivity of residues, we can assume $\omega = t^i.dt$.

The case $i \geq 0$ is easy because then ω can be extended to the valuation ring of $k[[t]]$ and therefore its image by ∂_v^{dR} is 0 according to the Gysin long exact sequence (3.1.1.b). The case $i = -1$ follows from the residual Riemann-Roch formula as explained in the previous example. For the remaining case, we use the reciprocity formula for $\mathbb{P}_k^1$ and ∂_v^{dR} (according to [Dég08b, 5.2.1] and [Ros96, 2.2]). According to this formula, we get:

$$\sum_{x \in \mathbb{P}_{k(0)}^1} \partial_x(\omega) = 0.$$

Because $\omega = t^i.dt$, the above equality gives:

$$\partial_0^{dR}(t^i.dt) = -\partial_\infty^{dR}(t^i.dt) \overset{(1)}{=} -\partial_\infty^{dR}(-u^{-(i+2)}.du) \overset{(2)}{=} 0$$

where equality (1) is given by the substitution $u = t^{-1}$, and (2) is true because $i + 2 \leq 0$. This concludes.

$\square$

Remark 5.4.6 The above proof also gives a tool to compute residues in rigid cohomology. Given any cohomology class ω in $H^1_{rig}(K/E)$, we can extend it to the completion of K thus it corresponds to an element in $H^1_{rig}(k((t))/E)$, in other words an overconvergent differential $\hat{\omega}$ form over $V((t))$. Then $\partial_v(\omega)$ is the residue in the usual sense: write $\omega = f(t).dt$ with $f(t) \in V\{t\}$, then

$$\partial_v(\omega) = \mathrm{res}_t(f(t))$$

seen as an element of K. More generally, our construction of residues should be linked with that of [Ber74, VII, 1.2].

Example 5.4.7 *Functoriality of coniveau spectral sequences.*— Assume that $\mathscr{S}$ is the category of excellent regular schemes. Let $\mathbb{E}$ be any absolute oriented ring spectrum.

Using the method of Bloch-Ogus in [BO74] and Gysin long exact sequences of the form (3.1.1.b), we get for any regular excellent scheme X and any integer $n \in \mathbb{Z}$, a *coniveau* spectral sequence of the form:

$$E_1^{p,q} = \bigoplus_{x \in X^{(n)}} \mathbb{E}^{q-p,n-p}(\kappa(x)) \Rightarrow \mathbb{E}^{p+q,n}(X)$$

which converges to the coniveau filtration on $\mathbb{E}^{**}(X)$.

Recall that this spectral sequence can be defined using the exact couple:

$$D^{p,q} = \varinjlim_{Z^*} \mathbb{E}^{p+q,n}(X - Z^{p+1}), \; E^{p,q} = \varinjlim_{Z^*} \mathbb{E}^{q-p,n-p}(Z^p - Z^{p+1})$$

where the limit is taken over the sequences $(Z^p)_{p \in \mathbb{N}}$ such that Z^p is a closed subscheme of codimension $\geq p$ in X satisfying the condition that $(Z^p - Z^{p+1})$ is regular. Indeed, using the fact X is excellent, we obtain that the corresponding set, ordered by term-wise inclusion, is filtering.

Then the maps of the exact couples are given by considering localisation long exact in sequences in cohomology — which come from Gysin triangles:

$$\begin{array}{ccc}
E^{p,q+1} \rightsquigarrow \mathbb{E}^{q-p+1,n-p}(Z^p - Z^{p+1}) & & \mathbb{E}^{q-p,n-p}(Z^p - Z^{p+1}) \rightsquigarrow E^{p,q} \\
\partial_{X-Z^{p+1},Z^p-Z^{p+1}} \uparrow & & \downarrow i_{p*} \\
D^{p,q} \rightsquigarrow \mathbb{E}^{p+q,n}(X - Z^{p+1}) \xleftarrow[\;j_p^*\;]{} \mathbb{E}^{p+q,n}(X - Z^p) \rightsquigarrow D^{p-1,q+1}
\end{array}$$

(see also the presentation of [Dég14]). According to this presentation, the
fact that this spectral sequence, and especially the differentials in the E_1-
term, is functorial with respect to any pseudo-morphism $\mathbb{E} \to \mathbb{F}$ of oriented
ring spectra follows from the Riemann-Roch formula applied to the mor-
phisms in the above diagram and the fact that it is enough to consider
sequences Z^* as above and such that the normal bundle of $Z^p - Z^{p+1}$ in
$X - Z^{p+1}$ is trivial.

Note that this phenomena was already observed in a particular case in
[Gil05, Th. 3.9].

5.5 Residual Riemann-Roch formula

5.5.1 So far, we have only worked out the trivial form of the residual
Riemann-Roch formula, when the Todd class involved is 1. Let us express
the general residual Riemann-Roch formula in the case of the usual Chern
character, as introduced in Paragraph 5.3.3.

Theorem 5.5.2 *Consider a closed regular pair* (X, Z) *of codimension* c.
Let $N_Z X$ *be the normal bundle of* Z *in* X *and put* $U = X - Z$.
Then the following diagram is commutative:

$$
\begin{array}{ccc}
K_r(U) & \xrightarrow{\ \partial_{X,Z}\ } & K_{r-1}(Z) \\
{\scriptstyle \mathrm{ch}^U_{r,n}}\big\downarrow & & \big\downarrow{\scriptstyle \sum_{i+j=n-c} \mathrm{Td}_i(-N_Z X)\cdot \mathrm{ch}^Z_{r-1,j}} \\
\mathrm{H}^{2n-r,n}_{\mathrm{B}}(U) & \xrightarrow{\ \partial_{X,Z}\ } & \mathrm{H}^{2(n-c)-r+1,n-c}_{\mathrm{B}}(Z)
\end{array}
$$

Remark 5.5.3 Once again, using the universality of Beilinson motivic cohomology among the absolute oriented cohomology with additive formal
group law (Prop. 5.3.1), the preceding formula gives also a formula for the
classical (mixed Weil) cohomologies.

Example 5.5.4 Let us first consider the case $r = 1$. Then one gets an explicit description of the residue morphism for K-theory, when $X = \mathrm{Spec}\,(A)$
and $Z = \mathrm{Spec}\,(A/I)$, assuming A and A/I are regular rings.

Indeed, one knows that $K_1(A_I) = \mathrm{GL}(A_I)^{ab}$, the abelianization of the
group of invertible matrices of arbitrary dimensions. Assume we are given
an endomorphism $u : A^r \to A^r$ such that $u \otimes_A A_I$ is an automorphism
of A^r_I. We will denote by $[u]$ the class of this isomorphism in $K_1(A)$. By
assumption, u is a monomorphism whose cokernel is supported on I. We
denote by $[\mathrm{coKer}(u)]$ the class of the corresponding (finitely presented) A/I-
module in $K_0(A/I)$. With these notations, one has the following formula:

$$\partial_{X,Z}([u]) = [\mathrm{coKer}(u)].$$

Assume furthermore $n = c = 1$. Recall that the following part of the higher Chern character $\mathrm{ch}_{1,1} : K_1(A) \to \mathbf{H}_{\mathrm{B}}^{1,1}(A) = A^\times \otimes \mathbb{Q}$ sends any matrix of $GL(A)$ to its determinant.

Assume Z is connected. By assumption, $I = (\pi)$ for a prime divisor π: A_I is a discrete valuation ring. We let v_π denote its valuation. Then, giving the above notations, the residual Riemann-Roch formula lands in $\mathbf{H}_{\mathrm{B}}^{0,0}(Z) = \mathbb{Q}$ and reads:

$$\boxed{v_\pi\big(\det(u)\big) = \mathrm{rk}_{A/I}\big([\mathrm{coKer}(u)]\big).}$$

Note that in fact, it is an integral formula as all the members are integers.

Consider now the case $n > 1$, r arbitrary: According to the coniveau spectral sequence of 5.4.7 applied for $\mathbb{E} = \mathbf{H}_{\mathrm{B}}$, we get that the group $\mathbf{H}_{\mathrm{B}}^{2n-1,n}(U)$ is the cohomology, tensored with $\mathbb{Q}$, in the middle of the following complex:

$$\bigoplus_{y \in U^{(n-2)}} K_2^M(\kappa(y)) \longrightarrow \bigoplus_{x \in U^{(n-1)}} \kappa(x)^\times \xrightarrow{\mathrm{div}} Z^n(U)$$

where the last map is the usual divisor class map (computation of Quillen). Thus, any element f of $\mathbf{H}_{\mathrm{B}}^{2n-1,n}(U)$ can be described as the class of a finite sum:

$$\sum_{x \in U^{(n-1)}} f_x$$

where f_x is a unit of $\kappa(x)$, which is the identity for almost all x, and such that the following n-codimensional cycle of U is zero:

$$\sum_{x \in U^{(n-1)}} \mathrm{div}_U(f_x) = 0.$$

Moreover, using this description of the group $\mathbf{H}_{\mathrm{B}}^{2n-1,n}(U)$, the residue map $\partial_{X,Z} : \mathbf{H}_{\mathrm{B}}^{2n-1,n}(U) \to CH^{n-c}(Z)_{\mathbb{Q}}$ can be described as follows:

$$\partial_{X,Z}(f) = \sum_{x \in U^{(n-1)}} \mathrm{div}_X(f_x)$$

where div_X denotes the divisor of the rational function of f_x seen as a cycle in X. Indeed, by assumption on (f_x), this cycle has support in Z.

Thus, we will represent the element $\mathrm{ch}_{1,n}([u]) \in \mathbf{H}_{\mathrm{B}}^{2n-1,n}(U)$ as the class of a sum: $\sum_{x \in U^{(n-1)}} f_{u,x}$ satisfying the conditions above.[54]

---

[54]It is rather delicate to give a formula for the $f_{u,x}$. One can only say that they describe the part of the element $[u] \in K_1(U)$ (topologically) supported in codimension $(n-1)$.

Assume $n = c$ and Z is connected with generic point η. Then the residual Riemann-Roch formula lands again in $\mathbf{H}_E^{0,0}(Z) = \mathbb{Q}$ and reads:

$$\sum_{x \in U^{(n-1)}, \eta < x} \operatorname{ord}_\eta^x (f_{u,x}) = \operatorname{rk}_{A/I}\big([\operatorname{coKer}(u)]\big)$$

where $\operatorname{ord}_\eta^x$ denotes the order of the rational function at x. Observe this is again an integral formula.

Let us finally consider the case where the codimension c of Z in X is arbitrary less than n. Then the residual formula can be stated as the following equality of cycles in $CH^{n-c}(Z)_\mathbb{Q}$:

$$\sum_{x \in U^{(n-1)}} \operatorname{div}_X (f_{u,x}) = \sum_{i+j=n-c} \operatorname{Td}_i(-N_Z X). \operatorname{ch}_{0,j}\big([\operatorname{coKer}(u)]\big).$$

(Recall that by assumption, the cycle on the left has support in Z.)

6 The axiomatic of Panin revisited

6.1 Axioms for (arithmetic) cohomologies

Our theory is obviously linked with the more classical theory of *oriented cohomology theory* developed by Panin and Smirnov (see [Pan03, Pan04, Pan09] and [Smi06a, Smi06b]). Our axioms are more restrictive as we ask for representability of a cohomology theory. Nevertheless, one can extract from Section 1 the following generalization of the axioms used by Panin:

Definition 6.1.1 A *ringed cohomology theory (with supports)* $\mathbb{E}$ on $\mathscr{S}$ is the datum for each closed pair (X, Z) in $\mathscr{S}$ of a bigraded abelian group $\mathbb{E}_Z^{**}(X)$ equipped with the following structures:

- contravariant functoriality as described in 1.2.6,
- covariant functoriality as described in 1.2.7,
- refined products as described in 1.2.8,
- for each closed pair (X, Z), a *boundary morphism*

$$\mathbb{E}^{n,m}(X - Z) \xrightarrow{\delta_{X,Z}} \mathbb{E}_Z^{n+1,m}(X)$$

  contravariantly natural and fitting in a Gysin long exact sequence of the form (3.1.1.a), which satisfies the axioms (E1)–(E7) described in Prop. 1.2.10 together with the following additional properties:

- *Homotopy*: for any scheme X, $\mathbb{E}^{**}(X) \to \mathbb{E}^{**}(\mathbb{A}_X^1)$ is an isomorphism,
- *Stability*: For any scheme S, let $\tilde{\mathbb{E}}^{2,1}(\mathbb{P}_S^1) := \mathbb{E}^{2,1}(\mathbb{P}_S^1)/\mathbb{E}^{2,1}(\{\infty\})$. There exists a family of classes $\eta_S \in \tilde{\mathbb{E}}^{2,1}(\mathbb{P}_S^1)$ indexed by schemes in $\mathscr{S}$ which is stable by pullbacks and such that for any scheme S and any integers (n,m) the following map is an isomorphism:

$$\mathbb{E}^{n,m}(S) \to \tilde{\mathbb{E}}^{n+2,m+1}(\mathbb{P}_S^1), x \mapsto \eta_S.p^*(x).$$

- *Excision*: for any excisive morphism of closed pairs $f : (Y,T) \to (X,Z)$ (see Def. 1.4.2), the pullback $f^* : \mathbb{E}_Z^{**}(X) \to \mathbb{E}_T^{**}(Y)$ is an isomorphism.

A morphism of ringed cohomology theories with support is a natural transformation compatible with contravariant and covariant functorialities, with refined products and with the operator $\delta_{X,Z}$ for closed pairs (X,Z) in $\mathscr{S}$.

We will say that $\mathbb{E}$ is *oriented* if there exists a natural transformation of presheaves of sets on $\mathscr{S}$:

$$c : \mathrm{Pic} \to \mathbb{E}^{2,1}$$

such that for any scheme S, $c_{\mathbb{P}_S^1}(\lambda) = \eta_S$ where $\lambda = \mathcal{O}(-1)$ is the canonical line bundle on $\mathbb{P}_S^1$.

We will say that a closed pair (X,Z) is $\mathbb{E}$-pure if the morphisms

$$\mathbb{E}^{**}(X,Z) \xleftarrow{\sigma_1^*} \mathbb{E}^{**}(D_Z X, \mathbb{A}_Z^1) \xrightarrow{\sigma_0^*} \mathbb{E}^{**}(N_Z, Z)$$

induced by the deformation diagram (1.3.1.a) are isomorphisms. We say that $\mathbb{E}$ is *absolutely pure* if any regular closed pair in $\mathscr{S}$ is $\mathbb{E}$-pure.

For short, we will say *arithmetic cohomology* for a ringed cohomology with support which is oriented and absolutely pure. A morphism of arithmetic cohomology is defined likewise, but beware we do not require the compatibility with the given orientations.

6.1.2 With this definition, we can extend all the results of sections 2, 3 and 4 as follows:

(1) One has to pay a special attention to the projective bundle formula (2.1.13) and realize that the lemma of Morel 2.1.14 can be stated and proven using cohomology with support (instead of working in the unstable homotopy category). Then one gets the working theory of Chern classes and formal group laws as established in Section 2.1.

(2) For Sections 2.3, 2.4, 3 and 4, the arguments just go through as we
have been careful to rely only on the axiomatic described in Section 1
and restated in the previous definition.

The results obtained here cover the one proved earlier by Panin.

Remark 6.1.3 Note that the axiomatic described here differs especially
with that of Panin because of two points:

- we have devised another axiom, *absolute purity*, especially relevant in
  the arithmetic case;
- we asked for the existence of a refined product.

Both properties are very strong and the natural examples are given by
representable cohomology theories — but see also the next section. Note
however that in the case of algebraic K-theory, they should be obtained
without using representability: the case of absolute purity is of course the
localization theorem of Quillen but the case of refined products is less obvious.

6.2   Étale cohomology

Example 6.2.1   (1) Let Λ be a torsion ring of characteristic exponent N.
Let $\mathscr{S}$ be the category of regular schemes on which N is invertible.
Then it follows from [AGV73], using the method described in Section 1.2 together with the functoriality of étale sheaves established
in [AGV73], that for any closed immersion $i : Z \to X$, the bigraded
cohomology groups:

$$H_Z^n(X_{\text{ét}}, \Lambda(m))$$

of the twisted sheaf $i^!\Lambda(m)$, computed in the small étale site of X, is
a ringed cohomology with support over $\mathscr{S}$ in the sense of the above
definition. Recall that according to [AGV73, IX, Th. 3.3], this cohomology theory is oriented with an additive formal group law. Moreover, it is absolutely pure over $\mathscr{R}$eg according to the (absolute purity)
theorem of Gabber.

Thus our constructions apply to this cohomology, which has an additive formal group law. In particular, we get maps for projective
morphisms of regular $\mathbb{Z}[1/N]$-schemes on étale cohomology with coefficients in Λ.[55]

---

[55]This Gysin morphisms agree with the one constructed by Gabber-Riou: see Remark 6.2.5.

(2) Let l be a prime number and $\mathscr{S}$ be the category of $\mathbb{Z}[1/l]$-regular schemes.

Then we can apply the construction of [Jan88] to get that *continuous l-adic étale cohomology with support*:

$$H^n_{\mathrm{cont},Z}(X,\mathbb{Z}_l(m))$$

defined in *loc. cit.*, Section 3, (after Remark 3.5) is an arithmetic cohomology in the previous sense. In fact, homotopy, stability and excision follow from the known results of [AGV73]. The refined product can be defined using the method of Section 6 given that we have a pairing

$$\Gamma_T(Z,-)\otimes\Gamma_Z(X,-)\to\Gamma_T(X,-)$$

of the functors of global sections with support for torsion sheaves (as in the above example). Axioms (E1)–(E7) then follow. Note this theory is oriented: this is *loc. cit.* (3.26). Finally, using the absolute purity theorem of Gabber in the form of the computation of $\Gamma_Z(X,-)$ for $Z\subset X$ regular schemes, we get that this cohomology theory with support is absolutely $\mathscr{R}eg$-pure. The same construction works for $\mathbb{Q}_l$-coefficients.

Thus we can also apply the constructions of this paper to l-adic étale cohomology (integral and rational).

6.2.2 Using the more sophisticated theory of [CD12b], we can get many examples as follows.

Let $\mathscr{T}$ be a motivic triangulated category over $\mathscr{S}$ in the sense of [CD12b, Def. 2.4.45]: in other words, this is a category fibered over the category $\mathscr{S}$ which satisfies the axioms (A1)–(A4) of Par. 1.1.1 together with the homotopy and stability property (in fact, we will not use the adjoint properties of *loc. cit.*). Then we can associate with $\mathscr{T}$ a ringed cohomology theory with support: for any closed immersion $i:Z\to X$ in $\mathscr{S}$, we put:

$$H^{n,m}_Z(X,\mathscr{T}):=\mathrm{Hom}_{\mathscr{T}(X)}(i_*(\mathbb{1}_Z),\mathbb{1}_X(m)[n])$$

where $\mathbb{1}_?$ is the cartesian section of $\mathscr{T}$ made by the unit for the tensor product and $\mathbb{1}_X(m)$ denotes the m-th Tate twist ([CD12b, 2.4.17]). Then exactly the same arguments as in the proof of Prop. 1.2.10 shows that this theory satisfies axioms (E1)–(E7).

Assume moreover that one has a premotivic adjunction ([CD12b, Def. 1.4.2])

$$\varphi^*:\mathscr{T}\to T'$$

where $\mathscr{T}$ and $\mathscr{T}'$ are motivic categories. Then, according to [CD12b, 2.3.11 or 2.4.53], for any closed immersion i, $\varphi^* i_* \simeq i_* \varphi^*$ through a canonical isomorphism (called an *exchange isomorphism*). Thus, given any closed pair (X, Z) in $\mathscr{S}$, one gets by applying φ^* a morphism:

$$H_Z^{n,m}(X, \mathscr{T}) \xrightarrow{\varphi} H_Z^{n,m}(X, \mathscr{T}')$$

which is compatible with contravariant functoriality (resp. covariant functoriality, refined product, boundary) because φ^* commutes with f^* for any morphism f (resp. i_* for any closed immersion i, tensor product, localization triangle).

Example 6.2.3   (1) *(Motivic) étale cohomology.*– Let R be any ring. For any scheme S, Cisinski and the author have introduced in [CD14, 5.1.3], following Voevodsky, the category $\mathrm{DM}_\mathrm{h}(S, R)$ of h-motives. We proved in *loc. cit.*, Th. 5.6.2, that it forms, for various S, a motivic triangulated category. In particular, the cohomology theory

$$H_{\mathrm{\acute{e}t}, Z}^{n,m}(X, R) := \mathrm{Hom}_{\mathrm{DM}_\mathrm{h}(X, R)}(i_*(\mathbb{1}_Z), \mathbb{1}_X(m)[n])$$

is a ringed cohomology with support, defined over the category of all schemes. According to *loc. cit.*, 5.6.2, it is even an arithmetic cohomology. According to Voevodsky, this cohomology theory is called the *étale motivic cohomology* with coefficients in R. In view of the second computation below, we think that it should simply be called the *étale cohomology* with coefficients in R.

According to the fundamental results of *loc. cit.*, one gets for any regular scheme X:

- if R is a $\mathbb{Q}$-algebra,

$$H_{\mathrm{\acute{e}t}}^{n,m}(X, R) = H_{\mathrm{B}}^{n,m}(X, R) = K_{2m-n}^{(m)}(X)$$

  is Beilinson motivic cohomology;
- if R is a torsion ring with characteristic exponent N, for any scheme X,

$$H_{\mathrm{\acute{e}t}}^{n,m}(X, R) = H_{\mathrm{\acute{e}t}}^n(X[1/N], R(m)),$$

  where $X[1/N]$ is the open part of X where N is invertible, and the right hand side is the usual étale cohomology of $X[1/N]$ with coefficients in R twisted m-times.

Moreover, for any ring extension R'/R, there is a premotivic adjunction:

$$\varphi : \mathrm{DM}_{\mathrm{h}}(S, R) \to \mathrm{DM}_{\mathrm{h}}(S, R')$$

so that we get a morphism of arithmetic cohomologies:

$$\varphi : H_{\text{ét}}^{**}(-, R) \to H_{\text{ét}}^{**}(-, R'). \qquad (6.2.3.\text{a})$$

(2) *Continuous étale cohomology.*– Let R be any valuation ring with parameter ℓ. Then, according to *loc. cit.*, 7.2.11, the homotopy ℓ-adic completion of $\mathrm{DM}_{\mathrm{h}}(-, R)$ gives a motivic triangulated category $\mathrm{DM}_{\mathrm{h}}(-, \hat{R}_\ell)$ and in particular a ringed cohomology theory with support, defined over the category of all schemes:

$$H_{\mathrm{cont}, Z}^{n,m}(X, \hat{R}_\ell) := \mathrm{Hom}_{\mathrm{DM}_{\mathrm{h}}(X, \hat{R}_\ell)}(i_*(\mathbb{1}_Z), \mathbb{1}_X(m)[n]).$$

According to *loc. cit.*, this is an arithmetic cohomology,.

Note that when R is a discrete valuation ring, and X a scheme such that the exponent characteristic of R/ℓ is invertible on X, according to *loc. cit.*, 7.2.21, the triangulated category $\mathrm{DM}_{\mathrm{h}}(X, \hat{R}_\ell)$ agree with Ekedahl category of ℓ-adic complexes. Thus the cohomology $H_{\mathrm{cont}, Z}^{**}(X, \hat{R}_\ell)$ is Jannsen continuous étale ℓ-adic cohomology and deserves the name of *continuous étale cohomology* with coefficients in $\hat{R}_l$.

From the obvious premotivic adjunction $\hat{\rho}_\ell^* : \mathrm{DM}_{\mathrm{h}}(S, R) \to \mathrm{DM}_{\mathrm{h}}(S, \hat{R}_\ell)$ (see [CD14, 7.2.4]), we get a morphism of arithmetic cohomologies:

$$\rho_\ell : H_{\text{ét}}^{**}(-, R) \to H_{\mathrm{cont}}^{**}(-, \hat{R}_\ell). \qquad (6.2.3.\text{b})$$

Let Q be the fraction field of R. We now easily get the rational version of continuous étale cohomology by taking tensor product by Q over R:

$$H_{\mathrm{cont}, Z}^{n,m}(X, Q_\ell) := H_{\mathrm{cont}, Z}^{n,m}(X, \hat{R}_\ell) \otimes_R Q.$$

which is again an arithmetic cohomology theory. And finally, a rational version of the previous morphism:

$$\rho_\ell : H_{\text{ét}}^{**}(-, Q) \to H_{\mathrm{cont}}^{**}(-, Q_\ell). \qquad (6.2.3.\text{c})$$

(3) For a prime number ℓ, combining (6.2.3.a) and (6.2.3.c), we get a morphism of ringed cohomologies with support on the category of regular $\mathbb{Z}[1/\ell]$-schemes:

$$\rho_\ell : H_{\mathrm{B}}^{**}(-) \to H_{\mathrm{cont}}^{**}(-, \mathbb{Q}_\ell) \qquad (6.2.3.\text{d})$$

from Beilinson motivic cohomology to continuous rational ℓ-adic cohomology.

As a corollary of the preceding examples and the constructions of this paper, we thus obtain:

Corollary 6.2.4 *Assume $\mathscr{S}$ is one of the following categories of schemes:*

(a) *regular noetherian schemes of finite dimension;*

(b) *smooth schemes over a noetherian (singular) scheme of finite dimension.*

Let R be a ring (resp. discrete valuation ring with parameter ℓ). In the respective case, we also denote by Q_l the fraction field of $\hat{R}_\ell$.

(1) *The ring cohomology theory $H^{**}_{\acute{e}t}(-, R)$ (resp. $H^{**}_{cont}(-, \hat{R}_\ell)$, $H^{**}_{cont}(-, Q_\ell)$) admits Chern classes, Gysin morphisms for any projective morphism of schemes in $\mathscr{S}$, and residue morphisms associated with a closed immersion $i : Z \to S$ of schemes in $\mathscr{S}$ which fit into the usual localization long exact sequence. These residues and Gysin morphisms satisfy the following properties: compatibility with transversal pullback, excess of intersection, projection formula.*

(2) *The five natural transformations of Example 6.2.3 are functorial with respect to Gysin morphisms and localization long exact sequences.*

(3) *For any integer $r \geq 0$, there exists a well defined higher Chern character:*

$$\mathrm{ch}_r : K_r(X) \to \oplus_{n \geq 0} H^{2n-r,n}_{\acute{e}t}(Y, Q_\ell)$$

from Quillen (resp. Thomason-Trobaugh in case (b)) algebraic K-theory such that for any projective morphism $f : Y \to X$ in $\mathscr{S}$, the following diagram commutes:

$$
\begin{array}{ccc}
K_r(Y)_{\mathbb{Q}} & \xrightarrow{\quad f_* \quad} & K_r(X)_{\mathbb{Q}} \\
{\scriptstyle \mathrm{Td}(\tau_f).\,\mathrm{ch}_t} \downarrow & & \downarrow {\scriptstyle \mathrm{ch}_t} \\
\oplus_{n \geq 0} H^{2n-r,n}_{\acute{e}t}(Y, Q_\ell) & \xrightarrow{\quad f_* \quad} & \oplus_{n \geq 0} H^{2n-r,n}_{\acute{e}t}(X, Q_\ell)
\end{array}
$$

where on the top line, f_* is the usual covariant functoriality of algebraic K-theory. Moreover for any closed immersion $i : Z \to X$ in $\mathscr{S}$, one gets:

$$
\begin{array}{ccc}
K_r(X - Z)_{\mathbb{Q}} & \xrightarrow{\quad \partial_{X,Z} \quad} & K_r(Z)_{\mathbb{Q}} \\
{\scriptstyle \mathrm{Td}(-N_Z X).\,\mathrm{ch}_t} \downarrow & & \downarrow {\scriptstyle \mathrm{ch}_t} \\
\oplus_{n \geq 0} H^{2n-r,n}_{\acute{e}t}(Y, Q_\ell) & \xrightarrow{\quad \partial_{X,Z} \quad} & \oplus_{n \geq 0} H^{2(n-c)-r+1,n-c}_{\acute{e}t}(X, Q_\ell),
\end{array}
$$

where $N_Z X$ is the normal bundle of Z in X.

Under assumption (b), the functor K_r can be replaced by Weibel homotopy invariant K-theory KH_r in the two previous diagram.

As explained in 6.1.2, Point (1) is a compact form of the results of Sections 2, 3 (recall excess of intersection: 3.2.12, projection formula: 3.2.10(b)). Point (2) follows from Th. 4.2.3 and Th. 4.3.2 because all theories have additive formal group law and there is only one strict isomorphism of formal group law: this implies the Todd class involved in each formulas is necessarily equal to 1 (see Section 5.1). Point (3) finally follows from Prop. 5.3.4 and Prop. 5.3.6.

Remark 6.2.5 When R is a torsion ring with characteristic exponent N, in [Rio14], Riou following a construction of Gabber has defined Gysin morphisms on étale cohomology of $\mathbb{Z}[1/N]$-schemes with coefficients in R, with respect to all lci projective maps between any noetherian schemes. If one restricts to regular schemes, we obtain using Th. 3.3.1 that our Gysin maps coincide with the construction of Gabber-Riou.

Let us be more precise. First, let us compare our conventions with that of *op. cit.* Let X be a scheme and $\mathcal{E}$ be a locally free $\mathcal{O}_X$-module. Then the vector bundle associated with $\mathcal{E}$ is $E = V(\mathcal{E}^\vee)$, the spectrum over X of the symmetric algebra induced by the *dual* of $\mathcal{E}$. Because of this convention one relates Chern classes used in [Rio14] with ours by the formula:

$$c_r(\mathcal{E}) = (-1)^r . c_r(E)$$

(compare with relation (5.2.9.a)).

Once this convention is settled, one can apply Theorem 3.3.1 as Riou proved the excess intersection formula in [Rio14, Prop. 2.3.2]. Note also that, because of Cor. 5.1.10, Chern classes are uniquely determined by the choice of a stability isomorphism:

$$H^2_{\text{ét}}(\mathbb{P}^1_S, R(1)) \simeq H^0_{\text{ét}}(S, R(0)) = R$$

(which necessarily appears as a particular case of the projective bundle formula).

References

[AGV73] M. Artin, A. Grothendieck, and J.-L. Verdier, *Théorie des topos et cohomologie étale des schémas*, Lecture Notes in Mathematics, **269, 270, 305**, Springer-Verlag, 1972–1973, Séminaire de Géométrie Algébrique du Bois–Marie 1963–64 (SGA 4).

[AH10]   J. Scholbach, A. Holmstrom, *Arakelov motivic cohomology I*, J. Algebraic Geom. **24** (2015), no. 4, 719–754.

[Art69]   M. Artin, *Algebraic approximation of structures over complete local rings*, Inst. Hautes Études Sci. Publ. Math., no.**36** (1969), 23–58.

[Ayo07a] J. Ayoub, *Les six opérations de Grothendieck et le formalisme des cycles évanescents dans le monde motivique (I)*, Astérisque, **314** (2007), Soc. Math. France.

[Ayo07b] ———, *Les six opérations de Grothendieck et le formalisme des cycles évanescents dans le monde motivique. II*, Astérisque (2007), no. 315, vi+364 pp. (2008).

[Beĭ84]   A. A. Beĭlinson, *Higher regulators and values of L-functions*, Current problems in mathematics, **24** (1984), Itogi Nauki i Tekhniki, Akad. Nauk SSSR Vsesoyuz. Inst. Nauchn. i Tekhn. Inform., Moscow, pp. 181–238.

[Ber74]   P. Berthelot, *Cohomologie cristalline des schémas de caractéristique p > 0*, Lecture Notes in Mathematics, **407**, Springer-Verlag, Berlin-New York, 1974.

[BGI71]   P. Berthelot, A. Grothendieck, and L. Illusie, *Théorie des intersections et théorème de Riemann-Roch*, Lecture Notes in Mathematics, **225**, Springer-Verlag, 1971, Séminaire de Géométrie Algébrique du Bois–Marie 1966–67 (SGA 6).

[BK86]   S. Bloch and K. Kato, *p-adic étale cohomology*, Inst. Hautes Études Sci. Publ. Math., no. **63** (1986), 107–152.

[BO74]   S. Bloch and A. Ogus, *Gersten's conjecture and the homology of schemes*, Ann. Sci. École Norm. Sup. (4) **7** (1974), 181–201 (1975).

[Bor03]   S. Borghesi, *Algebraic Morava K-theories*, Invent. Math. **151** (2003), no. 2, 381–413.

[BS58]   A. Borel and J.-P. Serre, *Le théorème de Riemann-Roch*, Bull. Soc. Math. France **86** (1958), 97–136.

[CD12a]   D.-C. Cisinski and F. Déglise, *Mixed weil cohomologies*, Adv. in Math. **230** (2012), no. 1, 55–130.

[CD12b]   ———, *Triangulated categories of mixed motives*, arXiv:0912.2110v3, (2012).

[CD14]   ——, *Étale motives*, Compos. Math. **152** (2016), no. 3, 556666.

[Cis13]   D.-C. Cisinski, *Descente par éclatements en K-théorie invariante par homotopie*, Ann. of Math. (2) **177** (2013), no. 2, 425–448.

[Dég08a]   F. Déglise, *Around the Gysin triangle II*, Doc. Math. **13** (2008), 613–675.

[Dég08b]   ——, *Motifs génériques*, Rend. Semin. Mat. Univ. Padova **119** (2008), 173–244.

[Dég11]   ——, *Modules homotopiques*, Doc. Math. **16** (2011), 411–455.

[Dég13]   ——, *Orientable homotopy modules*, Amer. Journ. of Math. **135** (2013), no. 2, 519–560.

[Dég14]   ——, *Suite spectrale du coniveau et t-structure homotopique*, Ann. Fac. Sci. Toulouse Math. (6) **23** (2014), no. 3, 591–609.

[Del77]   P. Deligne, *Cohomologie étale*, Lecture Notes in Mathematics, **569**, Springer-Verlag, Berlin, 1977, Séminaire de Géométrie Algébrique du Bois-Marie SGA 4 1/2, Avec la collaboration de J. F. Boutot, A. Grothendieck, L. Illusie et J. L. Verdier.

[Del87]   ——, *Le déterminant de la cohomologie*, Current trends in arithmetical algebraic geometry (Arcata, Calif., 1985), Contemp. Math., **67**, Amer. Math. Soc., 1987, pp. 93–177.

[DI05]   D. Dugger and D. C. Isaksen, *Motivic cell structures*, Algebr. Geom. Topol. **5** (2005), 615–652.

[Die74]   J. Dieudonné, *Cours de géométrie algébrique. I: Aperçu historique sur le développement de la géométrie algébrique.*, Paris: Presses Universitaires de France, 1974 (French).

[DRØ03]   B. Ian Dundas, O. Röndigs, and P. A. Østvær, *Motivic functors*, Doc. Math. **8** (2003), 489–525.

[Dye62]   E. Dyer, *Relations between cohomology theories*, Colloquium on Algebraic Topology, NATO, (1962).

[Fuj02]   K. Fujiwara, *A proof of the absolute purity conjecture (after Gabber)*, Algebraic geometry 2000, Azumino (Hotaka), Adv. Stud. Pure Math., **36** (2002), Math. Soc. Japan, Tokyo, pp. 153–183.

[Ful98] W. Fulton, *Intersection theory*, second ed., Ergebnisse der Mathematik und ihrer Grenzgebiete. 3. Folge. A Series of Modern Surveys in Mathematics [Results in Mathematics and Related Areas. 3rd Series. A Series of Modern Surveys in Mathematics], **2**, Springer-Verlag, Berlin, 1998.

[GD67] A. Grothendieck and J. Dieudonné, *Éléments de géométrie algébrique. IV. Étude locale des schémas et des morphismes de schémas IV*, Publ. Math. IHES **20, 24, 28, 32** (1964-1967).

[Gil81] H. Gillet, *Riemann-Roch theorems for higher algebraic K-theory*, Adv. in Math. **40** (1981), no. 3, 203–289.

[Gil05] ———, *K-theory and intersection theory*, Handbook of K-theory. **1, 2**, Springer, Berlin, 2005, pp. 235–293.

[Gra98] Jeremy J. Gray, *The Riemann-Roch theorem and geometry, 1854–1914*, Proceedings of the International Congress of Mathematicians, **III** (Berlin, 1998), no. Extra **III**, (1998), pp. 811–822 (electronic).

[Gro61] A. Grothendieck, *Éléments de géométrie algébrique. II. Étude globale élémentaire de quelques classes de morphismes*, Inst. Hautes Études Sci. Publ. Math. no. **8** (1961), 222.

[Gro77] ———, *Cohomologie ℓ-adique et fonctions L*, Lecture Notes in Mathematics, vol. 589, Springer-Verlag, 1977, Séminaire de Géométrie Algébrique du Bois–Marie 1965–66 (SGA 5).

[Hir66] F. Hirzebruch, *Topological methods in algebraic geometry*, Third enlarged edition. Die Grundlehren der Mathematischen Wissenschaften, Band 131, Springer-Verlag New York, Inc., New York, (1966).

[Hoy15] M. Hoyois, *From algebraic cobordism to motivic cohomology*, J. Reine Angew. Math. **702** (2015), 173–226. MR 3341470

[Jan88] U. Jannsen, *Continuous étale cohomology*, Math. Ann. **280** (1988), no. 2, 207–245.

[Lev09] M. Levine, *Comparison of cobordism theories*, J. Algebra **322** (2009), no. 9, 3291–3317.

[LM07] M. Levine and F. Morel, *Algebraic cobordism*, Springer Monographs in Mathematics, Springer, Berlin, (2007).

[LYZ13] M. Levine, Y. Yang, and G. Zhao, *Algebraic elliptic cohomology theory and flops, I*, arXiv:1311.2159v2, (2013).

[ML98] S. Mac Lane, *Categories for the working mathematician*, second ed., Graduate Texts in Mathematics, vol. 5, Springer-Verlag, New York, (1998).

[MV99] F. Morel and V. Voevodsky, $\mathbf{A}^1$-*homotopy theory of schemes*, Inst. Hautes Études Sci. Publ. Math., no. **90** (1999), 45–143 (2001).

[NSØ09a] N. Nauman, M. Spitzweck, and P.A. Østvær, *Motivic Landweber exactness*, Doc. Math. **14** (2009), 551–593.

[NSØ09b] N. Naumann, M. Spitzweck, and P. A. Østvær, *Motivic Landweber exactness*, Doc. Math. **14** (2009), 551–593.

[Ouw14] D. Ouwehand, *Local rigid cohomology of singular points*, arXiv:1401.1656, (2014).

[Pan03] I. Panin, *Oriented cohomology theories of algebraic varieties*, K-Theory **30** (2003), no. 3, 265–314, Special issue in honor of Hyman Bass on his seventieth birthday. Part III.

[Pan04] ———, *Riemann-Roch theorems for oriented cohomology*, Axiomatic, enriched and motivic homotopy theory, NATO Sci. Ser. II Math. Phys. Chem., **131** (2004), Kluwer Acad. Publ., Dordrecht, pp. 261–333.

[Pan09] ———, *Oriented cohomology theories of algebraic varieties. II (After I. Panin and A. Smirnov)*, Homology, Homotopy Appl. **11** (2009), no. 1, 349–405.

[PPR08] I. Panin, K. Pimenov, and O. Röndigs, *A universality theorem for Voevodsky's algebraic cobordism spectrum*, Homology, Homotopy Appl. **10** (2008), no. 2, 211–226.

[PS00] I. Panin and A. Smirnov, *Push-forwards in oriented cohomology theories of algebraic varieties*, K-theory preprint archive 0459, 2000.

[Qui69] D. Quillen, *On the formal group laws of unoriented and complex cobordism theory*, Bull. Amer. Math. Soc. **75** (1969), 1293–1298.

[Qui73] ———, *Higher algebraic K-theory. I*, Algebraic K-theory, I: Higher K-theories (Proc. Conf., Battelle Memorial Inst., Seattle, Wash., 1972), Springer, Berlin, 1973, pp. 85–147. Lecture Notes in Math., Vol. 341.

[Rie57] B. Riemann, *Theorie des Abel'schen functionen*, J. Reine Angew. Math. **54** (1857).

[Rio10] J. Riou, *Algebraic K-theory, $\mathbb{A}^1$-homotopy and Riemann-Roch theorems*, J. Topol. **3** (2010), no. 2, 229–264.

[Rio14] ———, *Classes de Chern, morphismes de Gysin, pureté absolue*, Travaux de Gabber sur l'uniformisation locale et la cohomologie des schémas quasi-excellents, no. **361-362** (2014), pp. 303–351.

[Roc65] G. Roch, *Ueber die anzahl der willkürlichen constanten in algebraischen functionen*, J. Reine Angew. Math. **64** (1865).

[Ros96] M. Rost, *Chow groups with coefficients*, Doc. Math. **1** (1996), No. 16, 319–393 (electronic).

[Shi07] E. Shinder, *The mishchenko and quillen formulas for oriented cohomology pretheories*, St. Petersburg Math. J. **18** (2007), 671–678.

[Smi06a] A. L. Smirnov, *Orientations and transfers in the cohomology of algebraic varieties*, Algebra i Analiz **18** (2006), no. 2, 167–224.

[Smi06b] ———, *The Riemann-Roch theorem for operations in the cohomology of algebraic varieties*, Algebra i Analiz **18** (2006), no. 5, 210–236.

[Sou85] C. Soulé, *Opérations en K-théorie algébrique*, Canad. J. Math. **37** (1985), no. 3, 488–550.

[Sou86] ———, *Régulateurs*, Astérisque (1986), no. 133-134, 237–253, Seminar Bourbaki, Vol. 1984/85.

[Spi13] M. Spitzweck, *A commutative $\mathbb{P}^1$-spectrum representing motivic cohomology over dedekind domains*, arXiv:1207.4078, (2013).

[Swa98] Richard G. Swan, *Néron-Popescu desingularization*, Algebra and geometry (Taipei, 1995), Lect. Algebra Geom., vol. 2, Int. Press, Cambridge, MA, (1998), pp. 135–192.

[Tat68] J. Tate, *Residues of differentials on curves*, Ann. Sci. École Norm. Sup. (4) **1** (1968), 149–159.

[Tho84] R. W. Thomason, *Absolute cohomological purity*, Bull. Soc. Math. France **112** (1984), no. 3, 397–406.

[Tho93] ———, *Les K-groupes d'un schéma éclaté et une formule d'intersection excédentaire*, Invent. Math. **112** (1993), no. 1, 195–215.

[TT90]   R. W. Thomason and T. Trobaugh, *Higher algebraic K-theory of schemes and of derived categories*, The Grothendieck Festschrift, Vol. III, Progr. Math., vol. 88, Birkhäuser Boston, Boston, MA, (1990), pp. 247–435.

[vdK71] W. van der Kallen, *Le K_2 des nombres duaux*, C. R. Acad. Sci. Paris Sér. A-B **273** (1971), A1204–A1207.

[Vez01] G. Vezzosi, *Brown-Peterson spectra in stable $\mathbb{A}^1$-homotopy theory*, Rend. Sem. Mat. Univ. Padova **106** (2001), 47–64.

[Vis07] A. Vishik, *Symmetric operations in algebraic cobordism*, Adv. Math. **213** (2007), no. 2, 489–552.

[Voe96] V. Voevodsky, *The Milnor conjecture*, K-theory Preprint Archives **170**, (1996).

[Voe98] ———, *$\mathbb{A}^1$-homotopy theory*, Proceedings of the International Congress of Mathematicians, Vol. I (Berlin, 1998), no. Extra Vol. I, (1998), pp. 579–604 (electronic).

[Voe02] ———, *Open problems in the motivic stable homotopy theory. I*, Motives, polylogarithms and Hodge theory, Part I (Irvine, CA, 1998), Int. Press Lect. Ser., vol. 3, Int. Press, Somerville, MA, (2002), pp. 3–34.

[Wil06] J. Wildeshaus, *The boundary motive: definition and basic properties*, Compos. Math. **142** (2006), no. 3, 631–656.

Acknowledgement

I want to warmly thank D.C. Cisinski, O. Gabber, H. Gillet, M. Levine, J. Riou and A. Vishik, for discussions and ideas that motivated and made this work possible as well as A. Navarro for a very useful proof-reading. Last but not least, I will never thank enough the anonymous referee of the current version for his incredibly careful reading and his many comments. In particular, he indicated to me a sign mistake in [Dég08a, Cor. 5.31] which is corrected in formula (3.2.14.a) of the present paper. He helped me to get a truly neater version of this paper.

FRÉDÉRIC DÉGLISE, E.N.S. LYON, UMPA, 46, ALLÉE D'ITALIE, 69364 LYON CEDEX 07 FRANCE
E-mail: frederic.deglise@ens-lyon.fr
URL: http://perso.ens-lyon.fr/frederic.deglise/

K-Theory
Copyright ©2018 Tata Institute of Fundamental Research
Publisher: Hindustan Book Agency, New Delhi, India

Non rationalité stable
d'hypersurfaces cubiques
sur des corps non algébriquement clos

J.-L. Colliot-Thélène

Introduction

Soit F un corps. Soit X une F-variété projective, lisse, géométriquement connexe X, telle que $X(F) \neq \emptyset$. On s'intéresse aux propriétés suivantes.

(i) Rationalité : La F-variété X est F-birationnelle à un espace projectif $\mathbf{P}_F^d$.

(ii) Rationalité stable : Il existe un entier r tel que la F-variété $X \times_F \mathbf{P}_F^r$ est F-birationnelle à $\mathbf{P}_F^{r+d}$.

(iii) Rétracte rationalité : Il existe un ouvert non vide $U \subset X$, un entier s, un ouvert non vide $V \subset \mathbf{P}_F^s$ et un F-morphisme $V \to U$ qui admet une section.

(iv) R-trivialité : Pour tout corps L contenant F, l'ensemble $X_L(L)/R$ quotient de $X(L)$ par la R-équivalence sur X_L a exactement un élément.

(v) CH_0-trivialité : La F-variété X est universellement CH_0- triviale, c'est-à-dire que pour tout corps L contenant F, l'application degré $deg_L :$ $CH_0(X_L) \to \mathbb{Z}$ est un isomorphisme (voir [1, 6]).

(vi) Trivialité de la cohomologie non ramifiée : Pour tout module fini galoisien M sur F d'ordre premier à la caractéristique de F, pour tout entier $i \geq 0$ et tout corps L contenant F, l'application naturelle $H^i(L, M) \to H_{nr}^i(L(X)/L, M)$ est un isomorphisme.

Chacune des propriétés implique la suivante. Pour le passage de (ii) à (iii) sur un corps quelconque, voir [18, Cor. 3.3]. Qu'en toute caractéristique l'hypothèse (iii) implique (iv) est établi par Kahn et Sujatha [10, Thm. 8.5.1 et Prop. 8.6.2]. Pour les définitions des différentes notions et des références pour les autres implications, on consultera [10, 1, 21].

Soit $X \subset \mathbf{P}_F^n$, avec $n = d + 1 \geq 3$ une hypersurface cubique lisse de dimension d possédant un point F-rationnel. Il est connu qu'une telle hypersurface est F-unirationnelle [11].

Pour F algébriquement clos, X est rationnelle si $d = 2$, non rationnelle si $d = 3$ et $\mathrm{char}(F) \neq 2$ (Clemens–Griffiths, Mumford, Murre). Pour $d \geq 4$, certaines hypersurfaces cubiques lisses de dimension paire sont rationnelles. Pour les hypersurfaces cubiques lisses de dimension impaire, on ne sait rien sur la rationalité, la rationalité stable, ou même la rétracte rationalité. Claire Voisin [25] a montré qu'il existe des hypersurfaces cubiques de dimension $d = 3$ sur $\mathbb{C}$ qui sont universellement CH_0-triviales. Elle a aussi établi cette dernière propriété pour de larges classes d'hypersurfaces cubiques de dimension $d = 4$.

Pour F non algébriquement clos, que peut-on dire ?

Pour $F = \mathbb{R}$ le corps des réels, B. Segre [23] observa qu'une surface cubique lisse $X/\mathbb{R}$ telle que l'espace topologique $X(\mathbb{R})$ ait deux composantes connexes n'est pas $\mathbb{R}$-rationnelle. Cette observation se généralise. Ceci est discuté au §1, où l'on montre que pour tout entier $n \geq 3$ et tout corps $F \subset \mathbb{R}$ il existe des hypersurfaces cubiques lisses $X \subset \mathbf{P}_F^n$ qui ne sont pas CH_0-triviales, et en particulier ne sont pas stablement rationnelles.

Que peut-on dire lorsque le corps F n'est pas formellement réel, i.e. lorsque -1 est une somme de carrés dans F ?

Pour F un corps de caractéristique différente de 2, on note $u(F) \leq \infty$ le plus grand entier $n \leq \infty$ tel qu'il existe une forme quadratique anisotrope de rang n sur F. Rappelons que l'on a $u(F((t))) = 2u(F)$.

Pour $X \subset \mathbf{P}_F^3$, on peut utiliser le groupe de Brauer pour donner des exemples de surfaces cubiques non stablement rationnelles [15]. On donne facilement de tels exemples déjà sur un corps fini $\mathbb{F}$, sur $\mathbb{C}((x))$, et sur $\mathbb{C}(x)$.

Sur $F = \mathbb{C}((x))((y))$, D. Madore [13, Proposition 2.1] a construit, par un argument de spécialisation élaboré, une hypersurface cubique diagonale $X \subset \mathbf{P}_F^4$ telle que l'application degré $CH_0(X) \to \mathbb{Z}$ ne soit pas un isomorphisme. Ceci implique que X n'est pas rétracte rationnelle.

Sur F un corps p-adique quelconque et sur $F = \mathbb{F}((x))$ (avec $\mathbb{F}$ fini de caractéristique différente de 3), par spécialisation à une hypersurface cubique singulière sur un corps fini, A. Pirutka et l'auteur [6, Théorème 1.19] ont construit des hypersurfaces cubiques lisses $X \subset \mathbf{P}_F^4$ avec un F-point qui ne sont pas universellement CH_0 triviales. On en déduit de tels exemples sur tout corps global de caractéristique différente de 3.

Dans la prépublication récente [2], par spécialisation à une hypersurface cubique produit d'une quadrique et d'un hyperplan, Chatzistamatiou et Levine construisent des exemples d'hypersurfaces cubiques lisses $X \subset \mathbf{P}_F^n$ avec un F-point, non universellement CH_0-triviales, et donc non rétractes rationnelles, dans la situation suivante : k un corps de caractéristique différente de 2, $F = k((x))$, $u(k) \geq 2^\ell + 1$ et $n = 2^\ell$. Une inspection de leur argument (voir §2 ci-dessous) montre qu'il suffit en fait de supposer $u(k) \geq 2^\ell$. On obtient ainsi de tels exemples $X \subset \mathbf{P}_F^{2^\ell}$ sur le corps $F = \mathbb{C}((\lambda_1)) \cdots ((\lambda_{\ell+1}))$.

Je propose ici deux autres méthodes pour obtenir des hypersurfaces cubiques lisses, avec un point rationnel, qui ne sont pas rétractes rationnelles.

Au §3, sur $F = k((x))$, avec k un corps de caractéristique différente de 2, possédant une forme de Pfister anisotrope de dimension 2^ℓ, avec $\ell \geq 2$, par exemple sur $F = \mathbb{C}((\lambda_1)) \cdots ((\lambda_{\ell+1}))$, je construis des exemples d'hypersurfaces cubiques lisses $X \subset \mathbf{P}_F^n$ avec un F-point, non rétractes rationnelles, pour tout entier n avec $3 \leq n \leq 2^{\ell-1} + 1$. Ceci ne couvre pas le cas $n = 2^\ell$, obtenu par la méthode de Chatzistamatiou et Levine. L'argument donné utilise la spécialisation de la R-équivalence sur une fibre spéciale géométriquement intègre mais singulière en codimension 2 et un succédané de cohomologie non ramifiée sur la désingularisation de cette fibre.

Au §4, pour k un corps de caractéristique différente de 3 possédant un élément $a \notin k^{*3}$, par exemple pour $k = \mathbb{C}((\lambda_1))$, sur le corps $F = k((\lambda_2)) \cdots ((\lambda_{\ell+1}))$ je construis des exemples d'hypersurfaces cubiques lisses diagonales $X \subset \mathbf{P}_F^n$ avec un F-point, non universellement CH_0-triviales, donc non rétractes rationnelles, pour tout entier n avec $3 \leq n \leq \ell + 3$. L'argument donné utilise la cohomologie non ramifiée, dont on démontre par spécialisations successives qu'elle n'est pas constante.

Au §5, on compare les résultats des §2, 3 et 4 sur les corps de séries formelles itérées, d'abord sur les complexes puis sur les corps p-adiques.

1 Composantes connexes réelles

Théorème 1.1 *Soit k un sous-corps de $\mathbb{R}$. Soit X une k-variété projective, lisse, géométriquement connexe. Dans chacun des cas suivants :*

(a) *la k-variété X est k-rationnelle,*

(b) *la k-variété X est k-rétracte rationnelle,*

(c) *la k-variété X est universellement CH_0-triviale,*

l'espace topologique $X(\mathbb{R})$ est non vide et connexe.

Démonstration Il suffit d'établir le résultat dans le cas $k = \mathbb{R}$. Le cas (a) est un cas particulier de (b) qui d'après [6, Lemme 1.5] est un cas particulier de (c). Si la $\mathbb{R}$-variété X est universellement CH_0-triviale, en particulier l'application degré $CH_0(X) \to \mathbb{Z}$ est un isomorphisme. Ainsi X possède un zéro-cycle de degré 1, et donc un point réel. Soit $s \geq 1$ le nombre de composantes connexes de $X(\mathbb{R})$. D'après [5, Prop. 3.2] (voir aussi [4, Thm. 3.1]), pour toute $\mathbb{R}$-variété projective, lisse, géométriquement connexe avec $X(\mathbb{R}) \neq \emptyset$, on a $CH_0(X)/2 = (\mathbb{Z}/2)^s$. Ainsi $s = 1$.

$\square$

Remarque 1.2 Sous l'hypothèse (a), on peut établir la connexité de $X(\mathbb{R})$ par des méthodes plus classiques. On montre directement que le nombre de composantes connexes de $X(\mathbb{R})$ est un invariant birationnel des $\mathbb{R}$-variétés projectives, lisses, géométriquement connexes (ce qui résulte aussi du résultat sur le groupe de Chow mentionné ci-dessus).

Proposition 1.3 *Soit k un sous-corps de $\mathbb{R}$. Pour tout entier $n \geq 2$, il existe une hypersurface cubique lisse $X \subset \mathbf{P}_k^n$ telle que l'espace topologique $X(\mathbb{R})$ ait deux composantes connexes. Une telle hypersurface n'est pas rétracte rationnelle.*

Démonstration   Soit $X \subset \mathbf{P}_k^{n+1}$ donnée par l'annulation de

$$R(x_1, \ldots, x_n, u, v) := (\sum_{i=1}^{n} x_i^2)v - u(u-v)(u+v).$$

Le lieu singulier est donné par $u = v = 0 = \sum_{i=1}^{n} x_i^2 = 0$, donc ne possède par de $\mathbb{R}$-point. Ainsi $X(\mathbb{R})$ est une variété C^∞. On vérifie que cette variété possède deux composantes connexes. Soit en effet $H \subset \mathbf{P}_{\mathbb{R}}^{n+1}$ l'hyperplan à l'infini défini par $v = 0$, et soit $\mathbf{A}_{\mathbb{R}}^{n+1}$ son complémentaire. La trace de $X_{\mathbb{R}}$ sur $\mathbf{A}_{\mathbb{R}}^{n+1}$ est donnée par l'équation affine

$$\sum_{i=1}^{n} x_i^2 = u(u-1)(u+1)$$

dont le lieu réel est la réunion disjointe d'une partie bornée satisfaisant $-1 \leq u \leq 0$ et d'une partie non bornée satisfaisant $u \geq 1$. Ceci montre que $X(\mathbb{R})$ est disconnexe, donc a deux composantes connexes, car c'est le maximum possible pour une hypersurface cubique X dans $\mathbf{P}_{\mathbb{R}}^n$ avec $n \geq 2$.

Soit $S(x_1, \ldots, x_n, u, v)$ une forme cubique sur $\mathbb{Q}$ définissant une hypersurface cubique lisse, par exemple

$$S(x_1, \ldots, x_n, u, v) = \sum_{i=1}^{n} x_i^3 + u^3 + v^3.$$

Pour $t \in k \subset \mathbb{R}$ non nul et très proche de 0, l'hypersurface cubique X_t de $\mathbf{P}_k^{n+1}$ définie par $R + tS = 0$ est lisse. Pour tout $t \in \mathbb{R}$ assez proche de 0, les variétés C^∞ données par $X_t(\mathbb{R})$ et $X_0(\mathbb{R}) = X(\mathbb{R})$ sont difféomorphes (théorème d'Ehresmann), et en particulier ont le même nombre de composantes connexes. La deuxième partie de l'énoncé résulte du théorème 1.1.

$\square$

2 Résultats de Chatzistamatiou et Levine

Commençons par rappeler un résultat de Totaro [24, Lemme 2.4 et argument subséquent]. Pour X une k-variété propre, on note $A_0(X)$ le noyau de l'application degré $deg_k : CH_0(X) \to \mathbb{Z}$.

Lemme 2.1 *Soit R un anneau de valuation discrète hensélien de corps résiduel k et de corps des fractions K. Soit $\mathcal{X}$ un R-schéma intègre propre et plat. Soit $X = \mathcal{X} \times_R K$. Supposons que la fibre spéciale Y/k est la réunion de deux diviseurs Y_1 et Y_2 sans composante commune. Soit Z le k-schéma $Y_1 \cap Y_2$. Supposons Y_1/k lisse et $A_0(X) = 0$. Alors :*

(i) *L'application $A_0(Z) \to A_0(Y_1)$ est surjective.*

(ii) *Si de plus l'indice de Z est égal à celui de Y_1, par exemple si Z possède un zéro-cycle de degré 1, alors l'application $CH_0(Z) \to CH_0(Y_1)$ est surjective.*

Démonstration On a une suite exacte

$$CH_0(Z) \to CH_0(Y_1) \oplus CH_0(Y_2) \to CH_0(Y) \to 0$$

où la seconde flèche envoie un couple (z_1, z_2) sur $z_1 - z_2$.

Comme Y_1 est lisse, tout zéro-cycle de degré zéro sur Y_1 est rationnellement équivalent sur Y_1 à un zéro-cycle z de degré zéro à support dans le complémentaire de Z dans Y_1. Un tel zéro-cycle z se relève en un zéro-cycle de degré zéro sur X, dont l'image par la flèche de spécialisation $CH_0(X) \to CH_0(Y)$ est l'image du couple $(z, 0) \in CH_0(Y_1) \oplus CH_0(Y_2)$. L'hypothèse $A_0(X) = 0$ assure que cette image est nulle. La suite exacte ci-dessus établit alors l'existence d'une classe $\zeta \in CH_0(Z)$ dont l'image est $(z, 0) \in CH_0(Y_1) \oplus CH_0(Y_2)$. En particulier le degré de ζ est zéro, ce qui établit l'assertion (i). L'assertion (ii) est alors claire. $\square$

Voici une version du résultat utilisé par Chatzistamatiou et Levine [2].

Lemme 2.2 *Soit R un anneau de valuation discrète de corps résiduel k et de corps des fractions K. Soit $\mathcal{X}$ un R-schéma intègre propre et plat. Soit $X = \mathcal{X} \times_R K$. Supposons que la fibre spéciale Y/k est, comme diviseur, la somme de deux diviseurs effectifs Y_1 et Y_2, qui comme k-variétés sont lisses et géométriquement intègres. Soit Z le k-schéma $Y_1 \cap Y_2$.*

Supposons la K-variété X géométriquement intègre et universellement CH_0-triviale. Si l'indice de $Y_{2k(Y_1)}$ sur le corps $k(Y_1)$ est égal à 1, alors l'application $CH_0(Z_{k(Y_1)}) \to CH_0(Y_{1k(Y_1)})$ est surjective, et l'indice de $Z_{k(Y_1)}$ sur $k(Y_1)$ est égal à 1.

Démonstration Il existe un homomorphisme local $R \to S$ d'anneaux de valuation discrète, avec S hensélien, de corps des fractions L induisant l'inclusion $k \subset k(Y_1)$ au niveau des corps résiduels. On considère alors la suite exacte

$$CH_0(Z_{k(Y_1)}) \to CH_0(Y_{1\,k(Y_1)}) \oplus CH_0(Y_{2\,k(Y_1)}) \to CH_0(Y_{k(Y_1)}) \to 0.$$

Soit z un zéro-cycle sur $Y_{1\,k(Y_1)}$. Comme Y_1 est lisse, ce zéro-cycle est rationnellement équivalent à un zéro-cycle z_1 à support étranger à Z. Par hypothèse, il existe un zéro-cycle w sur $Y_{2\,k(Y_1)}$ de degré égal à celui de z_1. Comme Y_2 est lisse, ce zéro-cycle w est rationnellement équivalent sur $Y_{2\,k(Y_1)}$ à un zéro-cycle z_2 dont le support est étranger à Z. Le zéro-cycle $z_1 - z_2$ sur $Y_{k(Y_1)}$ est à support dans le lieu lisse du morphisme $\mathcal{X} \to \mathrm{Spec}\,(R)$. Il se relève donc en un zéro-cycle de degré zéro sur X_L (ceci vaut même sur un corps non parfait). Comme la K-variété X est universellement CH_0-triviale, par spécialisation [8, Prop. 2.6], $z_1 - z_2$ est rationnellement équivalent à zéro sur $Y_{k(Y_1)}$. De la suite exacte ci-dessus on tire l'existence d'une classe de zéro-cycle sur $Z_{k(Y_1)}$ d'image la classe de z dans $CH_0(Y_{1\,k(Y_1)})$. Ceci établit la première partie de l'énoncé. Appliquant le résultat au zéro-cycle z de $Y_{1\,k(Y_1)}$ de degré 1 défini par le point générique de Y_1, on obtient la seconde partie de l'énoncé.

$\square$

Remarque 2.3 On peut donner des variantes de l'énoncé ci-dessus. Supposons Y_1/k et Y_2/k géométriquement intègres mais non nécessairement lisses. Supposons qu'il existe un zéro-cycle de degré 1 à support dans le lieu lisse de $Y_2 \setminus Z$. Alors l'indice de $Z_{k(Y_1)}$ sur $k(Y_1)$ est égal à 1. En effet, le point générique η de Y_1 définit un $k(Y_1)$-point lisse de $Y_{1\,k(Y_1)} \setminus Z_{k(Y_1)}$. L'argument ci-dessus établit alors l'existence d'un zéro-cycle sur $Z_{k(Y_1)}$ d'image la classe de η dans $CH_0(Y_{1\,k(Y_1)})$.

Théorème 2.4 (*Chatzistamatiou et Levine*) *Soit k un corps de caractéristique différente de 2. Si sur le corps k il existe une forme quadratique anisotrope en $n = 2^\ell$ variables, alors il existe une hypersurface cubique lisse $X \subset \mathbf{P}^n_{k((t))}$ qui possède un $k((t))$-point et qui n'est pas universellement CH_0-triviale.*

Démonstration Soit $\ell \geq 1$, soit k un corps, $\mathrm{car}(k) \neq 2$, et soit $q(x_1, \ldots, x_n)$ une forme quadratique anisotrope de rang exactement $n = 2^\ell$. Soit

$$q'(x_0, \ldots, x_n) := q(x_1, \ldots, x_n) + x_0^2.$$

Un théorème de Hoffmann [9] (dont une nouvelle démonstration fut donnée par Merkurjev [16] au moyen de formules du degré à la Rost) assure qu'il n'y a pas d'application rationnelle de la quadrique défine par $q' = 0$ dans $\mathbf{P}_k^n$ vers la quadrique définie par $q = 0$ dans $\mathbf{P}_k^{n-1}$. Il n'y a pas besoin ici de supposer que la forme quadratique q' est anisotrope. Soit $Y_1 \subset \mathbf{P}_k^n$ la quadrique lisse définie par $q' = 0$. Soit $Y_2 \subset \mathbf{P}_k^n$ l'hyperplan défini par $x_0 = 0$. Alors $Z = Y_1 \cap Y_2 \subset Y_2$ est la quadrique définie par $q = 0$ dans $Y_2 \simeq \mathbf{P}_k^{n-1}$. Il existe une forme cubique lisse en $n+1$ variables sur $k((t))$ qui se spécialise en $t = 0$ sur la forme cubique $q'(x_0, \ldots, x_n).x_0$, et qui possède un zéro non trivial sur $k((t))$, car aucun k-point de Y_2 n'est situé sur Z. Soit $X \subset \mathbf{P}_{k((t))}^n$ l'hypersurface cubique lisse qu'elle définit. Si l'indice de Z sur $k(Y_1)$ était égal à 1, alors par un théorème bien connu de Springer, la quadrique Z aurait un $k(Y_1)$-point, donc il y aurait une application rationnelle de Y_1 vers Z, contredisant le théorème d'Hoffmann. Le lemme 2.2 permet alors de conclure que l'hypersurface cubique $X \subset \mathbf{P}_{k((t))}^n$ n'est pas universellement CH_0-triviale.

$\square$

3 Formes quadratiques multiplicatives et R-équivalence

3.1 Certaines hypersurfaces cubiques

Soit k un corps de caractéristique différente de 2. Soit $q(x_1, \ldots, x_n)$, $n \geq 2$ une forme quadratique non dégénérée sur k. Soit $\rho \in k^*$, $\rho \neq 0, \rho \neq 1$.

Soit $X = X(q, \rho) \subset \mathbf{P}_k^{n+1}$ l'hypersurface cubique donnée par l'équation homogène[1]

$$q(x_1, \ldots, x_n)v = u(u - v)(u - \rho v).$$

Le lieu de non lissité de X est donné par $u = v = 0 = q = 0$.

Suivant [3, §5], soit $Y = Y(q, \rho)$ le fibré en quadriques sur $\mathbf{P}_k^1$ obtenu par recollement de la variété Y_1 définie par

$$q(X_1, \ldots, X_n) = U(U - 1)(U - \rho)T^2$$

dans $\mathbf{P}_k^n \times_k \mathbf{A}_k^1$ muni des coordonnées $(X_1, \ldots, X_n, T; U)$ et de la variété Y_2 définie par

$$q(X_1', \ldots, X_n') = V(1 - V)(1 - \rho V)T^2$$

dans $\mathbf{P}_k^n \times_k \mathbf{A}_k^1$ muni des coordonnées $(X_1', \ldots, X_n', T; V)$, le recollement se faisant via $V = 1/U$ et $(X_1', \ldots, X_n', T) = (U^{-2}X_1, \ldots, U^{-2}X_n, T)$.

---

[1] Pour $n = 2$, on obtient une surface de Châtelet.

Les k-variétés X et Y sont k-birationnellement isomorphes, elles contiennent toutes deux la k-variété affine lisse W d'équation

$$q(x_1, \ldots, x_n) = u(u-1)(u-\rho).$$

Plus précisément, on définit un morphisme birationnel de $X \setminus \{v = 0\}$ vers Y_1 par

$$(X_1, \ldots, X_n, T; U) = (x_1, \ldots, x_n, v; u/v).$$

On définit un morphisme birationnel de $X \setminus \{u = 0\}$ vers Y_2 par

$$(X'_1, \ldots, X'_n, T; V) = (x_1 v, \ldots, x_n v, u^2; v/u).$$

Ceci définit un morphisme de $X \setminus \{u = v = 0\}$ vers Y.

Montrons que ce morphisme s'étend en un morphisme de

$$X_{lisse} = X \setminus \{u = v = q = 0\}$$

vers Y. Il suffit d'établir cela au voisinage des points de $u = v = 0$ qui satisfont $q \neq 0$. Sans perte de généralité, on peut alors se placer sur l'ouvert affine $x_1 \neq 0$ de X. On s'intéresse donc à la variété affine définie par l'équation

$$q(1, x_2, \ldots, x_n) = u(u-v)(u-\rho v).$$

Cette équation se réécrit

$$[q(1, x_2, \ldots, x_n) + (1+\rho)u^2 - \rho uv].v = u^3.$$

Au voisinage d'un point de $u = v = 0$ avec $q \neq 0$ la fonction

$$f(x_2, \ldots, x_n, u, v) := q(1, x_2, \ldots, x_n) + (1+\rho)u^2 - \rho uv$$

est inversible. L'application rationnelle de X vers Y_2 est donnée sur l'ouvert affine défini par $x_1 \neq 0$ par

$$(X'_1, \ldots, X'_n, T; U) = (v, x_2 v, \ldots, x_n v, u^2; v/u).$$

Comme on a $fv = u^3$, cette application est donc aussi donnée par

$$(X'_1, \ldots, X'_n, T; U) = (u, x_2 u, \ldots, x_n u, f; u^2.f^{-1})$$

qui est un morphisme là où f est inversible, donc dans le voisinage de tout point avec $u = v = 0$ et $q \neq 0$.

On notera que l'image de $u = v = 0$, $q \neq 0$ est le point $(0, \ldots, 0, 1; 0)$ de $Y_2 \subset Y$.

Soit désormais $\theta : X_{lisse} \to Y$ le k-morphisme birationnel construit ci-dessus.

Soient $A \in W(k) \subset X(k)$ le k-point lisse défini par $(x_1, \ldots, x_n, u, v) = (0, \ldots, 0, 1, 1)$ et $B \in W(k) \subset X(k)$ le k-point lisse défini par $(x_1, \ldots, x_n, u, v) = (0, \ldots, 0, \rho, 1)$. Sur $Y(k)$, les images de ces points par θ sont donnés respectivement par $A_1 = \theta(A) \in Y_1(k)$ de coordonnées $(X_1, \ldots, X_n, T; u) = (0, \ldots, 0, 1; 1)$ et $B_1 = \theta(B) \in Y_1(k)$ de coordonnées $(X_1, \ldots, X_n, T; u) = (0, \ldots, 0, 1; \rho)$.

Lemme 3.1 *Supposons la forme quadratique q anisotrope sur k. Si les points $A, B \in W(k)$ sont R-équivalents sur X, alors $A_1 = \theta(A)$ et $B_1 = \theta(B)$ sont R-équivalents sur Y.*

Démonstration L'hypothèse sur q garantit que le lieu singulier de X, qui est donné par $u = v = q = 0$, ne contient aucun k-point. On a donc $X_{lisse}(k) = X(k)$. Soient P et Q deux points rationnels de $X(k)$ R-équivalents sur X. Par définition, il existe une famille finie de k-morphismes $\sigma_i : \mathbf{P}^1_k \to X$, $i = 1, \ldots, m-1$ avec $\sigma_i(0) = P_i, \sigma_i(\infty) = P_{i+1}$, et $P_0 = P$ et $P_m = Q$. Comme $X_{lisse}(k) = X(k)$, les points P_i sont tous dans X_{lisse}, les images des morphismes σ_i rencontrent X_{lisse}. La composition avec θ définit des applications rationnelles ρ_i de $\mathbf{P}^1_k$ vers Y, qui, puisque Y est projectif, sont en fait des morphismes. On a $\rho_i(0) = \sigma(P_i)$ et $\rho_i(\infty) = \sigma(P_{i+1})$. Ainsi $\theta(P)$ et $\theta(Q)$ sont R-équivalents sur Y.

$\square$

3.2 Rappel

Soit k un corps, car.$(k) \neq 2$, et soit ϕ une n-forme de Pfister anisotrope sur k. On sait (Pfister, voir [12]) que l'ensemble $N_\phi(k) \subset k^*$ des valeurs non nulles de ϕ est un sous-groupe de k^*. Soit Y une k-variété projective, lisse, géométriquement connexe. Soit $k(Y)$ son corps des fonctions.

Dans [3, 4], on a considéré le sous-groupe de $k(Y)^*$ formé des fonctions qui partout localement sur Y peuvent s'écrire comme le produit d'une unité locale et d'un élément de $N_\phi(k(Y))$. On a défini et étudié le quotient $D^\phi(Y)$ de ce groupe par le sous-groupe $N_\phi(k(Y))$.

Il y a en particulier un accouplement

$$Y(k) \times D^\phi(Y) \to D^\phi(k) = k^*/N_\Phi(k)$$

qui passe au quotient par la R-équivalence sur $Y(k)$ ([3, Cor. 4.1.5], conséquence de la fonctorialité et de l'invariance homotopique des groupes $D^\phi(Y)$).

L'invariance birationnelle des groupes $D^\phi(Y)$ fut établie par M. Rost [22].

3.3   Sur un corps de u-invariant fini

On renvoie à [12] pour les propriétés de base des formes quadratiques sur les corps.

Théorème 3.2 *Soit k un corps de caractéristique différente de 2 tel que $I^{n+1}k \neq 0$, i.e. tel qu'il existe une $(n+1)$-forme de Pfister anisotrope. Soit $F = k((t))$. Pour tout entier m avec $2 \leq m \leq 2^n$, il existe une hypersurface cubique lisse $X \subset \mathbf{P}_F^{m+1}$ avec un F-point qui n'est pas rétracte rationnelle, en particulier n'est pas stablement rationnelle.*

Démonstration   La condition sur k équivaut à l'existence d'une n-forme de Pfister ϕ, de dimension 2^n et d'un élément $\rho \in k^*$ non représenté par la forme ϕ sur k. Elle implique aussi que le u-invariant de k est au moins égal à 2^{n+1}.

Soit $q(x_1, \ldots, x_m)$, avec $m \geq 2$, une forme quadratique non dégénérée sur k, de dimension $m \leq 2^n$, qui est une sous-forme de la forme ϕ sur k. En particulier q est anisotrope.

Soit $Y = Y(q, \rho)$ comme au §3.1. On vérifie facilement (cf. [3, §5]) que la fonction rationnelle $u \in k(W)^* = k(Y)^*$ définit un élément de $D^\phi(Y)$, et que l'évaluation de cet élément sur A_1, resp. B_1, est $1 \in k^*/D^\phi(k)$, resp. $\rho \in k^*/D^\phi(k)$. Par hypothèse, on a $\rho \notin D^\phi(k)$. Ainsi les k-points A_1 et B_1 ne sont pas R-équivalents sur Y. (On voit aussi que l'application naturelle $D^\phi(k) \to D^\phi(Y)$ n'est pas un isomorphisme.)

Le lemme 3.1 garantit alors que les points A et B de $X(k)$ ne sont pas R-équivalents sur l'hypersurface cubique singulière X.

Soit $\Psi(x_1, \ldots, x_m, u, v) = 0$ une forme cubique sur le corps k définissant une hypersurface lisse dans $\mathbf{P}_k^{m+1}$. Soit $\mathcal{X} \subset \mathbf{P}_{k[[t]]}^{m+1}$ défini par

$$q(x_1, \ldots, x_m)v - u(u - v)(u - \rho v) + t\Psi(x_1, \ldots, x_m, u, v) = 0.$$

La fibre spéciale $\mathcal{X}_0$ est l'hypersurface cubique X sur le corps k discutée au §3.1. Elle contient les points k-rationnels lisses A et B. Par le lemme de Hensel, le point A se relève en des $k((t))$-points $\tilde{A}$ de la fibre générique $\mathcal{X}_t$ sur le corps $F = k((t))$. En outre, l'ensemble de tels $\tilde{A}$ est Zariski dense sur la F-variété $\mathcal{X}_t$. On a le même énoncé pour les relevés $\tilde{B}$ de B.

Supposons la F-variété $\mathcal{X}_t$ rétracte rationnelle. Alors il existe des ouverts non vides $U \subset \mathcal{X}_t$ et $V \subset \mathbf{P}_F^n$ et des F-morphismes $s : U \to V$ et $p : V \to U$ dont le composé $p \circ s$ est l'identité de U. Soient alors $\tilde{A}$ et $\tilde{B}$ des relevés de A et B dans U. Il existe une droite $\mathbf{P}_F^1 \subset \mathbf{P}_F^n$ contenant $s(\tilde{A})$ et $s(\tilde{B})$, et qui donc rencontre V. En composant avec $p : V \to U \subset \mathcal{X}_t$, puisque $\mathcal{X}_t$ est propre et que $\mathbf{P}_F^1$ est une courbe régulière, on obtient un F-morphisme $\mathbf{P}_F^1 \to \mathcal{X}_t$ tel que $\tilde{A}$ et $\tilde{B}$ soient dans l'image $\mathbf{P}^1(F)$. Ainsi $\tilde{A}$ et $\tilde{B}$ sont

R-équivalents sur la F-variété $\mathcal{X}_t$. On sait [14] que la R-équivalence se spécialise (ceci vaut pour tout $k[[t]]$-schéma propre): l'application

$$\mathcal{X}_t(F) = \mathcal{X}(k[[t]]) \to X_0(k) = X(k)$$

induit une application $\mathcal{X}_t(F)/R \to X(k)/R$. Mais A et B ne sont pas R-équivalents sur X. On conclut (voir l'introduction de l'article) que la F-hypersurface cubique lisse $\mathcal{X}_t$ n'est pas rétracte rationnelle.

$\square$

Remarque 3.3 On peut envisager diverses variantes de la démonstration.

(1) Tout en utilisant l'argument de spécialisation de la R-équivalence comme ci-desssus, on peut remplacer les groupes $D^\phi(Y)$ par la cohomologie non ramifiée. Soit $\phi = <1, -a_1> \otimes \cdots \otimes <1, -a_n>$. Le cup-produit

$$(a_1) \cup \cdots (a_n) \cup (u) \in H^{n+1}(k(Y), \mathbb{Z}/2),$$

où pour $b \in k(Y)^*$ on note $(b) \in k(Y)^*/k(Y)^{*2} = H^1(k(Y), \mathbb{Z}/2)$ la classe de b, est une classe non ramifiée qui est non constante car elle prend des valeurs distinctes dans $H^{n+1}(k, \mathbb{Z}/2)$. Ceci utilise le théorème ([20]) qu'une $(n+1)$-forme de Pfister non triviale sur le corps k a une image non nulle dans $H^{n+1}(k, \mathbb{Z}/2)$.

(2) On peut envisager une méthode qui ignore la R-équivalence, et utilise seulement l'équivalence rationnelle sur les zéro-cycles.

Supposons que l'on ait :
(*) Il existe un k-morphisme de désingularisation $f : X' \to X$ qui est universellement CH_0-trivial.

Si la F-variété $\mathcal{X}_t$ est rétracte rationnelle, alors elle est universellement CH_0-triviale [6, Lemme 1.3]. D'après [6, Théorème 1.12], sous l'hypothèse (*), ceci implique que la k-variété X' est universellement CH_0-triviale. Ceci implique alors [17] que les groupes de cohomologie non ramifiée à coefficients $\mathbb{Z}/2$ de X', qui sont des invariants k-birationnels des k-variétés projectives et lisses, sont réduits à la cohomologie du corps de base. Mais on sait (via le modèle Y) que ce n'est pas le cas, car on a une classe de cohomologie non ramifiée qui prend des valeurs distinctes en deux points – ce dernier point utilisant comme ci-dessus [20].

4 Hypersurfaces cubiques diagonales et cohomologie non ramifiée

Théorème 4.1 *Soit k un corps de caractéristique différente de 3, possédant un élément a qui n'est pas un cube. Soient $0 \leq n \leq m$ des entiers. Soit F*

un corps avec

$$k(\lambda_1, \ldots, \lambda_m) \subset F \subset F_m := k((\lambda_1)) \cdots ((\lambda_m)).$$

L'hypersurface cubique $X := X_{n,F}$ *de* $\mathbf{P}_F^{n+3}$ *définie par l'équation*

$$x^3 + y^3 + z^3 + aw^3 + \sum_{i=1}^{n} \lambda_i t_i^3 = 0$$

possède un point rationnel et n'est pas universellement CH_0-*triviale, en particulier elle n'est pas rétracte rationnelle.*

Démonstration Pour établir le résultat, on peut supposer que k contient une racine cubique primitive de l'unité, soit j, et que $F = F_m$. Le lemme 4.2 ci-dessous permet de supposer $n = m$. On fixe un isomorphisme $\mathbb{Z}/3 = \mu_3$ et on considère la cohomologie étale à coefficients $\mathbb{Z}/3$. On ignore les torsions à la Tate dans les notations. Etant donnés un corps L contenant k et des éléments $b_i, i = 1, \ldots, s$, de L^*, on note $(b_1, \ldots, b_s) \in H^s(L, \mathbb{Z}/3)$ le cup-produit, en cohomologie galoisienne, des classes $(b_i) \in L^*/L^{*3} = H^1(L, \mathbb{Z}/3)$.

On va démontrer par récurrence sur $n \neq 0$ l'assertion suivante, qui implique la proposition.

(A_n) Soient k, a, F_n et X_n/F_n comme ci-dessus. Le cup-produit

$$\alpha_n := ((x + jy)/(x + y), a, \lambda_1, \ldots, \lambda_n) \in H^{n+2}(F_n(X_n), \mathbb{Z}/3)$$

définit une classe de cohomologie non ramifiée (par rapport au corps de base F_n) qui ne provient pas d'une classe dans $H^{n+2}(F_n, \mathbb{Z}/3)$.

Le cas $n = 0$ est connu ([15, Chap. VI, §5] [7, §2.5.1]). Supposons l'assertion démontrée pour n.

La classe α_{n+1} sur la F_{n+1}-hypersurface $X_{n+1} \subset \mathbf{P}_{F_{n+1}}^{n+4}$ a ses résidus triviaux en dehors des diviseurs définis par $x + y = 0$ et $x + jy = 0$. Soit $\Delta \subset X_{n+1}$ le diviseur $x + y = 0$. Ce diviseur est défini par les équations

$$x + y = 0, z^3 + aw^3 + \sum_{i=1}^{n+1} \lambda_i t_i^3 = 0.$$

Le résidu de α_{n+1} au point générique de Δ est

$$\partial_\Delta(\alpha_{n+1}) = \pm(a, \lambda_1, \ldots, \lambda_{n+1}) \in H^{n+2}(F_{n+1}(\Delta), \mathbb{Z}/3).$$

Mais dans le corps des fonctions de Δ, on a

$$1 + a(w/z)^3 + \sum_{i=1}^{n+1} \lambda_i(t_i/z)^3 = 0$$

et cette égalité implique (cf. [19, Lemma 1.3]) :

$$(a, \lambda_1, \ldots, \lambda_{n+1}) = 0 \in H^{n+2}(F_{n+1}(\Delta), \mathbb{Z}/3).$$

Le même argument s'applique pour le diviseur défini par $x + jy = 0$. Ainsi α_{n+1} est une classe de cohomologie non ramifiée sur la F_{n+1}-hypersurface X_{n+1}.

Soit $\mathcal{X}_{n+1}$ le $F_n[[\lambda_{n+1}]]$-schéma défini par

$$x^3 + y^3 + z^3 + aw^3 + \sum_{i=1}^{n+1} \lambda_i t_i^3 = 0.$$

Le diviseur Z défini par $\lambda_{n+1} = 0$ sur $\mathcal{X}$ est le cône d'équation

$$x^3 + y^3 + z^3 + aw^3 + \sum_{i=1}^{n} \lambda_i t_i^3 = 0$$

dans $\mathbf{P}_{F_n}^{n+4}$, cône qui est birationnel au produit de $\mathbf{P}_{F_n}^1$ et de l'hypersurface cubique lisse $X_n \subset \mathbf{P}_{F_n}^{n+3}$ définie par

$$x^3 + y^3 + z^3 + aw^3 + \sum_{i=1}^{n} \lambda_i t_i^3 = 0.$$

Le corps des fonctions rationnelles de $\mathcal{X}_{n+1}$ est $F_{n+1}(X_{n+1})$.

On a

$$\partial_Z(\alpha_{n+1}) = \pm((x+jy)/(x+y), a, \lambda_1, \ldots, \lambda_n) \in H^{n+2}(F_n(Z), \mathbb{Z}/3).$$

Par l'hypothèse de récurrence

$$((x+jy)/(x+y), a, \lambda_1, \ldots, \lambda_n) \in H^{n+2}(F_n(X_n), \mathbb{Z}/3)$$

n'est pas dans l'image de $H^{n+2}(F_n, \mathbb{Z}/3)$. Ceci implique que

$$((x+jy)/(x+y), a, \lambda_1, \ldots, \lambda_n) \in H^{n+2}(F_n(Z)), \mathbb{Z}/3)$$

n'est pas dans l'image de $H^{n+2}(F_n, \mathbb{Z}/3)$. Du diagramme commutatif

$$\partial_Z : \qquad H^{n+3}(F_{n+1}(X), \mathbb{Z}/3) \longrightarrow H^{n+2}(F_n(Z), \mathbb{Z}/3)$$
$$\uparrow \qquad\qquad\qquad\qquad \uparrow$$
$$\partial_{\lambda_{n+1}=0} : \qquad H^{n+3}(F_{n+1}, \mathbb{Z}/3) \longrightarrow H^{n+2}(F_n, \mathbb{Z}/3)$$

on conclut que

$$\alpha_{n+1} := ((x+jy)/(x+y), a, \lambda_1, \ldots, \lambda_{n+1}) \in H^{n+3}(F_{n+1}(X), \mathbb{Z}/3)$$

n'est pas dans l'image de $H^{n+3}(F_{n+1}, \mathbb{Z}/3)$.

Ceci établit (A_n) pour tout entier n et implique (cf. [17]) que la F_n-variété X_n n'est pas universellement CH_0-triviale et n'est pas rétracte rationnelle.

□

Lemme 4.2 *Soit F un corps. Si une F-variété X projective, lisse, géométriquement connexe n'est pas universellement CH_0-triviale, alors la $F((t))$-variété $X \times_F F((t))$ n'est pas universellement CH_0-triviale, et donc n'est pas rétracte rationnelle.*

Démonstration Sur tout corps L contenant F, on dispose de l'application de spécialisation $CH_0(X_{L((t))}) \to CH_0(X_L)$, et cette application est surjective et respecte le degré.

□

Remarque 4.3 Il serait intéressant de comprendre la généralité de la construction faite dans le théorème 4.1. On utilise une classe de cohomologie non ramifiée non constante sur un modèle birationnel de la fibre spéciale d'une $k[[t]]$-schéma propre à fibres intègres, et on en tire une classe de cohomologie non ramifiée non constante de degré un de plus sur la fibre générique sur $k((t))$, essentiellement par cup-produit avec la classe d'une uniformisante de l'anneau $k[[t]]$.

On laisse au lecteur le soin d'établir l'analogue suivant du théorème 4.1.

Théorème 4.4 *Soient $p \neq 3$ un nombre premier et k un corps p-adique dont le corps résiduel contient les racines cubiques primitives de 1. Soit $a \in k^*$ une unité qui n'est pas un cube. Soit π une uniformisante de k. Soient $0 \leq n \leq m$ des entiers. Soit F un corps avec*

$$\mathbb{Q}(a)(\lambda_1, \ldots, \lambda_m) \subset F \subset k((\lambda_1)) \cdots ((\lambda_m)).$$

L'hypersurface cubique X_n de $\mathbf{P}_F^{n+4}$ définie par l'équation

$$x^3 + y^3 + z^3 + aw^3 + \pi t^3 + \sum_{i=1}^{n} \lambda_i t_i^3 = 0,$$

qui possède un point rationnel, n'est pas universellement CH_0-triviale et donc n'est pas rétracte rationnelle.

Exemples

En appliquant le théorème 4.1, on trouve $X_n \subset \mathbf{P}_F^{n+3}$ non rétracte rationnelle avec

$$k(\lambda_1, \ldots, \lambda_n) \subset F \subset k((\lambda_1)) \ldots ((\lambda_n))$$

dans les situations suivantes.

(i) Le corps $k = \mathbb{F}$ est un corps fini de caractéristique différente de 3 contenant les racines cubiques de 1.

(ii) Le corps k, de caractéristique différente de 3, possède une valuation discrète, par exemple k est le corps des fonctions d'une variété complexe de dimension au moins 1, ou est un corps p-adique, ou est un corps de nombres.

On trouve ainsi des hypersurfaces cubiques lisses non rétractes rationnelles dans $\mathbf{P}_{\mathbb{C}(x_1, \ldots, x_m)}^n$, avec un point rationnel, pour tout entier n avec $3 \leq n \leq m + 2$.

En appliquant le théorème 4.4, sur un corps k p-adique $(p \neq 3)$ contenant une racine cubique de 1, on trouve des hypersurfaces cubiques lisses non rétractes rationnelles dans $\mathbf{P}_{k(x_1, \ldots, x_m)}^n$, avec un point rationnel, pour tout entier n avec $4 \leq n \leq m + 4$.

5   Comparaison des résultats obtenus par les diverses méthodes

Les hypothèses faites sur le corps k dans chacun des trois derniers paragraphes diffèrent. Pour comparer la qualité des résultats qu'ils produisent, on considère la situation sur le corps $E_n := \mathbb{C}((t_1)) \cdots ((t_n))$. Comme on va voir, aucune des trois méthodes ne donne des résultats entièrement couverts par les deux autres.

En outre les méthodes des §1, 2 et 3 donnent des hypersurfaces cubiques avec un indice de torsion (comme défini dans [2]) égal à 2 et celle du §4 donne des exemples avec un indice de torsion égal à 3.

La méthode du §2 (Chatzistamatiou et Levine) fournit des hypersurfaces cubiques $X \subset \mathbf{P}_{E_n}^N$, avec un point rationnel, non rétractes rationnelles, pour N entier avec $N \geq 3$ de la forme $N = 2^\ell \leq 2^{n-1}$.

La méthode du §3 fournit des hypersurfaces cubiques $X \subset \mathbf{P}_{E_n}^N$, avec un point rationnel, non rétractes rationnelles, pour tout N entier avec $3 \leq N \leq 2^{n-2} + 1$.

La méthode du §4 fournit des hypersurfaces cubiques $X \subset \mathbf{P}_{E_n}^N$, avec un point rationnel, non rétractes rationnelles, pour tout N entier avec $3 \leq N \leq n + 2$.

On obtient ainsi des exemples de telles hypersurfaces cubiques $X \subset \mathbf{P}^{N}_{E_n}$ pour les valeurs suivantes de N.

Pour $n = 1$, les méthodes classiques donnent des exemples non universellement CH_0-triviaux avec $N = 3$. On a donc de tels exemples avec $N = 3$ pour tout $n \geq 1$.

Pour $n = 2$, on a déjà $N = 3$. La méthode du §4 donne des exemples non universellement CH_0-triviaux avec $3 \leq N \leq 4$. Un exemple avec $N = 4$ avait été obtenu par Madore [13], qui a montré que pour l'hypersurface cubique X d'équation

$$x^3 + y^3 + \lambda z^3 + \mu u^3 + \lambda \mu v^3 = 0$$

sur le corps $\mathbb{C}((\lambda))((\mu))$ le groupe de Chow réduit $A_0(X)$ n'est pas nul.

On a donc de tels exemples avec $N = 4$ pour tout $n \geq 2$.

Pour $n = 3$, on a déjà $N = 3, 4$. La méthode du §2 donne des exemples non universellement CH_0-triviaux pour $N = 4$, celle du §3 donne des exemples non rétractes rationnels avec $N = 3$, celle du §4 donne des exemples non universellement CH_0-triviaux pour $N \leq 5$. Le cas $N = 5$ est nouveau. On a donc de tels exemples avec $N = 5$ pour tout $n \geq 3$.

Pour $n = 4$, on a déjà $N = 3, 4, 5$. La méthode du §2 donne des exemples non universellement CH_0-triviaux pour $N = 4$ et $N = 8$ (nouveau cas) celle du §3 donne des exemples non rétractes rationnels avec $N \leq 5$, cas déjà obtenu, celle du §4 donne des exemples non universellement CH_0-triviaux avec $N \leq 6$. Le cas $N = 6$ est nouveau. La situation reste ouverte pour $N = 7$ et $N > 8$. On a donc de tels exemples avec $N = 6, 8$ pour tout $n \geq 4$.

A partir de $n = 5$, on a $n + 2 \leq 2^{n-2} + 1$. On obtient des exemples non rétractes rationnels avec $N \leq 2^{n-2} + 1$ par la méthode du §3 et des exemples non universellement CH_0-triviaux $N = 2^{\ell} \leq 2^{n-1}$ par la méthode du §2.

Soit maintenant k un corps p-adique et $F = k((t_1)) \cdots ((t_n))$. On obtient $X \subset \mathbf{P}^{N}_{F}$, $N \geq 3$, hypersurface cubique lisse avec un point rationnel et non rétracte rationnelle pour $N \geq 3$ satisfaisant l'une des conditions suivantes :

$N = 2^{\ell} \leq 2^{n+1}$, par la méthode du §2 (Chatzistamatiou et Levine);

$N \leq 2^n + 1$, par la méthode du §3 ;

$N \leq n + 4$, pour $p \neq 3$ et k contenant les racines cubiques de 1 par la méthode du §4 (Théorème 4.4).

Le cas des hypersurfaces cubiques lisses dans $\mathbf{P}^4_k$ sur k un corps p-adique quelconque [6, Théorème 1.19] n'est pas couvert par les résultats du présent article.

Question : Pour un corps F donné, l'ensemble des entiers $n \geq 3$ tels qu'il existe une hypersurface cubique lisse dans $\mathbf{P}^n_F$ avec un F-point et non universellement CH_0-triviale est-il un intervalle dans les entiers ?

Remerciements. L'exposé de Marc Levine au Colloque International de K-Théorie à Mumbai, en janvier 2016, et l'article de Chatzistamatiou et Levine [2] m'ont amené à ce travail. Le contenu du §4 a été trouvé à l'occasion de la rencontre EDGE 2016 à Edimbourg (juin 2016), rencontre où j'ai exposé les résultats de l'article. Je remercie Alena Pirutka pour diverses remarques. Je remercie l'IRSES Moduli et l'Institut Tata (Mumbai) pour leur soutien à l'occasion du colloque de Mumbai. Ce travail a bénéficié d'une aide de l'Agence Nationale de la Recherche portant la référence ANR-12-BL01-0005.

Bibliographie

[1] A. Auel, J.-L. Colliot-Thélène and Parimala, *Universal unramified cohomoloqy of cubic fourfolds containing a plane*, in *Brauer Groups and Obstruction Problems: Moduli Spaces and Arithmetic* (Palo Alto, 2013), Birkhäuser Progress in Mathematics **320** (2017), p. 29–55.

[2] André Chatzistamatiou et Marc Levine, *Torsion orders of complete intersections*, Algebra & Number Theory **11** (2017), no. 8, 17791835.

[3] J.-L. Colliot-Thélène, *Formes quadratiques multiplicatives et variétés algébriques*, Bulletin Soc. Math. France **106** (1978) 113–151.

[4] J.-L. Colliot-Thélène, *Formes quadratiques multiplicatives et variétés algébriques : deux compléments*, Bulletin Soc. Math. France **108** (1980) 213–227.

[5] J.-L. Colliot-Thélène et F. Ischebeck, *L'équivalence rationnelle sur les cycles de dimension zéro des variétés algébriques réelles*, C. R. Acad. Sc. Paris, t. **292** (1981) 723–725.

[6] J.-L. Colliot-Thélène et A. Pirutka, *Hypersurfaces quartiques de dimension 3 : non rationalité stable*, Ann. Sc. Éc. Norm. Sup., 4ème série, t. **49**, fasc. 2 (2016) 371–397.

[7] J.-L. Colliot-Thélène et J.-J. Sansuc, *La descente sur les variétés rationnelles, II.* Duke Math. J. **54** no. 2 (1987) 375–492.

[8] W. Fulton, *Intersection theory*, Ergebnisse der Math. und ihrer Grenzg. 3. Folge, Bd. **2**, Springer, Berlin, 1998.

[9] D. W. Hoffmann, *Isotropy of quadratic forms over the function field of a quadric*, Math. Zeitschrift **220** (1995), no. 3, 461–476.

[10] B. Kahn et R. Sujatha, *Birational geometry and localisation of categories*, Documenta Math. Extra Volume: Alexander S. Merkurjev's sixtieth birthday (2015) 277–334.

[11] J. Kollár, *Unirationality of cubic hypersurfaces*, J. Inst. Math. Jussieu **1** (2002), no. 3, 467–476.

[12] T. Y. Lam, *The algebraic theory of quadratic forms*, Benjamin/Cummings, 1973.

[13] D. Madore, *Équivalence rationnelle sur les hypersurfaces cubiques de mauvaise réduction*, J. Number Theory **128** (2008), 926–944.

[14] D. Madore, *Sur la spécialisation de la R-équivalence*, prépublication http:// perso.telecom-paristech.fr/~madore/specialz.pdf

[15] Yu. I. Manin, *Formes cubiques : algèbre, géométrie, arithmétique (en russe)*, Nauka, Moscou, 1972.

[16] A. S. Merkurjev, *Steenrod operations and degree formulas*, J. für die reine u. angew. Math. (Crelle) **565** (2003) 13–26.

[17] A. S. Merkurjev, *Unramified elements in cycle modules*, J. Lond. Math. Soc. **78** (2008), 51–64.

[18] A. S. Merkurjev, *Invariants of algebraic groups and retract rationality of classifying spaces*, in *Algebraic Groups: Structure and Actions*, Proceedings of Symposia in Pure Mathematics **94**, Amer. Math. Soc., Providence, RI, 2017, p. 277–294.

[19] J. Milnor, *Algebraic K-theory and quadratic forms*, Invent. Math. **9** (1970) 318–344.

[20] D. Orlov, A. Vishik, V. Voevodsky, *An exact sequence for $K_*^M/2$ with applications to quadratic forms*, Ann. of Math. (2) **165** (2007), no. 1, 1–13.

[21] A. Pirutka, *Varieties that are not stably rational, zero-cycles and unramified cohomology*, in *Algebraic Geometry: Salt Lake City 2015*, Proceedings of Symposia in Pure Mathematics **97-2**, Amer. Math. Soc., Providence, RI, 2018, p. 459–484.

[22] M. Rost, *Durch Normengruppen definierte birationale Invarianten*, C. R. Acad. Sci. Paris Sér. I Math. **310** (1990), no. 4, 189–192.

[23] B. Segre, *Sull'esistenza, sia nel campo razionale che nel campo reale, di involuzioni piane non birazionali*, Rend. Acc. Naz. Lincei, Sc. fis. mat. e nat. **10** (1951), 94–97.

[24] B. Totaro, *Hypersurfaces that are not stably rational*, J. Amer. Math. Soc. **29** (2016), 883–891.

[25] C. Voisin, *On the universal CH_0 group of cubic hypersurfaces*, J. Eur. Math. Soc. (JEMS) **19** (2017), no. 6, 1619–1653.

CNRS ET UNIVERSITÉ PARIS SUD MATHÉMATIQUES, BÂTIMENT 307, 91405, ORSAY CEDEX, FRANCE
E-mail: jean-louis.colliot-thelene@math.u-psud.fr

K-Theory
Copyright ©2018 Tata Institute of Fundamental Research
Publisher: Hindustan Book Agency, New Delhi, India

Computing topological Hochschild homology using twisted theories

Samik Basu

Abstract

The study of twisted theories generalizes earlier definitions of twisted K theory and cohomology with local coefficients on one hand and the Thom spectra associated to spherical fibrations on the other hand. We briefly recall this formulation for arbitrary E_∞-ring spectra. Apart from geometric considerations, the theory is amenable to detection of ring structures in module spectra and computation of invariants like topological Hochschild homology. We describe a computation using the 3-sphere.

1 Introduction

This paper studies twisted homology and cohomology theories and uses them to make some computations with ring spectra. Twisted theories are naturally defined groups associated to 1-cocycles on a space X with values in a group of automorphisms of a cohomology theory (see Section 2 for a precise construction). Curiously, this definition closely adheres to the classical idea of Thom spectra.

Thom spectra are a central concept in stable homotopy theory defined from Thom spaces of universal bundles over classifying spaces. The Thom spaces of the universal n-plane bundle over $BO(n)$ fit together to form a spectrum MO, and the Thom spaces of the universal complex n-plane bundle over $BU(n)$ form a spectrum MU. Similar constructions may be defined on other infinite sequences of groups such as SO and Sp to yield Thom spectra such as MSO and MSp.

Classically, the Thom spectra were defined and used to compute cobordism groups. The celebrated result of Thom ([26]) states that the set of n-manifolds up to cobordism is isomorphic to the nth homotopy group of the spectrum MO which in turn is used to compute the cobordism groups. The argument generalizes to cobordism with a fixed structure on the stable normal bundle. Consequently, the homotopy groups of MU computes complex cobordism and the homotopy groups of MSO computes oriented cobordism.

The computation of homotopy groups of spectra such as MO, MU or MSO relies on the Adams spectral sequence. Such spectral sequences are induced by ring objects in spectra. Using recent constructions, they may be defined as monoids and commutative monoids in the symmetric monoidal category of spectra using the smash product ([9] and [13]). Previously these were identified as algebras over certain operads : E_∞ ring spectra defined as algebras over an E_∞ operad, correspond to the commutative objects and A_∞ ring spectra, algebras over an A_∞ operad, are the associative objects.

It is also important to form modules over the ring spectra in order to carry out constructions such as forming quotients, mapping spaces, derived smash product, and others inspired from homological algebra (see [9]). A particular example is the Hochschild homology $HH_*(A; M)$ of a ring A with coefficients in a bi-module M over A, defined as the homology of the cyclic bar resolution. The analogous construction in spectra replaces tensor product by the smash product (see Section 4 for a definition in a general symmetric monoidal category). If R is a commutative ring spectrum, A an associative R-algebra, and M an A-bimodule, the cyclic bar construction gives a simplicial spectrum whose geometric realization is the topological Hochschild homology, $THH^R(A; M)$[1].

In view of the above, constructing ring structures on spectra are crucial in homotopy theory. The question whether a spectrum which possesses a ring structure up to homotopy is a monoid in the strict sense, is usually difficult to address. Techniques generally used to solve this are Segal's infinite loop space machine ([16], [25]), realizing spectra as Thom spectra ([14]), and an obstruction theory approach (see [12],[19],[21]).

For viewing a Thom spectrum as a ring spectrum, one considers a general construction. Given a map $f : X \to BO$, the classifying space for virtual bundles of dimension 0, or even more generally a map $f : X \to BF$ the classifying space for spherical fibrations, one may construct a Thom spectrum X^f([15]). In [14], the authors also show that if f is a map of loop spaces (respectively a map of infinite loop spaces), then X^f has an A_∞ structure (respectively E_∞ structure). This construction displays E_∞ structures on a large number of examples. Spectra like MU, MSO, MO, which arise from cobordism theories are all Thom spectra of infinite loop maps, and thus, are all E_∞.

The classical definition of Thom spectra for spherical fibrations has been extended in [2] and [3] to maps $f : X \to BGL_1R$, where GL_1R is the space of units of a ring spectrum R. This recovers the usual Thom spectra when

---

[1]This definition works only under certain technical conditions on R, A and M. For details, refer to [9].

one considers $R = S^0$, the sphere spectrum. The generalized Thom spectra $f \mapsto X^f$ take values in R-modules. The homotopy groups of the generalized Thom spectrum X^f are precisely the twisted R-homology groups and the homotopy groups of the R-module mapping spectrum $Map_{R-mod}(X^f, R)$ are the twisted R-cohomology groups. In the special case $R = HA$ the Eilenberg MacLane spectrum for a commutative ring, one obtains the co-homology with local coefficients (see Section 2.1). Another twisted coho-mology theory that precisely adheres to this definition is twisted K-theory (see Section 2.2).

In this paper we survey and extend results concerning computations with generalized Thom spectra. We note that if $X \simeq S^1$, the Thom spec-trum looks like a quotient of R. A similar result holds if $X \simeq S^3$. We use these spheres for they are the only ones with A_∞-structures, which are used to construct A_∞-structures on quotients of R. In particular, special-izing to $R = K_p^\wedge$ the spectrum of K-theory with p-adic coefficients, there are $\zeta : S^1 \to BGL_1 K_p^\wedge$ such that $(S^1)^\zeta \simeq K/p$ and $\zeta' : S^3 \to BGL_1 K_p^\wedge$ such that $(S^3)^{\zeta'} \simeq K/p$ (K/p stands for the spectrum of mod-p-K-theory). Quotients by ideals with $n-1$ generators are obtained when one considers $X \simeq SU(n)$.

In [6], the Thom spectrum is expressed as a functor which is symmetric monoidal. This allows us to view R-algebra structures on the R-module Thom spectra using space-level geometric data. The examples above arise from loop maps which implies a general result about realizing quotients as R-algebra spectra. For A_∞-R-algebra structures that arise in this way, the topological Hochschild homology is again equivalent to a Thom spectrum. The precise result in this framework states that for a loop map $\Omega\zeta : \Omega X \to BGL_1 R$, the Thom spectrum $\Omega X^{\Omega\zeta}$ has an A_∞-R-algebra structure. In this case, the topological Hochschild homology is equivalent to the Thom spectrum of a map $L^\eta\zeta : LX \to BGL_1 R$ (Theorem 4.4) (Here, LX stands for the free loop space of X). This is proved for the sphere spectrum, that is $R = S^0$ in [7] and for general R in [6].

Using the above result the groups $\pi_* THH^{K_p^\wedge}(K/p)$ have been computed in [5] using the space $X \simeq S^1$ and $\zeta : S^1 \to BGL_1 K_p^\wedge$ so that $(S^1)^\zeta \simeq K/p$. In this paper we make an analogous computation using $X \simeq S^3$ to prove

Theorem 1.1 *For odd primes p, there are there are A_∞-$K_p^\wedge$-algebra struc-tures on K/p for which*

$$\pi_k(THH^{K_p^\wedge}(K/p)) = \begin{cases} 0 & \text{if } k \text{ is odd} \\ (\mathbb{Z}/(p^\infty))^{\frac{p-1}{2}} & \text{if } k \text{ is even} \end{cases}$$

(see Theorem 5.5). It is worthwhile to note that in the computations of [5], there are homotopically different loop maps which induce the same values of

THH. This computation was accomplished earlier by Angeltveit in [1] using
a different approach. The approach with Thom spectra is a intuitive way of
observing the same results. Using the language of Thom spectra, we write
down homotopy classes of ring structures in a rather explicit manner from
which we may observe how the different values of topological Hochschild
homology stack up for various cases.

The key technique used throughout the computation is a splitting of the
spectrum $gl_1 K_p^\wedge$ for odd primes p. This is proved by applying logarithmic
cohomology operations defined in [20] to the spectrum $K_p^\wedge$ (Proposition 4.4
of [5]). The splitting allows us to write down all possible loop maps from
S^1 and S^3 to the space $BGL_1 K_p^\wedge$.

The paper is organized as follows. In section 2, we recall the classical
definition of cohomology with local coefficients and twisted K-theory. In
section 3, we define generalized Thom spectra, its associated twisted theories, and discuss some examples. In section 4, we sketch the main theorems
about ring structures on the R-module Thom spectra. In section 5, we
work out a computation of the topological Hochschild homology of K/p
using the Thom spectrum on the space S^3.

2   Examples of Twisted theories

This section deals with some examples of twisted cohomology theories. Perhaps a good motivation for these lie in the idea of identifying cohomology
with the sheaf cohomology of the constant sheaf. More precisely one has

$$H^*(X; A) \cong \hat{H}^*(X; \underline{A})$$

where the left-hand side is the cohomology of the singular cochain complex
and the right-hand side is the sheaf cohomology of the constant sheaf $\underline{A}$
associated to the abelian group A.

Recall that two sections of a bundle are said to be vertically homotopic
if the relevant homotopy becomes the constant identity on projection down
to the base. The definition may be extended to the case where $P \to Y$
is a fibration and $\alpha : X \to Y$ is a fixed map. In this case, the homotopy
classes of lifts of α to P (so that the homotopy becomes constant α on
projection to Y) are called the vertical homotopy classes of lifts. The groups
$\hat{H}^n(X; \underline{A})$ may be interpreted as vertical homotopy classes of sections of a
trivial bundle over X with fiber Eilenberg MacLane spaces $K(A, n)$.

Recall that a generalized cohomology theory R comprises functors R^n,
$n \in \mathbb{Z}$, satisfying the Eilenberg-Steenrod axioms except for the dimension
axiom. The Brown representability theorem implies that for such a theory
there are spaces R_n such that $R^n(X) \cong [X, R_n]$. As in the case of ordinary

cohomology, one may compute $R^n(X)$ again as vertical homotopy classes of maps over a trivial fibration over X with fiber R_n.

Suppose that one has a generalized cohomology theory R, and a group G^2 of automorphisms of R (so that G acts on R_n for each $n \in \mathbb{Z}$). A Čech 1-cocycle with values in G is defined as : for an open cover $\mathcal{U} = \{U_\alpha | \alpha \in I\}^3$ of X, associate each non-trivial intersection $U_\alpha \cap U_\beta \neq \Phi$ with an element $g_{\alpha\beta} \in G$ and for each non-trivial triple intersection $U_\alpha \cap U_\beta \cap U_\gamma \neq \Phi$ one has the equation $g_{\alpha\beta}g_{\beta\gamma} = g_{\alpha\gamma}$. Čech 1-cocycles (up to coboundaries) are classified by

$$\hat{H}^1(X; G) \cong [X, BG].$$

Given such a 1-cocycle α one may construct bundles R_α^n over X by taking the G-bundle over X classified by the above map and forming the associated bundle using the G-action on R_n. We let $R_\alpha^n(X)$ denote the vertical homotopy classes of sections from X to R_α^n. This is identified with the vertical homotopy classes of lifts in the diagram

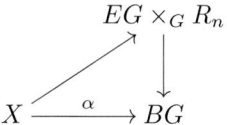

where $EG \to BG$ is the universal principal G-bundle associated to a topological group G. We call these groups R_α^n twisted R-cohomology theories (see [17, §22.1]).

2.1   Cohomology with local coefficients

The earliest example of a twisted cohomology theory as above is the cohomology with local coefficients. For an abelian group A the cohomology theory $H^*(-; A)$ is represented by the Eilenberg-MacLane spectrum HA. The nth-space of the spectrum is the Eilenberg-MacLane space $K(A, n)$.

Let π be a group which acts on A, that is, we have a homomorphism $\pi \to Aut(A)$. A Čech 1-cocycle α lies in $[X, B\pi] = Hom(\pi_1 X, \pi)$ for a connected space X. Then such an α induces a local coefficients $\mathcal{L}_\alpha$: $\pi_1 X \xrightarrow{\alpha} \pi \to Aut(A)$ on X whose value at each point is A and the action of the fundamental groupoid is given by the induced homomorphism from $\pi_1(X) \to Aut(A)$. For the local coefficient system $\mathcal{L}_\alpha$, the cohomology

---

[2]For a topological group these are defined to be 1-cocycles in for the sheaf of continuous functions to G.

[3]This is defined only on passing to a sufficiently small refinement because the Čech cocycles are defined to be elements of the limit of the cocycles over open coverings of X. The limit over the collection of open coverings is taken with respect to the ordering induced by refinement.

with local coefficients $H^*(X; \mathcal{L}_\alpha)$ is defined and one proves that this may be represented as vertical homotopy classes of lifts ([11])

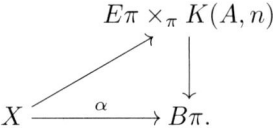

According to the definition above, the twisted cohomology groups $H^n_\alpha(X)$ are given by the vertical classes of sections for a fibration $P \times_\pi K(A, n) \to X$ where P is the π-bundle over X associated to the map $\alpha : X \to B\pi$. Note that P is obtained as the pullback of the fibration $E\pi \to B\pi$ via α. Hence, $P \times_\pi K(A, n)$ is the pullback of $E\pi \times_\pi K(A, n)$ via the map α. It follows that the vertical homotopy classes of sections are isomorphic to the homotopy classes of lifts above.

2.2   Twisted K-theory

Twisted K-theory was introduced by Donovan and Karoubi in 1970 [8] and generalized by Rosenberg for the K-theory of C^*-algebras [22]. Interest in the subject was rejuvenated in the 1990s after certain applications were found in string theory [18], [27]. A remarkable result relates the twisted equivariant K-theory of compact Lie groups G to the K group of positive energy representations of the loop group $\mathcal{L}G$ ([10]). The construction can be described as follows. We follow the discussion in [4] in relation to the context described at the beginning of the section.

One of the many spaces which are homotopic to the space $\mathbb{Z} \times BU$ is $Fred(\mathcal{H})$, the space of Fredholm operators on an infinite dimensional Hilbert space $\mathcal{H}$. Therefore, we have $K(X)(= K^0(X)) \cong [X, Fred(\mathcal{H})]$. We interpret this as vertical homotopy classes of sections on a trivial bundle over X with fiber $Fred(\mathcal{H})$.

Let $PU(\mathcal{H})$ be the projective unitary group, given by the space of unitary operators $U(\mathcal{H})$ modulo the transformations given by scalar multiplication. There is an action of $U(\mathcal{H})$ on $Fred(\mathcal{H})$ via conjugation of transformations which descends to $PU(\mathcal{H})$.

Now as in the above context, we consider twists ζ given by 1-cocycles over $PU(\mathcal{H})$ and use these to define twisted K-groups $K^\zeta(X)$. In order to make a precise definition one proves that the space $U(\mathcal{H})$ is contractible for an infinite dimensional Hilbert space $\mathcal{H}$ and thus, the space $PU(\mathcal{H})$ is a $K(\mathbb{Z}, 2)$. Therefore, we have,

$$\zeta \in [X, BPU(\mathcal{H})] = [X, K(\mathbb{Z}, 3)] = H^3(X; \mathbb{Z})$$

which implies that the twists above are parametrized by $H^3(X;\mathbb{Z})$. Our definition gives twisted K-groups $K^\zeta(X)$ for $\zeta \in H^3(X;\mathbb{Z})$. We may extend this action of $PU(\mathcal{H})$ to the other spaces of the K-theory spectrum to form the groups $K^{i,\zeta}(X)$ for $i \in \mathbb{Z}$.

Atiyah and Segal ([4]) extend this definition one step further. They consider instead of principal $PU(\mathcal{H})$-bundles, the associated bundle of infinite projective spaces with fiber $\mathcal{P}(\mathcal{H})$. Thus, a twist in the above sense is given by a bundle of infinite projective spaces $P \to X$. Then they extend the set of twists to the case of a bundle of projective spaces with involution, that splits each fiber P_x as a disjoint union of two projective spaces P_x^+ and P_x^-. These twists are now classified by $H^1(X;\mathbb{Z}/2) \oplus H^3(X;\mathbb{Z})$ with the first factor coming from 1-cocycles of involutions. They define for $\eta, \zeta \in H^1(X;\mathbb{Z}/2) \oplus H^3(X;\mathbb{Z})$ twisted K-groups $K^{i,\eta,\zeta}(X)$ analogous to the above situation.

3 Generalized Thom spectra

This section discusses the general framework for twisted homology theories. Given a commutative ring spectrum R, one defines the infinite loop space of units GL_1R. The space GL_1R may be identified with the invertible endomorphisms of R as a module over itself. Given a 1-cocycle over a space X with values in GL_1R that is a map $X \to BGL_1R$, one may define associated twisted R-homology and cohomology groups. The homology groups are computed as the homotopy groups of an R-module Thom spectrum, and the cohomology groups are homotopy groups of a mapping spectrum. We also discuss some examples and relate to the examples of the previous section.

3.1 Thom spectra from cocycles valued in the space of units

Let R be a commutative ring spectrum. The space of units, GL_1R generalizes the idea of the group of units of a ring. It is an H-space representing the functor which takes X to the group of units in $R^0(X)$, that is, it satisfies the universal property that $[X, GL_1R] = R^0(X)^\times$. We define

Definition 3.1 Let R be an E_∞-ring spectrum. Define $GL_1(R) \subset \Omega^\infty(R)$

as the pullback

$$GL_1(R) \longrightarrow \Omega^\infty(R) \qquad\qquad (3.1)$$

$$\downarrow \qquad\qquad\qquad \downarrow$$

$$\pi_0(R)^\times \longrightarrow \pi_0(R).$$

From the pullback square above, we see that $[X, GL_1 R]$ equals the elements in $[X, \Omega^\infty R](= R^0(X))$ which give units in $\pi_0 R$ on restriction to the 0-skeleton. This is precisely the set of units in $R^0(X)$, thus verifying the stated universal property. Substituting S^n for X one obtains

$$[S^n, GL_1(R)] = R^0(S^n)^\times.$$

Since R is a cohomology theory, $R^*(S^n) = R^*[\epsilon]/(\epsilon^2)$ where $deg(\epsilon) = n$ for $n > 0$. So for $n > 0$,

$$[S^n, GL_1(R)]_* = \widetilde{R}^0(S^n)^\times = 1 + \epsilon\pi_n(R)$$

and for $n = 0$, $R^0(S^0) = R^0 \times R^0$. So, $\pi_0(GL_1(R)) = \widetilde{R}^0(S^0)^\times = R^{0\times}$. Therefore, the homotopy groups of the space $GL_1(R)$ are given by

$$\pi_n(GL_1(R)) = \begin{cases} \pi_n(R) & \text{if } n > 0 \\ \pi_0(R)^\times & \text{if } n = 0. \end{cases}$$

Note that all the above definitions go through as long as R is an E_2-ring spectrum. If R is E_∞, then the space $\Omega^\infty R$ is an E_∞-ring space (see [15]). With respect to the multiplicative structure on $\Omega^\infty R$ and $\pi_0 R$ the pullback (3.1) is a pullback of E_∞-spaces. Hence, the space $GL_1 R$ is an E_∞-space and it follows that there is a spectrum $gl_1(R)$ whose 0th-space is $GL_1 R$.

Now we proceed to the definition of the Thom spectrum. The construction resembles a twisted version of the group ring. Given an extension τ

$$(\tau): 1 \to k^* \to E \to G \to 1$$

for a group G and a field k, one may form the twisted group ring $k^\tau[G] = \mathbb{Z}[E] \otimes_{\mathbb{Z}[k^*]} k$. If the extension τ is trivial, then $E = k^* \times G$, and $k^\tau[G]$ is the group ring $k[G]$. Imitating this definition for spectra leads to the construction of the Thom spectrum. In the case of spectra extensions such as τ are replaced by $GL_1 R$-fibrations induced by maps $X \to BGL_1 R$.

Definition 3.2 Let X be a space and let ζ be a map from X to $BGL_1(R)$. Let P be the $GL_1(R)$ bundle associated to the map ζ described by the homotopy pullback[4]

$$
\begin{array}{ccc}
P & \longrightarrow & EGL_1(R) \\
\downarrow & & \downarrow \\
X & \longrightarrow & BGL_1(R)
\end{array}
$$

and define the Thom spectrum X^ζ as

$$X^\zeta = \Sigma^\infty P_+ \wedge^L_{\Sigma^\infty GL_1(R)_+} R.$$

The Thom spectrum is a functor from the category of spaces over $BGL_1 R$ to the category of R-module spectra ([6]). We deduce from [3], which writes the functor as a colimit, that the Thom spectrum functor commutes with colimits and homotopy colimits.

3.2 Twisted homology theories

We may define twisted homology theories using Thom spectra as below.

Definition 3.3 The group $[X, BGL_1 R]$ for a commutative ring spectrum R is defined to be the set of R-twistings on a space X. For $\zeta \in [X, BGL_1 R]$, the ζ-twisted R-homology is defined as

$$R^\zeta_k(X) = \pi_0 Map_{R-mod}(\Sigma^k R, X^\zeta) \cong \pi_k X^\zeta$$

and the ζ-twisted R-cohomology is defined as

$$R^k_\zeta(X) = \pi_0 Map_{R-mod}(X^\zeta, \Sigma^k R).$$

Suppose that R is an Ω-spectrum given by the sequence of spaces $\{R_n\}$ and the $GL_1 R$-action on R induces a $GL_1 R$-action on each of the spaces R_n. Then, one may identify the ζ-twisted R-cohomology as

$$
\begin{aligned}
Map_{R-mod}(X^\zeta, \Sigma^k R) &\simeq Map_{R-mod}(\Sigma^\infty P_+ \wedge^L_{\Sigma^\infty GL_1(R)_+} R, \Sigma^k R) \\
&\simeq Map_{\Sigma^\infty GL_1 R_+ - mod}(\Sigma^\infty P_+, \Sigma^k R) \\
&\simeq Map_{GL_1 R}(P, R_k).
\end{aligned}
$$

The space $Map_{GL_1 R}(P, R_k)$ can in turn be identified with the space of lifts

$$
\begin{array}{ccc}
 & & EGL_1 R \times_{GL_1 R} R_k \\
 & \nearrow & \downarrow \\
X & \xrightarrow{\quad \zeta \quad} & BGL_1 R.
\end{array}
$$

---

[4]One notes that for E_∞-spaces G, there exists a quasi-fibration $G \to EG \to BG$.

Therefore, the group $R^k_\zeta(X)$ may be identified with the vertical homotopy classes of sections above.

Let $R = HA$, the Eilenberg MacLane spectrum for a commutative ring A. Then $\Omega^\infty HA$ is the discrete space A and so the space $GL_1 HA$ is the discrete space $A^\times$. It follows that the spectrum $gl_1 HA$ is $HA^\times$. A map $f : X \to BGL_1 HA \simeq BA^\times$ is equivalent to a homomorphism $\hat{f} : \pi_1 X \to A^\times$. Via the action of $A^\times$ on A this induces a local coefficient $\mathcal{L}_f$. Then the above identifications imply

$$HA^f_k(X) \cong H_k(X; \mathcal{L}_f), \quad HA^k_f(X) \cong H^k(X; \mathcal{L}_f).$$

Therefore, for $R = HA$ twisted R-homology and cohomology recovers respectively the homology and cohomology with local coefficients.

Next consider $R = KU$ the complex K-theory spectrum. For the units we have

$$\pi_n(GL_1 KU) = \begin{cases} \mathbb{Z} & \text{if } n = 2k > 0 \\ \mathbb{Z}/2 & \text{if } n = 0 \\ 0 & \text{otherwise.} \end{cases}$$

The units spectrum $gl_1 KU$-theory splits off into two wedge summands, one part having the homotopy groups up to dimension 2 and the other part having the homotopy groups in dimension ≥ 3. It follows that as a space $GL_1 KU \simeq \mathbb{Z}/2 \times K(\mathbb{Z}, 2) \times BSU$ and as a space $BGL_1 KU \simeq B\mathbb{Z}/2 \times K(\mathbb{Z}, 3) \times B^2 SU$ (see [17], section 22.3). For the splitting at odd primes also see [5], section 4. Given an element in $(\zeta, \eta) : H^1(X; \mathbb{Z}/2) \oplus H^3(X; \mathbb{Z})$ we obtain a map $X \to B\mathbb{Z}/2 \times K(\mathbb{Z}, 3)$ which we call $f_{\zeta,\eta}$. Composing with the inclusion $B\mathbb{Z}/2 \times K(\mathbb{Z}, 3) \to BGL_1 KU$ yields a map $\widehat{(\zeta, \eta)} : X \to BGL_1 KU$. The twisted K-groups of Atiyah and Segal are computed as

$$KU^r_{\widehat{(\zeta,\eta)}}(X) \cong K^{r,\zeta,\eta}(X).$$

3.3   Examples of certain Thom spectra

We construct some quotient modules as Thom spectra. The results of this section appear in sections 2.1 and 2.2 of [5]. We repeat the proofs here to introduce the basic maneuvers to be used in later sections.

Note that the Thom spectrum over the trivial inclusion $* \to BGL_1 R$ is weakly equivalent to R. If the map $\zeta : X \to BGL_1 R$ is null-homotopic, one readily observes that the pullback $P \simeq X \times GL_1 R$. It follows that

$$X^\zeta \simeq \Sigma^\infty P_+ \wedge^L_{\Sigma^\infty GL_1(R)_+} R \simeq (\Sigma^\infty X_+ \wedge \Sigma^\infty GL_1 R_+) \wedge^L_{\Sigma^\infty GL_1(R)_+} R \simeq R \wedge X_+.$$

The spectra $R \wedge X_+$ play the role of free R-modules. Next we consider Thom spectra over suspensions. Assume that $X \simeq \Sigma Y$, so that a map

$\zeta : X \to BGL_1R$ induces a unit $u_\zeta \in [Y, GL_1R] \cong R^0(Y)^\times$. The unit gives a map of R-modules $R \wedge Y_+ \to R$ which we also denote by u_ζ. In terms of this notation,

Proposition 3.4 *The Thom spectrum X^ζ is equivalent to the homotopy pushout of*

$$(R \xleftarrow{p} R \wedge Y_+ \xrightarrow{u_\zeta} R)$$

where the map p is induced by the projection $Y_+ \to S^0$.

Proof The space X is the union of two copies of CY, the cone on Y, whose intersection is Y, inducing the following homotopy pushout square of Thom spectra

$$\begin{array}{ccc} Y^\zeta & \longrightarrow & CY^\zeta \\ \downarrow & & \downarrow \\ CY^\zeta & \longrightarrow & (\Sigma Y)^\zeta. \end{array}$$

Since CY is contractible, $CY^\zeta \simeq R$. The map ζ restricted to Y factors through CY. Therefore, it is also trivial and $Y^\zeta \simeq R \wedge Y_+$. Hence, we obtain the following homotopy pushout

$$\begin{array}{ccc} R \wedge Y_+ & \longrightarrow & R \\ \downarrow & & \downarrow \\ R & \longrightarrow & (\Sigma Y)^\zeta. \end{array}$$

In order to compute the two maps $R \wedge Y_+ \to R$ in the diagram, one goes back to the definition of the Thom spectrum. The GL_1R-fibration over ΣY is equivalent to two copies of $GL_1R \times CY$ identified over Y via the map $u_\zeta : Y \to GL_1R$. Therefore, in appropriate notation, one of the maps in the diagram above is p and the other u_ζ.

$\square$

Now we use the above Proposition to describe a couple of explicit computations which will be used throughout the paper.

Proposition 3.5 *Let $X = S^1$, $R = K_p^\wedge$, the spectrum of p-adic K-theory, and $\zeta : S^1 \to BGL_1K_p^\wedge$ be given by $1 - p \in \pi_1 BGL_1(K_p^\wedge) = \pi_0 GL_1(K_p^\wedge) = \mathbb{Z}_p^\times$. Then $X^\zeta \simeq K/p$.*

Proof The circle, S^1, is the suspension of S^0. So, as in Proposition 3.4 one has a homotopy pushout of the Thom spectra

$$
\begin{array}{ccc}
K_p^\wedge \wedge S_+^0 & \longrightarrow & K_p^\wedge \\
\downarrow & & \downarrow \\
K_p^\wedge & \longrightarrow & (S^1)^\varsigma.
\end{array}
$$

The spectrum $K_p^\wedge \wedge S_+^0 \simeq K_p^\wedge \vee K_p^\wedge$ as a $K_p^\wedge$-module. Therefore, the homotopy pushout above writes $(S^1)^\varsigma$ as a homotopy cofiber of the map

$$K_p^\wedge \vee K_p^\wedge \to K_p^\wedge \vee K_p^\wedge$$

represented by the matrix

$$
\begin{pmatrix}
1 & 1 \\
1 & 1-p
\end{pmatrix}.
$$

This may be rewritten as the homotopy cofiber of the map $K_p^\wedge \xrightarrow{\times p} K_p^\wedge$. Therefore, $(S^1)^\varsigma \simeq K_p^\wedge/p = K/p$.

$\square$

Remark 3.6 In fact, the proof above may be carried out for any R. If $\zeta \in [S^1, BGL_1 R] = \pi_0 R^\times$, we deduce that $(S^1)^\varsigma$ is the cofiber of $(1 - \zeta)$ from R to itself.

Another case of interest for us is the 3-sphere.

Proposition 3.7 *Let* $X = S^3$, *and* $\zeta : S^3 \to BGL_1 K_p^\wedge$ *be given by* $\beta p \in \pi_3(BGL_1 K_p^\wedge) = \pi_2 GL_1 K_p^\wedge = \mathbb{Z}_p\{\beta\}$ (β *stands for the Bott element*). *Then,* $(S^3)^\varsigma \simeq K/p$.

Proof The space S^3 is the suspension on the space S^2, so once again by Proposition 3.4 one has a homotopy pushout

$$
\begin{array}{ccc}
K_p^\wedge \wedge S_+^2 & \longrightarrow & K_p^\wedge \\
\downarrow & & \downarrow \\
K_p^\wedge & \longrightarrow & (S^3)^\varsigma.
\end{array}
$$

Now we have $K_p^\wedge \wedge S_+^2 \simeq K_p^\wedge \vee \Sigma^2 K_p^\wedge$ and thus, one obtains that $(S^3)^\varsigma$ is the homotopy cofiber of

$$K_p^\wedge \vee \Sigma^2 K_p^\wedge \to K_p^\wedge \vee K_p^\wedge$$

represented by the matrix

$$\begin{pmatrix} 1 & 0 \\ 1 & \beta p \end{pmatrix}.$$

We readily deduce from the above that

$$(S^3)^\zeta \simeq cofiber(\Sigma^2 K_p^\wedge \xrightarrow{\beta p} K_p^\wedge) \simeq K/p.$$

$\square$

Remark 3.8 Computations such as the above fit into a general scheme of calculations of quotients as Thom spectra. In [6], it is proved that for a ring spectrum R whose homotopy is concentrated in even dimensions, and a sequence $\{x_1, \ldots, x_{n-1}\}$ with $x_i \in \pi_{2i}R$, there is a map $f : SU(n) \to BGL_1R$ such that the Thom spectrum $SU(n)^f \simeq$ the quotient $R/(x_1, \ldots, x_{n-1})$.

4 Ring structures and topological Hochschild homology

The Thom spectrum functor may be used to produce ring structures. In this section, we recall this formulation. We briefly outline how the topological Hochschild homology of these rings are computed as Thom spectra. Details and proofs appear in [6].

The R-module Thom spectrum induces a functor

$$\{spaces \ over \ BGL_1R\} \to \{R - modules\}.$$

For a commutative ring spectrum R, the category of R-modules has a symmetric monoidal structure induced by the smash product of R-modules (denoted by $\wedge_R$). The space of units, GL_1R, is an infinite loop space and hence, so is BGL_1R. Using either the category of I-spaces ([23]) or $*$-modules ([7], Section 4) one obtains a symmetric monoidal structure on the category of spaces in which the E_∞ spaces are commutative monoids. We fix such a structure $\otimes$ so that both GL_1R and BGL_1R are represented as commutative monoids. This induces a symmetric monoidal structure on $\{spaces \ over \ BGL_1R\}$ given by

$$(X \to BGL_1R, Y \to BGL_1R) \mapsto (X \otimes Y \to BGL_1R \otimes BGL_1R \to BGL_1R)$$

the latter map being the monoidal structure on BGL_1R. We have the following result from [6]

Theorem 4.1 *The R-module Thom spectrum functor is a symmetric monoidal functor.*

A consequence of the above Theorem is that once we have a loop map to BGL_1R, the induced Thom spectrum carries an A_∞-R-algebra structure. That is, if $\zeta : G \to BGL_1R$ is a map of loop spaces, then there is an induced A_∞ structure on the Thom spectrum. This produces a family of A_∞ structures on the Thom spectrum parametrized by the space of loop structures on the map ζ. A map f which loops down to ζ fits into the commutative diagram

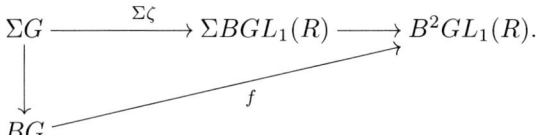

Example 4.2 Let $G = S^1$, $R = K_p^\wedge$ and $\zeta = 1 - p$ so that Proposition 3.5 implies that $(S^1)^\zeta \simeq K/p$. In this case, $BG \simeq \mathbb{C}P^\infty$. The space of extensions is the diagram

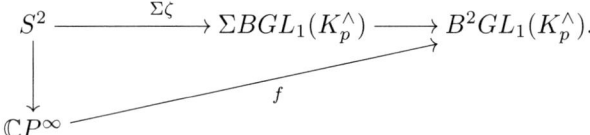

We tackle this extension problem by looking at the CW structure of $\mathbb{C}P^\infty$. It is made up of even dimensional cells, so that, all the attaching maps involve attaching cells along odd dimensional spheres. The spectrum, $K_p^\wedge$, has homotopy only in even dimensions, and hence, so does the space, $B^2GL_1K_p^\wedge$. Thus, all the obstructions to extending the map ζ must vanish. Therefore, we see that K/p is an A_∞ ring in the category of $K_p^\wedge$-modules. In fact, we get a family of ring structures parametrized by the above space of extensions.

Example 4.3 Let $G = S^3$, $R = K_p^\wedge$ and $\zeta = \beta p$ so that $(S^3)^\zeta \simeq K/p$ by Proposition 3.7. The classifying space of S^3 is the infinite quarternionic projective space, $\mathbb{H}P^\infty$, which has a cell in every dimension that is a multiple of 4. So, again there is no obstruction to extending $\Sigma\zeta$ to f in the

diagram

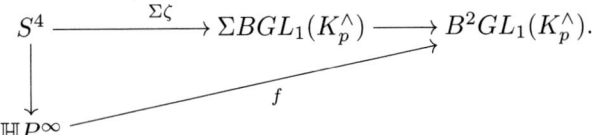

A curious fact about the A_∞-R-algebra structures obtained as Thom spectra of loop maps is that the topological Hochschild homology can be expressed as a Thom spectrum ([6]). For the sphere spectrum, this was proved in [7].

The Thom spectrum X^ζ of a map $\zeta : X \to BGL_1 R$ is a twisted R-module generated by X which becomes an analog of a twisted group ring when ζ is a loop map. The Hochschild Homology of group rings is given by the isomorphism

$$HH_*(k[G]) \cong k \otimes H_*(G, G^c)$$

where for the group homology of G^c, G acts on itself by conjugation.

The analog of Hochschild homology of algebras is topological Hochschild homology. For a monoid A, this is defined as the realization of the simplicial object given by the cyclic bar construction $B^{cy}(A)$. The cyclic bar construction $B^{cy}(A)$ may be defined for a monoid A in a symmetric monoidal category $(C, \otimes, \mathbb{1})$ as the simplicial object $B^{cy}(A)_n = A^{\otimes n+1}$ with face operators

$$d_i(a_0 \otimes a_1 \otimes \cdots \otimes a_n) \begin{cases} a_0 \otimes \cdots \otimes a_i a_{i+1} \otimes \cdots a_n & i = 0, 1, \cdots, n-1, \\ a_n a_0 \otimes \cdots a_{n-1} & i = n, \end{cases}$$

and degeneracy operators

$$s_i(a_0 \otimes a_1 \otimes \cdots \otimes a_n) = a_0 \otimes \cdots \otimes a_i \otimes \mathbb{1} \otimes a_{i+1} \otimes \cdots a_n \quad i = 0, 1, \cdots, n$$

One may use this to define the topological Hochschild homology of A_∞-ring spectra.

The analogous result for Hochschild homology of group rings in the case of topological Hochschild homology is the following result of Bökstedt and Waldhausen

$$THH(\Sigma^\infty \Omega X_+) \cong \Sigma^\infty LX_+$$

where LX is the free loop space of the space X given by $Map_*(S^1_+, X)$. The key idea governing the expressions above are the identifications $B^{cy}(M) \simeq LBM$ for group-like topological monoids M and $LBG \simeq EG \times_G G^c$ for topological groups G. In the latter identification, $EG \times_G G^c$ stands for the Borel construction on G with the group action of G acting on itself by conjugation whose homology computes the group homology $H_*(G, G^c)$.

The result of Bökstedt and Waldhausen for the category of R-modules is

$$THH^R(R \wedge \Omega X_+) \simeq R \wedge LX_+.$$

This is THH of the Thom spectrum of the constant map $* : \Omega X \to BGL_1R$ which is the Thom spectrum of the constant $* : LX \to B^2GL_1R$. In the general case the formula has the following form

$$THH^R(\Omega X^{\Omega \zeta}) \simeq LX^{L^\eta \zeta}.$$

The map $L^\eta \zeta$ is expressed as a composite of $L\zeta : LX \to LB^2GL_1R$ and a map $L^\eta : LB^2GL_1R \to BGL_1R$. We define the map L^η below.

Recall that there is a fibration

$$\Omega X \to LX \to X$$

where the right map is given by the evaluation at the basepoint of S^1. If X is a group-like H-space, then the fibration becomes trivial via a map α from $X \times \Omega X$ to LX given by the formula

$$\alpha(x, f)(t) = x.f(t).$$

Let η denote the Hopf map $S^3 \to S^2$. The Hopf map induces $\eta : B^2GL_1R \to BGL_1R$ as the composite

$$B^2GL_1R \simeq \Omega^2 B^4GL_1R \simeq Map_*(S^2, B^4GL_1R) \overset{\eta^*}{\to} Map_*(S^3, B^4GL_1R)$$
$$\simeq \Omega^3 B^4GL_1R$$
$$\simeq BGL_1R.$$

Now L^η is defined as the following composite using the H-space structure on the infinite loop space BGL_1R

$$LB^2GL_1(R) \simeq B^2GL_1(R) \times BGL_1(R) \overset{\eta \times id}{\to} BGL_1(R) \times BGL_1(R) \to BGL_1(R).$$

We have the Theorem

Theorem 4.4
$$THH^R(\Omega X^{\Omega \zeta}) = (LX)^{L^\eta \zeta}.$$

This was proved in the case $R = S^0$, the sphere spectrum in [7]. The general case is written out in [6]. We provide a sketch of the main idea below.

Sketch of proof: The topological Hochschild homology THH may be computed as the geometric realization of the cyclic bar construction[5]

$$THH(\Omega X^{\Omega \zeta}) \simeq \left| B^{cy}(\Omega X^{\Omega \zeta}) \right|.$$

The map $\Omega \zeta : \Omega X \to BGL_1 R$ is a map of loop spaces (and hence a map of monoids by using the I-space monoidal structure or $*$-modules). Hence, we obtain a map $B^{cy}(\Omega \zeta)$ from $B^{cy}(\Omega X)$ to $B^{cy}(BGL_1 R)$. Since the Thom spectrum functor is symmetric monoidal (Theorem 4.1) we have

$$
\begin{aligned}
THH^R(\Omega X^{\Omega \zeta}) &\simeq \left| B^{cy}(\Omega X^{\Omega \zeta}) \right| \\
&\simeq \left| Th(B^{cy}(\Omega \zeta)) \right| \\
&\simeq Th(\left| B^{cy}(\Omega \zeta) \right|).
\end{aligned}
$$

The map $\left| B^{cy}(\Omega \zeta) \right|$ is the composite

$$\left| B^{cy}(\Omega X) \right| \to \left| B^{cy}(BGL_1 R) \right| \to BGL_1 R$$

where the latter map $\left| B^{cy}(BGL_1 R) \right| \to BGL_1 R$ is a map of simplicial spaces induced by level wise multiplication using the H-space structure of $BGL_1 R$. Therefore, the theorem reduces to identifying the map $\left| B^{cy}(\Omega \zeta) \right|$ as $L^\eta \zeta$ or equivalently the map $B^{cy}(BGL_1 R) \to BGL_1 R$ as L^η. This is the content of Theorem 1.2 of [24].

5 Computations

In this section, we consider the Thom spectrum of $\zeta : S^3 \to BGL_1 K_p^\wedge$ defined by $\zeta = \beta p$ modulo the following identifications

$$[S^3, BGL_1(K_p^\wedge)] = \pi_3(BGL_1(K_p^\wedge)) = \pi_2(GL_1(K_p^\wedge)) = \pi_2(K_p^\wedge) = \mathbb{Z}_p\{\beta\}.$$

From Proposition 3.7 we know $(S^3)^\zeta \simeq K/p$, the spectrum for *mod p K*-theory.

As observed in Example 4.3, the map ζ above can always be written as a loop map. So, we get a space of A_∞ structures on K/p as a $K_p^\wedge$ module parametrized by the space of loop structures on ζ. This is the space of extensions

$$
\begin{array}{ccc}
S^4 & \xrightarrow{\Sigma \zeta} & \Sigma BGL_1 K_p^\wedge \longrightarrow B^2 GL_1 K_p^\wedge. \\
\downarrow & \overset{\hat{\zeta}}{\nearrow} & \\
\mathbb{H}P^\infty & &
\end{array}
\tag{5.1}
$$

---

[5]This is true only under certain hypothesis on the associated spectrum. Such technical details are completely checked in [6]. Essentially one needs to check similar technical criteria throughout the proof.

Our goal is to compute $THH^{K_p^\wedge}(K/p)$ for these A_∞ structures. Recall from Theorem 4.4 that for these A_∞-structures, $THH^{K_p^\wedge}(K/p)$ is equivalent to the Thom spectrum of a map from $L(\mathbb{H}P^\infty) \to BGL_1K_p^\wedge$. Since, η maps to 0 under the unit $S^0 \to K_p^\wedge$, we may write this map as $\widehat{L\zeta}$ in the following diagram

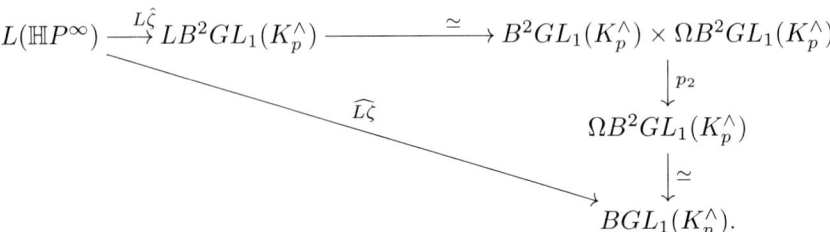

For the structure of $L\mathbb{H}P^\infty \simeq LBS^3$, we have the following well-known proposition

Proposition 5.1 $LBG \simeq G_{hG} = EG \times_G G^c$, where G^c denotes the space G acting on itself by conjugation.

Using the proposition, we deduce the following structure for $L\mathbb{H}P^\infty$.

Proposition 5.2 The space $L\mathbb{H}P^\infty$ may be expressed as a homotopy push-out of the diagram

$$\mathbb{H}P^\infty \xleftarrow{q} \mathbb{C}P^\infty \xrightarrow{q} \mathbb{H}P^\infty$$

where the maps $q : \mathbb{C}P^\infty \to \mathbb{H}P^\infty$ are the canonical quotient induced by the inclusion $S^1 \to S^3$ on classifying spaces.

Proof From Proposition 5.1, we realise that $L\mathbb{H}P^\infty \simeq L(BS^3) \simeq ES^3 \times_{S^3} S^3$ where S^3 acts on itself by conjugation. Let us study the various orbits of this action. Identifying S^3 with the unit quarternions, which gives S^3 the group structure, we see that 1 and -1 are the only points which commute with all of S^3. The orbits of these points are points, and the centralizer is all of S^3. The latitudes $(\alpha, *, *, *)$ are preserved under conjugation and hence these form the other orbits. In this case, the centralizer of these points is S^1.

Let U_+ and U_- be respectively the upper and lower hemispheres of S^3. Thus, we have the following S^3-equivariant pushout

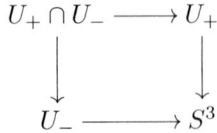

which is also a homotopy pushout. Taking homotopy coinvariants, we get a homotopy pushout

$$ES^3 \times_{S^3} (U_+ \cap U_-) \longrightarrow ES^3 \times_{S^3} U_+ \qquad (5.2)$$

$$ES^3 \times_{S^3} U_- \longrightarrow ES^3 \times_{S^3} S^3.$$

The spaces U_+ and U_- equivariantly deformation retracts to the north pole and south pole respectively and the space $U_+ \cap U_-$ deforms to S^2. So we have the equivalent homotopy pushout

$$ES^3 \times_{S^3} S^2 \longrightarrow ES^3 \times_{S^3} pt$$

$$ES^3 \times_{S^3} pt \longrightarrow ES^3 \times_{S^3} S^3.$$

In the above diagram we make the following identifications

$$ES^3 \times_{S^3} * \simeq BS^3 \simeq \mathbb{H}P^\infty$$

and,

$$ES^3 \times_{S^3} S^2 \simeq ES^3 \times_{S^3} S^3/S^1$$
$$\simeq ES^3 \times_{S^1} *$$
$$\simeq BS^1$$
$$\simeq \mathbb{C}P^\infty.$$

Therefore, we have the following homotopy pushout

$$\mathbb{C}P^\infty \longrightarrow \mathbb{H}P^\infty$$

$$\mathbb{H}P^\infty \longrightarrow L\mathbb{H}P^\infty$$

and the maps $\mathbb{C}P^\infty \to \mathbb{H}P^\infty$ are as stated. $\qquad \square$

Proposition 5.2 leads to the following result for

$$THH^{K_p^\wedge}(K/p) \simeq (L\mathbb{H}P^\infty)^{\widehat{L\zeta}}.$$

Proposition 5.3 *The spectrum* $(L\mathbb{H}P^\infty)^{\widehat{L\zeta}}$ *may be expressed as the homotopy cofiber*

$$K_p^\wedge \wedge \mathbb{C}P_+^\infty \to K_p^\wedge \wedge \mathbb{H}P_+^\infty \vee K_p^\wedge \wedge \mathbb{H}P_+^\infty$$

which implies the long exact sequence

$$\cdots \to K_{p\,*}^\wedge(\mathbb{C}P^\infty) \to K_{p\,*}^\wedge(\mathbb{H}P^\infty) \oplus K_{p\,*}^\wedge(\mathbb{H}P^\infty) \to \pi_*$$
$$(THH^{K_p^\wedge}(K/p)) \to \cdots$$

Proof Proposition 5.2 writes $L\mathbb{H}P^\infty$ as a homotopy pushout. Since the Thom spectrum functor respects homotopy colimits we have the following homotopy pushout

$$
\begin{array}{ccc}
(\mathbb{C}P^\infty)^{res(\widehat{L\zeta})} & \longrightarrow & (\mathbb{H}P^\infty)^{res(\widehat{L\zeta})} \\
\downarrow & & \downarrow \\
(\mathbb{H}P^\infty)^{res(\widehat{L\zeta})} & \longrightarrow & THH^{K_p^\wedge}(K/p).
\end{array}
$$

We want to understand this pushout square. The maps $\mathbb{H}P^\infty \to L\mathbb{H}P^\infty$ are induced by the inclusion of the constant loops. On composition with

$$L\mathbb{H}P^\infty \xrightarrow{\widehat{\zeta}} LB^2GL_1K_p^\wedge \to B^2GL_1K_p^\wedge \times BGL_1K_p^\wedge \xrightarrow{p_2} BGL_1K_p^\wedge$$

the map becomes trivial. Hence, the Thom spectrum $(\mathbb{H}P^\infty)^{res(\widehat{L\zeta})} \simeq K_p^\wedge \wedge \mathbb{H}P_+^\infty$. Since the map $\mathbb{C}P^\infty \to BGL_1K_p^\wedge$ factors through $\mathbb{H}P^\infty$, $res(\widehat{L\zeta})$ is also trivial on $\mathbb{C}P^\infty$. Thus, $(\mathbb{C}P^\infty)^{res(\widehat{L\zeta})} \simeq K_p^\wedge \wedge \mathbb{C}P_+^\infty$. Therefore, we obtain the homotopy pushout

$$
\begin{array}{ccc}
K_p^\wedge \wedge \mathbb{C}P_+^\infty & \longrightarrow & K_p^\wedge \wedge \mathbb{H}P_+^\infty \\
\downarrow & & \downarrow \\
K_p^\wedge \wedge \mathbb{H}P_+^\infty & \longrightarrow & THH^{K_p^\wedge}(K/p).
\end{array}
$$

On homotopy groups, this leads to the Mayer Vietoris sequence

$$\cdots \to K_{p\,*}^\wedge(\mathbb{C}P^\infty) \to K_{p\,*}^\wedge(\mathbb{H}P^\infty) \oplus K_{p\,*}^\wedge(\mathbb{H}P^\infty) \to \pi_*(THH^{K_p^\wedge}(K/p)) \to \cdots$$

$\square$

In order to compute THH, we need to identify the map

$$K_p^{\wedge} \wedge \mathbb{C}P_+^{\infty} \to K_p^{\wedge} \wedge \mathbb{H}P_+^{\infty} \vee K_p^{\wedge} \wedge \mathbb{H}P_+^{\infty}$$

in Proposition 5.3. We do this in the Proposition below

Proposition 5.4 *The two maps $K_p^{\wedge} \wedge \mathbb{C}P_+^{\infty} \to K_p^{\wedge} \wedge \mathbb{H}P_+^{\infty}$ may be expressed as q and $q \circ u^*$ for some choice of coordinates, where u^* is the map $K_p^{\wedge} \wedge \mathbb{C}P_+^{\infty} \to K_p^{\wedge} \wedge \mathbb{C}P_+^{\infty}$ induced by a unit $u \in K_p^{\wedge 0}(\mathbb{C}P^{\infty})^{\times}$.*

Moreover, the adjoint of the map $u : \Sigma \mathbb{C}P_+^{\infty} \to BGL_1R$, is homotopy equivalent to the composite

$$\Sigma^2 \mathbb{C}P_+^{\infty} \simeq S^2 \wedge \mathbb{C}P_+^{\infty} \overset{\mu}{\to} \mathbb{C}P^{\infty} \overset{res(\hat{\zeta})}{\to} B^2 GL_1 K_p^{\wedge}$$

where μ is the composition $S^2 \wedge \mathbb{C}P_+^{\infty} \to \mathbb{C}P^{\infty} \wedge \mathbb{C}P_+^{\infty} \to \mathbb{C}P^{\infty}$[6].

Proof Consider the following diagram

$$\mathbb{C}P^{\infty} \longrightarrow \mathbb{H}P^{\infty} \vee \mathbb{H}P^{\infty} \longrightarrow L\mathbb{H}P^{\infty} \longrightarrow \Sigma \mathbb{C}P^{\infty} \qquad (5.3)$$

with maps $*$, $\widehat{L\zeta}$, u to $BGL_1 K_p^{\wedge}$.

The top row is the Mayer Vietoris sequence obtained out of the homotopy pushout square above for $L\mathbb{H}P^{\infty}$. The maps $\mathbb{H}P^{\infty} \to L\mathbb{H}P^{\infty}$ are the inclusion of the constant loops, thus the maps $\mathbb{H}P^{\infty} \to BGL_1 K_p^{\wedge}$ are null-homotopic. Therefore, the map $\widehat{L\zeta} : L\mathbb{H}P^{\infty} \to BGL_1 K_p^{\wedge}$ factors through $\Sigma \mathbb{C}P^{\infty}$ which gives us a unit in $u \in K_p^{\wedge 0}(\mathbb{C}P^{\infty})$.

We use the above unit to write down the map in question for the Mayer Vietoris sequence. The map $q : \mathbb{C}P^{\infty} \to \mathbb{H}P^{\infty}$ is obtained by applying the classifying space functor B to the map of groups $S^1 \to S^3$. Now $L\mathbb{C}P^{\infty} \simeq S^1 \times \mathbb{C}P^{\infty}$ and by dividing the circle into upper and lower hemispheres we obtain a pushout square

$$\mathbb{C}P^{\infty} \sqcup \mathbb{C}P^{\infty} \simeq \mathbb{C}P^{\infty} \times U_+ \cap U_- \longrightarrow \mathbb{C}P^{\infty} \times U_+ \simeq \mathbb{C}P^{\infty} \qquad (5.4)$$

$$\mathbb{C}P^{\infty} \simeq \mathbb{C}P^{\infty} \times U_- \longrightarrow L\mathbb{C}P^{\infty}.$$

---

[6]The map is induced by multiplication on $\mathbb{C}P^{\infty}$. In terms of the identification $\mathbb{C}P^{\infty} \simeq K(\mathbb{Z}, 2)$, μ is represented by the cohomology class $\epsilon \otimes 1 + 1 \otimes x$ in $H^2(S^2 \times \mathbb{C}P^{\infty})$.

Note that the map of groups $S^1 \to S^3$ induces a map from the pushout (5.4) to (5.2). This allows us to extend the diagram (5.3) to

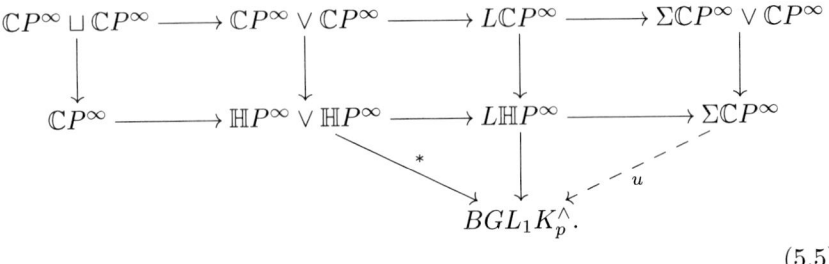

$$(5.5)$$

We pass to the $K_p^\wedge$ Thom spectra obtained by the composition of the associated maps to $BGL_1 K_p^\wedge$ and obtain the following diagram of Mayer Vietoris sequences

$$\cdots \longrightarrow K_{p*}^\wedge \mathbb{C}P^\infty \oplus K_{p*}^\wedge \mathbb{C}P^\infty \longrightarrow K_{p*}^\wedge \mathbb{C}P^\infty \oplus K_{p*}^\wedge \mathbb{C}P^\infty \longrightarrow \pi_*(L\mathbb{C}P^{\infty(L\hat{\zeta}\circ q)}) \longrightarrow \cdots$$

$$\cdots \longrightarrow K_{p*}^\wedge \mathbb{C}P^\infty \longrightarrow K_{p*}^\wedge \mathbb{H}P^\infty \oplus K_{p*}^\wedge \mathbb{H}P^\infty \longrightarrow \pi_*(L\mathbb{H}P^{\infty L\hat{\zeta}}) \longrightarrow \cdots$$

We are interested in computing the horizontal bottom left arrow. The left vertical arrow is induced by the fold map and the middle vertical arrow is $q_* \oplus q_*$. Now if we delete the middle row of the diagram (5.5) the diagram matches the analogous diagram in Proposition 3.3 of [5] computing the Thom spectrum $(L\mathbb{C}P^\infty)^{\hat{f}}$. As in the same proposition we deduce that the map $K_{p*}^\wedge \mathbb{C}P^\infty \oplus K_{p*}^\wedge \mathbb{C}P^\infty \to K_{p*}^\wedge \mathbb{C}P^\infty \oplus K_{p*}^\wedge \mathbb{C}P^\infty$ is given by the matrix

$$\begin{pmatrix} 1 & u \\ u & 1 \end{pmatrix}.$$

Hence, we deduce that the map $K_{p*}^\wedge \mathbb{C}P^\infty \to K_{p*}^\wedge \mathbb{H}P^\infty$ is given by $x \mapsto (q_*(x), q_*(u.x))$. Therefore, it remains to compute the unit u. From Proposition 3.4 of [5] we note that this may be computed as

$$(5.6)$$

$$
\begin{array}{ccc}
S^2 \wedge \mathbb{C}P_+^\infty & & \\
\downarrow{\scriptstyle \mu} & \searrow{\scriptstyle u} & \\
\mathbb{C}P_+^\infty & \xrightarrow{\hat{\zeta}\circ q} & B^2 GL_1 K_p^\wedge
\end{array}
$$

where $\mu =$ the multiplication induced by $S^2 \wedge \mathbb{C}P_+^\infty \to \mathbb{C}P^\infty \wedge \mathbb{C}P_+^\infty \to \mathbb{C}P^\infty$. This completes the proof of the Proposition. $\qquad\square$

Now we proceed towards the computation of THH using the Propositions above. Recall the structure of $gl_1 K_p^\wedge$ from Proposition 4.4 of [5]. For $p > 2$ we have the following split cofiber sequence

$$H\mu_{p-1} \vee \Sigma^2 H\mathbb{Z}_p \to gl_1(K_p^\wedge) \to K_p(\widehat{2}) \qquad (5.7)$$

where μ_{p-1} is the cyclic group of order $p - 1$ and the spectrum $K_p(\widehat{2})$ has the same homotopy groups as $K_p^\wedge$ in all dimensions other than 2 where $\pi_2 K_p(\widehat{2}) = 0$. We use this to calculate our possible A_∞ extensions. These are given by the diagram (5.1). Thus, we represent such elements as homotopy classes of maps in $[\mathbb{H}P^\infty, B^2 GL_1 K_p^\wedge]$. From (5.7) the space $B^2 GL_1 K_p^\wedge$ splits as follows

$$B^2 GL_1(K_p^\wedge) \simeq K(\mu_{p-1}, 2) \times K(\mathbb{Z}_p, 4) \times \Omega^\infty \Sigma^2 K_p(\widehat{2}).$$

The condition on the extensions is given by the restriction to S^4. These are elements in π_4. Note that

$$\pi_4(K(\mu_{p-1}, 2)) = \pi_2(\mu_{p-1}) = 0$$
$$\pi_4(\Omega^\infty \Sigma^2 K_p(\widehat{2})) = \pi_2(K_p(\widehat{2})) = 0$$
$$\pi_4(K(\mathbb{Z}_p, 4)) = \mathbb{Z}_p.$$

Thus, S^4 maps non trivially only to the factor $K(\mathbb{Z}_p, 4)$. Therefore, the condition implies that the map restricted from $\mathbb{H}P^\infty$ to S^4 must be equal to p. Since $H\mathbb{Z}_p^4(\mathbb{H}P^\infty) \to H\mathbb{Z}_p^4(S^4)$ is an isomorphism, the extension takes a fixed value on this factor. On the other factors there are no conditions on the extensions. Observe that

$$[\mathbb{H}P^\infty, K(\mu_{p-1}, 2)] = H\mu_{p-1}^2(\mathbb{H}P^\infty) = 0.$$

Hence, this factor does not contribute to our extensions. The factor $K_p(\widehat{2})$ does contribute, however, in this paper we assume for simplicity that the extension is 0 on that factor as well.

We compute THH for this extension by calculating the unit $u \in K_p^{\wedge 0}(\mathbb{C}P^\infty)$. The map $\mathbb{C}P^\infty \to B^2 GL_1(K_p^\wedge)$ is non trivial only on $\Sigma^4 H\mathbb{Z}_p$ where it equals p (actually px^2 where x denotes the generator of $H^2\mathbb{C}P^\infty$). We calculate this using the diagram (5.6) so that it remains only to compute u in the diagram

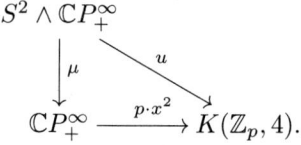

Now $\mu^*(x) = \epsilon \otimes 1 + 1 \otimes x$ (ϵ denotes the generator of $H^2 S^2$) so that $\mu^*(x^2) = 2\epsilon \otimes x$. Via the suspension isomorphism, the unit u corresponds to the class $2p$. Recall that the map $\Sigma^2 H\mathbb{Z}_p \to gl_1 K_p^\wedge$ is given by the inclusion of line bundles in the space of units. The group $H^2(X)$ is also the group of complex line bundles on X, which are units in $K^0(X)$. Therefore, the unit u corresponds to the K-theory class $(1+x)^{2p} \in K_p^{\wedge 0}(CP^\infty)$. This looks like $1 + 2px + \cdots$. The coefficients of each of the terms up to x^{p-1} is divisible by p and that of x^p is a unit. Then we may change the generator x of the K-theory ring of CP^∞ so that the unit looks like $1 + px + x^p$.

Now we compute (by using the 2-periodicity of $K_p^\wedge$, we put all the elements in degree 0, for which the equations below actually take place in degree 0)

$$K_p^\wedge{}_* CP^\infty \overset{q_*(u\cdot)}{\to} K_p^\wedge{}_* \mathbb{H}P^\infty.$$

Write

$$K_p^\wedge{}_*(\mathbb{C}P^\infty) = K_p^\wedge{}_* \{\beta_0, \beta_1, \ldots\}$$

where β_i is dual to x^i. Then,

$$\langle u(\beta_i), x^j \rangle = \langle \beta_i, x^j(1 + px + x^p) \rangle$$

$$= \begin{cases} 1 & \text{if } j = i, i - p \\ p & \text{if } j = i - 1 \\ 0 & \text{otherwise} \end{cases}$$

$$\implies u(\beta_i) = \begin{cases} \beta_0 & \text{if } i = 0 \\ \beta_i + p\beta_{i-1} & \text{if } 1 \le i \le p - 1 \\ \beta_i + p\beta_{i-1} + \beta_{i-p} & \text{if } i \ge p. \end{cases}$$

We finish the calculation using the Mayer Vietoris sequence

$$K_p^\wedge{}_*(\mathbb{C}P^\infty) \to K_p^\wedge{}_*(\mathbb{H}P^\infty) \oplus K_p^\wedge{}_*(\mathbb{H}P^\infty) \to \pi_*(THH^{K_p^\wedge}(K/p)).$$

The K-theory of $\mathbb{H}P^\infty$ is a free $K_p^\wedge{}_*$ on infinitely many generators a_i. The $K_p^\wedge$-theoretic Atiyah-Hirzebruch spectral sequence for $\mathbb{H}P^\infty$ collapses for dimension reasons. $H_*(\mathbb{H}P^\infty)$ is nonzero only in dimensions which are multiples of 4, and in each of these dimensions the group $\mathbb{Z}\alpha_i$ for some generators α_i. The elements a_i can be chosen to be the $\beta^{-2i}\alpha_i$ (where α_i is the corresponding element in $K_p^{\wedge 4i}(\mathbb{H}P^\infty)$ and β is the Bott class inducing the 2-periodicity of $K_p^\wedge$). Then, the map $q_* : K_p^\wedge{}_*(\mathbb{C}P^\infty) \to K_p^\wedge{}_*(\mathbb{H}P^\infty)$ is given by

$$\beta_{2i} \mapsto a_i$$
$$\beta_{2i-1} \mapsto 0.$$

The kernel of the map q_* is $K_{p\,*}^{\wedge}\{\beta_{2i-1}\}$. The second map from the kernel to $K_{p\,*}^{\wedge}(\mathbb{H}P^{\infty})$ is given by

$$\beta_{2i-1} \mapsto \begin{cases} pa_{i-1} & \text{if } 1 \leq i \leq \frac{p-1}{2} \\ pa_{i-1} + a_{i-\frac{p-1}{2}} & \text{if } i \geq \frac{p-1}{2}. \end{cases}$$

The cokernel is just $\frac{p-1}{2}$ copies of $Z/(p^{\infty}) \otimes K_{p\,*}^{\wedge}$. Therefore, we have for this case

$$\pi_k(THH^{K_p^{\wedge}}(K/p)) = \begin{cases} 0 & \text{if } k \text{ is odd} \\ (\mathbb{Z}/(p^{\infty}))^{\frac{p-1}{2}} & \text{if } k \text{ is even}. \end{cases}$$

We summarize this computation in

Theorem 5.5 *For odd primes p, there are ring structures on K/p as a $K_p^{\wedge}$ module for which*

$$\pi_k(THH^{K_p^{\wedge}}(K/p)) = \begin{cases} 0 & \text{if } k \text{ is odd} \\ (\mathbb{Z}/(p^{\infty}))^{\frac{p-1}{2}} & \text{if } k \text{ is even}. \end{cases}$$

Remark 5.6 One may also make computations allowing the extension on the $K_p(\hat{2})$ to be nonzero by proceeding as in section 5 of [5]. The space $K_p(\hat{2})$ may be expressed as a product of Adams summands for connective K-theory. In this case, one deduces that the unit takes the form

$$u_i = 1 + px + x^{2i-1}, 2 \leq i \leq p$$

for some i. For u_i, we compute as above to obtain

$$\pi_k(THH^{K_p^{\wedge}}(K/p)) = \begin{cases} 0 & \text{if } k \text{ is odd} \\ (\mathbb{Z}/(p^{\infty}))^{i-1} & \text{if } k \text{ is even}. \end{cases}$$

References

[1] V. Angeltveit, *Topological Hochschild homology and cohomology of A_{∞} ring spectra*, Geom. Topol. **12** (2008), 987–1032.

[2] M. Ando, A. J. Blumberg, D. Gepner, M. J. Hopkins, and C. Rezk, *Units of ring spectra, orientations and Thom spectra via rigid infinite loop space theory*, J. Topol. **7** (2014), 1077–1117.

[3] M. Ando, A. J. Blumberg, D. Gepner, M. J. Hopkins and C. Rezk, *An ∞-categorical approach to R-line bundles, R-module Thom spectra, and twisted R-homology*, J. Topol. **7** (2014), 869–893.

[4] M. F. Atiyah and G. Segal, *Twisted K-theory and cohomology*, Ukr. Mat. Vis. **1** (2004), 287–330.

[5] S. Basu, *Topological Hochschild homology of K/p as a $K_p^\wedge$-module*, Homology Homotopy Appl. **19** (2017), 253–280.

[6] S. Basu, S. Sagave and C. Schlichtkrull, *Generalized Thom spectra and their topological Hochschild homology*, to appear in J. Inst. Math. Jussieu.

[7] A. Blumberg, R. L. Cohen and C. Schlichtkrull, *Topological Hochschild homology of Thom spectra and the free loop space*, Geom. Topol. **14** (2010), 1165–1242.

[8] P. Donovan and M. Karoubi, *Graded Brauer groups and K-theory with local coefficients*, Inst. Hautes Études Sci. Publ. Math. **38** (1970), 5–25.

[9] A. D. Elmendorf, I. Kriz, M. Mandell and J. P. May, *Rings, modules and algebras in stable homotopy theory*, Math. Surverys Monogr. **47**, Amer. Math. Soc. Providence, RI, 1997. With an appendix by M. Cole.

[10] D. Freed, M. J. Hopkins and C. Teleman, *Loop groups and twisted K theory III*, Ann. Math (2) **174** (2011), 947–1007.

[11] S. Gitler, *Cohomology operations with local coefficients*, Amer. J. Math. **85** (1963), 156–188.

[12] P. Goerss and M. J. Hopkins, *Moduli spaces of commutative ring spectra*, London Math. Soc. Lecture Note Ser. **315** (2004), 151–200.

[13] M. Hovey, B. Shipley and J. Smith, *Symmetric Spectra*, J. Amer. Math. Soc. **13** (2000), 149–208.

[14] L. G. Lewis, J. P. May, M. Steinberger and J. E. Mcclure, *Equivariant stable homotopy theory*, Lecture Notes in Math. **1213**, Springer-Verlag, Berlin, 1986. With contributions by J. E. Mc-Clure.

[15] J. P. May, *E_∞ ring spaces and E_∞ ring spectra*, Lecture Notes in Math. **577**, Springer-Verlag, Berlin-New York, 1977. With contributions by Frank Quinn, Nigel Ray, and Jørgen Tornehave.

[16] J. P. May, *The construction of E_∞ ring spaces from bipermutative categories*, New topological contexts for Galois theory and algebraic geometry (BIRS 2008), 283–330, Geom. Topol. Monogr. **16**, Geom. Topol. Publ., Coventry, 2009.

[17] J. P. May and J. Sigurdsson, *Parametrized homotopy theory*, Math. Surveys Monogr. **132** Amer. Math. Soc. Providence, RI, 2006.

[18] R. Minasian and G. Moore, *K-theory and Ramond-Ramond charge*, J. High Energy Phys. (1997), no. 11, Paper 2, 7 pp.

[19] C. Rezk, *Notes on the Hopkins Miller Theorem,* Homotopy theory via algebraic geometry and group representations (Evanston, IL, 1997), 313–366, Contemp. Math., 220, Amer. Math. Soc., Providence, RI, 1998.

[20] C. Rezk, *The units of a ring spectrum and a logarithmic cohomology operation.* J. Amer. Math. **19** (2006), 969–1014.

[21] A. Robinson, *Obstruction theory and strict associativity of Morava K theories,* London Math. Soc. Lecture Note Ser. **139** (1989), 143–152.

[22] J. Rosenberg, *Continuous-trace algebras from the bundle theoretic point of view,* J. Austral. Math. Soc. Ser. A **47** (1989), 368–381.

[23] S. Sagave and C. Schlichtkrull, *Diagram spaces and symmetric spectra,* Adv. Math. **231** (2012), 2116–2193.

[24] C. Schlichtkrull, *Units of ring spectra and their traces in Algebraic K-theory.* Geom. Topol. **8** (2004), 645–673.

[25] G. Segal, *Categories and cohomology theories,* Topology **13** (1974), 293–312.

[26] R. Thom, *Quelques propriétés globales des variétés différentiables,* Comm. Math. Helv. **28** (1954), 17–86.

[27] E. Witten, *D-branes and K-theory,* J. High Energy Phys. (1998), No. 12, Paper 19, 41 pp.

STAT-MATH UNIT, INDIAN STATISTICAL INSTITUTE, KOLKATA 700108
E-mail: samik.basu2@gmail.com; samikbasu@isical.ac.in